LONDON MATHEMATICAL SOCIETY LECTURE NOTE SERIES

Managing Editor: Professor Endre Süli, Mathematical Institute, University of Oxford,
Woodstock Road, Oxford OX2 6GG, United Kingdom

The titles below are available from booksellers, or from Cambridge University Press at
www.cambridge.org/mathematics

London Mathematical Society Lecture Note Series: 472

Facets of Algebraic Geometry

A Collection in Honor of William Fulton's 80th Birthday

VOLUME 1

Edited by

PAOLO ALUFFI
Florida State University

DAVID ANDERSON
Ohio State University

MILENA HERING
University of Edinburgh

MIRCEA MUSTAȚĂ
University of Michigan, Ann Arbor

SAM PAYNE
University of Texas, Austin

CAMBRIDGE
UNIVERSITY PRESS

Shaftesbury Road, Cambridge CB2 8EA, United Kingdom

One Liberty Plaza, 20th Floor, New York, NY 10006, USA

477 Williamstown Road, Port Melbourne, VIC 3207, Australia

314–321, 3rd Floor, Plot 3, Splendor Forum, Jasola District Centre, New Delhi – 110025, India

103 Penang Road, #05–06/07, Visioncrest Commercial, Singapore 238467

Cambridge University Press is part of Cambridge University Press & Assessment,
a department of the University of Cambridge.

We share the University's mission to contribute to society through the pursuit of
education, learning and research at the highest international levels of excellence.

www.cambridge.org
Information on this title: www.cambridge.org/9781108792509

DOI: 10.1017/9781108877831

First published 2022

A catalogue record for this publication is available from the British Library

Library of Congress Cataloging-in-Publication data
Names: Aluffi, Paolo, 1960- editor.
Title: Facets of algebraic geometry : a collection in honor of William
Fulton's 80th birthday / edited by Paolo Aluffi, Florida State
University, David Anderson, Ohio State University, Milena Hering,
University of Edinburgh, Mircea Mustaţă, University of Michigan, Ann
Arbor, Sam Payne, University of Texas, Austin.
Description: Cambridge, United Kingdom ; New York, NY : Cambridge
University Press, 2022. | Series: London mathematical society lecture
note series ; 472, 473 | Includes bibliographical references.
Identifiers: LCCN 2021063035 (print) | LCCN 2021063036 (ebook) |
ISBN 9781108870061 (2 volume set ; paperback) | ISBN 9781108792509
(volume 1 ; paperback) | ISBN 9781108792516 (volume 2 ; paperback) |
ISBN 9781108877831 (volume 1 ; epub) | ISBN 9781108877855 (volume 2 ; epub)
Subjects: LCSH: Geometry, Algebraic. | Topology. | BISAC: MATHEMATICS / Topology
Classification: LCC QA564 .F33 2022 (print) | LCC QA564 (ebook) |
DDC 516.3/5–dc23/eng/20220126
LC record available at https://lccn.loc.gov/2021063035
LC ebook record available at https://lccn.loc.gov/2021063036

ISBN – 2 Volume Set 978-1-108-87006-1 Paperback
ISBN – Volume 1 978-1-108-79250-9 Paperback
ISBN – Volume 2 978-1-108-79251-6 Paperback

We dedicate these volumes with gratitude and admiration to William Fulton, on the occasion of his 80th birthday. We hope that the articles collected here, written by colleagues, friends, and students, may reflect the breadth of Bill's impact on mathematics, along with the depth of his enduring influence on all of us who have had the great good fortune to learn from and be inspired by him.

Contents of Volume 1

Contents of Volume 2

Contributors (Volume 1)

Paolo Aluffi
Florida State University

Asher Auel
Dartmouth College, New Hampshire

Michel Brion
Institut Fourier, Université Grenoble Alpes

Melody Chan
Brown University, Rhode Island

Alessio Corti
Imperial College London

Chiara Damiolini
Rutgers University, New Jersey

Dan Edidin
University of Missouri, Columbia

Lawrence Ein
University of Illinois, Chicago

Gavril Farkas
Humboldt-Universität zu Berlin

Matej Filip
University of Ljubljana

Søren Galatius
University of Copenhagen

Noah Giansiracusa
Bentley University, Massachusetts

Angela Gibney
Rutgers University, New Jersey

Eduardo González
University of Massachusetts, Boston

William Graham
University of Georgia

Johan P. Hansen
Aarhus Universitet, Denmark

Robert Lazarsfeld
Stony Brook University, New York

Leonardo C. Mihalcea
Virginia Tech University

Sam Payne
University of Texas, Austin

Andrea Petracci
Università di Bologna

Ryan Richey
University of Missouri, Columbia

Jörg Schürmann
Universität Münster

Changjian Su
University of Toronto

Nicola Tarasca
Virginia Commonwealth University

Chris T. Woodward
Rutgers University, New Jersey

Xian Wu
Jagiellonian University, Krakow

Contributors (Volume 2)

Milena Hering
University of Edinburgh

Philipp Jell
University of Regensburg

Allen Knutson
Cornell University, New York

Khazhgali Kozhasov
Technical University of Braunschweig

Thomas Lam
University of Michigan

Seung Jin Lee
Seoul National University

Joshua F. Lieber
California Institute of Technology

Yuri I. Manin
Max Planck Institute for Mathematics

Matilde Marcolli
California Institute of Technology

Mateusz Michałek
University of Konstanz

Benjamin Nill
Otto-von-Guericke-Universität Magdeburg

John Christian Ottem
University of Oslo

Ragni Piene
University of Oslo

Sutipoj Promtapan
University of North Carolina, Chapel Hill

Richárd Rimányi
University of North Carolina, Chapel Hill

Colleen Robichaux
University of Illinois at Urbana-Champaign

Mark Shimozono
Virginia Tech

Frank Sottile
Texas A & M University

Bernd Sturmfels
Max Planck Institute for Mathematics in the Sciences, Leipzig

Hendrik Süß
University of Manchester

Richard P. Thomas
Imperial College London

Robert Williams
Sam Houston State University

Harshit Yadav
Rice University, Houston

Li Ying
Vanderbilt University, Tennessee

Alexander Yong
University of Illinois at Urbana-Champaign

1

Positivity of Segre–MacPherson Classes

Paolo Aluffi[a], Leonardo C. Mihalcea[b], Jörg Schürmann[c]
and Changjian Su[d]

Dedicated to William Fulton on the occasion of his 80th birthday

Abstract. Let X be a complex nonsingular variety with globally generated tangent bundle. We prove that the signed Segre–MacPherson (SM) class of a constructible function on X with effective characteristic cycle is effective. This observation has a surprising number of applications to positivity questions in classical situations, unifying previous results in the literature and yielding several new results. We survey a selection of such results in this paper. For example, we prove general effectivity results for SM classes of subvarieties which admit proper (semi-)small resolutions and for regular or affine embeddings. Among these, we mention the effectivity of (signed) Segre–Milnor classes of complete intersections if X is projective and an alternation property for SM classes of Schubert cells in flag manifolds; the latter result proves and generalizes a variant of a conjecture of Fehér and Rimányi. Among other applications we prove the positivity of Behrend's Donaldson–Thomas invariant for a closed subvariety of an abelian variety and the signed-effectivity of the intersection homology Chern class of the theta divisor of a non-hyperelliptic curve; and we extend the (known) non-negativity of the Euler characteristic of perverse sheaves on a semi-abelian variety to more general varieties dominating an abelian variety.

1 Introduction

In this note, X will denote a nonsingular complex variety and $Z \subseteq X$ will be a closed subvariety; here (sub)varieties are by definition irreducible and reduced. We will assume that the tangent bundle of X is globally generated. In the

[a] Florida State University
[b] Virginia Tech University
[c] Universität Münster
[d] University of Toronto

projective case, this is equivalent to asking that X be a projective homogeneous variety – for example a projective space, a flag manifold, or an abelian variety; but our main results will hold in the non-complete case as well. We denote by $A_*(Z)$ the Chow group of cycles on Z modulo rational equivalence, and by $\mathsf{F}(Z)$ the group of constructible functions on Z; here we allow Z to be more generally a closed reduced subscheme of X. Our general aim is to investigate the *positivity* of certain rational equivalence classes associated with the embedding of Z in X, or more generally with suitable constructible functions on Z. We generalize several known positivity results on e.g., Euler characteristics, and provide a framework leading to analogous results in a broad range of situations.

Answering a conjecture of Deligne and Grothendieck, MacPherson [35] constructed a group homomorphism $c_* : \mathsf{F}(Z) \to A_*(Z)$ which commutes with proper push-forwards and satisfies a normalization property: if Z is nonsingular, then $c_*(\mathbb{1}_Z) = c(TZ) \cap [Z]$, where $c(TZ)$ is the total Chern class of Z. (MacPherson worked in homology; see [25, Example 19.1.7] for the refinement of the theory to the Chow group.) If $Y \subseteq Z$ is a constructible subset, the Chern–Schwartz–MacPherson (CSM) class $c_{SM}(Y) \in A_*(Z)$ is the image $c_*(\mathbb{1}_Y)$ of the indicator function of Y under MacPherson's natural transformation. Let $\varphi \in \mathsf{F}(Z)$. We will focus on the closely related Segre–MacPherson (SM) class

$$s_*(\varphi, X) := c(TX|_Z)^{-1} \cap c_*(\varphi) \in A_*(Z).$$

(The class $c(TX|_Z)$ is invertible in $A_*(Z)$, because it is of the form $1 + a$, where a is nilpotent.) In particular, we let $s_{SM}(Y, X)$ denote the Segre-Schwartz-MacPherson (SSM) class $s_*(\mathbb{1}_Y, X) \in A_*(Z)$; note that this class depends on both Y and the ambient variety X. If Y is a subvariety of Z, then the top-degree component of $s_{SM}(Y, X)$ in $A_{\dim Y}(Z)$ is the fundamental class $[\overline{Y}]$ of the closure of Y. Further, if $Y = Z$ is a *nonsingular* closed subvariety of X, then $s_{SM}(Y, X) \in A_* Y$ equals the ordinary Segre class $s(Y, X)$; in general, the two classes differ (as discussed in subsection 8.2 for Y a global complete intersection in a nonsingular projective variety X). See [1] and (in the equivariant case) [37] for general properties of SM classes, and [43] for their compability with transversal pullbacks.

There are 'signed' versions of both c_* and s_* (respectively, c_{SM} and s_{SM}), which appear naturally when relating them to characteristic cycles. If $c_*(\varphi) = c_0 + c_1 + \cdots$ is the decomposition into homogeneous components (i.e., $c_i \in A_i(Z)$) then the 'signed' class $\check{c}_*(\varphi)$ is defined by

$$\check{c}_*(\varphi) = c_0 - c_1 + c_2 - \cdots \quad \text{and} \quad \check{c}_{SM}(Y) := \check{c}_*(\mathbb{1}_Y);$$

that is, by changing the sign of each homogeneous component of odd dimension. One defines similarly the signed SM class

$$\check{s}_*(\varphi, X) := c(T^*X|_Z)^{-1} \cap \check{c}_*(\varphi) \quad \text{and} \quad \check{s}_{\mathrm{SM}}(Y, X) := \check{s}_*(\mathbb{1}_Y, X).$$

A basis for $\mathsf{F}(Z)$ consists of the *local Euler obstructions* Eu_Y for closed subvarieties Y of Z. In fact, the characteristic cycle of the (signed) local Euler obstruction is an irreducible Lagrangian cycle in T^*X, and from this perspective the functions Eu_Y are the 'atoms' of the theory; see equation (2) below. If Y is nonsingular, $\mathrm{Eu}_Y = \mathbb{1}_Y$. The local Euler obstruction is a subtle and well-studied invariant of singularities (see e.g., [35, 19, 31, 21, 13, 7]). The corresponding class $c_*(\mathrm{Eu}_Z) \in A_*(Z)$ for Z a subvariety of X is the *Chern–Mather class* of Z, $c_{\mathrm{Ma}}(Z) = \nu_*(c(\tilde{T}) \cap [\tilde{Z}])$, with $\nu \colon \tilde{Z} \to Z$ the *Nash blow-up* of Z and $\tilde{T}$ the tautological bundle on $\tilde{Z}$ extending TZ_{reg}, cf. [35] or [25, Example 4.2.9]. In particular $c_{\mathrm{Ma}}(Z) = c(TZ) \cap [Z]$ if Z is nonsingular. If Z is complete, we denote by $\chi_{\mathrm{Ma}}(Z) := \chi(Z, \mathrm{Eu}_Z)$ the degree of $c_{\mathrm{Ma}}(Z)$; so $\chi_{\mathrm{Ma}}(Z)$ equals the usual topological Euler characteristic $\chi(Z)$ if Z is nonsingular and complete. We also consider the corresponding Segre–Mather class $s_{\mathrm{Ma}}(Z, X) := s_*(\mathrm{Eu}_Z, X)$ as well as the signed classes

$$\check{c}_{\mathrm{Ma}}(Z) := \check{c}_*(\mathrm{Eu}_Z); \quad \check{s}_{\mathrm{Ma}}(Z, X) := c(T^*X|_Z)^{-1} \cap \check{c}_*(\mathrm{Eu}_Z).$$

With our conventions, we get $(-1)^{\dim Z} \check{c}_{\mathrm{Ma}}(Z) = \nu_*(c(\tilde{T}^*) \cap [\tilde{Z}])$ in terms of the dual tautological bundle on the Nash blow-up (which differs by the sign $(-1)^{\dim Z}$ from the definition of the signed Chern–Mather class used in some references like [40, 44]).

The main result in this paper is the (signed) effectivity of Segre–MacPherson classes in a large class of examples. By an *effective* class we mean a class which can be represented by a nonzero, non-negative, cycle.

Theorem 1.1 *Let X be a complex nonsingular variety, and assume that the tangent bundle TX is globally generated. Let $Z \subseteq X$ be a closed subvariety of X. Then the following hold:*

(a) The class $(-1)^{\dim Z} \check{s}_{Ma}(Z, X) \in A_*(Z)$ is effective.

(b) Assume that the inclusion $U \hookrightarrow Z$ is an affine morphism, where U is a locally closed smooth subvariety of Z. Then $(-1)^{\dim U} \check{s}_{SM}(U, X) \in A_*(Z)$ is effective.

If in addition X is assumed to be complete, then the requirement that TX is globally generated is equivalent to X being a homogeneous variety; cf. e.g., [17, Corollary 2.2]. Further, Borel and Remmert [11] (see also [17, Theorem 2.6]) prove that all complete homogeneous varieties are products

$(G/P) \times A$, where G is a semisimple Lie group, $P \subseteq G$ is a parabolic subgroup, and A is an abelian variety.

Theorem 1.1 is extended to more general constructible functions in Theorem 2.2 below. These theorems may be used to prove several positivity statements, unifying and generalizing analogous results from the existing literature. We list below the situations which we will highlight in this paper to illustrate applications of our methods, and the sections where these are discussed.

(a) Closed subvarieties of abelian varieties; primarily in §3.
(b) The proof of a generalization of a conjecture of Fehér and Rimányi [22] concerning SSM classes of Schubert cells in Grassmannians; §4.1.
(c) Complements of hyperplane arrangements; §4.2
(d) Positivity of certain Donaldson–Thomas type invariants; §5.
(e) Intersection homology Segre and Chern classes; §6.
(f) Semi-small resolutions; §8.1.
(g) Regular embeddings and Milnor classes; §8.2.
(h) Semi-abelian varieties and generalizations; §8.3.

Ultimately, these positivity statements follow from the effectivity of the associated characteristic cycles. In §7 we survey a more comprehensive list of situations in which the characteristic cycle is positive.

The proof of Theorem 1.1 will be given in §2.2 below. It is suprisingly easy, but not elementary; it is based on a classical formula by Sabbah [40], calculating the (signed) CSM class of a constructible function φ in terms of its characteristic cycle $CC(\varphi)$; see Theorem 2.1 below.

Acknowledgements. P. Aluffi acknowledges support from the Simons Collaboration Grant 625561; L. C. Mihalcea was supported in part by NSA Young Investigator Award H98320-16-1-0013 and the Simons Collaboration Grant 581675; J. Schürmann was funded by the Deutsche Forschungsgemeinschaft (DFG, German Research Foundation) under Germany's Excellence Strategy – EXC 2044–390685587, Mathematics Münster: Dynamics – Geometry – Structure.

P.A. thanks the University of Toronto for the hospitality. L.M. is grateful to R. Rimányi for many stimulating discussions about the CSM and SM classes, including the positivity conjecture from [22], and to the Math Department at UNC Chapel Hill for the hospitality during a sabbatical leave in the academic year 2017–18. The authors are grateful to an anonymous referee whose comments prompted a reorganization of this paper.

Finally, this paper is dedicated to Professor William Fulton in the occasion of his 80th birthday. His interest in positivity questions arising in algebraic geometry, and his influential ideas, continue to inspire us.

2 Characteristic Classes via Characteristic Cycles; Proof of the Main Theorem

2.1 Characteristic Cycles

Let X be a smooth complex variety. We recall a commutative diagram which plays a central role in seminal work of Ginzburg [27]; it is largely based on results from [8, 18, 30]. We also considered this diagram in our previous work [5, §4.2], and we use the notation from this reference.

$$\begin{array}{ccc} \mathrm{Perv}(X) & \xleftarrow{\quad \mathrm{DR} \quad}_{\sim} & \mathrm{Mod}_{rh}(\mathcal{D}_X) \\ {\scriptstyle \chi_{stalk}}\downarrow & & \downarrow{\scriptstyle \mathrm{Char}} \\ \mathsf{F}(X) & \xrightarrow[\sim]{\quad \mathrm{CC} \quad} & L(X) \end{array} \qquad (1)$$

Here $\mathrm{Mod}_{rh}(\mathcal{D}_X)$ denotes the abelian category of algebraic holonomic $\mathcal{D}_X$-modules with regular singularities, and $\mathrm{Perv}(X)$ is the abelian category of perverse (algebraically) constructible complexes of sheaves of $\mathbb{C}$-vector spaces on X; $\mathsf{F}(X)$ is the group of constructible functions on X and $L(X)$ is the group of conic Lagrangian cycles in T^*X. The functor DR is defined on $M \in \mathrm{Mod}_{rh}(\mathcal{D}_X)$ by

$$\mathrm{DR}(M) = R\mathcal{H}om_{\mathcal{D}_X}(\mathcal{O}_X, M)[\dim X],$$

that is, it computes the DeRham complex of a holonomic module (up to a shift), viewed as an *analytic* $\mathcal{D}_X$-module. This functor realizes the Riemann-Hilbert correspondence, and is an equivalence. We refer to e.g., [30, 27] for details. The left map χ_{stalk} computes the stalkwise Euler characteristic of a constructible complex, and the right map Char gives the characteristic cycle of a holonomic $\mathcal{D}_X$-module. The map CC is the characteristic cycle map for constructible functions; if $Z \subseteq X$ is closed and irreducible, then

$$\mathrm{CC}(\mathrm{Eu}_Z) = (-1)^{\dim Z}[T_Z^*X]; \qquad (2)$$

here Eu_Z is the local Euler obstruction (see §1), and $T_Z^*X := \overline{T_{Z^{reg}}^*X}$ is the conormal space of Z, i.e., the closure of the conormal bundle of the smooth locus of Z. The commutativity of diagram (1) is shown in [27] using deep $\mathcal{D}$-module techniques; it also follows from [41, Example 5.3.4, p. 359–360] (even for a holonomic $\mathcal{D}$-module without the regularity requirement). Also note that the upper transformations in (1) factor over the corresponding Grothendieck groups, so they also apply to complexes of such $\mathcal{D}$-modules. If $f: X \to Y$ is a proper map of smooth complex varieties, there are well-defined push-forwards for each of the objects in the diagram, denoted by f_*. Furthermore, all the maps commute with proper push forwards; cf. [27, Appendix]. For other proofs, see [29, Proposition 4.7.5] for the transformation DR, [41, §2.3] for the transformation χ_{stalk} and [42, §4.6] for the transformation CC (for the transformation Char it then follows from the commutativity of diagram (1)).

The next result, relating characteristic cycles to (signed) CSM classes, has a long history. See [40, Lemme 1.2.1], and more recently [38, (12)], [42, §4.5], [43, §3], especially diagram (3.1) in [43].

Theorem 2.1 *Let X be a complex nonsingular variety, and let $Z \subseteq X$ be a closed reduced subscheme. Let $\varphi \in \mathsf{F}(Z)$ be a constructible function on Z. Then*

$$\check{c}_*(\varphi) = c(T^*X|_Z) \cap \text{Segre}(CC(\varphi))$$

as elements in the Chow group $A_*(Z)$ of Z. Here $\text{Segre}(CC(\varphi))$ is the Segre class associated to the conic Lagrangian cycle $CC(\varphi) \subseteq T^*X|_Z$.

We recall the definition of the Segre class used in Theorem 2.1. Let $q: \mathbb{P}(T^*X|_Z \oplus \mathbb{1}) \to Z$ be the projection from the restriction of the projective completion of the cotangent bundle of X. If $C \subseteq T^*X|_Z$ is a cone supported over Z, and $\overline{C}$ is the closure in $\mathbb{P}(T^*X|_Z \oplus \mathbb{1})$, the Segre class is defined by

$$\text{Segre}(C) := q_* \left(\sum_{i \geq 0} c_1 \left(\mathcal{O}_{\mathbb{P}(T^*X|_Z \oplus 1)}(1) \right)^i \cap [\overline{C}] \right)$$

as an element of $A_*(Z)$; see [25, §4.1].

Every irreducible conic Lagrangian subvariety of T^*X is a conormal cycle T_Z^*X for $Z \subseteq X$ a closed subvariety; see e.g., [29, Theorem E.3.6]. From this it follows that every non-trivial characteristic cycle is a linear combination of conormal spaces:

$$CC(\varphi) = \sum_Y a_Y [T_Y^*X] \tag{3}$$

for uniquely determined closed subvarieties Y of Z and nonzero integer coefficients a_Y. By (2), the coefficients a_Y are determined by the equality of constructible functions

$$0 \neq \varphi = \sum_Y a_Y (-1)^{\dim Y} \mathrm{Eu}_Y. \tag{4}$$

2.2 Proof of the Main Theorem

The following result is at the root of all applications in this note.

Theorem 2.2 *Let X be a complex nonsingular variety such that TX is globally generated. Let $Z \subseteq X$ be a closed reduced subscheme of X and let $\varphi \in \mathsf{F}(Z)$ be a constructible function on Z such that the characteristic cycle $\mathrm{CC}(\varphi) \in A_*(T^*X|_Z)$ is effective.*

Then $\check{s}_*(\varphi, X)$ is effective in $A_*(Z)$. If TX is trivial (e.g., X is an abelian variety), then $\check{c}_*(\varphi)$ is effective.

Proof By Theorem 2.1,

$$\check{s}_*(\varphi, X) = c(T^*X|_Z)^{-1} \cap \check{c}_*(\varphi) = \mathrm{Segre}(\mathrm{CC}(\varphi)).$$

By hypothesis we have a decomposition (3) with positive coefficients a_Y. It follows from (4) that the Segre class of $\mathrm{CC}(\varphi)$ is a linear combination of Segre–Mather classes of subvarieties:

$$\mathrm{Segre}(\mathrm{CC}(\varphi)) = \sum_Y a_Y (-1)^{\dim Y} \mathrm{Segre}(\mathrm{CC}(\mathrm{Eu}_Y))$$

$$= \sum_Y a_Y (-1)^{\dim Y} \check{s}_{\mathrm{Ma}}(Y, X).$$

By definition, the top degree part of each signed Segre–Mather class $(-1)^{\dim Y} \check{s}_{\mathrm{Ma}}(Y, X)$ equals $[Y]$. Then the top degree part of $\mathrm{Segre}(\mathrm{CC}(\varphi))$ equals a positive linear combination of those fundamental classes $[Y]$ of maximal dimension, and in particular $\check{s}_*(\varphi, X) = \mathrm{Segre}(\mathrm{CC}(\varphi))$ is not zero. Since the tangent bundle TX is globally generated, it follows that the line bundle $\mathcal{O}_{\mathbb{P}(T^*X \oplus 1)}(1)$ is globally generated, as it is a quotient of $TX \oplus \mathbb{1}$. Therefore its first Chern class preserves non-negative classes. Since non-negativity is preserved by proper push-forwards, we can conclude that under the given hypotheses $\mathrm{Segre}(\mathrm{CC}(\varphi))$ is non-negative, and this completes the proof. □

Theorem 1.1 follows from Theorem 2.2:

Proof of Theorem 1.1 By Theorem 2.2, it suffices to show that the characteristic cycles for the constructible functions $(-1)^{\dim Z}\mathrm{Eu}_Z$ and $(-1)^{\dim U}\mathbb{1}_U$ are effective. If $\varphi = \check{\mathrm{Eu}}_Z := (-1)^{\dim Z}\mathrm{Eu}_Z$, then $\mathrm{CC}(\varphi)$ equals the conormal cycle $[T_Z^*X]$ of Z, and it is therefore trivially effective.

Consider then $(-1)^{\dim U}\mathbb{1}_U$, and let $j: U \hookrightarrow X$ be the inclusion. We use the Riemann–Hilbert correspondence (diagram (1)) to express the characteristic cycle. By definition,

$$(-1)^{\dim U}\mathbb{1}_U = \chi_{stalk}(j_!\mathbb{C}_U[\dim U]).$$

Since U is nonsingular, the sheaf $\mathbb{C}_U[\dim U]$ is perverse, and $\mathcal{O}_U$ is the corresponding regular holonomic $\mathcal{D}_U$-module. We have $j_!\mathbb{C}_U[\dim U] = \mathrm{DR}(j_!(\mathcal{O}_U))$; since j is an affine morphism, $j_!(\mathcal{O}_U)$ is a single regular holonomic $\mathcal{D}_X$-module (with support in Z); see [29, p. 95]. As pointed out in [29, p. 119], the characteristic cycles of non-trivial holonomic $\mathcal{D}_X$-modules are effective, and this finishes the proof. □

Remark 2.3 As the proof shows, the hypothesis that U is smooth in Theorem 1.1(ii) can be weakened, by only requiring that that $\mathbb{C}_U[\dim U]$ is a perverse sheaf. The proof of the effectivity then uses the fact that for an affine inclusion $j: U \hookrightarrow Z$, $j_!\mathbb{C}_U[\dim U]$ is a perverse sheaf on Z ([41, Lemma 6.0.2, p. 384 and Theorem 6.0.4, p. 409]). We will formalize this conclusion below, in Proposition 2.4. For the case in which $U = X \smallsetminus D$ is the open complement of a hypersurface D in $Z := X$, the result also follows from [41, Proposition 6.0.2, p. 404]. ⌋

2.3 Effective Characteristic Cycles (I)

The applications in the rest of the paper follow from Theorem 2.2: they represent situations when the characteristic cycle $\mathrm{CC}(\varphi)$ is effective.

As pointed out above, every non-trivial characteristic cycle is a linear combination of conormal spaces (3), and the coefficients a_Y in a linear combination are determined by the equality of constructible functions (4). The characteristic cycle of $\varphi \neq 0$ is effective if and only if the coefficients a_Y are positive. In particular, this condition is intrinsic to the constructible function $\varphi \in F(Z)$ and does not depend on the chosen closed embedding of Z into an ambient nonsingular variety X.

A key source of examples where $CC(\varphi)$ is effective, but possibly reducible, arises as follows. Constructible functions may be associated with (regular) holonomic $\mathcal{D}$-modules and perverse sheaves $\mathcal{F} \in \text{Perv}(Z)$, cf. diagram (1); for example, in the latter case the value of the constructible function $\varphi := \chi_{stalk}(\mathcal{F})$ at the point $z \in Z$ is the Euler characteristic $\varphi(z) = \chi(\mathcal{F}_z)$ of the stalk at z of the given complex of sheaves $\mathcal{F}$.

Proposition 2.4 *Let X be a complex nonsingular variety such that TX is globally generated, and let $Z \subseteq X$ be a closed reduced subscheme. Let $0 \neq \varphi \in F(Z)$ be a non-trivial constructible function associated with a regular holonomic $\mathcal{D}_X$-module supported on Z, or (equivalently) a perverse sheaf on Z. Then $\check{s}_*(\varphi, X)$ is effective in $A_*(Z)$.*

Proof This follows from the argument used in the proof of Theorem 2.2: the main observation is that the characteristic cycle of a non-trivial (regular) holonomic $\mathcal{D}$-module is effective; see e.g., [29, p. 119]. Further, perverse sheaves correspond to regular holonomic $\mathcal{D}$-modules by means of the Riemann–Hilbert correspondence (see e.g., [29, Theorem 7.2.5]), compatibly with the construction of the associated constructible functions and characteristic cycles; cf. diagram (1). □

There are situations where the characteristic cycle associated to a constructible sheaf is known to be irreducible: examples include characteristic cycles of the intersection cohomology sheaves of Schubert varieties in the Grassmannian [15], in more general minuscule spaces [10], of certain determinantal varieties [48], and of the theta divisors in the Jacobian of a non-hyperelliptic curve [14]. In all such cases, $\check{s}_*(\varphi, X)$ is effective provided that TX is globally generated, by Theorem 2.2. Also note that for the varieties Z listed above, the Chern–Mather class $c_{Ma}(Z)$ equals $c_{IH}(Z)$, the intersection homology class defined in §6 below. This follows because in this case the characteristic cycle of $\mathcal{IC}_Z$ is irreducible, thus it must agree with the conormal cycle of Z.

In the next few sections we discuss specific applications of Theorem 2.2 for various choices of the variety X or constructible function φ. The sections are mostly logically independent of each other, and the reader may skip directly to the case of interest. The only exception are the results concerning Abelian varieties; these will be mentioned throughout this note.

A more detailed discussion on effective characteristic cycles is given in §7 below, including a more comprehensive list of constructible functions φ for which $CC(\varphi)$ is effective, and operations on characteristic functions which preserve the effectivity of the corresponding characteristic cycles.

3 Abelian Varieties

If X is an abelian variety, then TX is trivial. (In fact, this characterizes abelian varieties among complete varieties, cf. [17, Corollary 2.3].) If TX is trivial, then for all constructible functions φ on Z the signed SM class agrees with the signed CM class: $\check{s}_*(\varphi, X) = \check{c}_*(\varphi) \in A_*(Z)$. In particular, $\check{s}_{Ma}(Z, X) = \check{c}_{Ma}(Z)$ for Z a subvariety of X. The following result follows then immediately from Theorems 1.1 and 2.2.

Corollary 3.1 *Let Z be a closed subvariety of a smooth variety X with TX trivial (for example, an abelian variety). Then $(-1)^{\dim Z} \check{c}_{Ma}(Z)$ is effective.*

More generally, let φ be a constructible function on Z such that $CC(\varphi)$ is effective. Then $\check{c}_*(\varphi) \in A_*(Z)$ is effective.

As an example, the total Chern–Mather class $c_{Ma}(\Theta)$ of the theta divisor in the Jacobian of a nonsingular curve must be signed-effective.

Corollary 3.1 implies that $\chi(Z, \varphi) \geq 0$, which also follows from [24, Theorem 1.3]. In particular, if Z is a closed subvariety of an abelian variety, then

$$(-1)^{\dim Z} \chi_{Ma}(Z) = (-1)^{\dim Z} \chi(Z, \mathrm{Eu}_Z) \geq 0.$$

For nonsingular subvarieties Z, the Euler obstruction Eu_Z equals $\mathbb{1}_Z$. Then the fact that $(-1)^{\dim Z} \chi(Z) \geq 0$ is proven (in the more general semi-abelian case) in [24, Corollary 1.5] (also see [20, (2)]). We note that the fact that $c(T^*Z) \cap [Z]$ is effective if Z is a *nonsingular* subvariety of an abelian variety X also follows immediately from the fact that T^*Z is globally generated, as it is a homomorphic image of the restriction of T^*X, which is trivial. Corollary 3.1 extends this result to *arbitrarily singular* closed subvarieties of a smooth variety X with trivial tangent bundle.

In fact, Corollary 3.1 also follows from Propositions 2.7 and 2.9 from [44], where explicit effective cycles representing $\check{s}_*(\varphi, X)$ in terms of suitable 'polar classes' are constructed.

4 Affine Embeddings

An important family of positivity statements arises from the indicator function $\mathbb{1}_U$ of a typically nonsingular and noncompact subvariety U of $Z \subseteq X$. In this case, among our main applications is the proof of a conjecture of Fehér and Rimányi about the effectivity of SSM classes of Schubert cells.

Throughout this section we impose the hypotheses of Theorem 1.1(b), i.e., X is a complex nonsingular variety with TX globally generated, $Z \subseteq X$ is a closed subvariety of X, and the inclusion $U \hookrightarrow Z$ is an affine morphism, with U a locally closed smooth subvariety of Z. Recall that

$$s_{\text{SM}}(U, X) := c(TX|_Z)^{-1} \cap c_*(\mathbb{1}_U); \quad \check{s}_{\text{SM}}(U, X) := c(T^*X|_Z)^{-1} \cap \check{c}_*(\mathbb{1}_U)$$

denote the SSM and the signed SSM classes associated to U. By Theorem 1.1 (b),

$$(-1)^{\dim U} \check{s}_{\text{SM}}(U, X) \in A_*(Z)$$

is effective. If TX is trivial, then $(-1)^{\dim U} \check{c}_{\text{SM}}(U)$ is effective, so in particular for X an abelian variety this implies that $(-1)^{\dim U} \chi(U) \geq 0$.

4.1 Schubert Cells in Flag Manifolds and a Conjecture of Fehér and Rimányi

Theorem 1.1 applies in particular if $U := X(u)^\circ$ is a Schubert cell in a flag manifold $X = G/P$, where G is a complex simple Lie group and P is a parabolic subgroup. For example, X could be a Grassmannian, or a complete flag manifold. Here $u \in W$ is a minimal length representative for its coset in W/W_P, where W is the Weyl group of G and W_P is the Weyl group of P. The Schubert *cell* $X(u)^\circ$ is defined to be BuB/P, where $B \subseteq P$ is a Borel subgroup. It is well known that $X(u)^\circ \cong \mathbb{C}^{\ell(u)}$, where $\ell(u)$ denotes the length of u. The closure $X(u)$ of $X(u)^\circ$ is the corresponding Schubert *variety,* and $X(u) = \bigsqcup_{w \leq u} X(w)^\circ$. We refer to e.g., [16] for further details on these definitions. Since the inclusion $X(u)^\circ \subseteq X$ is affine, we obtain the following result.

Corollary 4.1 *Let $X(u)^\circ$ be a Schubert cell in a generalized flag manifold G/P. Then the class $(-1)^{\ell(u)} \check{s}_{SM}(X(u)^\circ, G/P) \in A_*(X(u))$ is effective.*

Recall that $A_*(G/P)$ (resp., $A_*(X(u))$) has a $\mathbb{Z}$-basis given by fundamental classes $[X(v)]$ of Schubert varieties (with $X(v) \subseteq X(u)$, i.e., $v \leq u$). With this understood, Corollary 4.1 may be rephrased as follows.

Corollary 4.2 *Let $u \in W$ and consider the Schubert expansion*

$$s_{SM}(X(u)^\circ, G/P) = \sum a(w; u)[X(w)]$$

with $a(w; u) \in \mathbb{Z}$. Then $(-1)^{\ell(u)-\ell(w)} a(w; u) \geq 0$ for all w.

A similar positivity statement was conjectured by Fehér and Rimányi in §1.5 and Conjecture 8.4 of the paper [22]. Their conjecture is stated for certain degeneracy loci in quiver varieties, and in the 'universal' situation where the ambient space is a vector space with a group action. The Schubert cells and varieties in the flag manifolds of Lie type A are closely related to a compactified version of such quiver loci[1]. After passing to the compactified version of the statements from [22], Corollary 4.2 proves the conjecture from [22] in the Schubert instances; see [23, §6 and §7] for a comparison between the 'universal' and 'compactified' versions. A specific comparison between our calculations and those from [22] is included in the following example. We note that in arbitrary Lie type a description of Schubert varieties via quiver loci is not available.

Example 4.3   Let $X = \mathrm{Gr}(2,5)$ be the Grassmann manifold parametrizing subspaces of dimension 2 in $\mathbb{C}^5$. In this case one can index the Schubert cells by partitions included in the 2×3 rectangle, such that each cell has dimension equal to the number of boxes in the partition. With this notation, and using the calculation of CSM classes of Schubert cells from [3], one obtains the following matrix encoding Schubert expansions of SSM classes of Schubert cells:

$$
\begin{pmatrix}
1 & -4 & 5 & 4 & -2 & -10 & 5 & 4 & -4 & 1 \\
0 & 1 & -3 & -3 & 2 & 10 & -7 & -5 & 7 & -2 \\
0 & 0 & 1 & 0 & -2 & -3 & 7 & 3 & -9 & 3 \\
0 & 0 & 0 & 1 & 0 & -3 & 2 & 2 & -3 & 1 \\
0 & 0 & 0 & 0 & 1 & 0 & -3 & 0 & 3 & -1 \\
0 & 0 & 0 & 0 & 0 & 1 & -2 & -2 & 5 & -2 \\
0 & 0 & 0 & 0 & 0 & 0 & 1 & 0 & -2 & 1 \\
0 & 0 & 0 & 0 & 0 & 0 & 0 & 1 & -2 & 1 \\
0 & 0 & 0 & 0 & 0 & 0 & 0 & 0 & 1 & -1 \\
0 & 0 & 0 & 0 & 0 & 0 & 0 & 0 & 0 & 1
\end{pmatrix}.
$$

The columns, read left to right, and rows, read top to bottom, are indexed by:

$$\emptyset, \square, \square\square, \boxplus, \square\square\square, \boxplus\square, \square\boxplus, \boxplus\boxplus, \boxplus\square, \boxplus\boxplus\boxplus$$

After taking duals in the 2×3 rectangle, these give the same coefficients as in equation (3) from [22]. (The calculations in [22] are done in a stable limit,

---

[1] One example are the matrix Schubert varieties, regarded in the space of all matrices. The study of those of maximal rank is closely related to Schubert varieties in the Grassmannian.

therefore for our purposes one disregards partitions not included in the given rectangle.) Another example is given by the calculation of the SSM class for the partition $(3, 1)$ in $Gr(2, 6)$:

$$s_{SM}\left(\left(\boxed{}\right)^\circ\right) = \boxed{} - 3\,\boxed{} - 4\,\boxed{} + 13\,\boxed{} + 5\,\boxed{} - 22\,\square + 22\,\emptyset.$$

(Here λ denotes the Schubert class indexed by λ, and λ° is the indicator function of the Schubert cell.) This result is consistent with [22, Example 8.3]. ⌐

If the parabolic subgroup P is the Borel subgroup B, Corollary 4.2 is equivalent to the positivity of CSM classes of Schubert cells, [5, Corollary 1.4]. Indeed, in this case

$$s_{SM}(X(u)^\circ, G/B) = (-1)^{\ell(u)} \check{c}_*(\mathbb{1}_{X(u)^\circ})$$

as shown in [5, Theorem 7.5]. This equality does not hold for more general flag manifolds G/P, and its proof relies on additional properties relating the CSM/SSM classes to Demazure–Lusztig operators from the Hecke algebra [4, 5]. From this prospective, SSM classes appear to have a simpler behavior than the CSM classes, and one can obtain positivity-type statements for a larger class of varieties.

4.2 Complements of Hyperplane Arrangements

A typical example of an affine embedding $U \subseteq X$ is the complement $U = X \setminus D$ of a hypersurface $D \subseteq Z := X$. In particular, one can consider a projective hyperplane arrangement $\mathcal{A}$ in complex projective space $X = \mathbb{P}^n$, with $A := D$ the union of the hyperplanes and $U := \mathbb{P}^n \setminus A$ its complement. In this particular case Theorem 1.1(b) recovers a consequence of the following result of [2, Corollary 3.2]:

$$c_{SM}(U) = \pi_{\widehat{\mathcal{A}}}\left(\frac{-h}{1+h}\right) \cap \left(c(T\mathbb{P}^n) \cap [\mathbb{P}^n]\right).$$

Here h denotes the hyperplane class in $\mathbb{P}^n$, $\widehat{\mathcal{A}}$ is the corresponding 'central arrangement' in $\mathbb{C}^{n+1}$ with $\widehat{A}$ its union of linear hyperplanes, and $\pi_{\widehat{\mathcal{A}}}$ denotes the corresponding 'Poincaré polynomial' of $\widehat{\mathcal{A}}$ (see e.g., [2, p. 1880]):

$$\pi_{\widehat{\mathcal{A}}}(t) = \sum_{k=0}^{n+1} rk\, H^k\left(\mathbb{C}^{n+1} \setminus \widehat{A}, \mathbb{Q}\right) t^k.$$

In particular the Poincaré polynomial $\pi_{\hat{\mathcal{A}}}$ has non-negative coefficients and constant term one, and

$$(-1)^{\dim U} \check{s}_{\mathrm{SM}}(U, \mathbb{P}^n) = \pi_{\hat{\mathcal{A}}}\left(\frac{h}{1-h}\right) \cap [\mathbb{P}^n]$$

is effective.

5 Donaldson–Thomas Type Invariants

Let $Z \subseteq X$ be a closed reduced subscheme of X as before. K. Behrend ([7, Definition 1.4, Proposition 4.16]) defines a constructible function ν_Z and proves that if Z is proper, then the dimension-0 component of $c_*(\nu_Z)$ equals the corresponding virtual fundamental class $[Z]^{\mathrm{vir}}$, a 'Donaldson–Thomas type invariant' in the terminology of [7, p. 1308].

The characteristic cycle of Behrend's constructible function ν_Z is effective because the intrinsic normal cone of Z is effective. More explicitly, ν_Z is defined as a weighted sum

$$\nu_Z := \sum_Y (-1)^{\dim Y} \mathrm{mult}(Y) \cdot \mathrm{Eu}_Y,$$

where the summation is over the supports Y of the components of the intrinsic normal cone of Z and $\mathrm{mult}(Y)$ is the multiplicity of the corresponding component. (Cf. [7, Definition 1.4].) Since these multiplicities are positive, $\mathrm{CC}(\nu_Z)$ is effective, cf. (4).

Corollary 5.1 *Let X be a complex nonsingular variety with globally generated tangent bundle, and let $Z \subseteq X$ be a closed reduced subscheme. Then $\check{s}_*(\nu_Z, X) \in A_*(Z)$ is effective. If TX is trivial, then $\check{c}_*(\nu_Z)$ is effective.*

In particular for X an abelian variety this implies that $[Z]^{\mathrm{vir}}$ is non-negative and hence

$$\chi_{\mathrm{vir}}(Z) := \chi(Z, \nu_Z) = \deg([Z]^{\mathrm{vir}}) \geq 0.$$

6 Characteristic Classes from Intersection Cohomology

Another particular case of interest is the characteristic class of the intersection cohomology sheaf complex. If $Z \subseteq X$ is a closed subvariety, let $\mathcal{IC}_Z \in \mathrm{Perv}(Z)$ denote the intersection cohomology complex associated to Z. This is the key example of a perverse sheaf on Z, cf. [28], [29, Definition 8.2.13] or [41, p. 385]. The associated constructible function is $\mathrm{IC}_Z := \chi_{stalk}(\mathcal{IC}_Z)$. We let

$$c_{\mathrm{IH}}(Z) := (-1)^{\dim Z} c_*(\mathrm{IC}_Z),$$

be the *intersection homology Chern class* of Z. (Note that $c_{\mathrm{IH}}(Z)$ is an element of the Chow group of Z, not of its intersection homology.) The sign is introduced in order to ensure that $c_{\mathrm{IH}}(Z) = c(TZ) \cap [Z]$ if Z is nonsingular[2].

Similarly, if Z has a *small* resolution of singularities $f : Y \to Z$, then $\mathcal{IC}_Z \simeq Rf_* \mathbb{C}_Y[\dim Y]$ (see e.g., [41, Example 6.0.9, p. 400]) so that

$$c_{\mathrm{IH}}(Z) = f_*(c(TY) \cap [Y]).$$

This class corresponds to the constructible function $(-1)^{\dim Z} f_*(\mathbb{1}_Y)$. We also consider the signed version, $\check{c}_{\mathrm{IH}}(Z)$. For Z complete, the degree of the zero-dimensional component of $c_{\mathrm{IH}}(Z)$ equals the 'intersection homology Euler characteristic' of Z, $\chi_{\mathrm{IH}}(Z) = (-1)^{\dim Z} \chi(Z, \mathrm{IC}_Z)$. By the functoriality of c_* and χ_{stalk}, $\chi_{\mathrm{IH}}(Z)$ agrees with the intersection homology Euler characteristic defined as alternating sum of ranks of intersection homology groups, as in e.g., [20].

More generally, let $f : Y \to Z$ be a proper morphism. Using that χ_{stalk} commutes with Rf_* [41, §2.3] one may define $\varphi := \chi_{stalk}(Rf_*\mathbb{C}_Y) = f_*(\mathbb{1}_Y)$; then $c_*(\varphi) = f_* c_*(\mathbb{1}_Y)$ by the functoriality of c_*.

Theorem 2.2 and Proposition 2.4 imply the following result.

Corollary 6.1 *Let Z be a closed subvariety of a smooth variety X with TX globally generated. Then $\check{s}_*(\mathrm{IC}_Z, X)$ is effective. If TX is trivial (e.g., X is an abelian variety), then $(-1)^{\dim Z} \check{c}_{\mathrm{IH}}(Z) = \check{c}_{\mathrm{SM}}(\mathrm{IC}_Z)$ is effective.*

In particular for X an abelian variety this implies the inequality $(-1)^{\dim Z} \chi_{\mathrm{IH}}(Z) \geq 0$, recovering [20, Theorem 5.3]. In case Z has a *small resolution* $f : Y \to Z$,

$$(-1)^{\dim Z} \check{c}_{\mathrm{IH}}(Z) = f_*(c(T^*Y) \cap [Y]) = (-1)^{\dim Y} \check{c}_*(f_* \mathbb{1}_Y)$$

and this class is effective by Corollary 6.1.

7 Effective Characteristic Cycles (II)

In this section we collect several instances of positive characteristic cycles. Some of these were already used in the previous sections, and they will be reproduced here for completeness.

---

[2] More generally, $c_{\mathrm{IH}}(Z) = c_{\mathrm{SM}}(Z)$ if Z is a rational homology manifold (e.g., Z has only quotient singularities). In fact a quasi-isomorphism $\mathcal{IC}_Z \simeq \mathbb{C}_Z[\dim Z]$ characterizes a rational homology manifold Z [12, p. 34], see also [29, Proposition 8.2.21].

Proposition 7.1 *Let X be a complex nonsingular variety, and let $Z \subseteq X$ be a closed reduced subscheme. If $\varphi \in F(Z)$ is the constructible function in one of the cases listed below, then $CC(\varphi)$ is an effective cycle.*

(a) $\varphi = (-1)^{\dim Z} \mathrm{Eu}_Z$ *for Z a closed subvariety of X.*
(b) $\varphi = \nu_Z$ *(Behrend's constructible function, see §5).*
(c) $\varphi = \chi_{stalk}(\mathcal{F})$ *for a non-trivial perverse sheaf $\mathcal{F} \in \mathrm{Perv}(Z)$, e.g.:*

> *(c1)* $\varphi = \mathrm{IC}_Z$ *for Z a closed subvariety of X, see §6.*
> *(c2)* $\varphi = (-1)^{\dim Z} \mathbb{1}_Z$ *for Z pure-dimensional and smooth, or more generally a rational homology manifold.*
> *(c3)* $\varphi = (-1)^{\dim Z} \mathbb{1}_Z$ *for Z pure-dimensional with only local complete intersection singularities (i.e., $Z \hookrightarrow X$ is a regular embedding).*
> *(c4)* $\varphi = (-1)^{\dim Y} f_* \mathbb{1}_Y$ *for a proper surjective semi-small morphism of varieties $f : Y \to Z$, with Y a rational homology manifold and Z a closed subvariety of X. (See §8.1 for more on semi-small maps.)*
> *(c5)* $\varphi = (-1)^{\dim U} \mathbb{1}_U$, *where $U \subseteq Z$ is a (not necessarily closed) subvariety, such that the inclusion $U \hookrightarrow Z$ is an affine morphism and $\mathbb{C}_U[\dim U]$ is a perverse sheaf on U (e.g., U is smooth, a rational homology manifold, or with only local complete intersection singularities).*

Proof Part (a) follows from equation (2); part (b) from the discussion in §5; part (c) from Proposition 2.4; part (c1) is discussed in §6; part (c5) combines Theorem 1.1 and Remark 2.3. The remaining statements are proved as follows:

(c2): $\mathbb{C}_Z[\dim Z]$ is a perverse sheaf for Z pure-dimensional and smooth, with

$$\chi_{stalk}(\mathbb{C}_Z[\dim Z]) = (-1)^{\dim Z} \mathbb{1}_Z$$

by definition. The corresponding regular holonomic $\mathcal{D}$-module is just $\mathcal{O}_Z$. Similarly, $\mathbb{C}_Z[\dim Z]$ is a perverse sheaf for Z pure-dimensional and a rational homology manifold, since then $\mathcal{IC}_Z \simeq \mathbb{C}_Z[\dim Z]$; cf. [12, p. 34] or [29, Proposition 8.2.21].

(c3): $\mathbb{C}_Z[\dim Z]$ is a perverse sheaf for Z pure-dimensional with only local complete intersection singularities [41, Example 6.0.11, p. 404].

(c4): In the given hypotheses, the push-forward $Rf_*\mathbb{C}_Y[\dim Y]$ is a perverse sheaf; cf. [41, Example 6.0.9, p. 400] or [29, Definition 8.2.30]. Of course

$$\chi_{stalk}(Rf_*\mathbb{C}_Y[\dim Y]) = (-1)^{\dim Y} f_* \mathbb{1}_Y,$$

since f is proper. $\square$

To check that for a constructible function φ the coefficients a_Y in the expansion $CC(\varphi) = \sum_Y a_Y [T_Y^* X]$ are nonnegative, one can also use the following description of $CC(\varphi)$ in terms of 'stratified Morse theory for constructible functions' from [42, 44] or [41, §5.0.3]:

$$CC(\varphi) = \sum_S (-1)^{\dim S} \cdot \chi(NMD(S), \varphi) \cdot [\overline{T_S^* X}]$$

if φ is constructible with respect to a complex algebraic Whitney stratification of Z with connected smooth strata S. Here $\chi(NMD(S), \varphi)$ is the Euler characteristic of a corresponding *normal Morse datum* $NMD(S)$ weighted by φ. Then $CC(\varphi)$ is non-negative (resp., effective) if and only if

$$(-1)^{\dim S} \cdot \chi(NMD(S), \varphi) \geq 0$$

for all S (and, resp., $(-1)^{\dim S'} \cdot \chi(NMD(S'), \varphi) > 0$ for at least one stratum S'). If the complex of sheaves $\mathcal{F}$ is constructible with respect to this complex algebraic Whitney stratification of Z, then one gets for $\varphi := \chi_{stalk}(\mathcal{F})$ and $x \in S$ [41, (5.38) on p. 294]:

$$\chi(NMD(\mathcal{F}, x)[-\dim S]) = (-1)^{\dim S} \cdot \chi(NMD(S), \varphi).$$

This leads to a direct proof of Proposition 2.4 in the case of perverse sheaves, without using $\mathcal{D}$-modules: $NMD(\mathcal{F}, x)[-\dim S]$ as above is concentrated in degree zero for all S if and only if $\mathcal{F}$ is a perverse sheaf [41, Remark 6.0.4, p. 389]. This argument also shows that the condition that $CC(\chi_{stalk}(\mathcal{F}))$ be *effective* is much weaker than the condition that $\mathcal{F}$ be perverse.

We end this section by listing some basic operations of constructible functions which preserve the property of having an effective characteristic cycle. These operations may be used to construct many more examples to which our Theorem 2.2 applies.

Proposition 7.2 *Let Z be a closed reduced subscheme of a nonsingular complex algebraic variety X, and assume that φ is a constructible function on Z with $CC(\varphi)$ effective.*

(1) Let Z' be a closed reduced subscheme of a nonsingular variety X', with $CC(\varphi')$ effective. Then $CC(\varphi \boxtimes \varphi')$ is also effective for the constructible function $\varphi \boxtimes \varphi'$ on $Z \times Z'$ defined by

$$(\varphi \boxtimes \varphi')(z, z') := \varphi(z) \cdot \varphi'(z').$$

(2) Let $f : Z \to Z'$ be a finite morphism, i.e., f is proper with finite fibers, with Z' a closed reduced subscheme of a nonsingular complex algebraic

variety X'. Then $f_*(\varphi)$ is a constructible function on Z' with $CC(f_*(\varphi))$ effective. Here

$$f_*(\varphi)(z') := \sum_{z \in f^{-1}(z')} \varphi(z).$$

(3) Let $f: X' \to X$ be a morphism of nonsingular complex algebraic varieties such that $f: Z' := f^{-1}(Z) \to Z$ is a smooth morphism of relative dimension d. Then $(-1)^d f^*(\varphi) = (-1)^d \varphi \circ f$ is a constructible function on Z' with effective characteristic cycle.

(4) Let $f: X \to \mathbb{C}$ be a morphism and let D be the hypersurface $\{f = 0\}$. Denote by $\psi_f: F(Z) \to F(Z \cap D)$ the corresponding specialization of constructible functions [42, §2.4.7]. Here

$$\psi_f(\varphi)(x) := \chi(M_{f|Z,x}, \varphi),$$

with $M_{f|Z,x}$ a local Milnor fiber of $f|Z$ at $x \in Z \cap D$. Then $CC(-\psi_f(\varphi))$ is non-negative. It is effective in case $\varphi \neq 0$ has a presentation as in (4) and at least one Y with $a_Y > 0$ is not contained in D.

(5) Let $f: X' \to X$ be a morphism of nonsingular complex algebraic varieties such that f is non-characteristic with respect to the support $supp(CC(\varphi))$ of the characteristic cycle of φ (e.g., f is transversal to all strata S of a complex algebraic Whitney stratification of Z for which φ is constructible). Let $d := \dim X' - \dim X$. Then $(-1)^d f^*(\varphi) = (-1)^d \varphi \circ f$ is a constructible function on $Z' := f^{-1}(Z)$ with effective characteristic cycle.

Proof　　These results can be deduced from the following facts:

(1) $CC(\varphi \boxtimes \varphi') = CC(\varphi) \boxtimes CC(\varphi')$, which follows from $Eu_Y \boxtimes Eu_{Y'} = Eu_{Y \times Y'}$ [35], or from stratified Morse theory for constructible functions or sheaves [41, (5.6) on p. 277]:

$$\chi(NMD(S), \varphi) \cdot \chi(NMD(S'), \varphi') = \chi(NMD(S \times S'), \varphi \boxtimes \varphi').$$

(2) Using the graph embedding, we can assume that the finite map $f: Z \to Z'$ is induced from a submersion $f: X \to X'$ of ambient nonsingular varieties. Consider the induced correspondence of cotangent bundles:

$$T^*X \xleftarrow{df} f^*T^*X' \xrightarrow{\tau} T^*X'.$$

Here df is a closed embedding (since f is a submersion), and

$$\tau: df^{-1}(supp(CC(\varphi))) \to T^*X'|Z'$$

is finite, since $f: Z \to Z'$ is finite. Now $\tau(df^{-1} \operatorname{supp}(\mathrm{CC}(\varphi)))$ is known to be contained in a conic Lagrangian subset of $T^*X'|Z'$ (e.g., coming from a stratification of f [41, (4.16) on p. 249]). Therefore its dimension is bounded from above by $\dim X'$. Then also the dimension of $df^{-1}(\operatorname{supp}(\mathrm{CC}(\varphi)))$ is bounded from above by $\dim X'$ by the finiteness of τ, so that

$$df^*(\mathrm{CC}(\varphi)) = \mathrm{CC}(\varphi) \cap [f^*T^*X']$$

is a *proper intersection*. But then $\tau_*(df^*(\mathrm{CC}(\varphi)))$ is an *effective* cycle on $T^*X'|Z'$, and [42, §4.6]:

$$\tau_*(df^*(\mathrm{CC}(\varphi))) = \mathrm{CC}(f_*(\varphi)).$$

(3) This follows from $f^*\mathrm{Eu}_Y = \mathrm{Eu}_{f^{-1}(Y)}$ for Y a closed subvariety in Z. This can be checked locally, e.g., for $f : Z \times Y' \to Z$ the projection along a smooth factor Y', with $f^*\mathrm{Eu}_Y = \mathrm{Eu}_Y \boxtimes \mathbb{1}_{Y'} = \mathrm{Eu}_Y \boxtimes \mathrm{Eu}_{Y'}$.

(4) Again it is enough to consider $\check{\mathrm{Eu}}_Y := (-1)^{\dim Y}\mathrm{Eu}_Y$ for some subvariety Y of Z. If $Y \subseteq \{f = 0\}$, then $\psi_f(\check{\mathrm{Eu}}_Y) = 0$ by definition. So we can assume $Y \nsubseteq \{f = 0\}$. Then $\mathrm{CC}(-\psi_f(\check{\mathrm{Eu}}_Y))$ is by [40, Theorem 4.3] the (Lagrangian) specialization of the relative conormal space $[T^*_{f|Z}X]$ along the hypersurface $\{f = 0\}$, so that it is also effective.

(5) Consider again the induced correspondence of cotangent bundles:

$$T^*X' \xleftarrow{\ df\ } f^*T^*X \xrightarrow{\ \tau\ } T^*X.$$

Then by definition, f is *non-characteristic* with respect to the support $\operatorname{supp}(\mathrm{CC}(\varphi))$ of the characteristic cycle of φ if and only if

$$df : \tau^{-1}(\operatorname{supp}(\mathrm{CC}(\varphi))) \to T^*X'$$

is *proper* and therefore finite, cf. [43, Lemma 3.2] or [41, Lemma 4.3.1, p. 255]. If f is non-characteristic, then

$$\mathrm{CC}((-1)^d f^*(\varphi)) = df_*(\tau^*(\mathrm{CC}(\varphi)))$$

by [43, Theorem 3.3], and this cycle is *effective* if $\mathrm{CC}(\varphi)$ is effective. Indeed the proof of [43, Theorem 3.3] is done in two steps: first for a submersion, where our claim follows from the case (3) above; then the case of a closed embedding of a nonsingular subvariety is (locally) reduced by induction to the case of a hypersurface of codimension one (locally) given by an equation $\{f = 0\}$. Here it is deduced from case (4) above, with $Y \nsubseteq \{f = 0\}$ by the 'non-characteristic' assumption if $[T^*_Y X]$ appears with positive multiplicity in $\mathrm{CC}(\varphi)$. $\qquad \square$

8 Further Applications

In this final section we explain further applications of Theorem 2.2 and Proposition 2.4 via the theory of perverse sheaves.

8.1 Semi-small Maps

Recall that a morphism $f: Y \to Z$ is called *semi-small* if for all $i > 0$,

$$\dim\{z \in Z \mid \dim f^{-1}(z) \geq i\} \leq \dim Z - 2i;$$

the morphism f is *small* if in addition all inequalities are strict for $i > 0$. See [12, p. 30], [41, Example 6.0.9, p. 400], or [29, Definition 8.2.29].

Proposition 8.1 *Let $f: Y \to Z$ be a proper surjective semi-small morphism of varieties, with Y a rational homology manifold and Z a closed subvariety of a smooth variety X with TX globally generated. Then $(-1)^{\dim Y} \check{s}_*(f_* \mathbb{1}_Y, X)$ is effective.*

In particular, if TX is trivial, then the class $(-1)^{\dim Y} \check{c}_*(f_* \mathbb{1}_Y) = (-1)^{\dim Y} f_* \check{c}_{SM}(Y)$ is effective. If moreover X is complete (i.e., X is an abelian variety), then $(-1)^{\dim Y} \chi(Y) \geq 0$.

The simplest example of a proper semi-small map $f: Y \to Z := f(Y) \subseteq X$ is a closed embedding. A smooth projective variety Y has a proper semi-small morphism (onto its image) into an abelian variety X if and only if its Albanese morphism $alb_X: X \to Alb(X)$ is semi-small (onto its image) [34, Remark 1.3]. The corresponding signed Euler characteristic bound $(-1)^{\dim Y} \chi(Y) \geq 0$ is further refined in [39, Corollary 5.2]. As an example, if C is a smooth curve of genus $g \geq 3$ and X is its Jacobian, then the induced Abel-Jacobi map $C^d \to C^{(d)} \to X$ (with $C^{(d)}$ the corresponding symmetric product) is semi-small (onto its image) for $1 \leq d \leq g - 1$ [47, Corollary 12].

Proof of Proposition 8.1 By the given hypotheses, the push-forward $Rf_* \mathbb{C}_Y[\dim Y]$ is a perverse sheaf; cf. [41, Example 6.0.9, p. 400] or [29, Definition 8.2.30]. Further

$$\chi_{stalk}(Rf_* \mathbb{C}_Y[\dim Y]) = (-1)^{\dim Y} f_* \mathbb{1}_Y,$$

since f is proper. The statement follows then from Proposition 2.4. □

8.2 Regular Embeddings and Milnor Classes

For this application, assume that $Z \subseteq X$ is a regular embedding, as in [25, Appendix B.7]. For instance, Z could be a smooth closed subvariety, a hypersurface, or a local complete intersection in X.

Proposition 8.2 *Let X be a complex nonsingular variety such that TX is globally generated, and let $Z \subseteq X$ be a regular embedding. Then $(-1)^{\dim Z} \check{s}_{SM}(Z, X)$ is effective. If TX is trivial, then $(-1)^{\dim Z} \check{c}_{SM}(Z)$ is effective.*

Proof The hypothesis imply that $\mathbb{C}_Z[\dim Z]$ is a perverse sheaf, by Proposition 7.1. Then the claim follows from Theorem 2.2 and Proposition 2.4. $\qquad\square$

In particular for X an abelian variety this implies the inequality $(-1)^{\dim Z} \chi(Z) \geq 0$, recovering [20, Theorem 5.4].

If $Z \subseteq X$ is a closed embedding, with X smooth, then the class

$$c_F(Z) := c(TX) \cap s(Z, X) \in A_*(Z)$$

is the *Chern–Fulton class* of Z; this is another intrinsic Chern class of Z [25, Example 4.2.6]. If Z is a local complete intersection in X, then the normal cone $C_Z X = N_Z X$ is a vector bundle, and we have

$$c_F(Z) = c(TX) \cap \left(c(N_Z X)^{-1} \cap [Z] \right) \in A_*(Z);$$

in this case, this class is also called the *virtual Chern class* of Z. If Z is smooth, $N_Z X$ is the usual normal bundle, so that $c_F(Z) = c(TZ) \cap [Z] = c_{SM}(Z)$. In general, for singular Z, these classes can be different, and their difference[3]

$$\mathrm{Mi}(Z) := (-1)^{\dim Z} \left(c_F(Z) - c_{SM}(Z) \right) \in A_*(Z)$$

is called the *Milnor class* of Z. Let $\mathrm{SMi}(Z, X)$ be the corresponding *Segre–Milnor class*

$$
\begin{aligned}
\mathrm{SMi}(Z, X) &:= c(TX)^{-1} \cap \mathrm{Mi}(Z) \\
&= (-1)^{\dim Z} \left(c(N_Z X)^{-1} \cap [Z] - s_{SM}(Z, X) \right) \\
&= (-1)^{\dim Z} \left(s(Z, X) - s_{SM}(Z, X) \right).
\end{aligned}
$$

As before consider the associated signed classes $\check{\mathrm{Mi}}(Z)$ and

$$\check{\mathrm{SMi}}(Z, X) := c(T^* X|_Z)^{-1} \cap \check{\mathrm{Mi}}(Z).$$

Assume now that X is projective, with very ample line bundle L, and that

$$Z = \{ s_j = 0 \mid j = 1, \ldots, r \}$$

is the global complete intersection of codimension $r > 0$ defined by sections $s_j \in \Gamma(X, L^{\otimes a_j})$ for suitable positive integers a_j. Then [36, Theorem 1

---

[3] There are different sign conventions in the literature. Here we adopt the convention used in the original definition of Milnor classes, [38].

and Corollary 1] implies that $\mathrm{Mi}(Z) = c_*(\varphi)$ for a constructible function φ associated to a perverse sheaf supported on the singular locus Z_{sing} of Z.[4]

Applying Proposition 2.4 we obtain:

Corollary 8.3 *Let X be a smooth projective variety with TX globally generated, and let the subvariety $Z = \{s_j = 0 \mid j = 1, \ldots, r\} \subseteq X$ be a global complete intersection as described above. Then $\check{S\mathrm{Mi}}(Z, X) \in A_*(Z)$ is non-negative. If X is an Abelian variety, then $\check{\mathrm{Mi}}(Z) \in A_*(Z)$ is non-negative.*

As an illustration, if Z has only isolated singularities, then one can consider the specialization $y = -1$ from [36, Corollary 2] to deduce:

$$\varphi = \sum_{z \in Z_{sing}} \mu_z \cdot \mathbb{1}_z \quad \text{so that} \quad \mathrm{Mi}(Z) = c_*(\varphi) = \sum_{z \in Z_{sing}} \mu_z \cdot [z],$$

with $\mu_z > 0$ the corresponding *Milnor number* of the isolated complete intersection singularity $z \in Z_{sing}$. Here the last formula for the Milnor class $\mathrm{Mi}(Z)$ is due to [45, 46]. Therefore in this case, $\mathrm{SMi}(Z, X) = \check{S\mathrm{Mi}}(Z) = \mathrm{Mi}(Z) = \check{\mathrm{Mi}}(Z)$.

8.3 Semi-abelian Varieties

Recall that a *semi-abelian* variety G is a group scheme given as an extension

$$0 \to \mathbb{T} \to G \to A \to 0$$

of an abelian variety A by a torus $\mathbb{T} \simeq (\mathbb{C}^*)^n$ $(n \geq 0)$, so that

$$G \simeq L_1^0 \times_A \cdots \times_A L_n^0 \to A$$

for some degree-zero line bundles L_i over A, with L_i^0 the open complement of the zero-section in the total space $L_i \to A$ (cf. [24, (5.5)] or [33, §4]). Then the projection $p \colon G \to A$ has the following important stability property:

(stab): The group homomorphism $p_* \colon \mathsf{F}(X') \to \mathsf{F}(X)$ induced by the morphism $p \colon X' \to X$ maps the image $\mathrm{im}(\chi_{stalk} \colon \mathrm{Perv}(X') \to \mathsf{F}(X'))$ to $\mathrm{im}(\chi_{stalk} \colon \mathrm{Perv}(X) \to \mathsf{F}(X))$.

---

[4] More precisely, [36] studies the *Hirzebruch–Milnor class* $M_y(Z)$ of Z, measuring the difference between the *virtual* and the *motivic Hirzebruch class* $T_{y*}^{vir}(Z)$ and $T_{y*}(Z)$ of Z. Specializing to $y = -1$ shows ([36, Corollary 1]) that $\mathrm{Mi}(Z) = c_*(\varphi)$ for φ the constructible function associated with the underlying perverse sheaf of a mixed Hodge module $\mathcal{M}(s_1', \ldots, s_r')$ determined by the sections $s_1, \ldots, s_r$.

Note that the constant morphism $p\colon X' \to pt$ satisfies the property (stab) if and only if X' has the following Euler characteristic property:

$$\chi(X', \mathcal{F}) \geq 0 \quad \text{for all perverse sheaves } \mathcal{F} \in \mathrm{Perv}(X'), \tag{1}$$

since $\mathbb{Z}_{\geq 0} = im(\chi_{stalk}\colon \mathrm{Perv}(pt) \to \mathsf{F}(pt)) \subseteq \mathsf{F}(pt) = \mathbb{Z}$. In particular an abelian variety satisfies (1) by Theorem 2.2 and Proposition 2.4.

Proposition 8.4 *The class of morphisms satisfying property (stab) is closed under composition. Further, the following morphisms satisfy property (stab):*

(a) *$p\colon X' \to X$ is an affine morphism with all fibers zero-dimensional (e.g., a finite morphism or an affine inclusion).*

(b) *$p\colon X' \to X$ is an affine smooth morphism of relative dimension one, with all fibers non-empty, connected and of the same non-positive Euler characteristic $\chi_p \leq 0$.*

Example 8.5 The following morphisms $p\colon X' \to X$ are affine smooth morphisms of relative dimension one, with all fibers non-empty, connected and constant non-positive Euler characteristic $\chi_p \leq 0$:

(1) $p\colon L^0 \to X$ is the open complement of the zero-section in the total space of a line bundle $L \to X$.

(2) p is the projection $p\colon X \times C \to X$ of a product with a smooth non-empty, connected affine curve C with non-positive Euler characteristic $\chi(C) \leq 0$.

(3) More generally, let $p\colon X' \to X$ be an *elementary fibration* in the sense of M. Artin (cf. [6, Definition 1.1, p. 105]), i.e., such that it can be factorized as an open inclusion $j\colon X' \to \overline{X'}$ followed by a projective smooth morphism of relative dimension one $\bar{p}\colon \overline{X'} \to X$ with irreducible (or connected) fibers, such that the induced map of the reduced complement $\bar{p}\colon Z := \overline{X'} \setminus X' \to X$ is a surjective étale covering. Then $p\colon X' \to X$ is an affine morphism [6, Lemma 1.1.2, p. 106]. If X is connected, then the genus $g \geq 0$ of the fibers of $\bar{p}$ and the degree $n \geq 1$ of the covering $\bar{p}\colon Z \to X$ are constant, so that all fibers of p have the same Euler characteristic $\chi_p = 2 - 2g - n$. The final assumption $\chi_p = 2 - 2g - n \leq 0$ just means $(g, n) \neq (0, 1)$, i.e., only the affine line $\mathbb{A}^1(\mathbb{C})$ (with $\chi(\mathbb{A}^1(\mathbb{C})) = 1$) is not allowed as a fiber of p.

The stabilization property (stab) is preserved by compositions, so that the projection

$$G \simeq L_1^0 \times_A \cdots \times_A L_n^0 \to L_2^0 \times_A \cdots \times_A L_n^0 \to \cdots \to L_n^0 \to A$$

for a semi-abelian variety has the property (stab) by the first example above. Similarly for the composition of 'elementary fibrations'

$$X'_n \xrightarrow{p_n} X'_{n-1} \xrightarrow{p_{n-1}} \cdots \xrightarrow{p_2} X'_1 \xrightarrow{p_1} X$$

over a connected base X, with all fiber Euler characteristics $\chi_{p_i} \leq 0$.

Corollary 8.6 *Assume that the morphism $p \colon X' \to X$ has the property (stab), with X a smooth variety such that TX is globally generated. Let $\mathcal{F} \in \text{Perv}(X')$ be a perverse sheaf on X', with $\varphi := \chi_{stalk}(\mathcal{F})$. Then $\check{s}_*(p_*(\varphi))$ is non-negative. If TX is trivial, then $\check{c}_*(p_*(\varphi))$ is non-negative. In particular $\chi(X', \mathcal{F}) = \chi(X', \varphi) = \chi(X, p_*(\varphi)) \geq 0$ if X is an abelian variety.*

The Euler characteristic property (1) for a semi-abelian variety G is due to [24, Corollary 1.4] (but the proof in this reference uses results about characteristic cycles on suitable compactifications and does not extend to the more general context considered above). The Euler characteristic property (1) for an algebraic torus $\mathbb{T} \simeq (\mathbb{C}^*)^n$ is due to [26, Corollary 3.4.4] in the ℓ-adic context as an application of the *generic vanishing theorem*:

$$H^i(\mathbb{T}, \mathcal{F} \otimes L) = 0 \quad \text{for } i \neq 0$$

for a given perverse sheaf $\mathcal{F} \in \text{Perv}(\mathbb{T})$ and a *generic* rank one local system L on $\mathbb{T}$. See also [32, Theorem 1.2] resp., [34, Theorem 1.1] and [33, Corollary 1.3] for the *generic vanishing theorem* for complex tori, resp., semi-abelian varieties and algebraically constructible perverse sheaves in the classical topology (as used in this paper). The following proof of Proposition 8.4(b) is an adaption to the language of constructible functions of techniques used in these references for their proofs of the generic vanishing theorem. In this way we can prove the Euler characteristic property (1) in a much more general context, e.g., for any product of connected smooth affine curves different from the affine line $\mathbb{A}^1(\mathbb{C})$ (instead of complex tori).

We close the paper with the proof of Proposition 8.4.

Proof of Proposition 8.4   Note that $\chi_{stalk} \colon \text{Perv}(X) \to \text{F}(X)$ is *additive* in the sense that $\chi_{stalk}(\mathcal{F}) = \chi_{stalk}(\mathcal{F}') + \chi_{stalk}(\mathcal{F}'')$ for any short exact sequence

$$0 \to \mathcal{F}' \to \mathcal{F} \to \mathcal{F}'' \to 0$$

in the abelian category $\text{Perv}(X)$. In particular we have $\chi_{stalk}(\mathcal{F}' \oplus \mathcal{F}'') = \chi_{stalk}(\mathcal{F}') + \chi_{stalk}(\mathcal{F}')$ and the zero-sheaf is mapped to the zero-function. Therefore χ_{stalk} induces a map from the corresponding Grothendieck group $\chi_{stalk} \colon K_0(\text{Perv}(X)) \to \text{F}(X)$, and

$$\mathrm{im}(\chi_{stalk} \colon \mathrm{Perv}(X) \to \mathsf{F}(X))$$

is a *submonoid* of the abelian group $\mathsf{F}(X)$. Moreover, χ_{stalk} commutes with both pushforwards $Rf_!, Rf_*$ for a morphism $f \colon X' \to X$ [41, §2.3], with

$$f_! = f_* \colon K_0(\mathrm{Perv}(X')) \to K_0(\mathrm{Perv}(X)) \quad \text{and} \quad f_! = f_* \colon \mathsf{F}(X') \to \mathsf{F}(X)$$

in this complex algebraic context [41, (6.41), (6.42), p. 413]. In particular, the pushforward for constructible functions is functorial with $\chi(X', \mathcal{F}) = \chi(X', \varphi) = \chi(X, f_*(\varphi))$ for $\varphi = \chi_{stalk}(\mathcal{F})$ and $\mathcal{F} \in \mathrm{Perv}(X')$. Also, this shows that the property (stab) is preserved by compositions, as claimed in Proposition 8.4. The other parts of Proposition 8.4 are proved as follows.

(a) An affine morphism $p \colon X' \to X$ with zero-dimensional fibers induces *exact* functors $Rp_!, Rp_* \colon \mathrm{Perv}(X') \to \mathrm{Perv}(X)$ [41, Corollary 6.0.5, p. 397 and Theorem 6.0.4, p. 409].

(b) Let $p \colon X' \to X$ be an affine smooth morphism of relative dimension one, with all fibers non-empty, connected and with the same non-positive Euler characteristic $\chi_p \le 0$. Then the shifted pullback $p^*[1] \colon \mathrm{Perv}(X) \to \mathrm{Perv}(X')$ is *exact*, since p is smooth of relative dimension one [41, Lemma 6.0.3, p. 386]. Note that Rp_* is not necessarily exact for the perverse t-structure. Nevertheless, since p is affine of relative dimension one, the perverse cohomology sheaves ${}^m\mathcal{H}^i(Rp_*\mathcal{F})$ vanish for $i \ne -1, 0$ for every perverse sheaf $\mathcal{F} \in \mathrm{Perv}(X')$ [41, Corollary 6.0.5, p. 397 and Theorem 6.0.4, p. 409]. Moreover, the abelian category $\mathrm{Perv}(X')$ is a *length category*, i.e., it is noetherian and artinian, so that $\mathcal{F} \in \mathrm{Perv}(X')$ is a finite iterated extension of *simple* perverse sheaves on X' [9, Theorem 4.3.1, p. 112]. By the additivity of χ_{stalk}, it is enough to consider a *simple* perverse sheaf $\mathcal{F}$ on X'. If ${}^m\mathcal{H}^{-1}(Rp_*\mathcal{F}) = 0$, then Rp_*F is also perverse, with

$$p_*(\chi_{stalk}(\mathcal{F})) = \chi_{stalk}(Rp_*\mathcal{F}) \in \mathrm{im}(\chi_{stalk} \colon \mathrm{Perv}(X) \to \mathsf{F}(X)).$$

Assume now that ${}^m\mathcal{H}^{-1}(Rp_*\mathcal{F}) \ne 0 \in \mathrm{Perv}(X)$. Then also

$$p^* \left({}^m\mathcal{H}^{-1}(Rp_*\mathcal{F}) \right)[1] \ne 0 \in \mathrm{Perv}(X')$$

by the surjectivity of p. Since the fibers of p are non-empty and connected, one gets by [9, Corollary 4.2.6.2, p. 111] a *monomorphism*

$$0 \to p^* \left({}^m\mathcal{H}^{-1}(Rp_*\mathcal{F}) \right)[1] \to \mathcal{F}.$$

This has to be an isomorphism $p^* \left({}^m\mathcal{H}^{-1}(Rp_*\mathcal{F}) \right)[1] \simeq \mathcal{F}$, since $\mathcal{F}$ is simple. As mentioned before, $Rp_!$ and Rp_* induce the same constructible function under χ_{stalk}, and the stalk of $Rp_!$ calculates the compactly supported cohomology in the corresponding fiber. But

$$\chi_{stalk}(\mathcal{F}) = \chi_{stalk}\left(p^*\left({}^m\mathcal{H}^{-1}(Rp_*\mathcal{F})\right)[1]\right) = -p^*(\varphi')$$

is constant along the fibers of p, with

$$\varphi' := \chi_{stalk}({}^m\mathcal{H}^{-1}(Rp_*\mathcal{F})) \in \text{im}(\chi_{stalk} : \text{Perv}(X) \to \mathsf{F}(X)).$$

Finally, all fibers of p have by assumption the same non-positive Euler characteristic $\chi_p \leq 0$, so that

$$p_*(\chi_{stalk}(\mathcal{F})) = -p_*(p^*(\varphi')) = -\chi_p \cdot \varphi' \in \text{im}(\chi_{stalk} : \text{Perv}(X) \to \mathsf{F}(X)),$$

since $\text{im}(\chi_{stalk} : \text{Perv}(X) \to \mathsf{F}(X))$ is a submonoid of the abelian group $\mathsf{F}(X)$. $\qquad\square$

References

[1] Aluffi, Paolo. 2003. Inclusion-exclusion and Segre classes. II. Pages 51–61 of: *Topics in algebraic and noncommutative geometry (Luminy/Annapolis, MD, 2001)*. Contemp. Math., vol. 324. Amer. Math. Soc., Providence, RI.

[2] Aluffi, Paolo. 2013. Grothendieck classes and Chern classes of hyperplane arrangements. *Int. Math. Res. Not. IMRN*, 1873–1900.

[3] Aluffi, Paolo, and Mihalcea, Leonardo Constantin. 2009. Chern classes of Schubert cells and varieties. *J. Algebraic Geom.*, **18**(1), 63–100.

[4] Aluffi, Paolo, and Mihalcea, Leonardo Constantin. 2016. Chern–Schwartz–MacPherson classes for Schubert cells in flag manifolds. *Compos. Math.*, **152**(12), 2603–2625.

[5] Aluffi, Paolo, Mihalcea, Leonardo C, Schürmann, Jörg, and Su, Changjian. 2017. Shadows of characteristic cycles, Verma modules, and positivity of Chern-Schwartz-MacPherson classes of Schubert cells. *arXiv preprint arXiv:1709.08697*.

[6] André, Yves, and Baldassarri, Francesco. 2001. *De Rham cohomology of differential modules on algebraic varieties*. Progress in Mathematics, vol. 189. Birkhäuser Verlag, Basel.

[7] Behrend, Kai. 2009. Donaldson-Thomas type invariants via microlocal geometry. *Ann. of Math. (2)*, **170**(3), 1307–1338.

[8] Beĭlinson, Alexandre, and Bernstein, Joseph. 1981. Localisation de g-modules. *C. R. Acad. Sci. Paris Sér. I Math.*, **292**(1), 15–18.

[9] Beĭlinson, A. A., Bernstein, J., and Deligne, P. 1982. Faisceaux pervers. Pages 5–171 of: *Analysis and topology on singular spaces, I (Luminy, 1981)*. Astérisque, vol. 100. Soc. Math. France, Paris.

[10] Boe, Brian D., and Fu, Joseph H. G. 1997. Characteristic cycles in Hermitian symmetric spaces. *Canad. J. Math.*, **49**(3), 417–467.

[11] Borel, A., and Remmert, R. 1961/1962. Über kompakte homogene Kählersche Mannigfaltigkeiten. *Math. Ann.*, **145**, 429–439.

[12] Borho, Walter, and MacPherson, Robert. 1983. Partial resolutions of nilpotent varieties. Pages 23–74 of: *Analysis and topology on singular spaces, II, III (Luminy, 1981)*. Astérisque, vol. 101. Soc. Math. France, Paris.

[13] Brasselet, J.-P., Lê, Dũng Tráng, and Seade, J. 2000. Euler obstruction and indices of vector fields. *Topology*, **39**(6), 1193–1208.

[14] Bressler, P., and Brylinski, J.-L. 1998. On the singularities of theta divisors on Jacobians. *J. Algebraic Geom.*, **7**(4), 781–796.

[15] Bressler, P., Finkelberg, M., and Lunts, V. 1990. Vanishing cycles on Grassmannians. *Duke Math. J.*, **61**(3), 763–777.

[16] Brion, Michel. 2005. Lectures on the geometry of flag varieties. Pages 33–85 of: *Topics in cohomological studies of algebraic varieties*. Trends Math. Birkhäuser, Basel.

[17] Brion, Michel. 2012. Spherical varieties. Pages 3–24 of: *Highlights in Lie algebraic methods*. Progr. Math., vol. 295. Birkhäuser/Springer, New York.

[18] Brylinski, J.-L., and Kashiwara, M. 1981. Kazhdan-Lusztig conjecture and holonomic systems. *Invent. Math.*, **64**(3), 387–410.

[19] Brylinski, Jean-Luc, Dubson, Alberto S., and Kashiwara, Masaki. 1981. Formule de l'indice pour modules holonomes et obstruction d'Euler locale. *C. R. Acad. Sci. Paris Sér. I Math.*, **293**(12), 573–576.

[20] Elduque, Eva, Geske, Christian, and Maxim, Laurentiu. 2018. On the signed Euler characteristic property for subvarieties of abelian varieties. *J. Singul.*, **17**, 368–387.

[21] Ernström, Lars. 1994. Topological Radon transforms and the local Euler obstruction. *Duke Math. J.*, **76**(1), 1–21.

[22] Fehér, László M., and Rimányi, Richárd. 2018. Chern-Schwartz-MacPherson classes of degeneracy loci. *Geom. Topol.*, **22**(6), 3575–3622.

[23] Fehér, László M., Rimányi, Richard, and Weber, Andrzej. 2018. Motivic Chern classes and K-theoretic stable envelopes. *Proc. London Math. Soc.*, **122**(1), 153–189.

[24] Franecki, J., and Kapranov, M. 2000. The Gauss map and a noncompact Riemann-Roch formula for constructible sheaves on semiabelian varieties. *Duke Math. J.*, **104**(1), 171–180.

[25] Fulton, William. 1984. *Intersection theory*. Berlin: Springer-Verlag.

[26] Gabber, Ofer, and Loeser, François. 1996. Faisceaux pervers *l*-adiques sur un tore. *Duke Math. J.*, **83**(3), 501–606.

[27] Ginzburg, Victor. 1986. Characteristic varieties and vanishing cycles. *Invent. Math.*, **84**(2), 327–402.

[28] Goresky, Mark, and MacPherson, Robert. 1983. Intersection homology. II. *Invent. Math.*, **72**(1), 77–129.

[29] Hotta, Ryoshi, Takeuchi, Kiyoshi, and Tanisaki, Toshiyuki. 2008. *D-modules, perverse sheaves, and representation theory*. Progress in Mathematics, vol. 236. Birkhäuser Boston, Inc., Boston, MA.

[30] Kashiwara, M., and Tanisaki, T. 1984. The characteristic cycles of holonomic systems on a flag manifold related to the Weyl group algebra. *Invent. Math.*, **77**(1), 185–198.

[31] Lê, Dũng Tráng, and Teissier, Bernard. 1981. Variétés polaires locales et classes de Chern des variétés singulières. *Ann. of Math. (2)*, **114**(3), 457–491.

[32] Liu, Yongqiang, Maxim, Laurentiu, and Wang, Botong. 2018. Mellin transformation, propagation, and abelian duality spaces. *Adv. Math.*, **335**, 231–260.

[33] Liu, Yongqiang, Maxim, Laurentiu, and Wang, Botong. 2021. Perverse sheaves on semi-abelian varieties. *Selecta Math. (N.S.)*, **27**(2), article no. 30.

[34] Liu, Yongqiang, Maxim, Laurentiu, and Wang, Botong. 2019. Generic vanishing for semi-abelian varieties and integral Alexander modules. *Math. Z.*, **293**(1-2), 629–645.

[35] MacPherson, R. D. 1974. Chern classes for singular algebraic varieties. *Ann. of Math. (2)*, **100**, 423–432.

[36] Maxim, Laurentiu, Saito, Morihiko, and Schürmann, Jörg. 2013. Hirzebruch-Milnor classes of complete intersections. *Adv. Math.*, **241**, 220–245.

[37] Ohmoto, Toru. 2006. Equivariant Chern classes of singular algebraic varieties with group actions. *Math. Proc. Cambridge Philos. Soc.*, **140**(1), 115–134.

[38] Parusiński, Adam, and Pragacz, Piotr. 2001. Characteristic classes of hypersurfaces and characteristic cycles. *J. Algebraic Geom.*, **10**(1), 63–79.

[39] Popa, Mihnea, and Schnell, Christian. 2013. Generic vanishing theory via mixed Hodge modules. *Forum Math. Sigma*, **1**, e1, 60.

[40] Sabbah, Claude. 1985. Quelques remarques sur la géométrie des espaces conormaux. *Astérisque*, 161–192.

[41] Schürmann, Jörg. 2003. *Topology of singular spaces and constructible sheaves*. Instytut Matematyczny Polskiej Akademii Nauk. Monografie Matematyczne (New Series) [Mathematics Institute of the Polish Academy of Sciences. Mathematical Monographs (New Series)], vol. 63. Birkhäuser Verlag, Basel.

[42] Schürmann, Jörg. 2005. Lectures on characteristic classes of constructible functions. Pages 175–201 of: *Topics in cohomological studies of algebraic varieties*. Trends Math. Birkhäuser, Basel. Notes by Piotr Pragacz and Andrzej Weber.

[43] Schürmann, Jörg. 2017. Chern classes and transversality for singular spaces. Pages 207–231 of: *Singularities in Geometry, Topology, Foliations and Dynamics*. Trends in Mathematics. Basel: Birkhäuser.

[44] Schürmann, Jörg, and Tibăr, Mihai. 2010. Index formula for MacPherson cycles of affine algebraic varieties. *Tohoku Math. J. (2)*, **62**(1), 29–44.

[45] Seade, José, and Suwa, Tatsuo. 1998. An adjunction formula for local complete intersections. *Internat. J. Math.*, **9**(6), 759–768.

[46] Suwa, Tatsuo. 1997. Classes de Chern des intersections complètes locales. *C. R. Acad. Sci. Paris Sér. I Math.*, **324**(1), 67–70.

[47] Weissauer, Rainer. 2006. Brill-Noether Sheaves. *arXiv preprint arXiv:math/0610923*.

[48] Zhang, Xiping. 2018. Chern classes and characteristic cycles of determinantal varieties. *J. Algebra*, **497**, 55–91.

2

Brill–Noether Special Cubic Fourfolds of Discriminant 14

Asher Auel[a]

To Bill Fulton, on the occasion of his 80th birthday.

Abstract. We study the Brill–Noether theory of curves on polarized K3 surfaces that are Hodge theoretically associated to cubic fourfolds of discriminant 14. We prove that any smooth curve in the polarization class has maximal Clifford index and deduce that a smooth cubic fourfold contains disjoint planes if and only if it admits a Brill–Noether special associated K3 surface of degree 14. As an application, we prove that the complement of the pfaffian locus, inside the Noether–Lefschetz divisor $\mathcal{C}_{14}$ in the moduli space of smooth cubic fourfolds, is contained in the irreducible locus of cubic fourfolds containing two disjoint planes.

1 Introduction

Let X be a *cubic fourfold*, i.e., a smooth cubic hypersurface $X \subset \mathbb{P}^5$ over the complex numbers. Determining the rationality of X is a classical question in algebraic geometry. Some classes of rational cubic fourfolds have been described by Fano [11] and Tregub [48], [49]. Beauville and Donagi [5] prove that *pfaffian* cubic fourfolds, i.e., those defined by pfaffians of skew-symmetric 6×6 matrices of linear forms, are rational. Hassett [19] describes, via lattice theory, Noether–Lefschetz divisors $\mathcal{C}_d$ in the moduli space $\mathcal{C}$ of smooth cubic fourfolds. A parameter count shows that $\mathcal{C}_{14}$ is the closure of the locus $\mathcal{Pf}$ of pfaffian cubic fourfolds; Hodge theory shows (see [50, §3 Prop. 2]) that $\mathcal{C}_8$ is the locus of cubic fourfolds containing a plane. Hassett [18] identifies countably many divisors of $\mathcal{C}_8$ consisting of rational cubic fourfolds. Recently, Addington, Hassett, Tschinkel, and Várilly-Alvarado [2] identify countably

[a] Dartmouth College, New Hampshire

many divisors of C_{18} consisting of rational cubic fourfolds, and Russo and Stagliano [45], [46] have shown that the very general cubic fourfolds in C_{26}, C_{38}, and C_{42} are rational. Nevertheless, it is expected that the very general cubic fourfold (as well as the very general cubic fourfold containing a plane) is not rational.

Short of a pfaffian presentation, how can one tell if a given cubic fourfold is pfaffian? Beauville [4] provides a homological criterion for a cubic hypersurface to be pfaffian, which for cubic fourfolds is equivalent to containing a quintic del Pezzo surface, but it is not clear how to translate this criterion into Hodge theory. More generally, how can one understand the complement $C_{14} \setminus Pf$ of the pfaffian locus? Such questions are implicit in [3] and [48], where cubic fourfolds with certain numerical properties are shown to be outside or inside, respectively, the pfaffian locus. In particular, Tregub studies the locus C_Π of cubic fourfolds that contain two disjoint planes, showing that this locus is irreducible of codimension 2 in C, and that the general member does not contain a smooth quartic rational normal scroll nor a quintic del Pezzo surface, hence cannot be pfaffian. Our main result is that this is essentially all of the complement of the pfaffian locus.

Theorem 1.1 *The complement of the pfaffian locus Pf, inside the Noether–Lefschetz divisor C_{14} of the moduli space of cubic fourfolds, is contained in the irreducible locus C_Π of cubic fourfolds containing two disjoint planes.*

In other words, any $X \in C_{14}$ is pfaffian or contains two disjoint planes (or both).

The proof combines several ingredients revolving around the Brill–Noether theory of special divisors on curves in K3 surfaces of degree 14. We use the determination, due to Mukai [40], [41] of the smooth projective curves C of genus 8 that are linear sections of the grassmannian $G(2,6) \subset \mathbb{P}^{14}$. This turns out to be equivalent to C lacking a g_7^2, equivalently, that C is Brill–Noether general. We also use a modified conjecture of Harris and Mumford, as proved by Green and Lazarsfeld [15], as well as the generalization due to Lelli-Chiesa [36], on line bundles on K3 surfaces computing the Clifford indices of smooth curves in a given linear system. We also need the earlier work of Saint-Donat [47] and Reid [43], [44], on hyperelliptic and trigonal linear systems on K3 surfaces, as well as useful refinements due to Knutsen [25], [26], [27] of the original result by Green and Lazarsfeld. Combining these results with lattice theory computations for cubic fourfolds and their associated K3 surfaces, as developed by Hassett [18], [19], we prove that the Clifford index of curves in the polarization class of any K3 surface of degree 14 associated to X must take the maximal value 3 (see Theorem 5.4), putting strong constraints on the geometry of cubic fourfolds in terms of the Brill–Noether theory of their associated K3 surfaces.

More generally, one might call a cubic fourfold X *Brill–Noether special* if X has an associated K3 surface S that is Brill–Noether special in the sense of Mukai [41, Def. 3.8], a condition implying that S has an ample divisor such that the general curve in its linear system is Brill–Noether special, see §2.2. Then our main result can be summarized by saying that a special cubic fourfold of discriminant 14 is Brill–Noether special if and only if it contains two disjoint planes. It would be interesting to study the Brill–Noether special loci in other divisors C_d of special cubic fourfolds, for example, in C_{26}, C_{38}, and C_{42}. In the context of discriminant 26, Farkas and Verra [12] also appeal to the Brill–Noether theory of some associated K3 surfaces.

Our result, and more generally the ability to detect a pfaffian cubic fourfold via Hodge theory, has two immediate applications. First, we obtain a new explicit proof that every cubic fourfold in C_{14} is rational: Beauville and Donagi [5] prove that any pfaffian cubic fourfold is rational, and by a much more classical construction going back to Fano, every cubic fourfold containing disjoint planes is rational; this covers all cubic fourfolds in C_{14}. This rationality result was initially obtained by Bolognesi, Russo, and Staglianò [6] using a much more classical approach involving one apparent double point surfaces, though this has been recently subsumed by the path-breaking work on the deformation invariance of rationality by Kontsevich and Tschinkel [29]. Second, we prove the existence of nonempty irreducible components of $\mathcal{Pf} \cap \Pi$, which are necessarily of codimension ≥ 3 in $\mathcal{C}$. This immediately implies that the pfaffian locus is not Zariski open in C_{14}. While this result was initially obtained in the course of conversations with M. Bolognesi and F. Russo based on the computer algebra calculations of G. Staglianò and earlier drafts of our respective papers, the proof presented in §6.1 does not require any explicit computer algebra computations (as opposed to the proof in [6]). However, it still seems plausible that the pfaffian locus is open inside the moduli space of marked cubic fourfolds of discriminant 14.

The author is indebted to Y. Tschinkel and F. Bogomolov for providing a stimulating work environment at the Courant Institute of Mathematical Sciences, where this project started in May 2013, and to B. Hassett, who first suggested the possibility of investigating the Brill–Noether theory of curves on K3 surfaces in the context of cubic fourfolds. The author also thanks M. Bolognesi and F. Russo for animated and productive conversations during the preparation of this manuscript in March 2015, while we were exchanging our respective drafts. We are grateful to M. Hoff, D. Jensen, A. L. Knutsen, and M. Lelli-Chiesa for detailed explanations of various aspects of their work;

to N. Addington, T. Johnsen, A. Kumar, R. Lazarsfeld, and H. Nuer for
helpful conversations; and to the anonymous referee for very constructive
comments on the manuscript. The author was partially supported by NSF grant
DMS-0903039 and an NSA Young Investigator Grant.

2 Brill–Noether Theory for Polarized K3 Surfaces of Degree 14

All varieties are assumed to be over the complex numbers and all K3 surfaces
are assumed to be smooth and projective.

2.1 Grassmannians and Curves of Genus 8

Let $G(2,6) \subset \mathbb{P}^{14}$ be the grassmannian of 2-planes in a 6-dimensional vector
space, embedded in $\mathbb{P}^{14}$ via the Plücker embedding. It was classically known
that a general flag of linear subspaces $P \subset Q$ of dimension 6 and 7 in $\mathbb{P}^{14}$
cut from $G(2,6)$ a K3 surface of degree 14 containing a canonical curve C of
genus 8.

Recall that a g_d^r on a smooth projective curve C is a line bundle A of degree
d with $h^0(C,A) \geq r + 1$; it is *complete* if $h^0(C,A) = r + 1$.

Theorem 2.1 (Mukai [40]) *A smooth projective curve C of genus 8 is a linear
section of the grassmannian $G(2,6) \subset \mathbb{P}^{14}$ if and only if C has no g_7^2.*

The Brill–Noether theorem states that when

$$\rho(g,r,d) = g - (r+1)(g-d+r)$$

is negative, the general curve of genus g has no g_d^r. A curve supporting
such a g_d^r is called *Brill–Noether special*. A curve not supporting any g_d^r
whenever $\rho(g,r,d) < 0$ is called *Brill–Noether general*. When $\rho(g,r,d) =
-1$, Eisenbud and Harris [10] proved that the locus of curves, in the moduli
space $\mathcal{M}_g$ of curves of genus g, that support such a g_d^r, is irreducible of
codimension 1. In particular, the locus of curves of genus 8 having a g_7^2 is
of codimension 1 in $\mathcal{M}_8$.

The *Clifford index* of a line bundle A on a smooth projective curve C is the
integer

$$\gamma(A) = \deg(A) - 2r(A),$$

where $r(A) = h^0(C,A) - 1$ is the *rank* of A. The Clifford index of C is

$$\gamma(C) = \min\{\gamma(A) : h^0(C,A) \geq 2 \text{ and } h^1(C,A) \geq 2\}$$

and a line bundle A on C is said to compute the Clifford index of C if $\gamma(A) = \gamma(C)$. Clifford's theorem states that $\gamma(C) \geq 0$ with equality if and only if C is hyperelliptic; similarly $\gamma(C) = 1$ if and only if C is trigonal or a smooth plane quintic. At the other end, $\gamma(C) \leq \lfloor (g-1)/2 \rfloor$ with equality whenever C is Brill–Noether general.

Up to taking the adjoint line bundle $\omega_C \otimes A^\vee$, which has the same Clifford index, we can always assume that nontrivial special divisors g_d^r satisfy $1 \leq r \leq \lfloor (g-1)/2 \rfloor$ and $2 \leq d \leq g-1$. For $g = 8$, we list them for the convenience of the reader:

γ	3		2		1			0		
g_d^r	g_5^1	g_7^2	g_4^1	g_6^2	g_3^1	g_5^2	g_7^3	g_2^1	g_4^2	g_6^3
ρ	0	-1	-2	-4	-4	-7	-8	-6	-10	-12

In genus 8, the Brill–Noether special locus is controlled by the existence of a g_7^2.

Lemma 2.2 *A smooth projective curve C of genus 8 is Brill–Noether special if and only if it has a complete g_7^2.*

Proof First note that if a curve has a complete g_d^r, then it has a complete g_d^k for all k between $d - g$ and r. Hence if C has a g_7^2 then it has a complete g_7^2. We can argue by the Clifford index. In Clifford index 3, the only special divisor is a g_7^2. For Clifford index 2, we use the facts that any genus 8 curve with a g_6^2 has a g_4^1 and any genus 8 curve with a g_4^1 has a g_7^2, see [40, Lemmas 3.4, 3.8]. In Clifford index 1, any genus 8 curve is trigonal, so taking twice the g_3^1 and adding a base point will result in a g_7^2. Finally, in Clifford index 0, the curve is hyperelliptic, so taking thrice the g_2^1 and adding a base point will result in a g_7^2. $\square$

2.2 Brill–Noether Theory for Polarized K3 Surfaces

A *polarized K3 surface* (S, H) of degree d is a smooth projective K3 surface S together with a primitive ample line bundle H of self-intersection $d \geq 2$. If $C \subset S$ is a smooth irreducible curve in the linear system $|H|$, so that $|H|$ is base point free (i.e., H is globally generated), then $d = 2g - 2$, where g is the genus of C. Following Mukai [41, Def. 3.8], we say that a polarized K3 surface (S, H) of degree $2g - 2$ is *Brill–Noether general* if $h^0(S, H') h^0(S, H'') < h^0(S, H) = g + 1$ for any nontrivial decomposition $H = H' \otimes H''$. Otherwise, we say *Brill–Noether special*. If a smooth irreducible curve $C \in |H|$ is

Brill–Noether general then it follows that (S, H) is Brill–Noether general, cf. [24, Rem. 10.2]. While the converse is an open question in general, for low degrees it was checked by Mukai, using a case-by-case analysis.

Theorem 2.3 *A polarized K3 surface (S, H), with H globally generated of degree ≤ 18 or 22, is Brill–Noether general if and only if some smooth irreducible $C \in |H|$ is Brill–Noether general.*

Of course, we are mainly interested in the degree 14 case, where the results assembled below will suffice to prove the theorem.

The existence of special divisors on curves in a K3 surface was considered by Saint-Donat [47] and Reid [43], [44]. Harris and Mumford conjectured that the gonality of a curve should be constant in a base point free linear system on a K3 surface. A counterexample was found by Donagi and Morrison [9] (in fact, this turned out to be the unique counterexample, cf. [7], [28]) and the conjecture was modified by Green [16, Conj. 5.8] to one about the constancy of the Clifford index in a linear system. In a similar spirit, one is interested in the question of when a given g_d^r on a curve in a K3 surface is the restriction of a line bundle from the K3. The conjecture of Green was proved in a celebrated paper by Green and Lazarsfeld.

Theorem 2.4 (Green–Lazarsfeld [15]) *Let S be a K3 surface and $C \subset S$ a smooth irreducible curve of genus $g \geq 2$. Then $\gamma(C') = \gamma(C)$ for every smooth curve $C' \in |C|$. Furthermore, if $\gamma(C) < \lfloor (g - 1)/2 \rfloor$ then there exists a line bundle L on S whose restriction to any $C' \in |C|$ computes the Clifford index of C'.*

We can thus define the *Clifford index* $\gamma(S, H)$ of a polarized K3 surface (S, H) with H globally generated to be the Clifford index of any smooth irreducible curve $C \in |H|$, which is well-defined by Theorem 2.4.

In the case where (S, H) has degree 14, so that a smooth curve $C \in |H|$ has genus 8, we have that $\gamma(S, H) \leq 3$. If (S, H) is Brill–Noether general, then by Theorem 2.3, some smooth curve $C \in |H|$ is Brill–Noether general (hence has maximal Clifford index), so that $\gamma(S, H) = 3$. When (S, H) is Brill–Noether special and $\gamma(S, H) < 3$, the result of Green and Lazarsfeld allows us to find a line bundle on S whose restriction to $C \in |H|$ computes the Clifford index. In fact, already for $\gamma(S, H) \leq 1$, results of Saint-Donat [47, Thm. 5.2] and Reid [43, Thm. 1] ensure that these line bundles can be chosen to be elliptic pencils, see §2.3 for details. Finally, when (S, H) is Brill–Noether special and $\gamma(S, H) = 3$, we would like to know if a g_7^2 on a smooth curve $C \in |H|$ is the restriction of a line bundle on S. Since a g_7^2 has the generic Clifford index, we cannot appeal to the result of Green and Lazarsfeld. This situation, of Clifford general but not Brill–Noether general polarized K3 surfaces, is discussed more generally in [24, §10.2].

To this end, we have the following much more powerful result of Lelli-Chiesa, concerning when a specific g_d^r on a curve $C \subset S$ lying in a K3 surface is the restriction of a line bundle on S.

Theorem 2.5 (Lelli-Chiesa [36]) *Let S be a K3 surface and $C \subset S$ a smooth irreducible curve of genus $g \geq 2$ that is neither hyperelliptic nor trigonal. Let A be a complete g_d^r such that $r > 1$, $d \leq g - 1$, $\rho(g, r, d) < 0$, and $\gamma(A) = \gamma(C)$. Assume that there is no irreducible genus 1 curve $E \subset S$ such that $E.C = 4$ and no irreducible genus 2 curve $B \subset S$ such that $B.C = 6$. Then A is the restriction of a globally generated line bundle L on S.*

This result comes from an in-depth study of generalized Lazarsfeld–Mukai bundles extending the original strategy of [15].

Remark 2.6   According to [36, Thm. 4.2*ff.*], the hypothesis on curves of genus 1 and 2 is completely satisfied as long as $\gamma(C) > 2$; otherwise, there is a list of seven exceptional cases when $\gamma(C) = 2$. We also remark that, according to the construction in the proof of [36, Thm. 4.2] (see also [35, Lemma 3.3] and [15, Lemma 3.1]), the line bundle L can be chosen to be globally generated, though this is not mentioned in the statement of the main theorem in [36].

When a K3 surface has Picard rank one, Lazarsfeld [34] has shown that the general curve in the linear system of the polarization class is Brill–Noether general. Hence Brill–Noether special K3 surfaces have higher Picard rank.

2.3 Brill–Noether Special K3 Surfaces
via Lattice-polarizations

Let Σ be an even nondegenerate lattice of signature $(1, \rho - 1)$ with a distinguished class H of even norm $d > 0$. A Σ-*polarized* K3 surface is a polarized K3 surface (S, H) of degree d together with a primitive isometric embedding $\Sigma \hookrightarrow \mathrm{Pic}(S)$ preserving H. For a general discussion of lattice-polarized K3 surfaces and their moduli, see [8]. In particular, there exists a quasi-projective coarse moduli space $\mathcal{K}_\Sigma$ of dimension $20 - \rho$ and a forgetful morphism $\mathcal{K}_\Sigma \to \mathcal{K}_d$ to the moduli space of polarized K3 surfaces of degree d. The main result of this section is the following characterization of Brill–Noether special K3 surfaces of degree 14 via lattice polarizations. The same result is obtained by Greer, Li, and Tian [17] using a different calculation.

Table 1. *Lattices embedded in Brill–Noether special K3 surfaces of degree 14 and Clifford index* γ. *Here, d and d_0 are the discriminants of the lattice and of* $\langle H \rangle^\perp$, *respectively. The pair (b,c) refers to the unique rank 3 cubic fourfold lattice, whose associated lattice is the given one, normalized as in Proposition 4.1.*

$\gamma = 0$		$\gamma = 1$		$\gamma = 2$				$\gamma = 3$	
	H	E		H	E		H	E		H	L		H	L

	H	E		H	E		H	E		H	L		H	L
H	14	2	H	14	3	H	14	4	H	14	6	H	14	7
E	2	0	E	3	0	E	4	0	L	6	2	L	7	2

$d_S = -4$	$d_S = -9$	$d_S = -16$	$d_S = -8$	$d_S = -21$
$d_S^0 = -14$	$d_S^0 = -14 \cdot 9$	$d_S^0 = -14 \cdot 4$	$d_S^0 = -14 \cdot 2$	$d_S^0 = -6$
$(b,c) = (6,8)$	$(b,c) = (5,6)$	$(b,c) = (2,2)$	$(b,c) = (4,4)$	$(b,c) = (7,12)$

Theorem 2.7 *If a polarized K3 surface (S, H), with H globally generated of degree 14, is Brill–Noether special then it admits a lattice polarization for one of the five rank 2 lattice appearing in Table 1 and $\gamma(S, H)$ is bounded above by the corresponding value of γ on the table.*

To round out the classification, we remark that elliptic K3 surfaces (S, H) of degree 14 with a section are also Brill–Noether special. These admit a lattice polarization with a class E such that $E^2 = 0$ and $E.H = 1$. However, in this case, the linear system $|H|$ contains a nontrivial fixed component in its base locus, and in particular, there is no Clifford index defined. Taken together with the five lattices listed in Table 1, this shows that the Brill–Noether special locus in $\mathcal{K}_{14}$ is the union of six Noether–Lefschetz divisors.

Before the proof of Theorem 2.7, we need some lemmas on elliptic pencils on K3 surfaces, which are mostly contained in the work of Saint-Donat [47] and Knutsen [26], [27]. By an *elliptic pencil* we mean a line bundle E on a K3 surface S such that the generic member of the linear system $|E|$ is a smooth genus one curve. A result of Saint-Donat [47, Prop. 2.6(ii)] says that if E is generated by global sections and $E^2 = 0$, then E is a multiple of an elliptic pencil. If E is an elliptic pencil then E is primitive in $\mathrm{Pic}(S)$ (cf. [23, Ch. 2, Remark 3.13(i)]), $E^2 = 0$, $h^0(S, E) = 2$, and $h^1(S, E) = 0$.

Lemma 2.8 *Let (S, H) be a polarized K3 surface of degree $2g - 2 \geq 2$, let $C \in |H|$ be a smooth irreducible curve, and let E be a globally generated line bundle on S with $E^2 = 0$ and $E.C = d < 2g - 2$. Then $E|_C$ is a g_d^1 if and only if E is an elliptic pencil such that $h^1(S, E(-C)) = 0$.*

Proof First remark that since H is globally generated and $(E - C).C = d - (2g - 2) < 0$ by hypothesis, we get that $h^0(S, E(-C)) = 0$, cf. [27, Proof of Prop. 2.1].

Now, assume that E is an elliptic pencil and that $h^1(S, E(-C)) = 0$. Then the long exact sequence in cohomology associated to the exact sequence of sheaves

$$0 \to E(-C) \to E \to E|_C \to 0$$

together with the fact that $h^0(S, E) = 2$, implies that $h^0(C, E|_C) = 2$. Since $\deg(E|_C) = E.C = d$, we have that $E|_C$ is a g_d^1.

Now assume that $E|_C$ is a g_d^1. By Saint-Donat [47, Prop. 2.6(ii)], $E = F^{\otimes k}$ for an elliptic pencil F and some $k \geq 1$ dividing d. Again considering the same long exact sequence as above, the last terms, when rewritten using Serre duality and the fact that $H^0(S, E^\vee) = 0$ since E is effective, read

$$H^1(S, E|_C) \to H^0(S, E(-C)^\vee) \to 0.$$

By Riemann–Roch on C, we have

$$h^1(S, E|_C) = h^1(C, E|_C) = 2 - (d - g + 1),$$

hence $h^0(S, E(-C)^\vee) \leq 2 - (d - g + 1)$. By Riemann–Roch on S, we have

$$h^0(S, E(-C)^\vee) - h^1(E(-C)) = 2 + \frac{1}{2}(E - C)^2 = 2 + \frac{1}{2}(-2d + 2g - 2)$$

$$= 2 - (d - g + 1),$$

using Serre duality and the fact that $H^0(S, E(-C)) = 0$. Hence we have $h^0(S, E(-C)^\vee) \geq 2 - (d - g + 1)$ and $h^1(S, E(-C)) = 0$. However, the beginning terms of the long exact sequence read

$$0 \to H^0(S, E) \to H^0(S, E|_C) \to H^1(E(-C))$$

implying that $h^0(S, E) = h^0(C, E|_C) = 2$ (since $E|_C$ is a g_e^1). But $h^0(S, E) = k + 1$ and thus we conclude that $k = 1$, i.e., E is an elliptic pencil. $\qquad\square$

Proof of Theorem 2.7 Let $C \subset S$ be a smooth irreducible curve (of genus 8) in the linear system of H. We argue by the Clifford index of (S, H), equivalently, of C.

If $\gamma(C) = 0$, i.e., C is hyperelliptic by Clifford's Theorem, then by Saint-Donat [47, Thm. 5.2] (cf. Reid [43, Prop. 3.1]), the g_2^1 on C is the restriction of an elliptic pencil E such that $E.H = 2$.

If $\gamma(C) = 1$, i.e., C is trigonal, then by Reid [44, Thm. 1] (cf. [47, Thm. 7.2]), after verifying $8 > \frac{1}{4}3^2 + 3 + 2$, the g_3^1 on C is the restriction of an elliptic pencil E such that $E.H = 3$.

In these first two cases, the sublattice of Pic(S) generated by H and E is primitive. Indeed, if not, then this sublattice admits a finite index overlattice contained in Pic(S). However, using the correspondence between finite index overlattices and isotropic subgroups of the discriminant form (cf., Nikulin [42, §1.4]), we find that, in this case, the only finite index overlattice would admit a class $F \in$ Pic(S), where $eF = E$, for $e = 2$ or 3, respectively. However, as E is an elliptic pencil on S, it is a primitive class in Pic(S), hence no such overlattice exists.

If $\gamma(C) = 2$, then by Green–Lazarsfeld [15] (since the generic value of the Clifford index is 3, see Theorem 2.4), there is a line bundle L on S such that $L|_C$ is a g_4^1 or a g_6^2. Then $L.H = \deg(L|_C) = 4$ or 6, respectively. Furthermore, by a result of Knutsen [25, Lemma 8.3], we can choose L satisfying

$$0 \le L^2 \le 4 \quad \text{and} \quad 2L^2 \le L.H \quad \text{and} \quad 2 = L.H - L^2 - 2$$

with $L^2 = 4$ or $2L^2 = L.H$ if and only if $H = 2L$. However, since 14 is squarefree, $H = 2L$ is impossible, hence the only possibilities are that $L^2 = 0$ and $L.H = 4$, $L^2 = 0$ and $L.H = 6$, or $L^2 = 2$ and $L.H = 6$. As a consequence of Martens' proof [38] of the main result of [15] (cf. proof of [25, Lemma 8.3]), we can also choose L generated by global sections and with $h^1(S, L(-C)) = 0$. Suggestively, in the two former cases, we denote L by E.

We now argue that the case $E^2 = 0$ and $E.H = 6$ is impossible. First assume that E is an elliptic pencil. Lemma 2.8 then implies that $E|_C$ is a g_6^1, contradicting the assumption that it is a g_6^2. Hence E cannot be an elliptic pencil. Thus by the result of Saint-Donat mentioned above, $E = kF$ for $k = 2$, 3, 6 and an elliptic pencil F. The case $k = 6$ is impossible, since $F^2 = 0$ and $F.H = 1$ contradicts the ampleness of H. For $k = 2, 3$, we have $F^2 = 0$ and $F.H = 6/k \le 3$, so that results of Saint-Donat [47, Prop. 5.2, 7.15] imply that $F|_C$ is a $g_{6/k}^1$, contradicting the fact that $\gamma(C) = 2$.

In the remaining two cases, we argue that the sublattice of Pic(S) generated by H and E (resp. H and L) is primitive. As before, we appeal to the correspondence between finite index overlattices and isotropic subgroups of the discriminant form (cf., Nikulin [42, §1.4]). In the case $E^2 = 0$ and $E.H = 4$, the only finite overlattice would contain a class dividing E, however since $E|_C$ must be a g_4^1, then by Lemma 2.8, E is an elliptic pencil and is thus a primitive class in Pic(S). Hence, the sublattice generated by H and E is primitive. In the case $L^2 = 2$ and $L.H = 6$, the only finite index overlattice would contain a class $F \in$ Pic(S) such that $2F = H - L$, however, such F would then satisfy $F^2 = (H - L)^2/4 = 1$, which is impossible since Pic(S) is an even lattice. Hence, the sublattice generated by H and L is primitive.

Finally, assume that $\gamma(C) = 3$. Then all the hypotheses of the results of Lelli-Chiesa [36] (see Theorem 2.5) are satisfied, hence there exists a line bundle L on S such that $L|_C$ is a g_7^2. In particular, we have that $L.C = \deg(L|_C) = 7$. As before, by Remark 2.6, L can be chosen to be globally generated, so that $2n = L^2 \geq 0$. Furthermore, by [24, Prop. 10.5], we can choose L so that $L^2 = 2$. The sublattice of $\mathrm{Pic}(S)$ generated by H and L is then primitive since its discriminant is squarefree.

Thus in each case, the polarized K3 surface (S, H) has a lattice-polarization with respect to one of the lattices on Table 1.   □

Remark 2.9 Every smooth curve C of genus 8 contains a finite number of g_5^1 linear systems. If $\gamma(C) = 3$ and C lies on a K3 surface S with a primitive degree 14 polarization H, then it could happen that none of the g_5^1 divisors are the restriction of a line bundle from S (e.g., the corresponding Lazarsfeld–Mukai bundles are simple). However, if a g_5^1 is the restriction of a line bundle on S, then arguing as in the proof of Theorem 2.7, one can verify that the Picard lattice of S admits a primitive sublattice generated by H and E, where E is an elliptic pencil such that $H.E = 5$ and $E|_C$ is the g_5^1.

3 Lattice Polarized Cubic Fourfolds

Let X be a smooth cubic fourfold and let $A(X)$ denote the lattice of codimension 2 algebraic cycles $\mathrm{CH}^2(X)$ with its usual intersection form. Then via the cycle class map, $A(X)$ is isomorphic to $H^4(X, \mathbb{Z}) \cap H^{2,2}(X)$ by the validity of the integral Hodge conjecture for cubic fourfolds proved by Voisin [51].

Given a positive definite lattice Λ containing a distinguished element h^2 of norm 3, a Λ-*polarized* cubic fourfold is a cubic fourfold X together with the data of a primitive isometric embedding $\Lambda \hookrightarrow A(X)$ preserving h^2. The main results of Looijenga [37] and Laza [33] on the description of the period map for cubic fourfolds imply that smooth Λ-polarized cubic fourfolds exist if and only if Λ admits a primitive embedding into $H^4(X, \mathbb{Z}) = \langle 1 \rangle^{\oplus 21} \oplus \langle -1 \rangle^{\oplus 2}$ and Λ contains no short roots (i.e., elements $v \in \Lambda$ with norm 2 such that $v.h^2 = 0$) nor long roots (i.e., elements $v \in \Lambda$ with norm 6 such that $v.h^2 = 0$ and $v.\langle h^2 \rangle^{\perp} \subset 3\mathbb{Z}$ where we compute $\langle h^2 \rangle^{\perp} \subset H^4(X, \mathbb{Z})$). We call any such lattice Λ a *cubic fourfold lattice*. We remark that the conditions defining short and long roots can be checked completely within the lattice Λ: for short roots, this is clear; for long roots $v \in \Lambda$, the conditions are equivalent to $v.v = 6$, $v.h^2 = 0$, and $v \pm h^2$ is divisible by 3 in Λ.

For a cubic fourfold lattice Λ of rank ρ, an adaptation of the argument of Hassett [19, Thm. 3.1.2] (see also [20, §2.3]) proves that the moduli space

C_Λ of Λ-polarized cubic fourfolds is a quasi-projective variety of dimension $21 - \rho$. There is a forgetful map $C_\Lambda \to C$, whose image we denote by $C_{[\Lambda]}$. In other words, $C_{[\Lambda]} \subset C$ is the locus of cubic fourfolds X such that $A(X)$ admits a primitive isometric embedding of Λ preserving h^2. We remark that the forgetful map $C_\Lambda \to C_{[\Lambda]}$ is generically finite to one, and whose degree depends on the number of automorphisms of Λ fixing h^2.

The possible rank 2 cubic fourfold lattices were classified by Hassett [19]; such a lattice K_d is uniquely determined by its discriminant, which can be any number $d > 6$ such that $d \equiv 0, 2 \pmod 6$. Then C_{K_d} coincides with the moduli space C_d^{mar} of *marked* special cubic fourfolds of discriminant d considered by Hassett [19, §5.2] and $C_{[K_d]}$ coincides with the Noether–Lefschetz divisor $C_d \subset C$. For various cubic fourfold lattices Λ of rank 3, the loci $C_{[\Lambda]}$ were considered in [1], [3], [6], [14], [48], [49].

Given a primitive embedding $\Lambda \hookrightarrow \Lambda'$ of cubic fourfold lattices preserving h^2, there is an induced morphism $C_{\Lambda'} \to C_\Lambda$ and an inclusion of subvarieties $C_{[\Lambda']} \subset C_{[\Lambda]}$. In particular, we have that $C_{[\Lambda]} \subset C_d$ whenever Λ admits a primitive embedding of K_d preserving h^2.

When $\Lambda = \Pi$ is the lattice with Gram matrix

$$
\begin{array}{c|ccc}
 & h^2 & T & P \\
\hline
h^2 & 3 & 4 & 1 \\
T & 4 & 10 & -1 \\
P & 1 & -1 & 3
\end{array}
\quad \cong \quad
\begin{array}{c|ccc}
 & h^2 & P & P' \\
\hline
h^2 & 3 & 1 & 1 \\
P & 1 & 3 & 0 \\
P' & 1 & 0 & 3
\end{array}
\tag{1}
$$

with the isomorphism defined by $T = 2h^2 - P - P'$, then $C_{[\Pi]}$ is one of the most well-studied codimension 2 loci in the moduli space of cubic fourfolds, cf. [11], [48], [50, §3, App.].

Proposition 3.1 *The subvariety $C_{[\Pi]} \subset C$ is an irreducible component of $C_8 \cap C_{14}$ and coincides with the locus of cubic fourfolds that contain disjoint planes.*

Proof The proof of the first statement is in [3, Thm. 4], cf. [11], [48]. The existence of two disjoint planes follows from the proof given in Voisin [50, §3, App., Prop.] and the refinement due to Hassett [19, §3].   □

Fix an *admissible discriminant* $d > 6$, i.e., such that $d \equiv 0, 2 \bmod 6$ and such that $4 \nmid d$, $9 \nmid d$, and $p \nmid d$ for any odd prime $p \equiv 2 \bmod 3$. Hassett [19, §5] proves that for any cubic fourfold X with a marking of discriminant d, the orthogonal complement $K_d^\perp$ of K_d inside $H^4(X, \mathbb{Z})$ is Hodge isometric to a twist $\mathrm{Pic}(S)_0(-1)$ of the primitive cohomology lattice of a polarized K3 surface (S, H) of degree d, and that such a Hodge-theoretic association gives rise to a choice of open immersion $\mathcal{C}_{K_d} = \mathcal{C}_d^{\mathrm{mar}} \hookrightarrow \mathcal{K}_d$ of moduli spaces (cf. [19, Corollary 5.2.4]). The choice of such an open immersion is determined by an isomorphism between the discriminant forms of the abstract lattices $K_d^{\perp}$ and $\mathrm{Pic}(S)_0(-1)$, modulo scaling by $\{\pm 1\}$; there are 2^{r-1} such choices, where r is the number of distinct odd primes dividing d, see [19, Corollary 5.2.4], [20, Proposition 26].

Now, given a cubic fourfold lattice Λ and a fixed primitive embedding $K_d \hookrightarrow \Lambda$ preserving h^2, we are interested in generalizing this open immersion to Λ-polarized cubic fourfolds. We can do this explicitly in the case of interest to us, namely when $d = 14$ and the rank of Λ is 3, due to the following lemma.

Lemma 3.2 *Let Λ be a rank 3 cubic fourfold lattice with a fixed primitive embedding $K_{14} \hookrightarrow \Lambda$ preserving h^2. Then, up to isometry, there is a unique rank 2 even indefinite lattice $\sigma(\Lambda)$ with discriminant $-d(\Lambda)$, a distinguished class H of norm d, and such that the orthogonal complement of K_{14} in Λ is isometric (up to twist) with the orthogonal complement of H in $\sigma(\Lambda)$.*

Proof As the sublattice $K_{14} = \langle h^2, T \rangle \subset \Lambda$ is primitive, there exists a class $J \in \Lambda$ and integers a, b, c such that

$$
\begin{array}{c|ccc}
 & h^2 & T & J \\
\hline
h^2 & 3 & 4 & a \\
T & 4 & 10 & b \\
J & a & b & c \\
\end{array}
$$

By translating J to $J - a(T - h^2)$, we can assume that $a = 0$. Directly computing the determinant of this Gram matrix, we then find that $d(\Lambda) = -3b^2 + 14c \equiv (5b)^2$ is a square modulo 14. Let $0 \le \alpha \le 7$ be such that $\alpha^2 \equiv d(\Lambda)$ modulo 14. Then we can write $d(\Lambda) = \alpha^2 - 14\beta$ for some integer β. Now we argue that β is even. Since J is orthogonal to h^2 and $\langle h^2 \rangle^{\perp}$ is an even lattice, we have that $J^2 = c$ must be even. Thus $d(\Lambda) \equiv 0, 1 \pmod 4$. From the equation $d(\Lambda) = \alpha^2 - 14\beta$ we see that $d(\Lambda)$ and α have the same parity, and by looking modulo 4, we finally find that β must be even.

We now define $\sigma(\Lambda)$ to be the rank 2 lattice $\langle H, L \rangle$ with Gram matrix

$$
\begin{array}{c|cc}
 & H & L \\
\hline
H & 14 & \alpha \\
L & \alpha & \beta \\
\end{array}
$$

Then $d(\sigma(\Lambda)) = 14\beta - \alpha^2 = -d(\Lambda)$ and hence $\sigma(\Lambda)$ is an indefinite even lattice since $d(\Lambda) > 0$ and β is even.

We now directly calculate that the orthogonal complement of K_{14} in Λ is generated by $(4bh^2 - 3bT + 14J)/\gcd(b, 14)$ and that the orthogonal complement of H in $\sigma(\Lambda)$ is generated by $(\alpha H - 14L)/\gcd(\alpha, 14)$. Computing the self-intersections of these generators yields $14d(\Lambda)/\gcd(b, 14)^2$ and $-14d(\sigma(\Lambda))/\gcd(\alpha, 14)^2$, respectively. Noting that $\alpha \equiv \pm 5b$ modulo 14, we have that $\gcd(b, 14) = \gcd(\alpha, 14)$, which proves the claim about the isometry of orthogonal complements.

Finally, we remark that $\sigma(\Lambda)$ is unique up to isometry with these properties. Indeed, given any rank 2 even indefinite lattice with Gram matrix as above, after a translation and a possible reflection, we can always choose $0 \le \alpha \le 7$. But then α, β are uniquely determined by the equation $14\beta - \alpha^2 = -d(\Lambda)$. So $\sigma(\Lambda)$ is unique up to isometry.   □

The proof of Lemma 3.2 provides an algorithm, given the Gram matrix of Λ, to calculate a Gram matrix of $\sigma(\Lambda)$. As an example, we calculate that the Gram matrix of $\sigma(\Pi)$ is

$$
\begin{array}{c|cc}
 & H & E \\
\hline
H & 14 & 7 \\
E & 7 & 2
\end{array}
\tag{2}
$$

where Π is the lattice in (1), with fixed primitive embedding $K_{14} = \langle h^2, T \rangle \hookrightarrow \Pi$.

Now, for any rank 3 cubic fourfold lattice Λ with a fixed choice of primitive embedding $K_{14} \hookrightarrow \Lambda$ as in Lemma 3.2, consider the moduli space $\mathcal{K}_{\sigma(\Lambda)}$ of $\sigma(\Lambda)$-polarized K3 surfaces and the forgetful morphism $\mathcal{K}_{\sigma(\Lambda)} \to \mathcal{K}_{14}$, whose image is a divisor $\mathcal{K}_{[\sigma(\Lambda)]} \subset \mathcal{K}_{14}$. For any Λ-polarized cubic fourfold X, the fixed primitive embedding $K_{14} \hookrightarrow \Lambda$ determines a discriminant 14 marking of X, which induces an associated polarized K3 surface (S, H) of discriminant 14 admitting a $\sigma(\Lambda)$-polarization by Lemma 3.2. We recall that for discriminant 14, there is a unique choice of open immersion $\mathcal{C}_{K_{14}} \hookrightarrow \mathcal{K}_{14}$, see [19, §6]. Then following Hassett [19, §5.2], we have the following.

Proposition 3.3 *Let Λ be a rank 3 cubic fourfold lattice with a fixed primitive embedding $K_{14} \hookrightarrow \Lambda$ preserving h^2. Then there exists an open immersion $\mathcal{C}_\Lambda \hookrightarrow \mathcal{K}_{\sigma(\Lambda)}$ of moduli spaces and a commutative diagram*

$$
\begin{array}{ccc}
\mathcal{C}_{K_{14}} & \hookrightarrow & \mathcal{K}_{14} \\
\uparrow & & \uparrow \\
\mathcal{C}_\Lambda & \hookrightarrow & \mathcal{K}_{\sigma(\Lambda)}
\end{array}
$$

where the vertical arrows are the forgetful maps and the top horizontal arrow is the (unique choice of) open immersion constructed by Hassett.

4 Cubic Fourfold Lattice Normal Forms

This section is devoted to establishing normal forms for cubic fourfold lattices of rank 3 with a discriminant 14 marking and their associated K3 surface Picard lattices.

Proposition 4.1 *Let Λ be a rank 3 cubic fourfold lattice with a fixed primitive embedding of K_{14} preserving h^2. Then there exists a basis h^2, T, J of Λ with respect to which Λ has Gram matrix*

$$
\begin{array}{c|ccc}
 & h^2 & T & J \\
\hline
h^2 & 3 & 4 & 0 \\
T & 4 & 10 & b \\
J & 0 & b & c
\end{array}
\tag{3}
$$

for some integers $0 \le b \le 7$ and $c > \max(2, 3b^2/14)$ even.

Proof Just as in the proof of Lemma 3.2, we can choose the primitive sublattice $K_{14} = \langle h^2, T \rangle$, and then there exists a class $J \in \Lambda$ and integers b, c such that that Gram matrix of Λ has the shape (3). Since we have $(3T - 4h^2).h^2 = 0$, we can further translate J to $\pm J - m(3T - 4h^2)$, which preserves $h^2.J = 0$ and allows us modify b modulo $14 = (3T - 4h^2).T$ and up to sign, so we can choose representatives $0 \le b \le 7$. Being a primitive sublattice of $A(X)$, we know that Λ is positive definite, hence its discriminant $-3b^2 + 14c$ must be positive, which forces $c > 3b^2/14$. Already in the proof of Lemma 3.2, we saw that c must be even; also c must be greater than 2, since $\langle h^2 \rangle^\perp$ is an even lattice with no vectors of norm 2. Note that a similar normal form analysis is carried out in [3, §2], [1, Lemma 4.2].   □

One application of the normal form in Proposition 4.1 is that the lattice $\sigma(\Lambda)$ can be even more explicitly computed from Λ. Given (b, c) that determine Λ, we compute that $\sigma(\Lambda)$ has Gram matrix

$$
\begin{array}{c|cc}
 & H & L \\
\hline
H & 14 & 2b \\
L & 2b & \frac{b^2}{2} - c
\end{array}
\quad \text{or} \quad
\begin{array}{c|cc}
 & H & L \\
\hline
H & 14 & 7 - 2b \\
L & 7 - 2b & \frac{b^2 - 4b + 7}{2} - c
\end{array}
$$

depending on whether b is even or odd, respectively.

A consequence of this calculation is that $\sigma(\Lambda)$ together with H determines the pair (b, c), and hence Λ together with the fixed primitively embedded K_{14} up to isomorphism. In the last line of Table 1, we have recorded, by listing the pair (b, c), the unique rank 3 cubic fourfold lattices Λ with a primitive embedding $K_{14} \hookrightarrow \Lambda$ whose associated lattice is the given $\sigma(\Lambda)$.

Finally, we are in a position to deduce the following.

Proposition 4.2   *If a smooth cubic fourfold X has an associated K3 surface (S, H) of degree 14 that is Brill–Noether special and with Clifford index 3, then $X \in C_{[\Pi]}$.*

Proof   The cubic fourfold X, together with the discriminant 14 marking $K_{14} \hookrightarrow A(X)$ whose associated K3 surface is (S, H), determines a point on the moduli space $C_{K_{14}}$. Under the open immersion $C_{K_{14}} \hookrightarrow K_{14}$ constructed by Hassett, this point maps to (S, H). Since (S, H) is Brill–Noether special with Clifford index 3, Theorem 2.7 implies that S admits a Σ-polarization where Σ is the lattice in the $\gamma = 3$ column of Table 1. Hence (S, H) determines a point on the moduli space K_Σ. Via the Hodge isometry $K_d^\perp \cong \mathrm{Pic}(S)_0(-1)$, we lift $\Sigma \cap \mathrm{Pic}(S)_0(-1)$ to a primitive rank 3 lattice $\Lambda \subset A(X)$; this is nothing but the saturation of the sum of K_d and the image of $\Sigma \cap \mathrm{Pic}(S)_0(-1)$ in $A(X)$. A calculation of this saturation shows that, in fact, $\Lambda \cong \Pi$. Recalling that $\Sigma = \sigma(\Pi)$ by (2), we thus have that the moduli point of (S, H) in $K_{\sigma(\Pi)}$ is in the image of the open immersion $C_\Pi \hookrightarrow K_{\sigma(\Pi)}$, which finishes the proof by the commutativity of the diagram in Proposition 3.3. $\square$

In fact, later on in Theorem 5.4, we will show that for any smooth cubic fourfold X with a discriminant 14 marking, the polarized K3 surface (S, H) of degree 14 Hodge theoretically associated to X always has Clifford index 3.

We end this section with some lattice computations that will be useful later on.

Let $\Lambda = (\mathbb{Z}^n, b)$ be an integral nondegenerate lattice with bilinear form $b \colon \mathbb{Z}^n \times \mathbb{Z}^n \to \mathbb{Z}$ having Gram matrix B and discriminant $d(\Lambda) = \det(B)$. Then with respect to the dual standard basis, the dual lattice $\Lambda^\vee = (\mathbb{Z}^n, b^\vee)$ can be considered as a bilinear form $b^\vee \colon \mathbb{Z}^n \times \mathbb{Z}^n \to \frac{1}{d}\mathbb{Z}$ having Gram matrix B^{-1}. The canonical isometric embedding $\Lambda \to \Lambda^\vee$ is then identified with the matrix multiplication map $B \colon (\mathbb{Z}^n, b) \to (\mathbb{Z}^n, b^\vee)$. In particular, the discriminant group $\Lambda^\vee/\Lambda$ can be identified with the cokernel of the matrix B, which aids in explicit computations.

We state some useful, if not easy, necessary conditions for a lattice to occur as the intersection lattice of a smooth cubic fourfold.

Lemma 4.3   *No lattice of the following type can arise as the intersection lattice $A(X)$ of a smooth cubic fourfold X:*

1. *A unimodular lattice.*
2. *A lattice with odd rank $\rho \leq 11$ and discriminant a prime $p \equiv \rho \bmod 4$.*
3. *A lattice with odd rank $\rho \leq 11$ and discriminant exactly divisible by 2.*

Proof For (1), it is a consequence of the classification of unimodular lattices of small rank that every unimodular lattice of rank ≤ 22 has short roots, hence cannot arise from a smooth cubic fourfold.

We recall the notion of the discriminant form $q_A \colon A^\vee/A \to \mathbb{Z}/2\mathbb{Z}$ of $A = A(X)$, as well as the modulo 8 signature $\operatorname{sign} q_A$ considered in [42, §1]. We remark that, in the notation of [42, Proposition 1.8.1], for any odd prime p, we have that $\operatorname{sign} q_\theta^p(p) \equiv 1 - p \bmod 4$ and $\operatorname{sign} q_\theta^p(p^2) \equiv 0 \bmod 8$ for any nonsquare class θ modulo p. For (2), by [42, Theorem 1.10.1], for a lattice A of odd rank ≤ 11 to be the intersection lattice of a smooth cubic fourfolds X, it is necessary that the signature satisfy $\operatorname{sign} q_A \equiv 11 - \rho \bmod 4$. If A has discriminant p, then $\operatorname{sign} q_A \equiv 1 - p \bmod 4$, hence we must have, $p \equiv \rho - 2 \bmod 4$. This is impossible if $p \equiv \rho \bmod 4$.

For (3), if 2 strictly divides $\operatorname{disc}(A)$, then $q_A = q_\theta^2(2) + q(\text{odd})$, where $q(\text{odd})$ means a finite quadratic form on a group of odd order. By [42, Prop. 1.11.2*], $\operatorname{sign} q_\theta^2(2) \not\equiv 0 \bmod 2$ while $\operatorname{sign} q(\text{odd}) \equiv 0 \bmod 2$. Hence no such cubic fourfold X exists. $\qquad\square$

5 Clifford Index Bounds for Cubic Fourfolds

In this section, we recall the constructions of pfaffian cubic fourfolds by Beauville and Donagi [5] and Brill–Noether general K3 surfaces of degree 14 by Mukai [41, Thms. 3.9, 4.7], working up to a proof of our main results. Throughout, denote by $\mathbb{P}(W)$ the projective space of lines in a vector space W.

Let V be a $\mathbb{C}$-vector space of dimension 6 and consider the subvarieties G and Δ of $\mathbb{P}(\bigwedge^2 V)$ of tensors of rank 2 and ≤ 4, respectively. Then G coincides with the image of the Plücker embedding $G(2, V) \hookrightarrow \mathbb{P}(\bigwedge^2 V)$, hence has dimension 8 and degree 14 by the Schubert calculus, see [13]. Also, Δ coincides with the vanishing locus of the pfaffian map $\mathrm{pf} \colon \bigwedge^2 V \to \bigwedge^6 V$, hence is a hypersurface of degree 3. We have that G is the singular locus of Δ and that Δ coincides with the secant variety of G, see [40, Rem. 1.5]. Similarly, we define $G^\vee \subset \Delta^\vee \subset P(\bigwedge^2 V^\vee)$. Here, $\mathbb{P}(\bigwedge^2 V^\vee)$ is the space of alternating bilinear forms on V up to homothety, and $\Delta^\vee$ is the subvariety of degenerate forms.

If $L \subset \mathbb{P}(\bigwedge^2 V)$ is a linear subspace of dimension 8 intersecting G transversally then $S = L \cap G \subset L \cong \mathbb{P}^8$ is the projective model of a smooth Brill–Noether general polarized K3 surface (S, H) of degree 14, see [41, Thm. 3.9].

Conversely, if (S, H) is a Brill–Noether general polarized K3 surface of degree 14, then S has a *rigid* vector bundle E, unique up to isomorphism, such that E is stable of rank 2 with det $E \cong H$ and $\chi(S, E) = h^0(S, E) = 6$, see [41, Thm. 4.5]. In particular, the evaluation morphism $H^0(S, E) \otimes \mathcal{O}_S \to E$ is surjective, hence there is a grassmannian embedding $\Phi_E \colon S \to G(2, H^0(S, E)^\vee)$ taking $x \mapsto E_x^\vee$. Here, we think of $E_x^\vee \subset H^0(S, E)^\vee$ as a 2-dimensional subspace dual to the quotient map $H^0(S, E) \to E_x$ defined by the evaluation morphism. We have that $E = \Phi_E^* \mathcal{E}$, where $\mathcal{E}$ is the tautological rank 2 vector (sub)bundle on $G(2, H^0(S, E)^\vee)$. Composing with the Plücker embedding, we have an embedding $S \to \mathbb{P}(\bigwedge^2 H^0(S, E)^\vee)$. On the other hand, the exterior square $\bigwedge^2 H^0(S, E) \otimes \mathcal{O}_S \to \bigwedge^2 E$ of the evaluation morphism defines a linear map $\lambda \colon \bigwedge^2 H^0(S, E) \to H^0(S, \bigwedge^2 E) \cong H^0(S, H)$. As λ is surjective, we arrive at a commutative square of morphisms

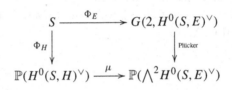

where μ is the linear embedding defined by λ. A result of Mukai is that this square is cartesian, see [41, Thm. 4.7], hence S can be written as an intersection $S = L \cap G \subset \mathbb{P}(\bigwedge^2 V)$ in our previous notation, where $L = \mathbb{P}(H^0(S, H)^\vee)$, $V = H^0(S, E)^\vee$, and $G = G(2, V)$. In conclusion, a polarized K3 surface (S, H) of degree 14 is Brill–Noether general if and only if its projective model is a transversal intersection $S = G \cap L \subset \mathbb{P}(\bigwedge^2 V)$ and any such (S, H) has a rigid vector bundle of rank 2, i.e., a rank 2 stable vector bundle E with det$(E) \cong H$ and $\chi(E) = 6$. By [39, Thm. 3.3(2)], Brill–Noether general polarized K3 surfaces $S = G \cap L$ and $S' = G \cap L'$ are projectively equivalent if and only if L and L' are equivalent under the action of $GL(V)$.

Still letting $L \subset \mathbb{P}(\bigwedge^2 V)$ be a linear subspace of dimension 8, if the projective dual linear subspace $L^\perp \subset \mathbb{P}(\bigwedge^2 V^\vee)$ of dimension 5 is not contained in $\Delta^\vee$, then $X = L^\perp \cap \Delta^\vee \subset L^\perp \cong \mathbb{P}^5$ is a pfaffian cubic fourfold, see [5, §2]. Conversely, writing $L^\perp = \mathbb{P}(W)$ for a subspace $W \subset \bigwedge^2 V^\vee$, then $L^\perp$ gives rise to a global section of the vector bundle $W^\vee \otimes \mathcal{O}_G(1)$, whose zero locus is precisely S.

Proposition 5.1 *Let V be a vector space of dimension 6 and let $L \subset$ $\mathbb{P}(\bigwedge^2 V)$ be a linear subspace of dimension 8. Assume that $S = L \cap G$ has* *dimension 2 and that $X = L^\perp \cap \Delta^\vee$ has dimension 4. If X is smooth then S is smooth and in this case there is a semiorthogonal decomposition*

$$\mathsf{D}^b(X) = \langle \mathcal{A}_X, \mathcal{O}_X, \mathcal{O}_X(1), \mathcal{O}_X(2) \rangle$$

and an equivalence of categories $\mathcal{A}_X \cong \mathsf{D}^b(S)$. More generally, if S is smooth, then the singular locus of X is contained in $L^\perp \cap G^\vee$.

Proof The smoothness statements, the semiorthogonal decomposition, and the equivalence of categories follow from Kuznetsov's theory of homological projective duality [31] applied to the duality between G and a noncommutative resolution of $\Delta^\vee$ that is supported along $G^\vee$ (cf. [30, Thms. 4.1, 10.1, 10.4] and also [32, Thm. 3.1]). □

Building on work of Mukai [40], it is proved in [21, Corollary 6.4] that if $S = L \cap G$ is smooth then $L^\perp \cap G^\vee \subset L^\perp \cap \Delta^\vee = X$, which contains the singular locus of X, is in bijection with the (finite) set of elliptic pencils E on S of degree 5.

By Hassett [19], any cubic fourfold of discriminant 14 has an associated K3 surface of degree 14. To link pfaffian cubic fourfolds and curves of genus 8 on the associated K3, we will need the following.

Proposition 5.2 *A smooth cubic fourfold X is pfaffian if and only if it has a discriminant 14 marking whose associated K3 surface (S, H) is Brill–Noether general.*

Proof First suppose that $X = L^\perp \cap \Delta^\vee$ is a smooth pfaffian cubic fourfold. By Proposition 5.1, $S = L \cap G$ is a K3 surface of degree 14 with a polarization H defined by the projective embedding $S \to L \cong \mathbb{P}^{14}$ and there is an equivalence $\mathcal{A}_X \cong \mathsf{D}^b(S)$. By Mukai [39, Thms. 3.10], (S, H) is Brill–Noether general. By Addington–Thomas [1] (cf. [22, Prop. 3.3]), the equivalence $\mathcal{A}_X \cong \mathsf{D}^b(S)$ induces a Hodge isometry of Mukai lattices $\widetilde{H}(\mathcal{A}_X, \mathbb{Z}) \cong \widetilde{H}(S, \mathbb{Z})$, which implies that X has a marking of discriminant 14 for which the associated polarized K3 surface is (S, H).

Now suppose that X is a smooth cubic fourfold with a marking of discriminant 14 whose associated polarized K3 surface (S, H) is Brill–Noether general. Then by Mukai [39, Prop. 4.7], H defines a projective embedding whose image is $S \cong L \cap G \subset L \cong \mathbb{P}^{14}$, for $L = \mathbb{P}(H^0(S, H)^\vee)$ as described above. Then $X' = L^\perp \cap \Delta^\vee \subset L^\perp \cong \mathbb{P}^5$ if a pfaffian cubic fourfold whose singular points are in bijection with elliptic pencils E on S such that $E.H = 5$. However, by Remark 5.5, (S, H) cannot admit any such elliptic pencils since it is associated to a smooth cubic fourfold of discriminant 14. Thus X' must be a smooth pfaffian cubic fourfold, so by Proposition 5.1, $\mathsf{D}^b(X')$ has a semiorthogonal decomposition $\langle \mathcal{A}_{X'}, \mathscr{O}_{X'}, \mathscr{O}_{X'}(1), \mathscr{O}_{X'}(2) \rangle$, and there is an equivalence $\mathcal{A}_{X'} \cong \mathsf{D}^b(S)$. In particular, by Addington–Thomas [1], X' has a marking of discriminant 14 whose associated polarized K3 surface is (S, H). By the injectivity of $\mathcal{C}_{K_{14}} \hookrightarrow \mathcal{K}_{14}$, we have that X and X' are isomorphic. In particular, X is pfaffian. $\qquad\square$

In terms of the open immersion of moduli spaces, Proposition 5.2 says that under the open immersion $\mathcal{C}_{K_{14}} \hookrightarrow \mathcal{K}_{14}$, the pfaffian locus coincides with the restriction of the Brill–Noether general locus to the image.

For the very general cubic fourfold X in $\mathcal{C}_{14}$, the (unique) associated polarized K3 surface (S, H) of degree 14 has Picard rank 1. The following is stated many times in the literature.

Corollary 5.3 *Any cubic fourfold X of discriminant 14 and with $A(X)$ of rank 2 is pfaffian.*

Proof Since the associated K3 surface (S, H) has Picard rank 1, the smooth curves $C \in |H|$ are Brill–Noether general, by Lazarsfeld [34]. Hence Proposition 5.2 applies to show that X is pfaffian. $\qquad\square$

As a further consequence, any cubic fourfold in $\mathcal{C}_{14} \smallsetminus \mathcal{P}f$ has $A(X)$ of rank ≥ 3.

By Proposition 5.2, X is not pfaffian if and only if (S, H) is Brill–Noether special for every degree 14 marking on X. Then by Theorem 2.7, we know that (S, H) must admit a lattice-polarization for some lattice in Table 1.

The main result of this section is the following.

Theorem 5.4 *Let X be a smooth cubic fourfold with a discriminant 14 marking and (S, H) an associated K3 surface of degree 14. Then $\gamma(S, H) = 3$.*

Proof As noted in §2.2, if (S, H) is Brill–Noether general, then we have $\gamma(S, H) = 3$. So we can assume that (S, H) is Brill–Noether special. In particular, $A(X)$ has rank ≥ 3 and let $\Lambda \subset A(X)$ be a primitive sublattice of rank 3 containing the marking K_{14}. Then by Theorem 2.7, (S, H) would admit an appropriate $\sigma(\Lambda)$-polarization for $\sigma(\Lambda)$ given on Table 1. By Proposition 3.3, any such cubic fourfold X would have a Λ-polarization, for the unique rank 3 cubic fourfold lattice Λ with specified $\sigma(\Lambda)$. We have enumerated these lattices Λ in Table 1. We now show that each such lattice Λ corresponding to $\gamma(S, H) < 3$ has roots, hence no smooth Λ-polarized cubic fourfolds exist in these cases.

When $\gamma(S, H) = 0$, the cubic fourfold lattice Λ with $(b, c) = (6, 8)$ has short root $4h^2 - 3T + 2J$. Hence $A(X) \supset \Lambda$ would contain a short root, thus no such smooth cubic fourfold exists.

When $\gamma(S, H) = 1$, the cubic fourfold lattice Λ with $(b, c) = (5, 6)$ has long root $4h^2 - 3T + 3J$. Hence $A(X) \supset \Lambda$ would contain a long root, thus no such smooth cubic fourfold exists.

When $\gamma(S, H) = 2$, we have two choices. The cubic fourfold lattice Λ with $(b, c) = (2, 2)$ has short root J. Hence $A(X) \supset \Lambda$ would contain a short root, thus no such smooth cubic fourfold exists. The cubic fourfold lattice Λ with $(b, c) = (4, 4)$ has long root $4h^2 - 3T + 3J$. Hence $A(X) \supset \Lambda$ would contain a long root, thus no such smooth cubic fourfold exists.

This rules out all possibilities with $\gamma(S, H) < 3$. Hence (S, H) is Brill–Noether special with $\gamma(S, H) = 3$. $\square$

Remark 5.5   This is a continuation of Remark 2.9. The rank 3 cubic fourfold lattice Λ, whose associated K3 surface lattice $\sigma(\Lambda)$ has degree 14 generated by H and E with $E.H = 5$ and $E^2 = 0$, must have $(b, c) = (1, 2)$. In this case, Λ has short root J. We conclude that there exist no smooth cubic fourfolds X whose associated K3 surface has a line bundle restricting to a g^1_5 on the smooth genus 8 curves in the polarization class.

Given an admissible $d = 2g - 2 > 6$, we wonder about the possible values of $0 \leq \gamma \leq \lfloor (g-1)/2 \rfloor$ that can be realized by Clifford indices polarized K3 surfaces associated to Brill–Noether special cubic fourfolds of discriminant d. For example, we expect that there are no "hyperelliptic" or "trigonal" special cubic fourfolds of any admissible discriminant.

6 Complement of the Pfaffian Locus

In this section, we can finally prove Theorem 1.1. We also show that cubic fourfolds can admit multiple discriminant 14 markings, some Brill–Noether special and some Brill–Noether general, implying that the pfaffian locus is not open inside $\mathcal{C}_{14}$.

Proof of Theorem 1.1   As a consequence of Theorem 5.4, the only component of the Brill–Noether special locus of $\mathcal{K}_{14}$ that intersects the image of $\mathcal{C}_{K_{14}}$ is the one corresponding to $\gamma = 3$ in Table 1. By Propositions 3.1 and 4.2, we have that the intersection of this component with the image of $\mathcal{C}_{K_{14}}$ in $\mathcal{K}_{14}$ coincides with the locus $\mathcal{C}_\Pi \subset \mathcal{C}_{K_{14}}$ (where we always consider Π with the fixed discriminant 14 marking derived from (1)). Applying the forgetful map, we see that the complement of the pfaffian locus in $\mathcal{C}_{14}$ is contained in $\mathcal{C}_{[\Pi]}$. $\square$

Example 6.1   Consider the lattice Λ:

	h^2	T	P	P'
h^2	3	4	1	1
T	4	10	0	0
P	1	0	3	0
P'	1	0	0	3

One can check, using the result of Laza [33] and Looijenga [37] on the image of the period map, that Λ is a cubic fourfold lattice, implying that the locus $\mathcal{C}_{[\Lambda]}$ of cubic fourfolds admitting a Λ-polarized has codimension 3 in the moduli space $\mathcal{C}$. We remark that the explicit example found by computers in [6, Ex. A.2] is contained in $\mathcal{C}_{[\Lambda]}$, and motivated its definition. By Proposition 3.1, the classes P and P' correspond to disjoint planes, and the class T generates a marking of discriminant 14, hence $\mathcal{C}_\Lambda$ is a divisor in $\mathcal{C}_\Pi$. Consider the four discriminant 14 markings generated by the classes:

$$T \qquad\qquad\qquad\qquad T' = 2h^2 - P - P'$$
$$T'' = 3h^2 - T - P \qquad\qquad T''' = 3h^2 - T - P'.$$

We compute that the polarized K3 surfaces (S, H) of degree 14 associated to these markings admit lattice-polarizations for the following lattices, respectively:

	H	C	C'
H	14	2	2
C	2	-2	1
C'	2	1	-2

	H	L	C
H	14	7	4
L	7	2	2
C	4	2	-2

	H	C	L
H	14	2	1
C	2	-2	0
L	1	0	-2

	H	C	L
H	14	2	1
C	2	-2	0
L	1	0	-2

Clearly T'' and T''' are permuted up to reordering the planes P and P', so the bottom two lattice polarizations are isomorphic.

   Now assume that X is very general in $\mathcal{C}_{[\Lambda]}$. Then the associated K3 surfaces have Picard lattices isomorphic to the ones above and in all four cases, one can verify that the degree 14 polarization class is very ample by an analysis of the ample cone. In the first, third, and fourth cases, one can also verify using Theorem 2.7 that the smooth genus 8 curves in the polarization class are Brill–Noether general, hence by Proposition 5.2, that X is pfaffian (in multiple ways). In the second case, one can verify that the second generator

C restricts to a g_7^2 on the smooth genus 8 curves in the polarization class (hence they are Brill–Noether special by Lemma 2.2). By Theorem 2.3, the polarized K3 surface associated to this second marking is Brill–Noether special.

References

[1] Addington, Nick, and Thomas, Richard. 2014. Hodge theory and derived categories of cubic fourfolds. *Duke Math. J.*, **163**(10), 1885–1927.

[2] Addington, Nicolas, Hassett, Brendan, Tschinkel, Yuri, and Várilly-Alvarado, Anthony. 2019. Cubic fourfolds fibered in sextic del Pezzo surfaces. *Amer. J. Math.*, **141**(6), 1479–1500.

[3] Auel, Asher, Bernardara, Marcello, Bolognesi, Michele, and Várilly-Alvarado, Anthony. 2014. Cubic fourfolds containing a plane and a quintic del Pezzo surface. *Alg. Geom.*, **1**(2), 181–193.

[4] Beauville, Arnaud. 2000. Determinantal hypersurfaces. *Michigan Math. J.*, **48**, 39–64. Dedicated to William Fulton on the occasion of his 60th birthday.

[5] Beauville, Arnaud, and Donagi, Ron. 1985. La variété des droites d'une hypersurface cubique de dimension 4. *C. R. Acad. Sci. Paris Sér. I Math.*, **301**(14), 703–706.

[6] Bolognesi, Michele, and Russo, Francesco. 2019. Some loci of rational cubic fourfolds. *Math. Annalen*, **373**, 165–190. with an appendix by Giovanni Staglianò.

[7] Ciliberto, Ciro, and Pareschi, Giuseppe. 1995. Pencils of minimal degree on curves on a $K3$ surface. *J. Reine Angew. Math.*, **460**, 15–36.

[8] Dolgachev, I. V. 1996. Mirror symmetry for lattice polarized $K3$ surfaces. *J. Math. Sci.*, **81**(3), 2599–2630. Translated from Itogi Nauki i Tekhniki, Seriya Sovremennaya Matematika i Ee Prilozheniya. Tematicheskie Obzory. Vol. 33, Algebraic Geometry-4, 1996.

[9] Donagi, Ron, and Morrison, David R. 1989. Linear systems on $K3$-sections. *J. Differential Geom.*, **29**(1), 49–64.

[10] Eisenbud, David, and Harris, Joe. 1989. Irreducibility of some families of linear series with Brill-Noether number -1. *Ann. Sci. École Norm. Sup. (4)*, **22**(1), 33–53.

[11] Fano, Gino. 1943. Sulle forme cubiche dello spazio a cinque dimensioni contenenti rigate razionali del 4° ordine. *Comment. Math. Helv.*, **15**, 71–80.

[12] Farkas, Gavril, and Verra, Alessandro. 2018. The universal K3 surface of genus 14 via cubic fourfolds. *J. Math. Pures Appl*, **11**, 1–20.

[13] Fulton, William. 1997. *Young tableaux*. London Mathematical Society Student Texts, vol. 35. Cambridge University Press, Cambridge. With applications to representation theory and geometry.

[14] Galluzzi, Federica. 2017. Cubic fourfolds containing a plane and K3 surfaces of Picard rank two. *Geometriae Dedicata*, **186**(1), 103–112.

[15] Green, Mark, and Lazarsfeld, Robert. 1987. Special divisors on curves on a $K3$ surface. *Invent. Math.*, **89**(2), 357–370.

[16] Green, Mark L. 1984. Koszul cohomology and the geometry of projective varieties. *J. Differential Geom.*, **19**(1), 125–171.

[17] Greer, Francois, Li, Zhiyuan, and Tian, Zhiyu. 2015. Picard groups on moduli of K3 surfaces with Mukai models. *Int. Math. Res. Not.*, **16**, 7238–7257.

[18] Hassett, Brendan. 1999. Some rational cubic fourfolds. *J. Algebraic Geom.*, **8**(1), 103–114.

[19] Hassett, Brendan. 2000. Special cubic fourfolds. *Compositio Math.*, **120**(1), 1–23.

[20] Hassett, Brendan. 2016. Cubic fourfolds, K3 surfaces, and rationality questions. Pages 26–66 of: Pardini, R., and Pirola, G.P. (eds), *Rationality Problems in Algebraic Geometry*. CIME Foundation Subseries, Lecture Notes in Mathematics, vol. 2172. Springer-Verlag.

[21] Hoff, Michael, and Knutsen, Andreas Leopold. 2021. Brill–Noether general K3 surfaces with the maximal number of elliptic pencils of minimal degree. *Geom. Dedicata*, **213**, 1–20.

[22] Huybrechts, Daniel. 2015. The K3 category of a cubic fourfold. *Compos. Math.*, **153**(3), 586–620.

[23] Huybrechts, Daniel. 2016. *Lectures on K3 surfaces*. Cambridge Studies in Advanced Mathematics, vol. 158. Cambridge University Press, Cambridge.

[24] Johnsen, Trygve, and Knutsen, Andreas Leopold. 2004. *K3 projective models in scrolls*. Lecture Notes in Mathematics, vol. 1842. Berlin: Springer-Verlag.

[25] Knutsen, Andreas Leopold. 2001. On kth-order embeddings of K3 surfaces and Enriques surfaces. *Manuscripta Math.*, **104**, 211–237.

[26] Knutsen, Andreas Leopold. 2002. Smooth curves on projective $K3$ surfaces. *Math. Scand.*, **90**(2), 215–231.

[27] Knutsen, Andreas Leopold. 2003. Gonality and Clifford index of curves on $K3$ surfaces. *Arch. Math. (Basel)*, **80**(3), 235–238.

[28] Knutsen, Andreas Leopold. 2009. On two conjectures for curves on $K3$ surfaces. *Internat. J. Math.*, **20**(12), 1547–1560.

[29] Kontsevich, Maxim, and Tschinkel, Yuri. 2019. Specialization of birational types. *Invent. Math.*, **217**(n), 415–432.

[30] Kuznetsov, Alexander. 2006. *Homological projective duality for Grassmannians of lines.* preprint arXiv:math/0610957.

[31] Kuznetsov, Alexander. 2007. Homological projective duality. *Publ. Math. Inst. Hautes Études Sci.*, 157–220.

[32] Kuznetsov, Alexander. 2010. Derived categories of cubic fourfolds. Pages 219–243 of: *Cohomological and geometric approaches to rationality problems.* Progr. Math., vol. 282. Boston, MA: Birkhäuser Boston Inc.

[33] Laza, Radu. 2010. The moduli space of cubic fourfolds via the period map. *Ann. of Math.*, **172**(1), 673–711.

[34] Lazarsfeld, Robert. 1986. Brill-Noether-Petri without degenerations. *J. Differential Geom.*, **23**(3), 299–307.

[35] Lelli-Chiesa, Margherita. 2013. Stability of rank-3 Lazarsfeld-Mukai bundles on K3 surfaces. *Proc. Lon. Math. Soc.*, **107**(2), 451–479.

[36] Lelli-Chiesa, Margherita. 2015. Generalized Lazarsfeld-Mukai bundles and a conjecture of Donagi and Morrison. *Adv. Math.*, **268**(2), 529–563.

[37] Looijenga, Eduard. 2009. The period map for cubic fourfolds. *Invent. Math.*, **177**, 213–233.

[38] Martens, Gerriet. 1989. On curves on K3 surfaces. Pages 174–182 of: *Algebraic curves and projective geometry (1988)*. Lecture Notes in Mathematics, vol. 1398. Springer.

[39] Mukai, Shigeru. 1988. Curves, K3 surfaces and Fano 3-folds of genus $\leq$ 10. Pages 357–377 of: *Algebraic Geometry and Commutative Algebra: In Honor of Masayoshi Nagata*, vol. 1. Tokyo: Kinokuniya.

[40] Mukai, Shigeru. 1993. Curves and Grassmannians. Pages 19–40 of: *Algebraic geometry and related topics (Inchon, 1992)*. Conf. Proc. Lecture Notes Algebraic Geom., I. Int. Press, Cambridge, MA.

[41] Mukai, Shigeru. 1995. New development of theory of Fano 3-folds: vector bundle method and moduli problem. *Sugaku*, **47**, 125–144.

[42] Nikulin, V. V. 1979. Integer symmetric bilinear forms and some of their geometric applications. *Izv. Akad. Nauk SSSR Ser. Mat.*, **43**(1), 111–177, 238.

[43] Reid, Miles. 1976a. Hyperelliptic linear systems on a K3 surface. *J. London Math. Soc. (2)*, **13**(3), 427–437.

[44] Reid, Miles. 1976b. Special linear systems on curves lying on a K3 surface. *J. London Math. Soc. (2)*, **13**(3), 454–458.

[45] Russo, Francesco, and Staglianò, Giovanni. 2019. Congruences of 5-secant conics and the rationality of some admissible cubic fourfolds. *Duke Math. J.*, **168**(5), 849–865.

[46] Russo, Francesco, and Staglianò, Giovanni. 2020. *Trisecant Flops, their associated K3 surfaces and the rationality of some Fano fourfolds*. arxiv:1909.01263.

[47] Saint-Donat, Bernard. 1974. Projective models of $K - 3$ surfaces. *Amer. J. Math.*, **96**, 602–639.

[48] Tregub, S. L. 1984. Three constructions of rationality of a cubic fourfold. *Vestnik Moskov. Univ. Ser. I Mat. Mekh.*, 8–14.

[49] Tregub, S. L. 1993. Two remarks on four-dimensional cubics. *Uspekhi Mat. Nauk*, **48**(2(290)), 201–202.

[50] Voisin, Claire. 1986. Théorème de Torelli pour les cubiques de $\mathbf{P}^5$. *Invent. Math.*, **86**(3), 577–601.

[51] Voisin, Claire. 2007. Some aspects of the Hodge conjecture. *Jpn. J. Math.*, **2**(2), 261–296.

3

Automorphism Groups of Almost Homogeneous Varieties

Michel Brion[a]

Dedicated to Bill Fulton on the occasion of his 80th birthday.

Abstract. Consider a smooth connected algebraic group G acting on a normal projective variety X with an open dense orbit. We show that $\text{Aut}(X)$ is a linear algebraic group if so is G; for an arbitrary G, the group of components of $\text{Aut}(X)$ is arithmetic. Along the way, we obtain a restrictive condition for G to be the full automorphism group of some normal projective variety.

1 Introduction

Let X be a projective algebraic variety over an algebraically closed field k. It is known that the automorphism group $\text{Aut}(X)$ has a natural structure of smooth k-group scheme, locally of finite type (see [20, p. 268]). This yields an exact sequence

$$1 \longrightarrow \text{Aut}^0(X) \longrightarrow \text{Aut}(X) \longrightarrow \pi_0 \, \text{Aut}(X) \longrightarrow 1,$$

where $\text{Aut}^0(X)$ is (the group of k-rational points of) a smooth connected algebraic group, and $\pi_0 \, \text{Aut}(X)$ is a discrete group.

To analyze the structure of $\text{Aut}(X)$, one may start by considering the connected automorphism group $\text{Aut}^0(X)$ and the group of components $\pi_0 \, \text{Aut}(X)$ separately. It turns out that there is no restriction on the former: every smooth connected algebraic group is the connected automorphism group of some normal projective variety X (see [4, Thm. 1]). In characteristic 0, we may further take X to be smooth by using equivariant resolution of singularities (see e.g. [28, Chap. 3]).

[a] Institut Fourier, Université Grenoble Alpes

By constrast, little seems to be known on the structure of the group of components. Every finite group G can be obtained in this way, as G is the full automorphism group of some smooth projective curve (see the main result of [33]). But the group of components is generally infinite, and it is unclear how infinite it can be.

The long-standing question whether this group is finitely generated has been recently answered in the negative by Lesieutre. He constructed an example of a smooth projective variety of dimension 6 having a discrete, non-finitely generated automorphism group (see [30]). His construction has been extended in all dimensions at least 2 by Dinh and Oguiso, see [14]. The former result is obtained over an arbitrary field of characteristic 0, while the latter holds over the complex numbers; it is further extended to odd characteristics in [35]. On the positive side, $\pi_0 \operatorname{Aut}(X)$ is known to be finitely presented for some interesting classes of projective varieties, including abelian varieties (see [1]) and complex hyperkähler manifolds (see [11, Thm. 1.5]).

In this article, we obtain three results on automorphism groups, which generalize recent work. The first one goes in the positive direction for *almost homogeneous* varieties, i.e., those on which a smooth connected algebraic group acts with an open dense orbit.

Theorem 1.1 *Let X be a normal projective variety, almost homogeneous under a linear algebraic group. Then $\operatorname{Aut}(X)$ is a linear algebraic group as well.*

This was first obtained by Fu and Zhang in the setting of compact Kähler manifolds (see [17, Thm. 1.2]). The main point of their proof is to show that the anticanonical line bundle is big. This relies on Lie-theoretical methods, in particular the g-anticanonical fibration of [25, I.2.7], also known as the Tits fibration. But this approach does not extend to positive characteristics, already when X is homogeneous under a semi-simple algebraic group: then any big line bundle on X is ample, but X is generally not Fano (see [23]).

To prove Theorem 1.1, we construct a normal projective variety X' equipped with a birational morphism $f : X' \to X$ such that the action of $\operatorname{Aut}(X)$ on X lifts to an action on X' that fixes the isomorphism class of a big line bundle. For this, we use a characteristic-free version of the Tits fibration (Lemma 3.1).

Our second main result goes in the negative direction, as it yields many examples of algebraic groups which cannot be obtained as the automorphism group of a normal projective variety. To state it, we introduce some notation.

Let G be a smooth connected algebraic group. By Chevalley's structure theorem (see [12] for a modern proof), there is a unique exact sequence of algebraic groups

$$1 \longrightarrow G_{\mathrm{aff}} \longrightarrow G \longrightarrow A \longrightarrow 1,$$

where G_{aff} is a smooth connected affine (or equivalently linear) algebraic group and A is an abelian variety. We denote by $\mathrm{Aut}_{\mathrm{gp}}^{G_{\mathrm{aff}}}(G)$ the group of automorphisms of the algebraic group G which fix G_{aff} pointwise.

Theorem 1.2 *With the above notation, assume that* $\mathrm{Aut}_{\mathrm{gp}}^{G_{\mathrm{aff}}}(G)$ *is infinite. If* $G \subset \mathrm{Aut}(X)$ *for some normal projective variety* X, *then* G *has infinite index in* $\mathrm{Aut}(X)$.

It is easy to show that $\mathrm{Aut}_{\mathrm{gp}}^{G_{\mathrm{aff}}}(G)$ is an arithmetic group, and to construct classes of examples for which this group is infinite, see Remark 4.3.

If G is an abelian variety, then $\mathrm{Aut}_{\mathrm{gp}}^{G_{\mathrm{aff}}}(G)$ is just its group of automorphisms as an algebraic group. In this case, Theorem 1.2 is due (in essence) to Lombardo and Maffei, see [32, Thm. 2.1]. They also obtain a converse over the field of complex numbers: given an abelian variety G with finite automorphism group, they construct a smooth projective variety X such that $\mathrm{Aut}(X) = G$ (see [32, Thm. 3.9]).

Like that of [32, Thm. 2.1], the proof of Theorem 1.2 is based on the existence of a homogeneous fibration of X over an abelian variety, the quotient of A by a finite subgroup scheme. This allows us to construct an action on X of a subgroup of finite index of $\mathrm{Aut}_{\mathrm{gp}}^{G_{\mathrm{aff}}}(G)$, which normalizes G and intersects this group trivially.

When X is almost homogeneous under G, the Albanese morphism provides such a homogeneous fibration, as follows from [3, Thm. 3]. A finer analysis of its automorphisms leads to our third main result.

Theorem 1.3 *Let* X *be a normal projective variety, almost homogeneous under a smooth connected algebraic group* G. *Then* $\pi_0 \mathrm{Aut}(X)$ *is an arithmetic group. In positive characteristics,* $\pi_0 \mathrm{Aut}(X)$ *is commensurable with* $\mathrm{Aut}_{\mathrm{gp}}^{G_{\mathrm{aff}}}(G)$.

(The second assertion does not hold in characteristic 0, see Remark 5.5.)

These results leave open the question whether every *linear* algebraic group is the automorphism group of a normal projective variety. Further open questions are discussed in the recent survey [9], in the setting of smooth complex projective varieties.

Acknowledgments. The above results have first been presented in a lecture at the School and Workshop on Varieties and Group Actions (Warsaw, September

23–29, 2018), with a more detailed and self-contained version of this article serving as lecture notes (see [7]). I warmly thank the organizers of this event for their invitation, and the participants for stimulating questions. Also, I thank Roman Avdeev, Yves de Cornulier, Fu Baohua, Hélène Esnault and Bruno Laurent for helpful discussions or email exchanges on the topics of this article, and the two anonymous referees for their valuable remarks and comments. Special thanks are due to Serge Cantat for very enlightening suggestions and corrections, and to Gaël Rémond for his decisive help with the proof of Theorem 1.3.

This work was partially supported by the grant 346300 for IMPAN from the Simons Foundation and the matching 2015–2019 Polish MNiSW fund.

2 Some Preliminary Results

We first set some notation and conventions, which will be valid throughout this article. We fix an algebraically closed ground field k of characteristic $p \geq 0$. By a *scheme*, we mean a separated scheme over k, unless otherwise stated; a *subscheme* is a locally closed k-subscheme. Morphisms and products of schemes are understood to be over k as well. A *variety* is an integral scheme of finite type. An *algebraic group* is a group scheme of finite type; a *locally algebraic group* is a group scheme, locally of finite type.

Next, we present some general results on automorphism groups, refering to [7, Sec. 2] for additional background and details. We begin with a useful observation:

Lemma 2.1 *Let $f : X \to Y$ be a birational morphism, where X and Y are normal projective varieties. Assume that the action of $\mathrm{Aut}(Y)$ on Y lifts to an action on X. Then the corresponding homomorphism $\rho : \mathrm{Aut}(Y) \to \mathrm{Aut}(X)$ is a closed immersion.*

Proof Since f restricts to an isomorphism on dense open subvarieties of X and Y, the scheme-theoretic kernel of ρ is trivial. Thus, ρ induces a closed immersion $\rho^0 : \mathrm{Aut}^0(Y) \to \mathrm{Aut}^0(X)$. On the other hand, we have $f_*(\mathcal{O}_X) = \mathcal{O}_Y$ by Zariski's Main Theorem; thus, Blanchard's lemma (see [8, Prop. 4.2.1]) yields a homomorphism

$$f_* : \mathrm{Aut}^0(X) \longrightarrow \mathrm{Aut}^0(Y).$$

Clearly, ρ^0 and f_* are mutually inverse; thus, the image of ρ contains $\mathrm{Aut}^0(X)$. Since $\mathrm{Aut}(X)/\mathrm{Aut}^0(X)$ is discrete, this yields the statement. □

We now discuss the action of automorphisms on line bundles. Given a projective variety X, the Picard group $\mathrm{Pic}(X)$ has a canonical structure of locally algebraic group (see [21]); its group of components, the Néron–Severi group $\mathrm{NS}(X)$, is finitely generated by [36, XIII.5.1]. The action of $\mathrm{Aut}(X)$ on $\mathrm{Pic}(X)$ via pullback extends to an action of the corresponding group functor, and hence of the corresponding locally algebraic group. As a consequence, for any line bundle $\pi : L \to X$ with class $[L] \in \mathrm{Pic}(X)$, the reduced stabilizer $\mathrm{Aut}(X,[L])$ is closed in $\mathrm{Aut}(X)$.

Given L as above, the polarization map

$$\mathrm{Aut}(X) \longrightarrow \mathrm{Pic}(X), \quad g \longmapsto [g^*(L) \otimes L^{-1}]$$

takes $\mathrm{Aut}^0(X)$ to $\mathrm{Pic}^0(X)$. Therefore, $\mathrm{Aut}^0(X)$ acts trivially on the quotient $\mathrm{Pic}(X)/\mathrm{Pic}^0(X) = \mathrm{NS}(X)$. This yields an action of $\pi_0 \mathrm{Aut}(X)$ on $\mathrm{NS}(X)$ and in turn, on the quotient of $\mathrm{NS}(X)$ by its torsion subgroup: the group of line bundles up to numerical equivalence, that we denote by $\mathrm{N}^1(X)$. Also, we denote by $[L]_{\mathrm{num}}$ the class of L in $\mathrm{N}^1(X)$, and by $\mathrm{Aut}(X,[L]_{\mathrm{num}})$ its reduced stabilizer in $\mathrm{Aut}(X)$. Then $\mathrm{Aut}(X,[L]_{\mathrm{num}})$ contains $\mathrm{Aut}^0(X)$, and hence is a closed subgroup of $\mathrm{Aut}(X)$, containing $\mathrm{Aut}(X,[L])$.

Further, recall that we have a central extension of locally algebraic groups

$$1 \longrightarrow \mathbb{G}_m \longrightarrow \mathrm{Aut}^{\mathbb{G}_m}(L) \longrightarrow \mathrm{Aut}(X,[L]) \longrightarrow 1,$$

where $\mathrm{Aut}^{\mathbb{G}_m}(L)$ denotes the group of automorphisms of the variety L which commute with the $\mathbb{G}_m$-action by multiplication on the fibers of π. For any integer n, the space $H^0(X, L^{\otimes n})$ is equipped with a linear representation of $\mathrm{Aut}^{\mathbb{G}_m}(L)$, and hence with a projective representation of $\mathrm{Aut}(X,[L])$. Moreover, the natural rational map

$$f_n : X \dashrightarrow \mathbb{P}\, H^0(X, L^{\otimes n})$$

(where the right-hand side denotes the projective space of hyperplanes in $H^0(X, L^{\otimes n})$) is equivariant relative to the action of $\mathrm{Aut}(X,[L])$.

Recall that L is *big* if f_n is birational onto its image for some $n \geq 1$. (See [29, Lem. 2.60] for further characterizations of big line bundles).

Lemma 2.2 *Let L a big and nef line bundle on a normal projective variety X. Then $\mathrm{Aut}(X,[L]_{\mathrm{num}})$ is an algebraic group.*

Proof It suffices to show that the locally algebraic group

$$G := \mathrm{Aut}(X,[L]_{\mathrm{num}})$$

has finitely many components. For this, we adapt the arguments of [31, Prop. 2.2] and [37, Lem. 2.23].

By Kodaira's lemma, we have $L^{\otimes n} \simeq A \otimes E$ for some positive integer $n \geq 1$, some ample line bundle A and some effective line bundle E on X (see [29, Lem. 2.60]). Since G is a closed subgroup of $\mathrm{Aut}(X, [L^{\otimes n}]_{\mathrm{num}})$, we may assume that $n = 1$.

Consider the ample line bundle $A \boxtimes A$ on $X \times X$. We claim that the degrees of the graphs $\Gamma_g \subset X \times X$, where $g \in G$, are bounded independently of g. This implies the statement as follows: the above graphs form a flat family of normal subvarieties of $X \times X$, parameterized by G (a disjoint union of open and closed smooth varieties). This yields a morphism from G to the Hilbert scheme $\mathrm{Hilb}_{X \times X}$. Since G is closed in $\mathrm{Aut}(X)$ and the latter is the reduced subscheme of an open subscheme of $\mathrm{Hilb}_{X \times X}$, this morphism is an immersion, say i. By [27, I.6.3, I.6.6.1], we may compose i with the Hilbert–Chow morphism to obtain a local immersion $\gamma : G \to \mathrm{Chow}_{X \times X}$. Clearly, γ is injective, and hence an immersion. Thus, the graphs Γ_g are the k-rational points of a locally closed subvariety of $\mathrm{Chow}_{X \times X}$. Since the cycles of any prescribed degree form a subscheme of finite type of $\mathrm{Chow}_{X \times X}$, our claim yields that G has finitely many components indeed.

We now prove this claim. Let $d := \dim(X)$ and denote by $L_1 \cdots L_d$ the intersection number of the line bundles $L_1, \ldots, L_d$ on X; also, we denote the line bundles additively. By [16, Thm. 9.6.3], $L_1 \cdots L_d$ only depends on the numerical equivalence classes of $L_1, \ldots, L_d$. With this notation, the degree of Γ_g relative to $A \boxtimes A$ is the self-intersection number $(A + g^*A)^d$. We have

$$(A + g^*A)^d = (A + g^*A)^{d-1} \cdot (L + g^*L) - (A + g^*A)^{d-1} \cdot (E + g^*E).$$

We now use the fact that

$$L_1 \cdots L_{d-1} \cdot E \geq 0$$

for any ample line bundles $L_1, \ldots, L_{d-1}$ (this is a very special case of [19, Ex. 12.1.7]), and hence for any nef line bundles $L_1, \ldots, L_{d-1}$ (since the nef cone is the closure of the ample cone). It follows that

$$(A + g^*A)^d \leq (A + g^*A)^{d-1} \cdot (L + g^*L)$$

$$= (A + g^*A)^{d-2} \cdot (L + g^*L)^2 - (A + g^*A)^{d-2} \cdot (L + g^*L) \cdot (E + g^*E).$$

Using again the above fact, this yields

$$(A + g^*A)^d \leq (A + g^*A)^{d-2} \cdot (L + g^*L)^2.$$

Proceeding inductively, we obtain $(A + g^*A)^d \leq (L + g^*L)^d$. Since g^*L is numerically equivalent to L, this yields the desired bound

$$(A + g^*A)^d \leq 2^d L^d. \qquad \square$$

Lemma 2.3 *Let L be a big line bundle on a normal projective variety X. Then $\mathrm{Aut}(X,[L])$ is a linear algebraic group.*

Proof Since $G := \mathrm{Aut}(X,[L])$ is a closed subgroup of $\mathrm{Aut}(X,[L^{\otimes n}])$ for any $n \geq 1$, we may assume that the rational map

$$f_1 : X \dashrightarrow \mathbb{P}\, H^0(X,L)$$

is birational onto its closed image Y_1. Note that f_1 is G-equivariant and the action of G on $\mathbb{P}\, H^0(X,L)$ stabilizes Y_1.

Consider the blowing-up of the base locus of L (defined as in [24, Ex. II.7.17.3]), and its normalization $\tilde{X}$. Denote by

$$\pi : \tilde{X} \longrightarrow X$$

the resulting birational morphism; then $\pi_*(\mathcal{O}_{\tilde{X}}) = \mathcal{O}_X$ by Zariski's Main Theorem. Let $\tilde{L} := \pi^*(L)$; then $H^0(\tilde{X}, \tilde{L}) \simeq H^0(X,L)$ and $\tilde{L} = L' \otimes E$, where L' is a line bundle generated by its subspace of global sections $H^0(X,L)$, and E is an effective line bundle. Thus, L' is big. Moreover, the action of G on X lifts to an action on $\tilde{X}$ which fixes $[\tilde{L}]$ and $[L']$. We now claim that the image of the resulting homomorphism $\rho : G \to \mathrm{Aut}(\tilde{X},[L'])$ is closed.

Note that ρ factors through a homomorphism

$$\eta : G \to \mathrm{Aut}(\tilde{X},[\tilde{L}]) = \mathrm{Aut}(\tilde{X},[\tilde{L}],[L']),$$

where the right-hand side is a closed subgroup of $\mathrm{Aut}(\tilde{X},[L'])$. Thus, it suffices to show that the image of η is closed. For this, we adapt the argument of Lemma 2.1. Consider the cartesian diagram

$$
\begin{array}{ccc}
\tilde{L} & \xrightarrow{\ \varphi\ } & L \\
\Big\downarrow{\scriptstyle \tilde{\pi}} & & \Big\downarrow{\scriptstyle \pi} \\
\tilde{X} & \xrightarrow{\ f\ } & X.
\end{array}
$$

Since $f_*(\mathcal{O}_{\tilde{X}}) = \mathcal{O}_X$, we have $\varphi_*(\mathcal{O}_{\tilde{L}}) = \mathcal{O}_L$. So Blanchard's lemma (see [8, Prop. 4.2.1]) yields a homomorphism

$$\varphi_* : \mathrm{Aut}^{\mathbb{G}_m}(\tilde{L})^0 \longrightarrow \mathrm{Aut}^{\mathbb{G}_m}(L)^0,$$

and hence a homomorphism

$$f_* : \mathrm{Aut}(\tilde{X}, \tilde{L})^0 \longrightarrow \mathrm{Aut}(X,[L])^0 = G^0$$

which is the inverse of $\eta^0 : G^0 \to \mathrm{Aut}(\tilde{X}, \tilde{L})^0$. Thus, the image of η contains $\mathrm{Aut}(\tilde{X}, \tilde{L})^0$. This implies our claim.

By this claim, we may replace the pair (X, L) with $(\tilde{X}, L')$; equivalently, we may assume that the big line bundle L is generated by its global sections. Then $\mathrm{Aut}(X, [L]_{\mathrm{num}})$ is an algebraic group by Lemma 2.2. Since G is a closed subgroup of $\mathrm{Aut}(X, [L]_{\mathrm{num}})$, it is algebraic as well. So the image of the homomorphism $G \to \mathrm{Aut}(\mathbb{P} H^0(X, L)) \simeq \mathrm{PGL}_N$ is closed. As f_1 is birational, the scheme-theoretic kernel of this homomorphism is trivial; thus, G is linear. $\qquad\square$

Finally, we record a classical bigness criterion, for which we could locate no reference in the generality that we need:

Lemma 2.4 *Consider an effective Cartier divisor D on a projective variety X and let $U := X \setminus \mathrm{Supp}(D)$. If U is affine, then $\mathcal{O}_X(D)$ is big.*

Proof Denote by $s \in H^0(X, \mathcal{O}_X(D))$ the canonical section, so that $U = X_s$ (the complement of the zero locus of s). Let $h_1, \ldots, h_r$ be generators of the k-algebra $\mathcal{O}(U)$. Then there exist positive integers $n_1, \ldots, n_r$ such that $h_i\, s^{n_i} \in H^0(X, \mathcal{O}_X(n_i D))$ for $i = 1, \ldots, r$. So $h_i\, s^n \in H^0(X, \mathcal{O}_X(nD))$ for any $n \geq n_1, \ldots, n_r$. It follows that the rational map $f_n : X \dashrightarrow \mathbb{P} H^0(X, \mathcal{O}_X(nD))$ restricts to a closed immersion $U \to \mathbb{P} H^0(X, \mathcal{O}_X(nD))_{s^n}$, and hence is birational onto its image. $\qquad\square$

3 Proof of Theorem 1.1

We first obtain a characteristic-free analogue of the Tits fibration:

Lemma 3.1 *Let G be a connected linear algebraic group, and H a subgroup scheme. For any $n \geq 1$, denote by G_n (resp. H_n) the n-th infinitesimal neighborhood of the neutral element e in G (resp. H).*

(i) *The union of the G_n $(n \geq 1)$ is schematically dense in G.*
(ii) *For $n \gg 0$, we have the equality $N_G(H^0) = N_G(H_n)$ of scheme-theoretic normalizers.*
(iii) *The canonical morphism $f : G/H \to G/N_G(H^0)$ is affine.*

Proof (i) Denote by $\mathfrak{m} \subset \mathcal{O}(G)$ the maximal ideal of e; then

$$G_n = \mathrm{Spec}(\mathcal{O}(G)/\mathfrak{m}^n)$$

for all n. Thus, the assertion is equivalent to $\bigcap_{n \geq 1} \mathfrak{m}^n = 0$. This is proved in [26, I.7.17]; we recall the argument for the reader's convenience. If G is smooth, then $\mathcal{O}(G)$ is a noetherian domain, hence the assertion follows from Nakayama's lemma. For an arbitrary G, we have an isomorphism of algebras

$$\mathcal{O}(G) \simeq \mathcal{O}(G_{\text{red}}) \otimes A,$$

where A is a local k-algebra of finite dimension as a k-vector space (see [13, III.3.6.4]). Thus, $\mathfrak{m} = \mathfrak{m}_1 \otimes 1 + 1 \otimes \mathfrak{m}_2$, where $\mathfrak{m}_1$ (resp. $\mathfrak{m}_2$) denotes the maximal ideal of e in $\mathcal{O}(G_{\text{red}})$ (resp. the maximal ideal of A). We may choose an integer $N \geq 1$ such that $\mathfrak{m}_2^N = 0$; then $\mathfrak{m}^n \subset \mathfrak{m}_1^{n-N} \otimes A$ for all $n \geq N$. Since G_{red} is smooth and connected, we have $\bigcap_{n \geq 1} \mathfrak{m}_1^n = 0$ by the above step; this yields the assertion.

(ii) Since $H_n = H^0 \cap G_n$ and G normalizes G_n, we have

$$N_G(H^0) \subset N_G(H_n).$$

To show the opposite inclusion, note that $(H_n)_{n-1} = H_{n-1}$, hence we have $N_G(H_n) \subset N_G(H_{n-1})$. This decreasing sequence of closed subschemes of G stops, say at n_0. Then $N_G(H_{n_0})$ normalizes H_n for all $n \geq n_0$. In view of (i), it follows that $N_G(H_{n_0})$ normalizes H^0.

(iii) We have a commutative triangle

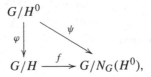

where φ is a torsor under H/H^0 (a finite constant group), and ψ is a torsor under $N_G(H^0)/H^0$ (a linear algebraic group). In particular, ψ is affine. Let U be an open affine subscheme of $G/N_G(H^0)$. Then $\psi^{-1}(U) \subset G/H^0$ is open, affine and stable under H/H^0. Hence $f^{-1}(U) = \varphi(\psi^{-1}(U))$ is affine. $\qquad\square$

Remark 3.2 (i) The first infinitesimal neighborhood G_1 may be identified with the Lie algebra $\mathfrak{g}$ of G; thus, $G_1 \cap H^0 = H_1$ is identified with the Lie algebra $\mathfrak{h}$ of H.

If $\text{char}(k) = 0$, then $N_G(H^0) = N_G(\mathfrak{h})$, since every subgroup scheme of G is uniquely determined by its Lie subalgebra (see e.g. [13, II.6.2.1]). As a consequence, the morphism $f : G/H \to G/N_G(H^0)$ is the $\mathfrak{g}$-anticanonical fibration considered in [25, I.2.7] (see also [22, §4]).

By contrast, if $\text{char}(k) > 0$ then the morphism $G/H \to G/N_G(\mathfrak{h})$ is not necessarily affine (see e.g. [3, Ex. 5.6]). In particular, the inclusion $N_G(H^0) \subset N_G(\mathfrak{h})$ may be strict.

(ii) If $\text{char}(k) = p > 0$, then G_{p^n} is the n-th Frobenius kernel of G, as defined for example in [26, I.9.4]; in particular, G_{p^n} is a normal infinitesimal subgroup scheme of G. Then the above assertion (i) just means that the union of the iterated Frobenius kernels is schematically dense in G.

We may now prove Theorem 1.1. Recall its assumptions: X is a normal projective variety on which a smooth connected linear algebraic group G acts with an open dense orbit. The variety X is unirational in view of [2, Thm. 18.2]; thus, its Albanese variety is trivial. By duality, it follows that the Picard variety $\text{Pic}^0(X)$ is trivial as well (see [16, 9.5.25]). Therefore, $\text{Pic}(X) = \text{NS}(X)$ is fixed pointwise by $\text{Aut}^0(X)$. Using Lemma 2.3, this implies that $\text{Aut}^0(X)$ is linear. We may thus assume that $G = \text{Aut}^0(X)$; in particular, G is a normal subgroup of $\text{Aut}(X)$.

Denote by $X_0 \subset X$ the open G-orbit; then X_0 is normalized by $\text{Aut}(X)$. Choose $x \in X_0(k)$ and denote by H its scheme-theoretic stabilizer in G. Then we have $X_0 = G \cdot x \simeq G/H$ equivariantly for the G-action. We also have $X_0 = \text{Aut}(X) \cdot x \simeq \text{Aut}(X)/\text{Aut}(X,x)$ equivariantly for the $\text{Aut}(X)$-action. As a consequence, $\text{Aut}(X) = G \cdot \text{Aut}(X,x)$.

Next, choose a positive integer n such that $N_G(H^0) = N_G(H_n)$ (see Lemma 3.1). The action of $\text{Aut}(X)$ on G by conjugation normalizes G_n and induces a linear representation of $\text{Aut}(X)$ in $V := \mathcal{O}(G_n)$, a finite-dimensional vector space. The ideal of H_n is a subspace $W \subset V$, and its stabilizer in G equals $N_G(H^0)$. We consider W as a k-rational point of the Grassmannian $\text{Grass}(V)$ parameterizing linear subspaces of V of the appropriate dimension. The linear action of $\text{Aut}(X)$ on V yields an action on $\text{Grass}(V)$. The subgroup scheme $\text{Aut}(X,x)$ fixes W, since it normalizes $\text{Aut}(X,x)^0 = H^0$. Thus, we obtain

$$\text{Aut}(X) \cdot W = G \cdot \text{Aut}(X,x) \cdot W = G \cdot W \simeq G/N_G(H^0).$$

As a consequence, the morphism $f : G/H \to G/N_G(H^0)$ yields an $\text{Aut}(X)$-equivariant morphism

$$\tau : X_0 \longrightarrow Y,$$

where Y denotes the closure of $\text{Aut}(X) \cdot W$ in $\text{Grass}(V)$.

We may view τ as a rational map $X \dashrightarrow Y$. Let X' denote the normalization of the graph of this rational map, i.e., of the closure of X_0 embedded diagonally in $X \times Y$. Then X' is a normal projective variety equipped with an action of $\text{Aut}(X)$ and with an equivariant morphism $f : X' \to X$ which restricts to an isomorphism above the open orbit X_0. By Lemma 2.1, the image of $\text{Aut}(X)$ in $\text{Aut}(X')$ is closed; thus, it suffices to show that $\text{Aut}(X')$ is a linear algebraic group. So we may assume that τ extends to a morphism $X \to Y$, that we will still denote by τ.

Next, consider the boundary, $\partial X := X \setminus X_0$, that we view as a closed reduced subscheme of X; it is normalized by $\text{Aut}(X)$. Thus, the action of $\text{Aut}(X)$ on X lifts to an action on the blowing-up of X along ∂X, and on its normalization. Using Lemma 2.1 again, we may further assume that ∂X is the support of an effective Cartier divisor Δ, normalized by $\mathrm{Aut}(X)$; thus, the line bundle $\mathcal{O}_X(\Delta)$ is $\mathrm{Aut}(X)$-linearized.

We also have an ample, $\mathrm{Aut}(X)$-linearized line bundle M on Y, the pull-back of $\mathcal{O}(1)$ under the Plücker embedding of $\mathrm{Grass}(V)$. Thus, there exist a positive integer m and a nonzero section $t \in H^0(Y, M^{\otimes m})$ which vanishes identically on the boundary $\partial Y := Y \setminus G \cdot W$. Then $L := \tau^*(M)$ is an $\mathrm{Aut}(X)$-linearized line bundle on X, equipped with a nonzero section $s := \tau^*(t)$ which vanishes identically on $\tau^{-1}(\partial Y) \subset \partial X$. Denote by D (resp. E) the divisor of zeroes of s (resp. t). Then $D + \Delta$ is an effective Cartier divisor on X, and we have

$$X \setminus \mathrm{Supp}(D + \Delta) = X_0 \setminus \mathrm{Supp}(D) = f^{-1}(G \cdot W \setminus \mathrm{Supp}(E)).$$

Since $\partial Y \subset \mathrm{Supp}(E)$, we have $G \cdot W \setminus \mathrm{Supp}(E) = Y \setminus \mathrm{Supp}(E)$. The latter is affine as M is ample. Since the morphism f is affine (Lemma 3.1), it follows that $X \setminus \mathrm{Supp}(D + \Delta)$ is affine as well. Hence $D + \Delta$ is big (Lemma 2.4). Also, $\mathcal{O}_X(D + \Delta) = L \otimes \mathcal{O}_X(\Delta)$ is $\mathrm{Aut}(X)$-linearized. Using Lemma 2.3, we conclude that $\mathrm{Aut}(X)$ is a linear algebraic group.

4 Proof of Theorem 1.2

By [3, Thm. 2], there exists a G-equivariant morphism

$$f : X \longrightarrow G/H$$

for some subgroup scheme $H \subset G$ such that $H \supset G_{\mathrm{aff}}$ and H/G_{aff} is finite; equivalently, H is affine and G/H is an abelian variety. Then the natural map $A = G/G_{\mathrm{aff}} \to G/H$ is an isogeny. Denote by Y the scheme-theoretic fiber of f at the origin of G/H; then Y is normalized by H, and the action map

$$G \times Y \longrightarrow X, \quad (g, y) \longmapsto g \cdot y$$

factors through an isomorphism

$$G \times^H Y \xrightarrow{\;\simeq\;} X,$$

where $G \times^H Y$ denotes the quotient of $G \times Y$ by the action of H via

$$h \cdot (g, y) := (gh^{-1}, h \cdot y).$$

This is the fiber bundle over G/H associated with the faithfully flat H-torsor $G \to G/H$ and the H-scheme Y. The above isomorphism identifies f with the morphism $G \times^H Y \to G/H$ obtained from the projection $G \times Y \to G$.

We now obtain a reduction to the case where G is *anti-affine*, i.e., $\mathcal{O}(G) = k$. Recall that G has a largest anti-affine subgroup scheme G_{ant}; moreover, G_{ant} is smooth, connected and centralizes G (see [13, III.3.8]). We have the Rosenlicht decomposition $G = G_{\text{ant}} \cdot G_{\text{aff}}$; in addition, the scheme-theoretic intersection $G_{\text{ant}} \cap G_{\text{aff}}$ contains $(G_{\text{ant}})_{\text{aff}}$, and the quotient $(G_{\text{ant}} \cap G_{\text{aff}})/(G_{\text{ant}})_{\text{aff}}$ is finite (see [8, Thm. 3.2.3]). As a consequence,

$$G = G_{\text{ant}} \cdot H \simeq (G_{\text{ant}} \times H)/(G_{\text{ant}} \cap H) \quad \text{and} \quad G/H \simeq G_{\text{ant}}/(G_{\text{ant}} \cap H).$$

Thus, we obtain an isomorphism of schemes

$$G_{\text{ant}} \times^{G_{\text{ant}} \cap H} Y \xrightarrow{\sim} X, \tag{1}$$

and an isomorphism of abstract groups

$$\text{Aut}^H_{\text{gp}}(G) \xrightarrow{\sim} \text{Aut}^{G_{\text{ant}} \cap H}_{\text{gp}}(G_{\text{ant}}). \tag{2}$$

Next, we construct an action of the subgroup $\text{Aut}^H_{\text{gp}}(G) \subset \text{Aut}^{G_{\text{aff}}}_{\text{gp}}(G)$ on X. Let $\gamma \in \text{Aut}^H(G)$. Then $\gamma \times \text{id}$ is an automorphism of $G \times Y$, equivariant under the above action of H. Moreover, the quotient map

$$\pi : G \times Y \longrightarrow G \times^H Y$$

is a faithfully flat H-torsor, and hence a categorical quotient. It follows that there is a unique automorphism δ of $G \times^H Y = X$ such that the diagram

$$
\begin{array}{ccc}
G \times Y & \xrightarrow{\gamma \times \text{id}} & G \times Y \\
\pi \downarrow & & \downarrow \pi \\
X & \xrightarrow{\quad \delta \quad} & X
\end{array}
$$

commutes. Clearly, the assignement $\gamma \mapsto \delta$ defines a homomorphism of abstract groups

$$\varphi : \text{Aut}^H_{\text{gp}}(G) \longrightarrow \text{Aut}(X).$$

We also have a natural homomorphism

$$\psi : \text{Aut}^H_{\text{gp}}(G) \longrightarrow \text{Aut}_{\text{gp}}(G/H).$$

By construction, f is equivariant under the action of $\text{Aut}^H_{\text{gp}}(G)$ on X via φ, and its action on G/H via ψ. Moreover, we have in $\text{Aut}(X)$

$$\varphi(\gamma) \, g \, \varphi(\gamma)^{-1} = \gamma(g)$$

for all $\gamma \in \text{Aut}^H_{\text{gp}}(G)$ and $g \in G$. In particular, the image of the homomorphism φ normalizes G.

Lemma 4.1 *With the above notation, the homomorphism ψ is injective. Moreover, $\mathrm{Aut}_{\mathrm{gp}}^{H}(G)$ is a subgroup of finite index of $\mathrm{Aut}_{\mathrm{gp}}^{G_{\mathrm{aff}}}(G)$.*

Proof To show both assertions, we may assume that G is anti-affine by using the isomorphism (2). Then G is commutative and hence its endomorphisms (of algebraic group) form a ring, $\mathrm{End}_{\mathrm{gp}}(G)$. Let $\gamma \in \mathrm{Aut}_{\mathrm{gp}}^{H}(G)$ such that $\psi(\gamma) = \mathrm{id}_{G/H}$; then $\gamma - \mathrm{id}_{G} \in \mathrm{End}_{\mathrm{gp}}(G)$ takes G to H, and H to the neutral element. Thus, $\gamma - \mathrm{id}_{G}$ factors through a homomorphism $G/H \to H$; but every such homomorphism is trivial, since G/H is an abelian variety and H is affine. So $\gamma - \mathrm{id}_{G} = 0$, proving the first assertion.

For the second assertion, we may replace H with any larger subgroup scheme K such that K/G_{aff} is finite. By [5, Thm. 1.1], there exists a finite subgroup scheme $F \subset G$ such that $H = G_{\mathrm{aff}} \cdot F$. Let n denote the order of F; then F is contained in the n-torsion subgroup scheme $G[n]$, and hence $H \subset G_{\mathrm{aff}} \cdot G[n]$.

We now claim that the subgroup scheme $G[n]$ is finite for any integer $n > 0$. If $\mathrm{char}(k) = p > 0$, then G is a semi-abelian variety (see [8, Prop. 5.4.1]) and the claim follows readily. If $\mathrm{char}(k) = 0$, then G is an extension of a semi-abelian variety by a vector group U (see [8, §5.2]). Since the multiplication map n_{U} is an isomorphism, we have $G[n] \simeq (G/U)[n]$; this completes the proof of the claim.

By this claim, we may replace H with the larger subgroup scheme $G_{\mathrm{aff}} \cdot G[n]$ for some integer $n > 0$. Then the restriction map

$$\rho : \mathrm{Aut}_{\mathrm{gp}}^{G_{\mathrm{aff}}}(G) \longrightarrow \mathrm{Aut}_{\mathrm{gp}}(G[n])$$

has kernel $\mathrm{Aut}_{\mathrm{gp}}^{H}(G)$. Thus, it suffices to show that ρ has a finite image.

Note that the image of ρ is contained in the image of the analogous map $\mathrm{End}_{\mathrm{gp}}(G) \to \mathrm{End}_{\mathrm{gp}}(G[n])$. Moreover, the latter image is a finitely generated abelian group (since so is $\mathrm{End}_{\mathrm{gp}}(G)$ in view of [8, Lem. 5.1.3]) and is n-torsion (since so is $\mathrm{End}_{\mathrm{gp}}(G[n])$). This completes the proof. □

Lemma 4.2 *With the above notation, the homomorphism φ is injective. Moreover, its image is the subgroup of $\mathrm{Aut}(X)$ which normalizes G and centralizes Y; this subgroup intersects G trivially.*

Proof Let $\gamma \in \mathrm{Aut}_{\mathrm{gp}}^{H}(G)$ such that $\varphi(\gamma) = \mathrm{id}_{X}$. In view of the equivariance of f, it follows that $\psi(\gamma) = \mathrm{id}_{G/H}$. Thus, $\gamma = \mathrm{id}_{G}$ by Lemma 4.1. So φ is injective; thus, we will identify $\mathrm{Aut}_{\mathrm{gp}}^{H}(G)$ with the image of φ.

As already noticed, this image normalizes G; it also centralizes Y by construction. Conversely, let $u \in \mathrm{Aut}(X)$ normalizing G and centralizing Y.

Since H normalizes Y, the commutator $uhu^{-1}h^{-1}$ centralizes Y for any schematic point $h \in H$. Also, $uhu^{-1}h^{-1} \in G$. But in view of (1), we have $X = G_{\mathrm{ant}} \cdot Y$, where G_{ant} is central in G. It follows that $uhu^{-1}h^{-1}$ centralizes X. Hence u centralizes H, and acts on G by conjugation via some $\gamma \in \mathrm{Aut}_{\mathrm{gp}}^{H}(G)$. For any schematic points $g \in G$, $y \in Y$, we have $u(g \cdot y) = ugu^{-1}u(y) = \gamma(g)u(y) = \gamma(g)y$, that is, $u = \varphi(\gamma)$.

It remains to show that $\mathrm{Aut}_{\mathrm{gp}}^{H}(G)$ meets G trivially. Let $\gamma \in \mathrm{Aut}_{\mathrm{gp}}^{H}(G)$ such that $\varphi(\gamma) \in G$. Then γ acts on G/H by a translation, and fixes the origin. So $\psi(\gamma) = \mathrm{id}_{G/H}$, and $\gamma = \mathrm{id}_G$ by using Lemma 4.1 again. $\qquad\square$

Now Theorem 1.2 follows by combining Lemmas 4.1 and 4.2.

Remark 4.3   (i) With the above notation, $\mathrm{Aut}_{\mathrm{gp}}^{H}(G)$ is the group of integer points of a linear algebraic group defined over the field of rational numbers. Indeed, we may reduce to the case where G is anti-affine, as in the beginning of the proof of Lemma 4.1. Then $\mathrm{Aut}_{\mathrm{gp}}^{H}(G)$ is the group of units of the ring $\mathrm{End}_{\mathrm{gp}}^{H}(G) =: R$; moreover, the additive group of R is free of finite rank (as follows from [8, Lem. 5.1.3]). So the group of units of the finite-dimensional $\mathbb{Q}$-algebra $R_{\mathbb{Q}} := R \otimes_{\mathbb{Z}} \mathbb{Q}$ is a closed subgroup of $\mathrm{GL}(R_{\mathbb{Q}})$ (via the regular representation), and its group of integer points relative to the lattice $R \subset R_{\mathbb{Q}}$ is just $\mathrm{Aut}_{\mathrm{gp}}^{H}(G)$.

In other terms, $\mathrm{Aut}_{\mathrm{gp}}^{H}(G)$ is an *arithmetic group*; it follows e.g. that this group is finitely presented (see [1]).

(ii) The commensurability class of $\mathrm{Aut}_{\mathrm{gp}}^{G_{\mathrm{aff}}}(G)$ is an isogeny invariant. Consider indeed an isogeny

$$u : G \longrightarrow G',$$

i.e., u is a faithfully flat homomorphism and its kernel F is finite. Then u induces isogenies $G_{\mathrm{aff}} \to G'_{\mathrm{aff}}$ and $G_{\mathrm{ant}} \to G'_{\mathrm{ant}}$. In view of Lemma 4.1, we may thus assume that G and G' are anti-affine. We may choose a positive integer n such that F is contained in the n-torsion subgroup scheme $G[n]$; also, recall from the proof of Lemma 4.1 that $G[n]$ is finite. Thus, there exists an isogeny

$$v : G' \longrightarrow G$$

such that $v \circ u = n_G$ (the multiplication map by n in G). Then we have a natural homomorphism

$$u_* : \mathrm{Aut}_{\mathrm{gp}}^{G_{\mathrm{aff}} \cdot F}(G) \longrightarrow \mathrm{Aut}_{\mathrm{gp}}^{G'_{\mathrm{aff}}}(G')$$

which lies in a commutative diagram

$$\mathrm{Aut}_{\mathrm{gp}}^{G_{\mathrm{aff}}\cdot G[n]}(G) \xrightarrow{u_*} \mathrm{Aut}_{\mathrm{gp}}^{G'_{\mathrm{aff}}\cdot \mathrm{Ker}(v)}(G') \xrightarrow{v_*} \mathrm{Aut}_{\mathrm{gp}}^{G_{\mathrm{aff}}}(G)$$

$$\downarrow \qquad\qquad\qquad \downarrow$$

$$\mathrm{Aut}_{\mathrm{gp}}^{G_{\mathrm{aff}}\cdot F}(G) \xrightarrow{u_*} \mathrm{Aut}_{\mathrm{gp}}^{G'_{\mathrm{aff}}}(G'),$$

where all arrows are injective, and the images of the vertical arrows have finite index (see Lemma 4.1 again). Moreover, the image of the homomorphism $v_* \circ u_* = (v \circ u)_* = (n_G)_*$ has finite index as well, since this homomorphism is identified with the inclusion of $\mathrm{Aut}_{\mathrm{gp}}^{G_{\mathrm{aff}}\cdot G[n]}(G)$ in $\mathrm{Aut}_{\mathrm{gp}}^{G_{\mathrm{aff}}}(G)$. This yields our assertion.

(iii) There are many examples of smooth connected algebraic groups G such that $\mathrm{Aut}_{\mathrm{gp}}^{G_{\mathrm{aff}}}(G)$ is infinite. The easiest ones are of the form $A \times G_{\mathrm{aff}}$, where A is an abelian variety such that $\mathrm{Aut}_{\mathrm{gp}}(A)$ is infinite. To construct further examples, consider a smooth connnected algebraic group G, denote by $\alpha : G \to A$ the quotient homomorphism by G_{aff}, and let $h : A \to B$ a non-zero homomorphism to an abelian variety. Then $G' := G \times B$ is a smooth connected algebraic group, and the assignment $(g,b) \mapsto (g, b + h(\alpha(g)))$ defines an automorphism of G' of infinite order, which fixes pointwise $G'_{\mathrm{aff}} = G_{\mathrm{aff}}$.

5 Proof of Theorem 1.3

By [3, Thm. 3], the Albanese morphism of X is of the form

$$f : X \longrightarrow \mathrm{Alb}(X) = G/H,$$

where H is an affine subgroup scheme of G containing G_{aff}. Thus, we have as in Section 4

$$X \simeq G \times^H Y \simeq G_{\mathrm{ant}} \times^K Y,$$

where $K := G_{\mathrm{ant}} \cap H$ and Y denotes the (scheme-theoretic) fiber of f at the origin of G/H. Then Y is a closed subscheme of X, normalized by H.

Lemma 5.1   (i) G_{ant} *is the largest anti-affine subgroup of* $\mathrm{Aut}(X)$. *In particular,* G_{ant} *is normal in* $\mathrm{Aut}(X)$.

(ii) $f_*(\mathcal{O}_X) = \mathcal{O}_{G/H}$.

(iii) *If K is smooth, then Y is a normal projective variety, almost homogeneous under the reduced neutral component* H_{red}^0.

Proof (i) By [8, Prop. 5.5.3], $\mathrm{Aut}(X)$ has a largest anti-affine subgroup $\mathrm{Aut}(X)_{\mathrm{ant}}$, and this subgroup centralizes G. Thus,

$$G' := G \cdot \mathrm{Aut}(X)_{\mathrm{ant}}$$

is a smooth connected subgroup scheme of $\mathrm{Aut}^0(X)$ containing G as a normal subgroup scheme. As a consequence, G' normalizes the open G-orbit in X. So $G/H = G'/H'$ for some subgroup scheme H' of G'; equivalently, $G' = G \cdot H'$. Also, H' is affine in view of [8, Prop. 2.3.2]. Thus, the quotient group $G'/G \simeq H'/H$ is affine. But, $G'/G = \mathrm{Aut}_{\mathrm{ant}}(X)/(\mathrm{Aut}_{\mathrm{ant}}(X) \cap G)$ is anti-affine. Hence $G' = G$, and $\mathrm{Aut}_{\mathrm{ant}}(X) = G_{\mathrm{ant}}$.

(ii) Consider the Stein factorization $f = g \circ h$, where $g : X' \to G/H$ is finite, and $h : X \to X'$ satisfies $h_*(\mathcal{O}_X) = \mathcal{O}_{X'}$. By Blanchard's lemma (see [8, Prop. 4.2.1]), the G-action on X descends to a unique action on X' such that g is equivariant. As a consequence, X' is a normal projective variety, almost homogeneous under this action, and h is equivariant as well. Let $H' \subset G$ denote the scheme-theoretic stabilizer of a k-rational point of the open G-orbit in X'. Then $H' \subset H$ and the homogeneous space H/H' is finite. It follows that $H' \supset G_{\mathrm{aff}}$, i.e., G/H' is an abelian variety. Thus, so is X'; then $X' = X$ and $h = \mathrm{id}$ by the universal property of the Albanese variety.

(iii) Since K is smooth, it normalizes the reduced subscheme $Y_{\mathrm{red}} \subset Y$. Thus, $G_{\mathrm{ant}} \times^K Y_{\mathrm{red}}$ may be viewed as a closed subscheme of $G_{\mathrm{ant}} \times^K Y = X$, with the same k-rational points (since $Y_{\mathrm{red}}(k) = Y(k)$). It follows that $X = G_{\mathrm{ant}} \times^K Y_{\mathrm{red}}$, i.e., Y is reduced.

Next, let $\eta : \tilde{Y} \to Y$ denote the normalization map. Then the action of K lifts uniquely to an action on $\tilde{Y}$. Moreover, the resulting morphism $G_{\mathrm{ant}} \times^K \tilde{Y} \to X$ is finite and birational, hence an isomorphism. Thus, η is an isomorphism as well, i.e., Y is normal. Since Y is connected and closed in X, it is a projective variety.

It remains to show that Y is almost homogenous under $H^0_{\mathrm{red}} =: H'$ (a smooth connected linear algebraic group). Since the homogeneous space H/H' is finite, so is the natural map $\varphi : G/H' \to G/H$; as a consequence, G/H' is an abelian variety and φ is an isogeny. We have a cartesian square

$$
\begin{array}{ccc}
X' := G \times^{H'} Y & \longrightarrow & G/H' \\
\downarrow & & \downarrow{\scriptstyle \varphi} \\
X = G \times^{H} Y & \longrightarrow & G/H.
\end{array}
$$

Thus, X' is a normal projective variety, almost homogeneous under G. Its open G-orbit intersects Y along an open subvariety, which is the unique orbit of H'. $\qquad\square$

We denote by $\mathrm{Aut}(X, Y)$ the normalizer of Y in $\mathrm{Aut}(X)$; this is the stabilizer of the origin for the action of $\mathrm{Aut}(X)$ on $\mathrm{Alb}(X) = G/H$. Likewise, we denote by $\mathrm{Aut}_{\mathrm{gp}}(G_{\mathrm{ant}}, K)$ (resp. $\mathrm{Aut}(Y, K)$) the normalizer of K in $\mathrm{Aut}_{\mathrm{gp}}(G_{\mathrm{ant}})$ (resp. $\mathrm{Aut}(Y)$).

Lemma 5.2   (i) *There is an exact sequence*

$$\pi_0(K) \longrightarrow \pi_0 \, \mathrm{Aut}(X) \longrightarrow \pi_0 \, \mathrm{Aut}(X, Y) \longrightarrow 0.$$

(ii) *We have a closed immersion*

$$\iota : \mathrm{Aut}(X, Y) \longrightarrow \mathrm{Aut}_{\mathrm{gp}}(G_{\mathrm{ant}}, K) \times \mathrm{Aut}(Y, K)$$

with image consisting of the pairs (γ, v) such that $\gamma|_K = \mathrm{Int}(v)|_K$.

Proof (i) Since the normal subgroup scheme G_{ant} of $\mathrm{Aut}(X)$ acts transitively on $G/H = \mathrm{Aut}(X)/\mathrm{Aut}(X, Y)$, we have

$$\mathrm{Aut}(X) = G_{\mathrm{ant}} \cdot \mathrm{Aut}(X, Y).$$

Moreover, $G \cap \mathrm{Aut}(X, Y) = H$, hence $G_{\mathrm{ant}} \cap \mathrm{Aut}(X, Y) = K$. Thus, we obtain $\mathrm{Aut}^0(X) = G_{\mathrm{ant}} \cdot \mathrm{Aut}(X, Y)^0$ and

$$\begin{aligned}
\pi_0 \, \mathrm{Aut}(X) &= \mathrm{Aut}(X, Y)/(G_{\mathrm{ant}} \cdot \mathrm{Aut}(X, Y)^0 \cap \mathrm{Aut}(X, Y)) \\
&= \mathrm{Aut}(X, Y)/(G_{\mathrm{ant}} \cap \mathrm{Aut}(X, Y)) \cdot \mathrm{Aut}(X, Y)^0 \\
&= \mathrm{Aut}(X, Y)/K \cdot \mathrm{Aut}(X, Y)^0.
\end{aligned}$$

This yields readily the desired exact sequence.

(ii) Let $u \in \mathrm{Aut}(X, Y)$, and v its restriction to Y. Since u normalizes G_{ant}, we have $u(g \cdot y) = \mathrm{Int}(u)(g) \cdot u(y) = \mathrm{Int}(u)(g) \cdot v(y)$ for all schematic points $g \in G_{\mathrm{ant}}$ and $y \in Y$. Moreover, u normalizes $K = G_{\mathrm{ant}} \cap \mathrm{Aut}(X, Y)$, hence we have $\mathrm{Int}(u) \in \mathrm{Aut}_{\mathrm{gp}}(G_{\mathrm{ant}}, K)$ and $\mathrm{Int}(v) \in \mathrm{Aut}(Y, K)$. In view of the invariance property $g \cdot y = gh^{-1} \cdot h \cdot y$ for any schematic point $h \in H$, we obtain $v(h \cdot y) = \mathrm{Int}(u)(h) \cdot v(y)$, i.e., $\mathrm{Int}(u) = \mathrm{Int}(v)$ on K. Thus, u is uniquely determined by the pair $(\mathrm{Int}(u), v)$, and this pair satisfies the assertion. Conversely, any pair (γ, v) satisfying the assertion yields an automorphism u of X normalizing Y, via $u(g \cdot y) := \gamma(g) \cdot v(y)$. $\square$

In view of the above lemma, we identify $\mathrm{Aut}(X, Y)$ with its image in $\mathrm{Aut}_{\mathrm{gp}}(G_{\mathrm{ant}}, K) \times \mathrm{Aut}(Y, K)$ via ι. Denote by

$$\rho : \mathrm{Aut}(X, Y) \longrightarrow \mathrm{Aut}_{\mathrm{gp}}(G_{\mathrm{ant}}, K)$$

the resulting projection.

Lemma 5.3 *The above map ρ induces an exact sequence*

$$\pi_0 \operatorname{Aut}^K(Y) \longrightarrow \pi_0 \operatorname{Aut}(X, Y) \longrightarrow I \longrightarrow 1,$$

where I denotes the subgroup of $\operatorname{Aut}_{\mathrm{gp}}(G_{\mathrm{ant}}, K)$ consisting of those γ such that $\gamma|_K = \operatorname{Int}(v)|_K$ for some $v \in \operatorname{Aut}(Y, K)$.

Proof By Lemma 5.2 (ii), we have an exact sequence

$$1 \longrightarrow \operatorname{Aut}^K(Y) \longrightarrow \operatorname{Aut}(X, Y) \overset{\rho}{\longrightarrow} I \longrightarrow 1.$$

Moreover, the connected algebraic group $\operatorname{Aut}(X, Y)^0$ centralizes G_{ant} in view of [8, Lem. 5.1.3]; equivalently, $\operatorname{Aut}(X, Y)^0 \subset \operatorname{Ker}(\rho)$. This readily yields the assertion. $\qquad\square$

We now consider the case where K is *smooth*; this holds for instance if $\operatorname{char}(k) = 0$. Then $\operatorname{Aut}(Y)$ is a linear algebraic group by Theorem 1.1 and Lemma 5.1 (iii). Thus, so is the subgroup scheme $\operatorname{Aut}^K(Y)$, and hence $\pi_0 \operatorname{Aut}^K(Y)$ is finite. Together with Lemmas 5.2 and 5.3, it follows that $\pi_0 \operatorname{Aut}(X)$ is commensurable with I.

To analyze the latter group, we consider the homomorphism

$$\eta : I \longrightarrow \operatorname{Aut}_{\mathrm{gp}}(K), \quad \gamma \longmapsto \gamma|_K$$

with kernel $\operatorname{Aut}^K_{\mathrm{gp}}(G_{\mathrm{ant}})$. Since K is a commutative linear algebraic group, it has a unique decomposition

$$K \simeq D \times U,$$

where D is diagonalizable and U is unipotent. Thus, we have

$$\operatorname{Aut}_{\mathrm{gp}}(G_{\mathrm{ant}}, K) = \operatorname{Aut}_{\mathrm{gp}}(G_{\mathrm{ant}}, D) \cap \operatorname{Aut}_{\mathrm{gp}}(G_{\mathrm{ant}}, U),$$

$$\operatorname{Aut}_{\mathrm{gp}}(Y, K) = \operatorname{Aut}_{\mathrm{gp}}(Y, D) \cap \operatorname{Aut}_{\mathrm{gp}}(Y, U),$$

$$\operatorname{Aut}_{\mathrm{gp}}(K) \simeq \operatorname{Aut}_{\mathrm{gp}}(D) \times \operatorname{Aut}_{\mathrm{gp}}(U).$$

Under the latter identification, the image of η is contained in the product of the images of the natural homomorphisms

$$\eta_D : \operatorname{Aut}(Y, D) \longrightarrow \operatorname{Aut}_{\mathrm{gp}}(D), \quad \eta_U : \operatorname{Aut}(Y, U) \longrightarrow \operatorname{Aut}_{\mathrm{gp}}(U).$$

The kernel of η_D (resp. η_U) equals $\operatorname{Aut}^D(Y)$ (resp. $\operatorname{Aut}^U(Y)$); also, the quotient $\operatorname{Aut}(Y, D)/\operatorname{Aut}^D(Y)$ is finite in view of the rigidity of diagonalizable group schemes (see [13, II.5.5.10]). Thus, the image of η_D is finite as well.

As a consequence, I is a subgroup of finite index of

$$J := \{\gamma \in \text{Aut}_{\text{gp}}(G_{\text{ant}}, K) \mid \gamma|_U = \text{Int}(v)|_U \text{ for some } v \in \text{Aut}(Y, K)\}. \quad (3)$$

Also, $\pi_0 \text{Aut}(X)$ is commensurable with J.

If $\text{char}(k) > 0$, then G_{ant} is a semi-abelian variety (see [8, Prop. 5.4.1]) and hence U is finite. Also, U is smooth since so is K. Thus, $\text{Aut}_{\text{gp}}(U)$ is finite, and hence the image of η is finite as well. Therefore, $\pi_0 \text{Aut}(X)$ is commensurable with $\text{Aut}_{\text{gp}}^K(G)$, and hence with $\text{Aut}_{\text{gp}}^{G^{\text{aff}}}(G)$ by Lemma 4.1. This completes the proof of Theorem 1.3 in the case where $\text{char}(k) > 0$ and K is smooth.

Next, we handle the case where $\text{char}(k) > 0$ and K is arbitrary. Consider the n-th Frobenius kernel $I_n := (G_{\text{ant}})_{p^n} \subset G_{\text{ant}}$, where n is a positive integer. Then $I_n \cap K$ is the n-th Frobenius kernel of K; thus, the image of K in G_{ant}/I_n is smooth for $n \gg 0$ (see [13, III.3.6.10]). Also, $\text{Aut}(X)$ normalizes I_n (since it normalizes G_{ant}), and hence acts on the quotient X/I_n. The latter is a normal projective variety, almost homogeneous under G/I_n (as follows from the results in [6, §2.4]). Moreover, $(G/I_n)_{\text{ant}} = G_{\text{ant}}/I_n$ and we have

$$\text{Alb}(X/I_n) \simeq \text{Alb}(X)/I_n \simeq G_{\text{ant}}/I_n K \simeq (G_{\text{ant}}/I_n)/(K/I_n \cap K),$$

where $K/I_n \cap K$ is smooth for $n \gg 0$.

We now claim that the homomorphism $\text{Aut}(X) \to \text{Aut}(X/I_n)$ is bijective on k-rational points. Indeed, every $v \in \text{Aut}(X/I_n)(k)$ extends to a unique automorphism of the function field $k(X)$, since this field is a purely inseparable extension of $k(X/I_n)$. As X is the normalization of X/I_n in $k(X)$, this implies the claim.

It follows from this claim that the induced map

$$\pi_0 \text{Aut}(X) \to \pi_0 \text{Aut}(X/I_n)$$

is an isomorphism. Likewise, every algebraic group automorphism of G_{ant} induces an automorphism of G_{ant}/I_n and the resulting map

$$\text{Aut}_{\text{gp}}(G_{\text{ant}}) \to \text{Aut}_{\text{gp}}(G_{\text{ant}}/I_n)$$

is an isomorphism, which restricts to an isomorphism

$$\text{Aut}_{\text{gp}}^K(G_{\text{ant}}) \xrightarrow{\simeq} \text{Aut}_{\text{gp}}^{K/I_n \cap K}(G_{\text{ant}}/I_n).$$

All of this yields a reduction to the case where K is smooth, and hence completes the proof of Theorem 1.3 when $\text{char}(k) > 0$.

It remains to treat the case where $\text{char}(k) = 0$. Consider the extension of algebraic groups

$$0 \longrightarrow D \times U \longrightarrow G_{\text{ant}} \longrightarrow A \longrightarrow 0,$$

where $A := G_{\mathrm{ant}}/K = G/H = \mathrm{Alb}(X)$ is an abelian variety. By [8, §5.5], the above extension is classified by a pair of injective homomorphisms

$$X^*(D) \longrightarrow \widehat{A}(k), \quad U^\vee \longrightarrow H^1(A, \mathcal{O}_A) = \mathrm{Lie}(\widehat{A}),$$

where $X^*(D)$ denotes the character group of D, and $\widehat{A}$ stands for the dual abelian variety of A. The images of these homomorphisms yield a finitely generated subgroup $\Lambda \subset \widehat{A}(k)$ and a subspace $V \subset \mathrm{Lie}(\widehat{A})$. Moreover, we may identify $\mathrm{Aut}_{\mathrm{gp}}(G_{\mathrm{ant}}, K)$ with the subgroup of $\mathrm{Aut}_{\mathrm{gp}}(\widehat{A})$ which stabilizes Λ and V. This identifies the group J defined in (3), with the subgroup of $\mathrm{Aut}_{\mathrm{gp}}(\widehat{A}, \Lambda, V)$ consisting of those γ such that $\gamma|_V \in \mathrm{Aut}(Y, K)|_V$, where $\mathrm{Aut}(Y, K)$ acts on V via the dual of its representation in U. Note that $\mathrm{Aut}(Y, K)|_V$ is an algebraic subgroup of $\mathrm{GL}(V)$. Therefore, the proof of Theorem 1.3 will be completed by the following result due to Gaël Rémond:

Lemma 5.4 *Assume that* $\mathrm{char}(k) = 0$. *Let A be an abelian variety. Let Λ be a finitely generated subgroup of $A(k)$. Let V be a vector subspace of $\mathrm{Lie}(A)$. Let G be an algebraic subgroup of* $\mathrm{GL}(V)$. *Let*

$$\Gamma := \{\gamma \in \mathrm{Aut}_{\mathrm{gp}}(A, \Lambda, V) \mid \gamma|_V \in G\}.$$

Then Γ is an arithmetic group.

Proof By the Lefschetz principle, we may assume that k is a subfield of $\mathbb{C}$.

As k is algebraically closed, there is no difference between the automorphisms of A and those of its extension to $\mathbb{C}$. Thus, we may assume that $k = \mathbb{C}$.

We denote $W := \mathrm{Lie}(A)$, $L \subset W$ its period lattice, and L' the subgroup of W containing L such that $\Lambda = L'/L$; then L' is a free abelian group of finite rank. Let

$$\Gamma' := \mathrm{Aut}(L) \times \mathrm{Aut}(L') \subset G' := \mathrm{Aut}(L \otimes \mathbb{Q}) \times \mathrm{Aut}(L' \otimes \mathbb{Q}).$$

If we choose bases of L and L' of rank r and s say then this inclusion reads $\mathrm{GL}_r(\mathbb{Z}) \times \mathrm{GL}_s(\mathbb{Z}) \subset \mathrm{GL}_r(\mathbb{Q}) \times \mathrm{GL}_s(\mathbb{Q})$. We see G' as the group of $\mathbb{Q}$-points of the algebraic group $\mathrm{GL}_r \times \mathrm{GL}_s$ over $\mathbb{Q}$. To show that Γ is arithmetic, it suffices to show that it is isomorphic with the intersection in G' of Γ' with the $\mathbb{Q}$-points of some algebraic subgroup G'' of $\mathrm{GL}_r \times \mathrm{GL}_s$ defined over $\mathbb{Q}$.

Now it is enough to ensure that $G''(\mathbb{Q})$ is the set of pairs $(\varphi, \psi) \in G'$ satisfying the following conditions (where $\varphi_{\mathbb{R}}$ stands for the extension $\varphi \otimes \mathrm{id}_{\mathbb{R}}$ of φ to $W = L \otimes \mathbb{R}$):

(1) $\varphi_{\mathbb{R}}$ is a $\mathbb{C}$-linear endomorphism of W,
(2) $\varphi_{\mathbb{R}}(L' \otimes \mathbb{Q}) \subset L' \otimes \mathbb{Q}$,
(3) $\varphi_{\mathbb{R}}|_{L' \otimes \mathbb{Q}} = \psi$,

(4) $\varphi_{\mathbb{R}}(V) \subset V$,

(5) $\varphi_{\mathbb{R}}|_V \in G$.

Indeed, if $(\varphi, \psi) \in \Gamma'$ satisfies these five conditions then $\varphi \in \text{Aut}(L)$ induces an automorphism of A thanks to (1), it stabilizes $\Lambda \otimes \mathbb{Q}$ because of (2) and then Λ itself by (3), since $\psi \in \text{Aut}(L')$. With (4) it stabilizes V and (5) yields that it lies in Γ.

It thus suffices to show that these five conditions define an algebraic subgroup of $\text{GL}_r \times \text{GL}_s$ (over $\mathbb{Q}$). But the subset of $M_r(\mathbb{Q}) \times M_s(\mathbb{Q})$ consisting of pairs (φ, ψ) satisfying (1), (2) and (3) is a sub-$\mathbb{Q}$-algebra, so its group of invertible elements comes indeed from an algebraic subgroup of $\text{GL}_r \times \text{GL}_s$ over $\mathbb{Q}$.

On the other hand, (4) and (5) clearly define an algebraic subgroup of $\text{GL}_r \times \text{GL}_s$ over $\mathbb{R}$. But as we are only interested in $\mathbb{Q}$-points, we may replace this algebraic subgroup by the (Zariski) closure of its intersection with the $\mathbb{Q}$-points $\text{GL}_r(\mathbb{Q}) \times \text{GL}_s(\mathbb{Q})$. This closure is an algebraic subgroup of $\text{GL}_r \times \text{GL}_s$ over $\mathbb{Q}$ and the $\mathbb{Q}$-points are the same. $\square$

Remark 5.5 Assume that $\text{char}(k) = 0$. Then by the above arguments, $\pi_0 \text{Aut}(X)$ is commensurable with $\text{Aut}_{\text{gp}}^{G_{\text{aff}}}(G)$ whenever K is *diagonalizable*, e.g., when G is a semi-abelian variety. But this fails in general. Consider indeed a non-zero abelian variety A and its universal vector extension,

$$0 \longrightarrow U \longrightarrow G \longrightarrow A \longrightarrow 0,$$

where $U \simeq H^1(A, \mathcal{O}_A)^\vee$ is a vector group of the same dimension as A. Then G is a smooth connected algebraic group, and $G_{\text{aff}} = U$. Moreover, G is anti-affine (see [8, Prop. 5.4.2]). Let $Y := \mathbb{P}(U \oplus k)$ be the projective completion of U. Then the action of U on itself by translation extends to an action on Y, and $X := G \times^U Y$ is a smooth projective equivariant completion of G by a projective space bundle over A. One may check that $\text{Aut}_{\text{gp}}^U(G)$ is trivial, and $\text{Aut}(X) \simeq \text{Aut}(A)$; in particular, the group $\pi_0 \text{Aut}(X) \simeq \text{Aut}_{\text{gp}}(A)$ is not necessarily finite.

References

[1] Borel, Armand. 1963. Arithmetic properties of linear algebraic groups. In: *Proc. Int. Congr. Math., Stockholm 1962*, 10–22.

[2] Borel, Armand. 1991. *Linear algebraic groups. Second enlarged edition*. Grad. Texts Math. **126**, Springer.

[3] Brion, Michel. 2010. Some basic results on actions of nonaffine algebraic groups. In: *Symmetry and Spaces*, Prog. Math. **278**, 1–20, Birkhäuser.

[4] Brion, Michel. 2014. On automorphisms and endomorphisms of projective varieties. In: *Automorphisms in birational and affine geometry*, Springer Proc. Math. Stat. **79**, 59–82, Springer.

[5] Brion, Michel. 2015. On extensions of algebraic groups with finite quotient. *Pacific J. Math.* **279**, 135–153.

[6] Brion, Michel. 2017. Algebraic group actions on normal varieties. *Trans. Moscow Math. Soc.* **78**, 85–107.

[7] Brion, Michel. 2019. Notes on automorphism groups of projective varieties. Available at www-fourier.univ-grenoble-alpes.fr/~mbrion/autos.pdf

[8] Brion, Michel; Samuel, Preena, and Uma, V. 2013. *Lectures on the structure of algebraic groups and geometric applications*. Hindustan Book Agency, New Dehli.

[9] Cantat, Serge. 2019. Automorphisms and dynamics: a list of open problems. In: *Proc. Int. Congr. Math., Rio 2018*, World Scientific, Singapore.

[10] Cantat, Serge and Dolgachev, Igor. 2012 Rational surfaces with a large group of automorphisms. *J. Amer. Math. Soc.* **25**, 863–905.

[11] Cattaneo, Andrea and Fu, Lie. 2019. Finiteness of Klein actions and real structures on compact hyperkähler manifolds, *Math. Ann.* **315**, 1783–1822.

[12] Conrad, Brian. 2002. A modern proof of Chevalley's theorem on algebraic groups. *J. Ramanujan Math. Soc.* **17**, 1–18.

[13] Demazure, Michel and Gabriel, Pierre. 1970. *Groupes algébriques*. Masson, Paris.

[14] Dinh, Tien-Cuong and Oguiso, Keiji. 2019. A surface with discrete and non-finitely generated automorphism group. *Duke Math. J.* **168**, 941–966.

[15] Grothendieck, Alexandre. 1961. *Éléments de géométrie algébrique (rédigés avec la collaboration de J. Dieudonné) : II. Étude globale élémentaire de quelques classes de morphismes.* Pub. Math. I.H.É.S. **8**, 5–222.

[16] Fantechi, Barbara; Göttsche, Lothar; Illusie, Luc; Kleiman, Steven; Nitsure, Nitin and Vistoli, Angelo. 2005. *Fundamental algebraic geometry: Grothendieck's FGA explained.* Math. Surveys Monogr. **123**, Amer. Math. Soc.

[17] Fu, Baohua and Zhang, De-Qi. 2013 A characterization of compact complex tori via automorphism groups. *Math. Ann.* **357**, 961–968.

[18] Fujiki, Akira. 1978. On automorphism groups of compact Kähler manifolds. *Inventiones math.* **44**, 225–258.

[19] Fulton, William. 1998. *Intersection theory*. Ergebnisse der Math. **2**, Springer.

[20] Grothendieck, Alexandre. 1961. Techniques de construction et théorèmes d'existence en géométrie algébrique IV : les schémas de Hilbert. *Sém. Bourbaki*, Vol. **6**, Exp. 221, 249–276.

[21] Grothendieck, Alexandre. 1962. Techniques de descente et théorèmes d'existence en géométrie algébrique. V. Les schémas de Picard : théorèmes d'existence. *Sém. Bourbaki*, Vol. 7, Exp. 232, 143–161.

[22] Haboush, William. 1974. The scheme of Lie sub-algebras of a Lie algebra and the equivariant cotangent map. *Nagoya Math. J.* **53**, 59–70.

[23] Haboush, William and Lauritzen, Niels. 1993. Varieties of unseparated flags. In: *Linear algebraic groups and their representations*. Contemp. Math. **153**, 35–57, Amer. Math. Soc.

[24] Hartshorne, Robin. 1977. *Algebraic geometry*. Grad. Texts in Math. **52**, Springer.

[25] Huckleberry, Alan and Oeljeklaus, Eberhard. 1984. *Classification theorems for almost homogeneous spaces.* Institut Élie Cartan **9**, Nancy.

[26] Jantzen, Jens Carsten. 2003. *Representations of algebraic groups. Second edition.* Math. Surveys Monogr. **107**, Amer. Math. Soc.

[27] Kollár, János. 1999. *Rational curves on algebraic varieties.* Ergebnisse der Math. **32**, Springer.

[28] Kollár, János. 2007. *Lectures on resolution of singularities.* Annals of Math. Stud. **166**, Princeton.

[29] Kollár, János and Mori, Shigefumi. 1998. *Birational geometry of algebraic varieties.* Cambridge Tracts Math. **134**, Cambridge University Press.

[30] Lesieutre, John. 2018. A projective variety with discrete, non-finitely generated automorphism group. *Inventiones Math.* **212**, 189–211.

[31] Lieberman, David I. 1978. Compactness of the Chow scheme: applications to automorphisms and deformations of Kähler manifolds. In: *Fonctions de plusieurs variables complexes, Sémin. François Norguet.* Lect. Notes Math. **670**, 140–186, Springer.

[32] Lombardo, Daniele and Maffei, Andrea. 2020. Abelian varieties as automorphism groups of smooth projective varieties. *Int. Math. Research Notices* (7), 1942–1956.

[33] Madden, Daniel J. and Valentini, Robert C. 1983. The group of automorphisms of algebraic function fields. *J. Reine Angew. Math.* **343**, 162–168.

[34] Mumford, David. 2008. *Abelian varieties. With appendices by C. P. Ramanujam and Yuri Manin. Corrected reprint of the 2nd ed. 1974.* Hindustan Book Agency, New Dehli.

[35] Oguiso, Keiji. 2020. A surface in odd characteristic with discrete and non-finitely generated automorphism group. *Adv. Math.* **375**, 107397, 20 pp.

[36] Berthelot, Pierre; Grothendieck, Alexandre, and Illusie, Luc. 1971. *Séminaire de Géométrie Algébrique du Bois Marie 1966–67. Théorie des intersections et théorème de Riemann-Roch (SGA 6).* Lect. Notes Math. **225**, Springer.

[37] Zhang, De-Qi. 2009. Dynamics of automorphisms on projective complex manifolds. *J. Differential Geom.* **82**, 691–722.

4

Topology of Moduli Spaces of Tropical Curves with Marked Points

Melody Chan[a], Søren Galatius[b] and Sam Payne[c]

Dedicated to William Fulton on the occasion of his 80th birthday

Abstract. This article is a sequel to [14]. We study a space $\Delta_{g,n}$ of genus g stable, n-marked tropical curves with total edge length 1. Its rational homology is identified with both top-weight cohomology of the complex moduli space $\mathcal{M}_{g,n}$ and with the homology of a marked version of Kontsevich's graph complex, up to a shift in degrees.

We prove a contractibility criterion that applies to various large subspaces of $\Delta_{g,n}$. From this we derive a description of the homotopy type of $\Delta_{1,n}$, the top weight cohomology of $\mathcal{M}_{1,n}$ as an S_n-representation, and additional calculations of $H_i(\Delta_{g,n}; \mathbb{Q})$ for small (g,n). We also deduce a vanishing theorem for homology of marked graph complexes from vanishing of cohomology of $\mathcal{M}_{g,n}$ in appropriate degrees, by relating both to $\Delta_{g,n}$. We comment on stability phenomena, or lack thereof.

1 Introduction

In [14], we studied the topology of the *tropical moduli space* Δ_g of stable tropical curves of genus g and total edge length 1. Here, a tropical curve is a metric graph with nonnegative integer vertex weights; it is said to be *stable* if every vertex of weight zero has valence at least 3. With appropriate degree shift, the rational homology of Δ_g is isomorphic to both Kontsevich's graph homology and the top weight cohomology of the complex algebraic moduli space $\mathcal{M}_g$. As an application of these identifications, we deduced that

[a] Brown University, Rhode Island
[b] University of Copenhagen
[c] University of Texas, Austin

$$\dim H^{4g-6}(\mathcal{M}_g; \mathbb{Q}) > 1.32^g + \text{constant},$$

disproving conjectures of Church, Farb, and Putman [12, Conjecture 9] and of Kontsevich [38, Conjecture 7C], which would have implied that these cohomology groups vanish for all but finitely many g.

In this article, we expand on [14], in two main ways.

1. We introduce marked points. Given $g, n \geq 0$ such that $2g - 2 + n > 0$, we study a space $\Delta_{g,n}$ parametrizing stable tropical curves of genus g with n labeled, marked points (not necessarily distinct). The case $n = 0$ recovers the spaces $\Delta_g = \Delta_{g,0}$ studied in [14].
2. We are interested here in $\Delta_{g,n}$ as a topological space, instead of studying only its rational homology. For example, we prove that several large subspaces of $\Delta_{g,n}$ are contractible, and determine the homotopy type of $\Delta_{1,n}$.

As for (1), the introduction of marked points to the basic setup of [14] poses no new technical obstacles. In particular, we note that $\Delta_{g,n}$ is the boundary complex of the Deligne–Mumford compactification $\overline{\mathcal{M}}_{g,n}$ of $\mathcal{M}_{g,n}$ by stable curves. This identification implies that there is a natural isomorphism

$$\widetilde{H}_{k-1}(\Delta_{g,n}; \mathbb{Q}) \xrightarrow{\cong} \mathrm{Gr}^W_{6g-6+2n} H^{6g-6+2n-k}(\mathcal{M}_{g,n}; \mathbb{Q}), \tag{1}$$

identifying the reduced rational homology of $\Delta_{g,n}$ with the top graded piece of the weight filtration on the cohomology of $\mathcal{M}_{g,n}$. See §6.

As for (2), studying the combinatorial topology of $\Delta_{g,n}$ in more depth is a new contribution compared to [14], where we only needed the rational homology of Δ_g for our applications. Note that $\Delta_{g,n}$ is a *symmetric Δ-complex* (§3). A new technical tool introduced in this article is a theory of collapses for symmetric Δ-complexes, roughly analogous to discrete Morse theory for CW complexes (Proposition 4.11). We apply this tool to prove that several natural subcomplexes of $\Delta_{g,n}$ are contractible. Recall that an edge in a connected graph is called a *bridge* if deleting it disconnects the graph.

Theorem 1.1 *Assume $g > 0$ and $2g - 2 + n > 0$. Each of the following subcomplexes of $\Delta_{g,n}$ is either empty or contractible.*

1. *The subcomplex $\Delta_{g,n}^w$ of $\Delta_{g,n}$ parametrizing tropical curves with at least one vertex of positive weight.*
2. *The subcomplex $\Delta_{g,n}^{lw}$ of $\Delta_{g,n}$ parametrizing tropical curves with loops or vertices of positive weight.*

3. The subcomplex $\Delta_{g,n}^{\text{rep}}$ of $\Delta_{g,n}$ parametrizing tropical curves in which at least two marked points coincide.
4. The closure $\Delta_{g,n}^{\text{br}}$ of the locus of tropical curves with bridges.

It is easy to classify when these loci are nonempty; see Remark 4.20. We refer to [21, §5.3] and [15] for related results on contractibility of spaces of graphs with bridges.

We use Theorem 1.1 to deduce a number of consequences, which we outline below.

The genus 1 case. When $g = 0$, the topology of the spaces $\Delta_{0,n}$ has long been understood; they are shellable simplicial complexes, homotopy equivalent to a wedge sum of $(n - 2)!$ spheres of dimension $n - 4$ [54]. Moreover, the character of $H_{n-4}(\Delta_{0,n}; \mathbb{Q})$ as an S_n-representation is computed in [49]. Our results below give an analogous complete understanding when $g = 1$. Namely, the spaces $\Delta_{1,1}$ and $\Delta_{1,2}$ are easily seen to be contractible (see Remark 5.1), and for $n \geq 3$, we have the following theorem.

Theorem 1.2 *For $n \geq 3$, the space $\Delta_{1,n}$ is homotopy equivalent to a wedge sum of $(n - 1)!/2$ spheres of dimension $n - 1$. The representation of S_n on $H_{n-1}(\Delta_{1,n}; \mathbb{Q})$ induced by permuting marked points is*

$$\text{Ind}_{D_n, \phi}^{S_n} \text{Res}_{D_n, \psi}^{S_n} \text{sgn}.$$

Here, $\phi \colon D_n \to S_n$ is the dihedral group of order $2n$ acting on the vertices of an n-gon, $\psi \colon D_n \to S_n$ is the action of the dihedral group on the edges of the n-gon, and sgn denotes the sign representation of S_n.

Note that the signs of these two permutation actions of the dihedral group are different when n is even. Reflecting a square across a diagonal, for instance, exchanges one pair of vertices and two pairs of edges. Moreover, calculating characters shows that these two representations of D_4 remain non-isomorphic after inducing along $\phi \colon D_4 \to S_4$.

Let us sketch how the expression in Theorem 1.2 arises; the complete proof is given in §5. The $(n-1)!/2$ spheres mentioned in the theorem are in bijection with the $(n - 1)!/2$ left cosets of D_n in S_n; each may be viewed as a way to place n markings on the vertices of an unoriented n-cycle. Choosing left coset representatives $\sigma_1, \ldots, \sigma_k$ where $k = (n - 1)!/2$, for any $\pi \in S_n$ we have $\pi \sigma_i = \sigma_j \pi'$ for some $\pi' \in D_n$. Then, writing $[\sigma_i]$ for the fundamental class of the corresponding sphere, it turns out that the S_n-action on top homology of $\Delta_{1,n}$ is described as $\pi \cdot [\sigma_i] = \pm[\sigma_j]$, where the sign depends on the sign of the permutation on the *edges* of the n-cycle induced by π'. This is because the ordering of edges determines the orientation of the corresponding sphere. This implies that the S_n-representation on $H_{n-1}(\Delta_{1,n};\mathbb{Q})$ is exactly $\operatorname{Ind}_{D_n,\phi}^{S_n} \operatorname{Res}_{D_n,\psi}^{S_n} \operatorname{sgn}$.

Combining (1) and Theorem 1.2, and noting S_n-equivariance, gives the following calculation for the top weight cohomology of $\mathcal{M}_{1,n}$.

Corollary 1.3 *The top weight cohomology of $\mathcal{M}_{1,n}$ is supported in degree n, with rank $(n-1)!/2$, for $n \geq 3$. Moreover, the representation of S_n on $\operatorname{Gr}_{2n}^W H^n(\mathcal{M}_{1,n};\mathbb{Q})$ induced by permuting marked points is*

$$\operatorname{Ind}_{D_n,\phi}^{S_n} \operatorname{Res}_{D_n,\psi}^{S_n} \operatorname{sgn}.$$

Corollary 1.3 is consistent with the recently proven formula for the S_n-equivariant top-weight Euler characteristic of $\mathcal{M}_{g,n}$ in [10]. See Remarks 6.3, 6.4, and 6.5 for further discussion of Corollary 1.3 and its context.

In the case $g \geq 2$, we no longer have a complete understanding of the homotopy type of $\Delta_{g,n}$. However, the contractibility results in Theorem 1.1 enable computer calculations of $H_i(\Delta_{g,n};\mathbb{Q})$ for small (g,n), presented in the Appendix.

Theorem 1.1 can also be used to deduce a lower bound on connectivity of the spaces $\Delta_{g,n}$. We do not pursue this here, but refer to [13, Theorem 1.3].

Marked graph complexes. In [14] we gave a cellular chain complex $C_*(X;\mathbb{Q})$ associated to any symmetric Δ-complex X, and we showed that it computes the reduced rational homology of (the geometric realization of) X. In the case $X = \Delta_{g,n}$, we are able to deduce that $C_*(\Delta_{g,n};\mathbb{Q})$ is quasi-isomorphic to the *marked graph complex* $G^{(g,n)}$, using Theorem 1.1. We shall define $G^{(g,n)}$ precisely in §2. Briefly, it is generated by isomorphism classes of connected, n-marked stable graphs Γ together with the choice of one of the two possible orientations of $\mathbb{R}^{E(\Gamma)}$. Kontsevich's graph complex $G^{(g)}$ [38, 39] occurs as the special case $n = 0$. The markings that we consider are elsewhere called hairs, half-edges, or legs; one difference between $G^{(g,n)}$ and many of the hairy graph complexes in the existing literature [18, 19, 40, 52] is that our markings are ordered rather than unordered. Hairy graphs with ordered markings do appear in the work of Tsopméné and Turchin on string links [51]; they study the more general situation where each marking carries a label from an ordered set $\{1,\ldots,r\}$, as well as the special case where multiple markings may carry the same label [50, §2.2.1]. Our $G^{(g,n)}$ agrees with the complex denoted $M_g(P_2^n)$ in [50].

Theorem 1.4 *For $g \geq 1$ and $2g - 2 + n \geq 2$, there is a natural surjection of chain complexes*

$$C_*(\Delta_{g,n};\mathbb{Q}) \to G^{(g,n)},$$

decreasing degrees by $2g - 1$, *inducing isomorphisms on homology*

$$\tilde{H}_{k+2g-1}(\Delta_{g,n}; \mathbb{Q}) \xrightarrow{\cong} H_k(G^{(g,n)})$$

for all k.

We recover [14, Theorem 1.3], in the special case where $n = 0$. For an analogous result with coefficients in a different local system, see [21, Proposition 27].

Combining (1) and Theorem 1.4 gives the following.

Corollary 1.5 *There is a natural isomorphism*

$$H_k(G^{(g,n)}) \xrightarrow{\sim} \mathrm{Gr}^W_{6g-6+2n} H^{4g-6+2n-k}(\mathcal{M}_{g,n}; \mathbb{Q}),$$

identifying marked graph homology with the top weight cohomology of $\mathcal{M}_{g,n}$.

Corollary 1.5 allows for an interesting application from moduli spaces back to graph complexes: applying known vanishing results for $\mathcal{M}_{g,n}$, we obtain the following theorem for marked graph homology.

Theorem 1.6 *The marked graph homology* $H_k(G^{(g,n)})$ *vanishes for* $k < \max\{0, n - 2\}$. *Equivalently,* $\tilde{H}_k(\Delta_{g,n}; \mathbb{Q})$ *vanishes for* $k < \max\{2g - 1, 2g - 3 + n\}$.

Theorem 1.6 generalizes a theorem of Willwacher for $n = 0$. See [55, Theorem 1.1] and [14, Theorem 1.4].

A transfer homomorphism. It may be deduced from [52, Theorem 1] and Theorem 1.4 above that $\tilde{H}_k(\Delta_g; \mathbb{Q})$ can be identified with a summand of $\tilde{H}_k(\Delta_{g,1}; \mathbb{Q})$. In the notation of op.cit., this is essentially the special case $n = m = 0$ and $h = 1$ (their $\mathrm{HGC}_{0,0}^{g,1}$ is then a cochain complex isomorphic to a shift of our $(G^{(g,1)})^\vee$, while their $\mathrm{GC}_{S^0 H_1}^g$ is isomorphic to our $(G^{(g)})^\vee$). In §5.3 we give a proof of this in our setup, including an explicit construction of the splitting on the level of cellular chains of the tropical moduli spaces.

Theorem 1.7 *For* $g \geq 2$, *there is a natural homomorphism of cellular chain complexes*

$$t : C_*(\Delta_g; \mathbb{Q}) \to C_*(\Delta_{g,1}; \mathbb{Q})$$

which descends to a homomorphism $G^{(g)} \to G^{(g,1)}$ *and induces injections* $\tilde{H}_k(\Delta_g; \mathbb{Q}) \hookrightarrow \tilde{H}_k(\Delta_{g,1}; \mathbb{Q})$ *and* $H_k(G^{(g)}) \hookrightarrow H_k(G^{(g,1)})$, *for all* k.

The homomorphism t is obtained as a weighted sum over all possible vertex markings, and may thus be seen as analogous to a transfer map. The resulting injection on homology is particularly interesting because $\bigoplus_g \tilde{H}_{2g-1}(\Delta_g; \mathbb{Q})$

is large: its graded dual is isomorphic to the Grothendieck-Teichmüller Lie algebra, as discussed in [14]. Combining with (1), we obtain the following.

Corollary 1.8 *We have*

$$\dim \mathrm{Gr}_{6g-4}^W H^{4g-4}(\mathcal{M}_{g,1}; \mathbb{Q}) > \beta^g + constant$$

for any $\beta < \beta_0$, where $\beta_0 \approx 1.32\ldots$ is the real root of $t^3 - t - 1 = 0$.

Corollary 1.8 can also be deduced purely algebro-geometrically from the analogous result for $\mathcal{M}_g$ proved in [14], without using the transfer homomorphism t. See Remark 6.6. The following result is deduced by an easy application of the topological Gysin sequence, see Corollary 6.7.

Corollary 1.9 *Let $\mathrm{Mod}_{g,1}$ denote the mapping class group of a genus g surface with one parametrized boundary component. Then*

$$\dim H^{4g-3}(\mathrm{Mod}_{g,1}; \mathbb{Q}) > \beta^g + constant$$

for any $\beta < \beta_0 \approx 1.32\ldots$ as above.

Acknowledgments. We are grateful to E. Getzler, A. Kupers, D. Petersen, O. Randal-Williams, O. Tommasi, K. Vogtmann, and J. Wiltshire-Gordon for helpful conversations related to this work. MC was supported by NSF DMS-1204278, NSF CAREER DMS-1844768, NSA H98230-16-1-0314, and a Sloan Research Fellowship. SG was supported by the European Research Council (ERC) under the European Union's Horizon 2020 research and innovation programme (grant agreement No 682922), the EliteForsk Prize, and by the Danish National Research Foundation (DNRF151 and DNRF151). SP was supported by NSF DMS-1702428 and a Simons Fellowship. He thanks UC Berkeley and MSRI for their hospitality and ideal working conditions.

2 Marked Graphs and Moduli of Tropical Curves

In this section, we recall the construction of the topological space $\Delta_{g,n}$ as a moduli space for genus g stable, n-marked tropical curves. The construction in [14, §2] is the special case $n = 0$. Following the usual convention, we write $\Delta_g = \Delta_{g,0}$.

2.1 Marked Weighted Graphs and Tropical Curves

All graphs in this paper are connected, with loops and parallel edges allowed. Let G be a finite graph, with vertex set $V(G)$ and edge set $E(G)$. A *weight function* is an arbitrary function $w\colon V(G) \to \mathbb{Z}_{\geq 0}$. The pair (G, w) is called a *weighted graph*. Its *genus* is

$$g(G, w) = b_1(G) + \sum_{v \in V(G)} w(v),$$

where $b_1(G) = |E(G)| - |V(G)| + 1$ is the first Betti number of G. The *core* of a weighted graph (G, w) is the smallest connected subgraph of G that contains all cycles of G and all vertices of positive weight, if $g(G, w) > 0$, or the empty subgraph if $g(G, w) = 0$.

An *n-marking* on G is a map $m\colon \{1, \ldots, n\} \to V(G)$. In figures, we depict the marking as a set of n labeled half-edges or legs attached to the vertices of G.

An *n-marked weighted graph* is a triple $\mathbf{G} = (G, m, w)$, where (G, w) is a weighted graph and m is an n-marking. The *valence* of a vertex v in a marked weighted graph, denoted $\mathrm{val}(v)$, is the number of half-edges of G incident to v plus the number of marked points at v. In other words, a loop edge based at v counts twice towards $\mathrm{val}(v)$, once for each end, an ordinary edge counts once, and a marked point counts once (as suggested by the interpretation of markings as half-edges). We say that $\mathbf{G}$ is *stable* if for every $v \in V(G)$,

$$2w(v) - 2 + \mathrm{val}(v) > 0.$$

Equivalently, $\mathbf{G}$ is stable if and only if every vertex of weight 0 has valence at least 3, and every vertex of weight 1 has valence at least 1.

2.2 The Category $\Gamma_{g,n}$

The stable n-marked graphs of genus g form the objects of a category which we denote $\Gamma_{g,n}$. The morphisms in this category are compositions of contractions of edges $G \to G/e$ and isomorphisms $G \to G'$. For the sake of removing any ambiguity about what that might mean, we now give a precise definition.

Formally, a graph G is a finite set $X(G) = V(G) \sqcup H(G)$ (of "vertices" and "half-edges"), together with two functions $s_G, r_G\colon X(G) \to X(G)$ satisfying $s_G^2 = \mathrm{id}$ and $r_G^2 = r_G$, such that

$$\{x \in X(G)\colon r_G(x) = x\} = \{x \in X(G)\colon s_G(x) = x\} = V(G).$$

Informally: s_G sends a half-edge to its other half, while r_G sends a half-edge to its incident vertex. We let $E(G) = H(G)/(x \sim s_G(x))$ be the set of edges. The definitions of n-marking, weights, genus, and stability are as stated in the previous subsection.

The objects of the category $\Gamma_{g,n}$ are all connected stable n-marked graphs of genus g. For an object $\mathbf{G} = (G, m, w)$ we shall write $V(\mathbf{G})$ for $V(G)$ and similarly for $H(\mathbf{G})$, $E(\mathbf{G})$, $X(\mathbf{G})$, $s_{\mathbf{G}}$ and $r_{\mathbf{G}}$. Then a morphism $\mathbf{G} \to \mathbf{G}'$ is a function $f: X(\mathbf{G}) \to X(\mathbf{G}')$ with the property that

$$f \circ r_{\mathbf{G}} = r_{\mathbf{G}'} \circ f \text{ and } f \circ s_{\mathbf{G}} = s_{\mathbf{G}'} \circ f,$$

and subject to the following three requirements:

- Noting first that $f(V(\mathbf{G})) \subset V(\mathbf{G}')$, we require $f \circ m = m'$, where m and m' are the respective marking functions of $\mathbf{G}$ and $\mathbf{G}'$.
- Each $e \in H(\mathbf{G}')$ determines the subset $f^{-1}(e) \subset X(\mathbf{G})$ and we require that it consists of precisely one element (which will then automatically be in $H(\mathbf{G})$).
- Each $v \in V(\mathbf{G}')$ determines a subset $S_v = f^{-1}(v) \subset X(\mathbf{G})$ and $\mathbf{S}_v = (S_v, r|_{S_v}, s|_{S_v})$ is a graph; we require that it be connected and have $g(\mathbf{S}_v, w|_{\mathbf{S}_v}) = w(v)$.

Composition of morphisms $\mathbf{G} \to \mathbf{G}' \to \mathbf{G}''$ in $\Gamma_{g,n}$ is given by the corresponding composition $X(\mathbf{G}) \to X(\mathbf{G}') \to X(\mathbf{G}'')$ in the category of sets.

Our definition of graphs and the morphisms between them is standard in the study of moduli spaces of curves and agrees, in essence, with the definitions in [1, X.2] and [2, §3.2], as well as those in [37] and [26].

Remark 2.1 We also note that any morphism $\mathbf{G} \to \mathbf{G}'$ can be alternatively described as an isomorphism following a finite sequence of *edge collapses*: if $e \in E(\mathbf{G})$ there is a canonical morphism $\mathbf{G} \to \mathbf{G}/e$, where $\mathbf{G}/e$ is the marked weighted graph obtained from $\mathbf{G}$ by collapsing e together with its two endpoints to a single vertex $[e] \in \mathbf{G}/e$. If e is not a loop, the weight of $[e]$ is the sum of the weights of the endpoints of e, and if e is a loop the weight of $[e]$ is one more than the old weight of the end-point of e. If $S = \{e_1, \ldots, e_k\} \subset E(\mathbf{G})$ there are iterated edge collapses $\mathbf{G} \to \mathbf{G}/e_1 \to (\mathbf{G}/e_1)/e_2 \to \ldots$ and any morphism $\mathbf{G} \to \mathbf{G}'$ can be written as such an iteration followed by an isomorphism from the resulting quotient of $\mathbf{G}$ to $\mathbf{G}'$.

We shall say that $\mathbf{G}$ and $\mathbf{G}'$ have the same *combinatorial type* if they are isomorphic in $\Gamma_{g,n}$. In fact there are only finitely many isomorphism classes of objects in $\Gamma_{g,n}$, since any object has at most $6g - 6 + 2n$ half-edges and $2g - 2 + n$ vertices and for each possible set of vertices and half-edges there are finitely many ways of gluing them to a graph, and finitely many possibilities for the n-marking and weight function. In order to get a *small* category $\Gamma_{g,n}$ we shall tacitly pick one object in each isomorphism class and pass to the full subcategory on those objects. Hence $\Gamma_{g,n}$ is a *skeletal* category. (However, we shall try to use language compatible with any choice of small equivalent subcategory of $\Gamma_{g,n}$.) It is clear that all Hom sets in $\Gamma_{g,n}$ are finite, so $\Gamma_{g,n}$ is in fact a finite category.

Replacing $\Gamma_{g,n}$ by some choice of skeleton has the effect that if $\mathbf{G}$ is an object of $\Gamma_{g,n}$ and $e \in E(\mathbf{G})$ is an edge, then the marked weighted graph $\mathbf{G}/e$ is likely not equal to an object of $\Gamma_{g,n}$. Given $\mathbf{G}$ and e, there is a unique *morphism* $q: \mathbf{G} \to \mathbf{G}'$ in $\Gamma_{g,n}$ factoring through an isomorphism $\mathbf{G}/e \to \mathbf{G}'$. As usual it is the pair $(\mathbf{G}', q)$ which is unique and not the isomorphism $\mathbf{G}/e \cong \mathbf{G}'$. By an abuse of notation, we shall henceforth write $\mathbf{G}/e \in \Gamma_{g,n}$ for the codomain of this unique morphism, and similarly G/e for its underlying graph.

Definition 2.2 Let us define functors

$$H, E: \Gamma_{g,n}^{\mathrm{op}} \to (\text{Finite sets, injections})$$

as follows. On objects, $H(\mathbf{G}) = H(G)$ is the set of half-edges of $\mathbf{G} = (G, m, w)$ as defined above. A morphism $f: \mathbf{G} \to \mathbf{G}'$ determines an injective function $H(f): H(\mathbf{G}') \to H(\mathbf{G})$, sending $e' \in H(\mathbf{G}')$ to the unique element $e \in H(\mathbf{G})$ with $f(e) = e'$. We shall write $f^{-1} = H(f): H(\mathbf{G}') \to H(\mathbf{G})$ for this map. This clearly preserves composition and identities, and hence defines a functor. Similarly for $E(\mathbf{G}) = H(\mathbf{G})/(x \sim s_G(x))$ and $E(f)$.

2.3 Moduli Space of Tropical Curves

We now recall the construction of moduli spaces of stable tropical curves, as the colimit of a diagram of cones parametrizing possible lengths of edges for each fixed combinatorial type, following [2, 5, 8].

Fix integers $g, n \geq 0$ with $2g - 2 + n > 0$. A *length function* on $\mathbf{G} = (G, m, w) \in \Gamma_{g,n}$ is an element $\ell \in \mathbb{R}_{>0}^{E(\mathbf{G})}$, and we shall think geometrically of $\ell(e)$ as the *length* of the edge $e \in E(\mathbf{G})$. An n-marked genus g *tropical curve* is then a pair $\Gamma = (\mathbf{G}, \ell)$ with $\mathbf{G} \in \Gamma_{g,n}$ and $\ell \in \mathbb{R}_{>0}^{E(\mathbf{G})}$. We shall say that $(\mathbf{G}, \ell)$ is *isometric* to $(\mathbf{G}', \ell')$ if there exists an isomorphism $\phi: \mathbf{G} \to \mathbf{G}'$ in $\Gamma_{g,n}$ such that $\ell' = \ell \circ \phi^{-1}: E(\mathbf{G}') \to \mathbb{R}_{>0}$. The *volume* of $(\mathbf{G}, \ell)$ is $\sum_{e \in E(\mathbf{G})} \ell(e) \in \mathbb{R}_{>0}$.

We can now describe the underlying set of the topological space $\Delta_{g,n}$, which is the main object of study in this paper. It is the set of isometry classes of n-marked genus g tropical curves of volume 1. We proceed to describe its topology and further structure as a closed subspace of the moduli space of tropical curves.

Definition 2.3   Fix $g, n \geq 0$ with $2g - 2 + n > 0$. For each object $\mathbf{G} \in \Gamma_{g,n}$ define the topological space

$$\sigma(\mathbf{G}) = \mathbb{R}_{\geq 0}^{E(\mathbf{G})} = \{\ell \colon E(\mathbf{G}) \to \mathbb{R}_{\geq 0}\}.$$

For a morphism $f \colon \mathbf{G} \to \mathbf{G}'$ define the continuous map $\sigma f \colon \sigma(\mathbf{G}') \to \sigma(\mathbf{G})$ by

$$(\sigma f)(\ell') = \ell \colon E(\mathbf{G}) \to \mathbb{R}_{\geq 0},$$

where ℓ is given by

$$\ell(e) = \begin{cases} \ell'(e') & \text{if } f \text{ sends } e \text{ to } e' \in E(\mathbf{G}'), \\ 0 & \text{if } f \text{ collapses } e \text{ to a vertex.} \end{cases}$$

This defines a functor $\sigma \colon \Gamma_{g,n}^{\mathrm{op}} \to$ Spaces and the topological space $M_{g,n}^{\mathrm{trop}}$ is defined to be the colimit of this functor.

In other words, the topological space $M_{g,n}^{\mathrm{trop}}$ is obtained as follows. For each morphism $f \colon \mathbf{G} \to \mathbf{G}'$, consider the map $L_f \colon \sigma(\mathbf{G}') \to \sigma(\mathbf{G})$ that sends $\ell' \colon E(\mathbf{G}') \to \mathbb{R}_{>0}$ to the length function $\ell \colon E(\mathbf{G}) \to \mathbb{R}_{>0}$ obtained from ℓ' by extending it to be 0 on all edges of $\mathbf{G}$ that are collapsed by f. So L_f linearly identifies $\sigma(\mathbf{G}')$ with some face of $\sigma(\mathbf{G})$, possibly $\sigma(\mathbf{G})$ itself. Then

$$M_{g,n}^{\mathrm{trop}} = \left(\coprod \sigma(\mathbf{G}) \right) \Big/ \{\ell' \sim L_f(\ell')\},$$

where the equivalence runs over all morphisms $f \colon \mathbf{G} \to \mathbf{G}'$ and all $\ell' \in \sigma(\mathbf{G}')$.

As we shall explain in more detail in later sections, $M_{g,n}^{\mathrm{trop}}$ naturally comes with more structure than just a topological space: $M_{g,n}^{\mathrm{trop}}$ is a *generalized cone complex*, as defined in [2, §2], and associated to a *symmetric Δ-complex* in the sense of [14]. This formalizes the observation that $M_{g,n}^{\mathrm{trop}}$ is glued out of the cones $\sigma(\mathbf{G})$.

The *volume* defines a function $v \colon \sigma(\mathbf{G}) \to \mathbb{R}_{\geq 0}$, given explicitly as $v(\ell) = \sum_{e \in E(\mathbf{G})} \ell(e)$, and for any morphism $\mathbf{G} \to \mathbf{G}'$ in $\Gamma_{g,n}$ the induced map $\sigma(\mathbf{G}) \to \sigma(\mathbf{G}')$ preserves volume. Hence there is an induced map $v \colon M_{g,n}^{\mathrm{trop}} \to \mathbb{R}_{\geq 0}$, and there is a unique element in $M_{g,n}^{\mathrm{trop}}$ with volume 0 which we shall denote $\bullet_{g,n}$. The underlying graph of $\bullet_{g,n}$ consists of a single vertex with weight g that carries all n marked points.

Definition 2.4   We let $\Delta_{g,n}$ be the subspace of $M_{g,n}^{\mathrm{trop}}$ parametrizing curves of volume 1, i.e., the inverse image of $1 \in \mathbb{R}$ under $v \colon M_{g,n}^{\mathrm{trop}} \to \mathbb{R}_{\geq 0}$.

Thus $\Delta_{g,n}$ is homeomorphic to the link of $M_{g,n}^{\mathrm{trop}}$ at the cone point $\bullet_{g,n}$.

2.4 The Marked Graph Complex

Fix integers $g, n \geq 0$ with $2g - 2 + n > 0$. The *marked graph complex* $G^{(g,n)}$ is a chain complex of rational vector spaces. As a graded vector space, it has generators $[\Gamma, \omega, m]$ for each connected graph Γ of genus g (Euler characteristic $1 - g$) with or without loops, equipped with a total order ω on its set of edges and a marking $m \colon \{1, \ldots, n\} \to V(G)$, such that at each vertex v, the number of half-edges incident to v plus $|m^{-1}(v)|$ is at least 3. These generators are subject to the relations

$$[\Gamma, \omega, m] = \mathrm{sgn}(\sigma)[\Gamma', \omega', m']$$

if there exists an isomorphism of graphs $\Gamma \cong \Gamma'$ that identifies m with m', and under which the edge orderings ω and ω' are related by the permutation σ. In particular this forces $[\Gamma, \omega, m] = 0$ when Γ admits an automorphism that fixes the markings and induces an odd permutation on the edges. A genus g graph Γ with v vertices and e edges is declared to be in homological degree $v - (g + 1) = e - 2g$. When $n = 0$, this convention agrees with [55]; it is shifted by $g + 1$ compared to [38].

For example, $G^{(1,1)}$ is 1-dimensional, supported in degree 0, with a generator corresponding to a single loop based at a vertex supporting the marking. On the other hand, $G^{(2,0)} = 0$, since all generators of $G^{(2,0)}$ are subject to the relation $[\Gamma, \omega, m] = -[\Gamma, \omega, m]$.

The differential on $G^{(g,n)}$ is defined as follows: given $[\Gamma, \omega, m] \neq 0$,

$$\partial[\Gamma, \omega, m] = \sum_{i=0}^{N} (-1)^i [\Gamma/e_i, \omega|_{E(\Gamma) \smallsetminus \{e_i\}}, \pi_i \circ m], \tag{2}$$

where $\omega = (e_0 < e_1 < \cdots < e_N)$ is the total ordering on the edge set $E(\Gamma)$ of Γ, the graph Γ/e_i is the result of collapsing e_i to a point, $\pi_i \colon V(\Gamma) \to V(\Gamma/e_i)$ is the resulting surjection of vertex sets, and $\omega|_{E(\Gamma/e_i)}$ is the induced ordering on the subset $E(\Gamma/e_i) = E(\Gamma) \smallsetminus \{e_i\}$. If e_i is a loop, we interpret the corresponding term in (2) as zero.

Remark 2.5 We may equivalently define $G^{(g,n)}$ as follows: take the graded vector space

$$\bigoplus_{(\Gamma, m)} \Lambda^{|E(\Gamma)|} \mathbb{Q}^{E(\Gamma)},$$

where the top exterior power $\Lambda^{|E(\Gamma)|} \mathbb{Q}^{E(\Gamma)}$ appears in homological degree $|E(\Gamma)| - 2g$, and impose the following relations. For any isomorphism $\phi \colon (\Gamma, m) \to (\Gamma', m')$, let

$$\phi_* \colon \Lambda^{|E(\Gamma)|} \mathbb{Q}^{E(\Gamma)} \to \Lambda^{|E(\Gamma')|} \mathbb{Q}^{E(\Gamma')}$$

denote the induced isomorphism of 1-dimensional vector spaces. Then we set $w = \phi_* w$ for each $w \in \Lambda^{|E(\Gamma)|} \mathbb{Q}^{E(\Gamma)}$. Viewing nonzero elements of $\Lambda^{|E(\Gamma)|} \mathbb{Q}^{E(\Gamma)}$ as orientations on $\mathbb{R}^{|E(\Gamma)|}$, this description accords with the rough definition of $G^{(g,n)}$ in terms of graphs and orientations given in the introduction.

3 Symmetric Δ-Complexes and Relative Cellular Homology

We briefly recall the notion of symmetric Δ-complexes and explain how $\Delta_{g,n}$ is naturally interpreted as an object in this category. We then recall the cellular homology of symmetric Δ-complexes developed in [14, §3], and extend this to a cellular theory of relative homology for pairs. This relative cellular homology of pairs will be applied in §5 to prove that there is a surjection $C_*(\Delta_{g,n}; \mathbb{Q}) \to G^{(g,n)}$ that induces isomorphisms in homology, as stated in Theorem 1.4.

3.1 The Symmetric Δ-Complex Structure on $\Delta_{g,n}$

Let I denote the category whose objects are the sets $[p] = \{0, \ldots, p\}$ for nonnegative integers p, together with $[-1] := \emptyset$, and whose morphisms are injections of sets.

Definition 3.1 A *symmetric Δ-complex* is a presheaf on I, i.e., a functor from I^{op} to **Sets**.

As in [14, §3] we write $X_p = X([p])$, except when that would create double subscripts. The *geometric realization* of X is the topological space

$$|X| = \left(\coprod_{p=0}^{\infty} X_p \times \Delta^p \right) / \sim , \tag{3}$$

where Δ^p is the standard p-simplex and $\sim$ is the equivalence relation that is generated as follows. For each $x \in X_p$ and each injection $\theta : [q] \hookrightarrow [p]$, the simplex $\{\theta^*(x)\} \times \Delta^q$ is identified with a face of $\{x\} \times \Delta^p$ via the linear map that takes the vertex $(\theta^*(x), e_i)$ to $(x, e_{\theta(i)})$. The cone CX is defined similarly, but with $\mathbb{R}_{\geq 0}^{p+1}$ in place of Δ^p. Note that our symmetric Δ-complexes include the data of an *augmentation*, that is, the locally constant map $|X| \to X_{-1}$ induced by the unique inclusion $[-1] \hookrightarrow [p]$.

Example 3.2 Most importantly for our purposes, $\Delta_{g,n}$ is naturally identified with the geometric realization of a symmetric Δ-complex (and $M_{g,n}^{\text{trop}}$ is the associated cone), as we now explain.

Consider the following functor $X = X_{g,n} \colon I^{\text{op}} \to \mathsf{Sets}$. The elements of X_p are equivalence classes of pairs $(\mathbf{G}, \tau)$ where $\mathbf{G} \in \Gamma_{g,n}$ and $\tau \colon E(\mathbf{G}) \to [p]$ is a bijective edge-labeling. Two edge-labelings are considered equivalent if they are in the same orbit under the evident action of $\text{Aut}(\mathbf{G})$. Here $\mathbf{G}$ ranges over all objects in $\Gamma_{g,n}$ with exactly $p+1$ edges. (Recall from §2.2 that we have tacitly picked one element in each isomorphism class in $\Gamma_{g,n}$.)

Next, for each injective map $\iota \colon [p'] \to [p]$, define the following map $X(\iota) \colon X_p \to X_{p'}$; given an element of X_p represented by $(\mathbf{G}, \tau \colon E(\mathbf{G}) \to [p])$, contract the edges of $\mathbf{G}$ whose labels are not in $\iota([p']) \subset [p]$, then relabel the remaining edges with labels $[p']$ as prescribed by the map ι. The result is a $[p']$-edge-labeling of some new object $\mathbf{G}'$, and we set $X(\iota)(\mathbf{G})$ to be the corresponding element of $X_{p'}$.

The geometric realization of X is naturally identified with $\Delta_{g,n}$, as follows. Recall that a point in the relative interior of Δ^p is expressed as $\sum_{i=0}^{p} a_i e_i$, where $a_i > 0$ and $\sum a_i = 1$. Then an element $x \in X_p$ corresponds to a graph in $\Gamma_{g,n}$ together with a labeling of its $p+1$ edges, and a point $(x, \sum a_i e_i)$ corresponds to the isomorphism class of stable tropical curve with underlying graph x in which the edge labeled i has length a_i. By abuse of notation, we will use $\Delta_{g,n}$ to refer to this symmetric Δ-complex, as well as its geometric realization.

Remark 3.3 We note that the cellular complexes of $\Delta_{g,n}$ have natural interpretations in the language of modular operads, developed by Getzler and Kapranov to capture the intricate combinatorial structure underlying relations between the S_n-equivariant cohomology of $\mathcal{M}_{g,n}$, for all g and n, and that of $\overline{\mathcal{M}}_{g',n'}$, for all g' and n' [26]. In particular, the cellular cochain complexes $C^*(\Delta_{g,n})$, with their S_n-actions, agree with the Feynman transform of the modular operad ModCom, assigning the vector space $\mathbb{Q}$, with trivial S_n-action, to each (g,n) with $2g-2+n > 0$, and assigning 0 otherwise. There is a quotient map ModCom $\to$ Com, which is an isomorphism for $g = 0$ and where the commutative operad Com is zero for $g > 0$. The Feynman transform of this quotient map is, up to a regrading, the map appearing in Theorem 1.4. Since the Feynman transform is homotopy invariant and has an inverse up to quasi-isomorphism, it cannot turn the non-quasi-isomorphism ModCom $\to$ Com into a quasi-isomorphism, and therefore the map in Theorem 1.4 cannot be a quasi-isomorphism for all (g,n). In fact it is not in the exceptional cases $g = 0$ and $(g,n) = (1,1)$, where $\Delta_{g,n}^w = \emptyset$.

See also [4], especially §3.2.1 and §6.2.

3.2 Relative Homology

We now present a relative cellular homology for pairs of symmetric Δ-complexes. This is a natural extension of the cellular homology theory for symmetric Δ-complexes developed in [14, §3], which we briefly recall.

Let $X\colon I^{\mathrm{op}} \to$ Sets be a symmetric Δ-complex. The group of cellular p-chains $C_p(X)$ is defined to be the co-invariants

$$C_p(X) = (\mathbb{Q}^{\mathrm{sign}} \otimes_{\mathbb{Q}} \mathbb{Q}X_p)_{S_{p+1}},$$

where $\mathbb{Q}X_p$ denotes the $\mathbb{Q}$-vector space with basis X_p on which $S_{p+1} = I^{\mathrm{op}}([p],[p])$ acts by permuting the basis vectors. This comes with a natural differential ∂ induced by $\sum (-1)^i (d_i)_*\colon \mathbb{Q}X_p \to \mathbb{Q}X_{p-1}$, and the homology of $C_*(X)$ is identified with the rational homology $\widetilde{H}_*(|X|,\mathbb{Q})$, reduced with respect to the augmentation $|X| \to X_{-1}$, cf. [14, Proposition 3.8].

This rational cellular chain complex has the following natural analogue for *pairs* of symmetric Δ-complexes: there is a relative cellular chain complex, computing the relative (rational) homology of the geometric realizations. Let us start by discussing what "subcomplex" should mean in this context.

Lemma 3.4 *Let* $X,Y\colon I^{\mathrm{op}} \to$ *Sets be symmetric* Δ-*complexes, and let* $f\colon X \to Y$ *be a map (i.e., a natural transformation of functors). If* $f_p\colon X_p \to Y_p$ *is injective for all* $p \geq 0$, *then* $|f|\colon |X| \to |Y|$ *is also injective.*

If f_p *is injective for all* $p \geq -1$, *then* $Cf\colon CX \to CY$ *is also injective, where* CX *and* CY *are the cones over* X *and* Y.

Proof sketch Let us temporarily write $H^X(x) < S_{p+1}$ for the stabilizer of a simplex $x \in X_p$, and similarly $H^Y(f(x))$ for the stabilizer of $f(x) \in Y_p$. The maps $f_p\colon X_p \to Y_p$ are equivariant for the S_{p+1} action, and injectivity of f_p implies that $H^X(x) = H^Y(f(x))$ and the induced map of orbit sets $X_p/S_{p+1} \to Y_p/S_{p+1}$ is injective.

As a set, $|X|$ is the disjoint union of $(\Delta^p \setminus \partial\Delta^p)/H^X(x)$ over all p and one $x \in X_p$ in each S_{p+1}-orbit, and similarly for $|Y|$. At the level of sets, the induced map $|f|\colon |X| \to |Y|$ then restricts to a bijection from each subset $(\Delta^p \setminus \partial\Delta^p)/H^X(x) \subset |X|$ onto the corresponding subset of $|Y|$, and these subsets of $|Y|$ are disjoint.

The statement about $CX \to CY$ is proved in a similar way. $\square$

Conversely, it is not hard to see that the geometric map $|f|: |X| \to |Y|$ is injective *only* when $f_p: X_p \to Y_p$ is injective for all $p \geq 0$. In this situation, in fact $|f|: |X| \to |Y|$ is always a homeomorphism onto its image. This is obvious in the case of finite complexes where both spaces are compact, which is the only case needed in this paper, and in general is proved in the same way as for CW complexes (the CW topology on a subcomplex agrees with the subspace topology).

Definition 3.5 A *subcomplex* $X \subset Y$ of a symmetric Δ-complex is a subfunctor X of $Y: I^{\mathrm{op}} \to$ Sets, in which for each p, X_p is a subset of Y_p, with the subfunctor being given by the canonical inclusions $X_p \to Y_p$. The inclusion $\iota: X \to Y$ then induces an injection $|\iota|: |X| \to |Y|$, which we shall use to identify $|X|$ with its image $|X| \subset |Y|$.

In particular, we emphasize that for each injection $\iota: [p'] \to [p]$ the map $X(\iota): X_p \to X_{p'}$ is a restriction of $Y(\iota): Y_p \to Y_{p'}$.

If $X \subset Y$ is a subcomplex of a symmetric Δ-complex, we obtain a map $\iota_*: C_*(X) \to C_*(Y)$ of cellular chain complexes which is injective in each degree, and we define a relative cellular chain complex $C_*(Y, X)$ by the short exact sequences

$$0 \to C_p(X) \xrightarrow{\iota_p} C_p(Y) \to C_p(Y, X) \to 0 \tag{4}$$

for all $p \geq -1$. Similarly for cochains and with coefficients in an abelian group A.

Proposition 3.6 *Let Y be a symmetric Δ-complex and $X \subset Y$ a subcomplex. Let $\iota: X \to Y$ be the inclusion, and let $C\iota: CX \to CY$ be the induced maps of cones over X and $|\iota|: |X| \to |Y|$ be the restriction of $C\iota$. Then there is a natural isomorphism*

$$H_{p+1}((CY/CX), (|Y|/|X|); \mathbb{Q}) \cong H_p(C_*(Y, X)).$$

In particular, if $f_{-1}: X_{-1} \to Y_{-1}$ is a bijection, we get a natural isomorphism

$$H_p(C_*(Y, X)) \cong H_p(|Y|, |X|; \mathbb{Q}).$$

Similarly for cohomology. A similar result also holds with coefficients in an arbitrary abelian group A, provided that S_{p+1} acts freely on X_p and Y_p for all p.

Proof sketch We prove the statement about homology. The short exact sequence (4) induces a long exact sequence in homology, which maps to the long exact sequence in singular homology of the geometric realizations. For each p, the first, second, fourth, and fifth vertical arrows

$$H_p(C_*(X)) \longrightarrow H_p(C_*(Y)) \longrightarrow H_p(C_*(Y,X)) \longrightarrow H_{p-1}(C_*(X)) \longrightarrow H_{p-1}(C_*(Y))$$

$$H_p(|X|;\mathbb{Q}) \longrightarrow H_p(|Y|;\mathbb{Q}) \longrightarrow H_p(|Y|,|X|;\mathbb{Q}) \longrightarrow H_{p-1}(|X|;\mathbb{Q}) \longrightarrow H_{p-1}(|Y|;\mathbb{Q})$$

are shown to be isomorphisms in [14, Proposition 3.8], so the middle vertical arrow is also an isomorphism. □

4 A Contractibility Criterion

In this section, we develop a general framework for contracting subcomplexes of a symmetric Δ-complex, loosely in the spirit of discrete Morse theory. We then apply this technique to prove Theorem 1.1, showing that three natural subcomplexes of $\Delta_{g,n}$ are contractible. See Figure 1 for a running illustration of the various definitions that follow.

4.1 A Contractibility Criterion

Let X be a symmetric Δ-complex, i.e., a functor $I^{op} \to$ Sets. For each injective map $\theta: [q] \to [p]$ we write

$$\theta^*: X_p \to X_q,$$

for the induced map on simplices, where X_i denotes the set of i-simplices $X([i])$.

For $\sigma \in X_p$ and $\theta: [q] \hookrightarrow [p]$, we say that $\tau = \theta^*\sigma \in X_q$ is a *face* of σ, with face map θ, and we write $\tau \precsim \sigma$. Thus $\precsim$ is a reflexive and transitive relation. It descends to a partial order $\preceq$ on the set $\coprod X_p/S_{p+1}$ of symmetric orbits of simplices. We write $[\sigma]$ for the S_{p+1}-orbit of a p-simplex σ.

Definition 4.1 A *property* on X is a subset of the vertices $P \subset X_0$.

One could call this a "vertex property," but we avoid that terminology since our motivating example is $X = \Delta_{g,n}$, when the vertices of X are 1-edge graphs. In this situation, we are interested in properties of *edges* of graphs in $\Gamma_{g,n}$ that are preserved by automorphisms and uncontractions, such as the property of being a bridge. See Example 4.3.

Let $P \subset X_0$ be a property, and let $\sigma \in X_p$. For $i = 0, \ldots, p$, we understand the ith vertex of σ to be $v_i = \iota^*(\sigma)$, where $\iota: [0] \to [p]$ sends 0 to i. We write

$$P(\sigma) = \{i \in [p]: v_i \text{ is in } P\}$$

for the set labeling vertices of σ that are in P, and call these *P-vertices* of σ. Similarly, we write

$$P^c(\sigma) = \{i \in [p]: v_i \text{ is not in } P\}$$

for the complementary set, and call these the *non-P-vertices* of σ.

We write $\text{Simp}(X) = \coprod_{p \geq 0} X_p$ for the set of all simplices of X, and define

$$P(X) = \{\sigma \in \text{Simp}(X): P(\sigma) \neq \emptyset\}.$$

We call the elements of $P(X)$ the *P-simplices* of X; they are the simplices with at least one P-vertex. If $P^c(\sigma) = \emptyset$ then we say σ is a *strictly P-simplex*.

Any collection of simplices of X naturally generates a subcomplex whose simplices are all faces of simplices in the collection. We write X_P for the subcomplex of X generated by $P(X)$. Let $P^*(X)$ denote the set of simplices of X_P. In other words,

$$P^*(X) = \text{Simp}(X_P) = \{\tau \in \text{Simp}(X): \tau \precsim \sigma \text{ for some } \sigma \in P(X)\}.$$

The set $\text{Simp}(X) \setminus P(X)$ is also the set of simplices in a subcomplex of X, as is $P^*(X) \setminus P(X)$.

Example 4.2   Figure 1 shows a symmetric Δ-complex X with five 0-simplices, 7 symmetric orbits of 1-simplices and 3 symmetric orbits of 2-simplices; we have chosen to illustrate an example where S_{p+1} acts freely on the p-simplices for each p.

There are two vertices in P. The subcomplex X_P is then all of X, since the maximal simplices all have at least one P-vertex. In this example, we therefore have $P^*(X) = \text{Simp}(X)$. The only simplices *not* in $P(X)$ are marked $P^* \setminus P$ in the figure. The strictly P-simplices are drawn in black.

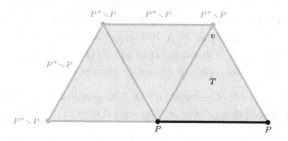

Figure 1   A symmetric Δ-complex X with two P-vertices. Here $X_P = X$, and v is a co-P-face of T.

Example 4.3 We pause to explain how these definitions apply to $X = \Delta_{g,n}$. For any $g, n \geq 0$ with $2g - 2 + n > 0$, recall that the p-simplices in $\Delta_{g,n}$ are pairs $(\mathbf{G}, t)$ where $\mathbf{G} \in \Gamma_{g,n}$ and $t \colon E(\mathbf{G}) \to [p]$ is a bijection. There is a bijection from $\coprod_{p \geq -1} \Delta_{g,n}([p])/S_{p+1}$ to the set of isomorphism classes of objects in $\Gamma_{g,n}$, sending $[(\mathbf{G}, t)]$ to $\mathbf{G}$.

The vertices X_0 are simply one-edge graphs $\mathbf{G} \in \Gamma_{g,n}$, since any such graph has a unique edge labeling. There is one such graph which is a loop; the others are bridges. For each g' satisfying $0 \leq g' \leq g$ and each subset $A \subset [n]$ with $2g' - 1 + |A| > 0$ and $2(g - g') - 1 + (n - |A|) > 0$, there is a unique one-edge graph in $\Gamma_{g,n}$ with vertices v_1 and v_2, such that $w(v_1) = g'$ and $m^{-1}(v_1) = A$. We write $\mathbf{B}(g', A) \in \Delta_{g,n}([0])$ for the corresponding vertex.

Note that $\mathbf{B}(g', A) = \mathbf{B}(g - g', [n] \smallsetminus A)$. We define the property

$$P_{g',n'} = \{\mathbf{B}(g', A) \colon |A| = n'\} \subset \Delta_{g,n}([0]).$$

A simplex $\sigma = (\mathbf{G}, t) \in \Delta_{g,n}$ is a $P_{g',n'}$-simplex if and only if $\mathbf{G}$ has a (g', n')-bridge, i.e., a bridge separating subgraphs of types (g', n') and $(g - g', n - n')$ respectively. Similarly, $(\mathbf{G}', t') \in P^*_{g',n'}(X)$ if $\mathbf{G}'$ admits a morphism in $\Gamma_{g,n}$ from some $\mathbf{G}$ with a (g', n')-bridge.

We return to the general case, where X is a symmetric Δ-complex and $P \subset X_0$ is a property, and define a co-P face as a face such that all complementary vertices lie in P.

Definition 4.4 Given $\sigma \in X_p$ and $\theta \colon [q] \hookrightarrow [p]$, we say that θ is a co-P *face map* if $[p] \smallsetminus \operatorname{im} \theta \subset P(\sigma)$. In this case, we say that $\tau = \theta^*(\sigma)$ is a co-P *face* of σ.

We write $\tau \precsim_P \sigma$ if τ is a co-P face of σ. Then $\precsim_P$ is a reflexive, transitive relation, and it induces a partial order $\preceq_P$ on $\coprod_{p \geq 0} X_p/S_{p+1}$, where $[\tau] \preceq_P [\sigma]$ if $\tau \precsim_P \sigma$.

Example 4.5 In Figure 1, the 0-simplex v is a co-P-face of T.

In our main example $X = \Delta_{g,n}$, for any property $P \subset X_0$, let us say that an edge $e \in E(\mathbf{G})$ is a P-*edge* if the graph obtained from $\mathbf{G}$ by collapsing each element of $E(\mathbf{G}) - \{e\}$ is in P. A P-*contraction* is a contraction of $\mathbf{G}$ by a subset, possibly empty, of P-edges. Then a face $(\mathbf{G}', t')$ of $(\mathbf{G}, t)$ is a co-P face if and only if $\mathbf{G}'$ is isomorphic to a P-contraction of $\mathbf{G}$.

The *automorphisms* of a simplex $\sigma \in X_p$, denoted $\operatorname{Aut}(\sigma)$, are the bijections $\psi \colon [p] \to [p]$ such that $\psi^*\sigma = \sigma$. The natural map $\{\sigma\} \times \Delta^p \to |X|$ factors through $\Delta^p/\operatorname{Aut}(\sigma)$.

A face $\tau \precsim \sigma$ is *canonical* (meaning canonical up to automorphisms) if, for any two injections θ_1 and θ_2 from $[q]$ to $[p]$ such that $\theta_i^*(\sigma) = \tau$, there exists $\psi \in \text{Aut}(\sigma)$ such that $\theta_1 = \psi \theta_2$. Note that the property of $\tau \precsim \sigma$ being canonical depends only on the respective S_{q+1} and S_{p+1} orbits; we will say that $[\tau] \preceq [\sigma]$ is canonical if $\tau \precsim \sigma$ is so.

Remark 4.6 In [14, §3.4] we defined a category J_X with object set $\coprod_{p \geq -1} X_p$. Automorphisms of σ in this category agree with automorphisms in the above sense, and the relation $\tau \precsim \sigma$ holds if and only if there exists a morphism $\sigma \to \tau$ in J_X. In op. cit. we also defined a Δ-complex $\text{sd}(X)$ called the subdivision of X, and a canonical homeomorphism $|\text{sd}(X)| \cong |X|$. Geometrically, $[\sigma]$ and $[\tau]$ are then 0-simplices of $\text{sd}(X)$, and they are related by $\preceq$ if and only if there exists a 1-simplex connecting them. The relation is canonical if and only if there is precisely one 1-simplex in $\text{sd}(X)$ between them.

Example 4.7 For any subgroup $G < S_{p+1}$, the quotient Δ^p / G carries a natural structure of symmetric Δ-complex in which every face is canonical. For an example of a face inclusion that is not canonical, consider the Δ-complex consisting of a loop formed by one vertex and one edge (viewed as a symmetric Δ-complex with one 0-simplex and two 1-simplices, cf. [14, §3]). The automorphism groups of all simplices are trivial, so neither of the vertex-edge inclusions is canonical. For an example of noncanonical face inclusions in $\Delta_{g,n}$, see Example 4.13.

The main technical result of this section, Proposition 4.11, involves canonical co-P-maximal faces and co-P-saturation, defined as follows. See Example 4.10 below.

Definition 4.8 Let $Z \subset \coprod X_p / S_{p+1}$, and let P be any property. We say that Z admits *canonical co-P maximal faces* if, for every $[\tau] \in Z$, the poset of those $[\sigma] \in Z$ such that $[\tau] \preceq_P [\sigma]$ has a unique maximal element $[\hat{\sigma}]$ and moreover $[\tau] \preceq [\hat{\sigma}]$ is canonical.

Definition 4.9 Let $Y \subset \text{Simp}(X)$ be any subset and let P be any property on X. We call Y *co-P-saturated* if $\tau \in Y$ and $\tau \precsim_P \sigma$ implies $\sigma \in Y$.

Example 4.10 We illustrate Definitions 4.8 and 4.9 for the symmetric Δ-complex X drawn in Figure 1. Here, the full set of symmetric orbits $\coprod_{p \geq 0} X_p / S_{p+1}$ admits canonical co-P maximal faces. On the other hand, the 1-skeleton $\coprod_{p=0,1} X_p / S_{p+1}$ does not: indeed, the poset $\{[\sigma] : [v] \precsim_P [\sigma]\}$ has two maximal elements. Finally, the set of simplices in the subcomplex generated by the 2-simplex T is co-P-saturated, while the vertex v taken by itself is not co-P-saturated.

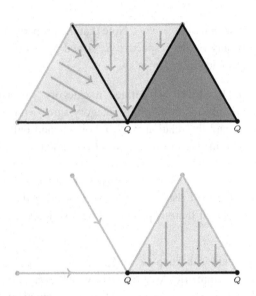

Figure 2 The deformation retractions $X = X_{Q,2} \searrow X_{Q,1}$ and $X_{Q,1} \searrow X_{Q,0}$ as in Proposition 4.11, with $P = \emptyset$.

Given $P \subset X_0$ and an integer $i \geq 0$, let $X_{P,i}$ denote the subcomplex of X generated by the set $V_{P,i}$ of P-simplices with at most i non-P vertices. When no confusion seems possible, we write $X_{P,i}$ for the image of the natural map

$$\coprod_{p \geq 0} \left(X_p \cap V_{P,i} \right) \times \Delta^p \longrightarrow |X|. \tag{5}$$

For example, for X and the property Q as shown in Figure 2, the subcomplexes $X_{Q,1}$ and $X_{Q,0}$ are shown in black in Figure 2 (top and bottom, respectively).

We use X_P to denote $X_{P,\infty}$, i.e., X_P is the subcomplex generated by all P-simplices. In the specific case where $X = \Delta_{g,n}$, we abbreviate the notation and write

$$\Delta_{P,i} = (\Delta_{g,n})_{P,i}, \quad \text{and} \quad \Delta_P = (\Delta_{g,n})_P.$$

Note that $\Delta_{P,i}$ parametrizes the closure of the locus of tropical curves with at least one P-edge and at most i non-P edges. For instance, if P is the property $P_{1,0}$ defined in Example 4.3, then the subspace $\Delta_P \subset \Delta_{g,n}$ is the locus of tropical curves with either a loop or a vertex of positive weight.

We now state the main technical result of this section, which is a tool for producing deformation retractions inside symmetric Δ-complexes.

Proposition 4.11 *Let X be a symmetric Δ-complex. Suppose $P, Q \subset X_0$ are properties satisfying the following conditions.*

1. The set of simplices $P^*(X)$ is co-Q-saturated.
2. The set of symmetric orbits of $X \setminus P^*(X)$ admits canonical co-Q maximal faces.

Then there are strong deformation retractions $(X_P \cup X_{Q,i}) \searrow (X_P \cup X_{Q,i-1})$ for each $i > 0$.

If, in addition, every strictly Q-simplex is in $P^*(X)$, then there is a strong deformation retraction $(X_P \cup X_Q) \searrow X_P$.

The basic ideas underlying Proposition 4.11 are discussed in Remark 4.14, below. We now give an example illustrating the conditions in Proposition 4.11 on $X = \Delta_{g,n}$.

Example 4.12 Suppose $P \subset \Delta_{g,n}([0])$ is a property. Because membership in $P(\Delta_{g,n})$ and $P^*(\Delta_{g,n})$ does not depend on edge-labeling, we say that $\mathbf{G}$ is in P if $(\mathbf{G}, t) \in P(\Delta_{g,n})$ for any, or equivalently every, edge-labeling t. Similarly, we say that $\mathbf{G}$ is in P^* if $(\mathbf{G}, t) \in P^*(\Delta_{g,n})$ for any, or equivalently every, edge-labeling t.

Suppose $g > 1$ and $n = 0$, and let $P = P_{1,0}$ and $Q = P_{2,0}$, as defined in Example 4.3. Note that $\mathbf{G} \in P_{1,0}^*(X)$ if and only if $\mathbf{G}$ has a loop or a positive vertex weight. One may then check that P and Q satisfy the conditions of Proposition 4.11. The content is that every graph with a loop or weight, upon expansion by a $(2,0)$-bridge, still has a loop or weight; and that any graph with no loops or weights has a *canonical* expansion by $(2,0)$-bridges. Furthermore, every strictly Q-simplex is in $P^*(\Delta_{g,n})$. Then Proposition 4.11 asserts the existence of a deformation retraction $X_{P_{1,0}} \cup X_{P_{2,0}} \searrow X_{P_{1,0}}$. In fact, this deformation retraction is the first step in the $n = 0$ case of Theorem 4.19(2).

Example 4.13 We give an example of a face inclusion in $\Delta_{g,n}$ that is *not* canonical. Let $\mathbf{G}$ and $\mathbf{G}'$ be the graphs shown in Figure 3, on the left and right, respectively. Note that $\mathbf{G}'$ is isomorphic to a contraction of $\mathbf{G}$. Let us consider the equivalence relation on morphisms $\mathbf{G} \to \mathbf{G}'$ given by $\alpha_1 \sim \alpha_2$ if $\alpha_2 = \theta \alpha_1$ for some $\theta \in \mathrm{Aut}(\mathbf{G})$. This equivalence relation partitions $\mathrm{Mor}(\mathbf{G}, \mathbf{G}')$ into exactly three classes, which are naturally in bijection with the three distinct unordered partitions of $E(\mathbf{G}')$ into two groups of two. In particular, there exist $\alpha_1, \alpha_2 \colon \mathbf{G} \to \mathbf{G}'$ such that there is no $\theta \in \mathrm{Aut}(\mathbf{G})$ with $\alpha_2 = \theta \alpha_1$. Finally, by equipping $\mathbf{G}$ and $\mathbf{G}'$ appropriately with edge labelings t and t', respectively, this example can be promoted to an example of a face map in $\Delta_{g,n}$ which is non-canonical. This example shows that the full set of symmetric orbits of

G   G'

Figure 3 The graphs **G** and **G'** on the left and right respectively. There is a morphism **G** → **G'** in $\Gamma_{4,0}$ but it is *not* canonical with respect to Aut(**G**).

simplices in $\Delta_{g,n}$ does *not* admit canonical co-$P_{2,0}$ maximal faces. However, the set of symmetric orbits of $\mathrm{Simp}(\Delta_{g,n}) \smallsetminus P_{1,0}^*(\Delta_{g,n})$ does, which is all that is required in Condition (2).

Remark 4.14   We now sketch the idea of the proof of Proposition 4.11, before proceeding to the proof itself. Let us first assume that $P = \emptyset$. In this case, Proposition 4.11 simplifies to the following: if Q is a property such that the symmetric orbits of X admit canonical co-Q maximal faces, then there is a strong deformation retraction $X_{Q,i} \searrow X_{Q,i-1}$ for each i. These retractions are drawn in an example in Figure 2.

Note that every p-simplex $\sigma \in V_{Q,i}$ which is not in $V_{Q,i-1}$ has precisely $p + 1 - i$ vertices in Q and i vertices not in Q. To such a simplex we shall associate a map $\Delta^p \to \partial \Delta^p$ by subtracting from the barycentric coordinates corresponding to vertices not in Q and adding to the remaining ones, in a way that glues to a retraction $X_{Q,i} \to X_{Q,i-1}$. Gluing the corresponding straight-line homotopies will give a homotopy from the identity to this retraction. In order to carry this out, the main technical task is to verify that the different homotopies may in fact be glued, which is where Condition (2) is used.

Now, dropping the condition that $P = \emptyset$ temporarily imposed in the previous paragraph, we obtain a *relative* version of the same argument. In this case, Condition (1) is needed in addition to guarantee that the relevant straight-line homotopies are constant on their overlap with X_P. This relative formulation is useful to apply the proposition repeatedly over a sequence of properties $P_1, \ldots, P_n$. This sequential use of the proposition is packaged below as Corollary 4.18.

The following definition and lemma will be used in the proof of Proposition 4.11. Let $P, Q \subset X_0$ be properties on X. Recall that $Q(\sigma)$ and $Q^c(\sigma)$ denote the labels of the Q-vertices and non-Q-vertices of σ, respectively. Let $i > 0$, and let

$$S_i = \{\sigma \in V_{Q,i} : \sigma \notin P^*(X)\}. \tag{6}$$

Let $T_i \subset S_i$ be the subset of simplices σ whose symmetric orbits $[\sigma]$ are maximal with respect to the partial order $\preceq_Q$, and let Y_i be the subcomplex of X generated by T_i.

Definition 4.15 Let σ be a p-simplex of X with $\sigma \in T_i$. We define a homotopy $\rho_{\sigma,i} \colon \Delta^p \times [0,1] \to \Delta^p$ as follows. If $|Q^c(\sigma)| < i$ then $\rho_{\sigma,i}$ is the constant homotopy. Otherwise, define a map

$$r_{\sigma,i} \colon \Delta^p \to \partial \Delta^p$$

as follows. Given $\ell \in \Delta^p$, let

$$\gamma = \min_{e \in Q^c(\sigma)} \ell(e);$$

note that $\gamma = 0$ is possible. Then define $r_{\sigma,i}(\ell)$ in coordinates by

$$r_{\sigma,i}(\ell)(e) = \begin{cases} \ell(e) - \gamma & \text{if } e \in Q^c(\sigma) \\ \ell(e) + \gamma i/|Q(\sigma)| & \text{if } e \in Q(\sigma). \end{cases} \tag{7}$$

(Note that $|Q(\sigma)| > 0$ since $\sigma \in T_i$.) Then let $\rho_{\sigma,i}$ be the straight line homotopy from the identity to $r_{\sigma,i}$.

Now we prove two lemmas demonstrating that the homotopies $\rho_{\sigma,i}$ glue appropriately. Write π for the quotient map

$$\pi \colon \left(\coprod_{p=0}^{\infty} X_p \times \Delta^p \right) \to |X| \tag{8}$$

as in Equation (3), and write $\sim$ for the equivalence relation $(\sigma_1, \ell_1) \sim (\sigma_2, \ell_2)$ if $\pi(\sigma_1, \ell_1) = \pi(\sigma_2, \ell_2)$. We prove first that points in the inverse image of X_P are fixed by each $\rho_{\sigma,i}$.

Lemma 4.16 Let X be a symmetric Δ-complex and $P, Q \subset X_0$ properties on X such that $P^*(X)$ is co-Q-saturated. Suppose $\sigma \in X_p$ is a simplex with $\sigma \in T_i$. Given $\ell \in \Delta^p$, if $\pi(\sigma, \ell) \in X_P$ then $r_{\sigma,i}(\ell) = \ell$. Thus $\rho_{\sigma,i}(\ell, t) = \ell$ for all $t \in [0,1]$.

Proof of Lemma 4.16 We prove the contrapositive, namely that if $r_{\sigma,i}(\ell) \neq \ell$ then $\pi(\sigma, \ell)$ is not in X_P. In general, there exists some q, some $\tau \in X_q$ (uniquely determined up to S_{q+1}-action) and some $\ell' \in (\Delta^q)^\circ$ such that $(\sigma, \ell) \sim (\tau, \ell')$. Moreover, the assumption $r_{\sigma,i}(\ell) \neq \ell$ implies that $\gamma > 0$ in Equation (7). Therefore, $\tau \precsim_Q \sigma$. Now $\sigma \notin P^*(X)$ since $\sigma \in T_i$, so $\tau \notin P^*(X)$ since $P^*(X)$ is co-Q saturated. Therefore $\pi(\sigma, \ell) = \pi(\tau, \ell') \notin X_P$. $\square$

Next, we prove that the homotopies $\rho_{\sigma,i}$ agree on overlaps, as σ ranges over T_i.

Lemma 4.17　*Let X be a symmetric Δ-complex and $P, Q \subset X_0$ properties such that*

1. *$P^*(X)$ is co-Q-saturated, and*
2. *the set of symmetric orbits of $X \smallsetminus P^*(X)$ admits canonical co-Q maximal faces.*

Given $\sigma_1 \in X_{p_1}$ and $\sigma_2 \in X_{p_2}$ with $\sigma_1, \sigma_2 \in T_i$, suppose $\ell_1 \in \Delta^{p_1}$ and $\ell_2 \in \Delta^{p_2}$ are such that $(\sigma_1, \ell_1) \sim (\sigma_2, \ell_2)$. Writing $r_1 = r_{\sigma_1,i}$ and $r_2 = r_{\sigma_2,i}$ for short, we have

$$(\sigma_1, r_1(\ell_1)) \sim (\sigma_2, r_2(\ell_2)).$$

Therefore $(\sigma_1, \rho_{\sigma_1,i}(\ell_1, t)) \sim (\sigma_2, \rho_{\sigma_2,i}(\ell_2, t))$ for all $t \in [0,1]$.

Proof of Lemma 4.17　Again, there exists some q, some $\tau \in X_q$ (uniquely defined up to S_{q+1}-action) and some $\ell \in (\Delta^q)^\circ$ such $(\tau, \ell) \sim (\sigma_1, \ell_1) \sim (\sigma_2, \ell_2)$. Moreover if $\tau \in P^*(X)$ then we are done by Claim 4.16, so we assume $\tau \notin P^*(X)$. There are two cases.

First, if $|Q^c(\tau)| < i$, then for each $j = 1, 2$, either $\tau \not\precsim_Q \sigma_j$ or $\tau \precsim_Q \sigma_j$; in the first case we have $\min_{e \in Q^c(\sigma_j)} \ell(e) = 0$, and in the second case we have $|Q^c(\sigma_j)| = |Q^c(\tau)| < i$. Thus in both cases, $r_1(\ell_1) = \ell_1$ and $r_2(\ell_2) = \ell_2$, which proves the claim.

Second, if $|Q^c(\tau)| = i$, we have $[\tau] \preceq_Q [\sigma_1]$ and $[\tau] \preceq_Q [\sigma_2]$. In fact, for $j = 1, 2$, we claim $[\sigma_j]$ is maximal such that $[\tau] \preceq_Q [\sigma_j]$. Indeed, if $[\tau] \preceq_Q [\sigma_j] \prec_Q [\sigma]$ for some $[\sigma]$, then

- $|Q(\sigma)| > 0$, since $\sigma_j \not\precsim_Q \sigma$;
- $|Q^c(\sigma)| \le i$, since $|Q^c(\sigma)| = |Q^c(\sigma_j)| \le i$;
- $\sigma \notin P^*(X)$, since $\sigma_j \notin P^*(X)$.

But this contradicts that $\sigma_j \in T_i$. We note again that $\tau \notin P^*(X)$ by assumption. Therefore $[\sigma_1] = [\sigma_2]$, by the hypothesis that the symmetric orbits of $X \smallsetminus P^*(X)$ admit canonical co-Q maximal faces.

Let us treat the special case $\sigma_1 = \sigma_2$; the general case will follow easily from it. Write $\sigma = \sigma_1 = \sigma_2$ and $r = r_1 = r_2$. For $j = 1, 2$, since $(\tau, \ell) \sim (\sigma, \ell_j)$, there exists $\alpha_j : [q] \hookrightarrow [p]$ such that $\alpha_j^* \sigma = \tau$ and $\ell_j = \alpha_{j*} \ell$. By canonicity of co-$Q$-maximal faces, there exists $\theta \in \mathrm{Aut}(\sigma)$ with $\theta \alpha_1 = \alpha_2$, so

$$\ell_2 = \alpha_{2*} \ell = \theta_* \alpha_{1*} \ell = \theta_* \ell_1 = \ell_1 \circ \theta.$$

Since $\theta \in \text{Aut}(\sigma)$, $\theta e \in Q(\sigma)$ if and only if $e \in Q(\sigma)$ for all $e \in [p]$. Therefore by Equation (7) we have

$$(\sigma, r(\ell_1)) \sim (\sigma, r(\ell_1 \circ \theta)) \sim (\sigma, r(\ell_2)),$$

as desired.

Finally, the general case follows from the previous one by replacing σ_2 with $\sigma_1 = \phi^* \sigma_2$ for some $\phi \colon [p] \to [p]$, and replacing ℓ_2 with $\ell_2 \circ \phi$. Indeed, we have $(\sigma_1, \ell_1) \sim (\sigma_2, \ell_2) \sim (\sigma_1, \ell_2 \circ \phi)$ and

$$(\sigma_2, r_2(\ell_2))) \sim (\phi^* \sigma_2, r_2(\ell_2) \circ \phi)$$
$$\sim (\sigma_1, r_1(\ell_2 \circ \phi))$$
$$\sim (\sigma_1, r_1(\ell_1)).$$

The second equivalence follows from the fact that $e \in Q(\sigma_2)$ if and only if $\phi(e) \in Q(\phi^* \sigma_2) = Q(\sigma_1)$ for every $e \in [p]$. The last equivalence follows from the previous computation for the case $\sigma_1 = \sigma_2$. This proves Claim 4.17. $\square$

We now proceed with the proof of Proposition 4.11.

Proof of Proposition 4.11   Let X be a symmetric Δ-complex, and let $P, Q \subset X_0$ be properties satisfying:

1. $P^*(X)$ is co-Q-saturated, and
2. the symmetric orbits of $X \smallsetminus P^*(X)$ admit canonical co-Q maximal faces.

We wish to exhibit a deformation retraction $X_P \cup X_{Q,i} \searrow X_P \cup X_{Q,i-1}$.

Recall that Y_i is the subcomplex of X generated by the set of simplices T_i; by the usual abuse of notation we will also write Y_i for the homeomorphic image of its geometric realization in $|X|$. First we note $X_P \cup X_{Q,i} = X_P \cup Y_i$. The inclusion $\supset$ is clear since $\text{Simp}(X_{Q,i}) \supset T_i$. The inclusion $\subset$ is also apparent: suppose $\tau \in \text{Simp}(X)$ has $|Q(\tau)| > 0$ and $|Q^c(\tau)| \leq i$. If $\tau \in P^*(X)$ then its image in $|X|$ is in X_P. Otherwise, $\tau \in S_i$, so $\tau \precsim_Q \sigma$ for some $\sigma \in T_i$, so the image of τ in $|X|$ lies in Y_i.

Therefore, we have a map

$$r_i \colon (X_P \cup X_{Q,i}) \to X_P \cup X_{Q,i}$$

that is obtained by gluing the maps $r_{\sigma,i}$ for $\sigma \in T_i$, together with the constant map on X_P. The fact that we may glue these maps together is the content of Lemmas 4.16 and 4.17. Moreover r_i restricts to the constant map on $X_P \cup X_{Q,i-1}$ by construction.

Next, we show that the image of r_i is $X_P \cup X_{Q,i-1}$. Let $\sigma \in T_i$ be a p-simplex and let $\ell \in \Delta^p$. Now there exists some q, some $\tau \in X_q$, and some

$\ell' \in (\Delta^q)^\circ$, such that $(\sigma, r_i(\ell)) \sim (\tau, \ell')$. Examining (7) shows that $|Q(\tau)| > 0$ and $|Q^c(\tau)| < |Q^c(\sigma)| = i$. Now if $\tau \in P^*(X)$, then $\pi(\sigma, r_i(\ell)) \in X_P$. Otherwise, if $\tau \notin P^*(X)$, then $\tau \in S_{i-1}$, so $\pi(\sigma, r_i(\ell)) \in X_{Q,i-1}$. This argument shows that the map $\rho_i : (X_P \cup X_{Q,i}) \times [0,1] \to X_P \cup X_{Q,i}$, defined to be the straight line homotopy associated to r_i, is a deformation retraction onto $X_P \cup X_{Q,i-1}$. Thus we have a strong deformation retraction $X_P \cup X_{Q,i} \searrow X_P \cup X_{Q,i-1}$ for each i, and hence a strong deformation retraction $X_P \cup X_Q \searrow X_P \cup X_{Q,0}$.

Finally, we check that if every strictly Q-simplex is in $P^*(X)$, then $X_{Q,0} \subset X_P$. Indeed, if this condition holds, then

$$\text{Simp}(X_{Q,0}) = \{\sigma \in \text{Simp}(X) \mid Q^c(\sigma) = \emptyset\} \subset \text{Simp}(X_P) = P^*(X),$$

so $X_{Q,0} \subset X_P$ as desired. Thus, under this condition there is a strong deformation retraction $X_P \cup X_Q \searrow X_P$, finishing the proof of the proposition. $\square$

We record an obvious corollary of Proposition 4.11, obtained by applying it repeatedly.

Corollary 4.18 *Let X be a symmetric Δ-complex, and let $P_1, \ldots, P_N$ be a sequence of properties.*

1. *Suppose that for $i = 2, \ldots, N$, the two properties $P = P_1 \cup \cdots \cup P_{i-1}$ and $Q = P_i$ satisfy that*

   - *$P^*(X)$ is co-Q-saturated,*
   - *the symmetric orbits of $X \setminus P^*(X)$ admit canonical co-Q maximal faces, and*
   - *every strictly Q-simplex is in $P^*(X)$.*

   Then there exists a strong deformation retraction

   $$X_{P_1 \cup \cdots \cup P_N} \searrow X_{P_1}.$$

2. *If in addition the symmetric orbits of X admit canonical co-P_1 maximal faces, then there exists a strong deformation retraction*

   $$X_{P_1 \cup \cdots \cup P_N} \searrow X_{P_1,0}.$$

Here the spaces X_P and $X_{P,0}$, for a property P, are the ones defined in (5).

4.2 Contractible Subcomplexes of $\Delta_{g,n}$

Here, we prove contractibility of three natural subcomplexes of $\Delta_{g,n}$. First, recall that a *bridge* of a connected graph G is an edge whose deletion disconnects G, and let $\Delta_{g,n}^{\mathrm{br}} \subset \Delta_{g,n}$ denote the closure of the locus of tropical curves with bridges. It is the geometric realization of the subcomplex generated by those $\mathbf{G} \in \Gamma_{g,n}$ with bridges. Next, we say that $\mathbf{G} = (G, m, w) \in \Gamma_{g,n}$ has *repeated markings* if the marking function m is not injective. Let $\Delta_{g,n}^{\mathrm{rep}} \subset \Delta_{g,n}$ be the locus of tropical curves with repeated markings. Let $\Delta_{g,n}^{\mathrm{w}}$ be the locus of tropical curves with at least one vertex of positive weight, and let $\Delta_{g,n}^{\mathrm{lw}}$ be the locus of tropical curves with loops or vertices of positive weights.

Note that $\Delta_{g,n}^{\mathrm{w}}$ is a closed subcomplex of $\Delta_{g,n}^{\mathrm{lw}}$ and both $\Delta_{g,n}^{\mathrm{rep}}$ and $\Delta_{g,n}^{\mathrm{lw}}$ are closed subcomplexes of $\Delta_{g,n}^{\mathrm{br}}$. For some purposes, it is most useful to contract the largest possible subcomplex. However, contractibility of smaller subcomplexes is also valuable; for instance, the contractibility of $\Delta_{g,n}^{\mathrm{w}}$ allows us to identify the reduced homology of $\Delta_{g,n}$ with graph homology in Theorem 1.4.

We recall the statement of Theorem 1.1.

Theorem 1.1 *Assume* $g > 0$ *and* $2g - 2 + n > 0$. *Each of the following subcomplexes of* $\Delta_{g,n}$ *is either empty or contractible.*

1. *The subcomplex* $\Delta_{g,n}^{\mathrm{w}}$ *of* $\Delta_{g,n}$ *parametrizing tropical curves with at least one vertex of positive weight.*
2. *The subcomplex* $\Delta_{g,n}^{\mathrm{lw}}$ *of* $\Delta_{g,n}$ *parametrizing tropical curves with loops or vertices of positive weight.*
3. *The subcomplex* $\Delta_{g,n}^{\mathrm{rep}}$ *of* $\Delta_{g,n}$ *parametrizing tropical curves in which at least two marked points coincide.*
4. *The closure* $\Delta_{g,n}^{\mathrm{br}}$ *of the locus of tropical curves with bridges.*

We will prove this theorem by applying Corollary 4.18 to a particular sequence of properties, as follows. Let $\mathbf{G} = (G, m, w) \in \Gamma_{g,n}$. Recall from Example 4.3 that an edge $e \in E(G)$ is a (g', n')-*bridge* if $\mathbf{G}/(E(G) - e) \cong \mathbf{B}(g', n')$, i.e., a (g', n')-bridge separates $\mathbf{G}$ into subgraphs of types (g', n') and $(g - g', n - n')$, respectively. We write $P_{g', n'}$ for the property $\{\mathbf{B}(g', n')\}$.

Theorem 4.19 *Let* $g > 0$ *and* $X = \Delta_{g,n}$.

1. *If* $n \geq 2$, *then the sequence of properties*

$$P_{0,n}, P_{0,n-1}, \ldots, P_{0,2}, P_{1,0}, P_{1,1}, \ldots, P_{1,n}, P_{2,0}, \ldots P_{2,n}, \ldots$$

satisfies both conditions of Corollary 4.18.
2. *If* $n = 0$ *or* 1 *and* $g > 1$, *then the sequence of properties*

$$P_{1,0}, \ldots, P_{1,n}, P_{2,0}, \ldots P_{2,n}, \ldots$$

satisfies both conditions of Corollary 4.18.

Each of these sequences of properties is finite. The last term of each of the two sequences above is chosen so that each type of bridge is named once. Precisely, if g is even, the last term is $P_{g/2, \lfloor n/2 \rfloor}$; if g is odd, the last term is $P_{(g-1)/2, n}$.

Remark 4.20   With the standing assumption that $g > 0$, the loci $\Delta_{g,n}^{lw}$ are never empty, and $\Delta_{g,n}^{w}$ is empty only when $(g,n) = (1,1)$. The locus $\Delta_{g,n}^{rep}$ is empty exactly when $n \leq 1$. The locus $\Delta_{g,n}^{br}$ is empty exactly when $(g,n) = (1,1)$. Otherwise, it contains $\Delta_{g,n}^{lw} \cup \Delta_{g,n}^{rep}$.

Proof that Theorem 4.19 implies Theorem 1.1   First we show Theorem 1.1(4). We treat two cases: if $n \geq 2$, let $P_1, \ldots, P_N$ denote the sequence of properties in part (1) of Theorem 4.19(1); if $n \leq 1$ and $g > 1$, let $P_1, \ldots, P_N$ denote the sequence of properties in part (2) of Theorem 4.19. In either case, $\cup P_i$ is the property of being a bridge, so $\Delta_{\cup P_i} = \Delta_{g,n}^{br}$. In the first case, $P_1 = P_{0,n}$ is the property of being a $(0,n)$-bridge, and note that $\Delta_{P_{0,n},0}$ is a point: there is a unique (up to isomorphism) tropical curve whose edges are all $(0,n)$-bridges. In the second case, $P_1 = P_{1,0}$ is the property of being a $(1,0)$-bridge, and $\Delta_{P_{1,0},0}$ is a $(g-1)$-simplex, parametrizing nonnegative edge lengths on a tree with g leaves of weight 1, and a central vertex supporting n markings. Then by Theorem 4.19, we may apply Corollary 4.18 to produce a deformation retraction from $\Delta_{\cup P_i} = \Delta_{g,n}^{br}$ to a contractible space. This shows Theorem 1.1(4).

We deduce Theorem 1.1(3) by considering only the subsequence of properties $P_1 = P_{0,n}, \ldots, P_{n-1} = P_{0,2}$. Indeed, $P_{0,n}, \ldots, P_{0,2}$, being an initial subsequence of the properties listed in Theorem 4.19(1), also satisfies both conditions of Corollary 4.18. Moreover $\Delta_{\cup P_i} = \Delta_{g,n}^{rep}$ and $\Delta_{P_{0,n},0}$ is a point. So by Corollary 4.18, we conclude that $\Delta_{g,n}^{rep}$ is contractible for all $g > 0$ and $n > 1$.

For Theorem 1.1(2), if $(g,n) = (1,1)$ the claim is trivial. Else, we verify directly that the properties $P = \emptyset$ and $Q = P_{1,0}$ satisfy the conditions (1) and (2) of Proposition 4.11. In other words, we verify directly that every graph admits a canonical maximal expansion by $(1,0)$-bridges. If **G** has no loops or weights, the expansion is trivial. Otherwise, the expansion is as follows: for any vertex v with

$$\mathrm{val}(v) + 2w(v) > 3,$$

replace every loop based at v with a bridge from v to a loop; add $w(v)$ bridges to vertices of weight 1, and set $w(v) = 0$. Contractibility of the loop-and-weight locus follows from Proposition 4.11, noting that this locus is exactly the subcomplex $(\Delta_{g,n})_{1,0}$ of $\Delta_{g,n}$ whenever $(g,n) \neq (1,1)$.

The proof of Theorem 1.1(1) is similar. If $(g,n) = (1,1)$ then $\Delta_{g,n}^{w}$ is empty. Otherwise, $\Delta_{g,n}^{w}$ itself is a symmetric Δ-complex, and we may consider the properties $P = \emptyset$ and $Q = P_{1,0}$ on it. This pair of properties still satisfies the conditions (1) and (2) of Proposition 4.11, since any $\mathbf{G}$ with a vertex of positive weight has a canonical maximal expansion of $\mathbf{G}$ by $(1,0)$-bridges that again has a vertex of positive weight whenever $\mathbf{G}$ does; this expansion is described above. So by Proposition 4.11, $\Delta^w_{g,n} = (\Delta^w_{g,n})_{P_{1,0}}$ deformation retracts down to the subcomplex of $\Delta^w_{g,n}$ consisting of graphs in which the only edges are $(1,0)$-edges, and this subcomplex is contractible. □

In order to prove Theorem 4.19, it will be convenient to develop a theory of *block decompositions* of stable weighted, marked graphs. Let us start with usual graphs, without weights or markings. If G is a connected graph, we say $v \in V(G)$ is a cut vertex if deleting it disconnects G. A *block* of G is a maximal connected subgraph with at least one edge and no cut vertices.

Example 4.21   If G is a graph on two vertices v_1, v_2 with a loop at each of v_1 and v_2 and n edges between v_1 and v_2, then G has three blocks: the loop at v_1, the loop at v_2, and the n edges between v_1 and v_2.

Returning to marked, weighted graphs, we define an *articulation point* of a stable, marked weighted graph $\mathbf{G} = (G, m, w)$ to be a vertex $v \in V(G)$ such that at least one of the following conditions holds:

(i) v is a cut vertex of G,
(ii) $w(v) > 0$, or
(iii) $|m^{-1}(v)| \geq 2$.

Articulation points are analogues of cut vertices for marked, weighted graphs. Let $\mathcal{A}$ denote the set of articulation points, and let $\mathcal{B}$ denote the set of blocks of the underlying graph G.

Definition 4.22   Let $\mathbf{G}$ be a weighted, marked graph. The *block graph* of $\mathbf{G}$, denoted $\mathrm{Bl}(\mathbf{G})$, is a graph defined as follows. The vertex set is $\mathcal{A} \cup \mathcal{B}$, and there is an edge $E = (v, B)$ from $v \in \mathcal{A}$ to $B \in \mathcal{B}$ if and only if $v \in B$.

In this way $\mathrm{Bl}(\mathbf{G})$ is naturally a tree, whose vertices are articulation points and blocks. The block graph of the graph G in Example 4.21 is drawn in Figure 4. The vertices of the block graph are depicted as the blocks and articulation points to which they correspond. The edges of the block graph are drawn in blue.

Figure 4 Block graph of G as in Example 4.21.

At this point, we will equip both the articulation points and the blocks with weights and markings on the vertices, according to the following conventions. If $v \in \mathcal{A}$ is an articulation point, we take it to have the weight and markings it has in **G**. That is, v has weight $w(v)$ and markings $m^{-1}(v)$. If $B \in \mathcal{B}$ is a block, then we give each vertex $x \in V(B)$ weights and markings according to the following rule. If $x \in \mathcal{A}$ then we equip it with weight 0 and no markings. Otherwise, we equip x with the same weights and markings as it had in **G**. In this way, we now regard each articulation point and each block as a weighted marked graph. We emphasize that these weighted marked graphs need not be stable. We note

$$\sum_{\mathbf{H} \in V(\mathrm{Bl}(\mathbf{G}))} g(\mathbf{H}) = g, \qquad \sum_{\mathbf{H} \in V(\mathrm{Bl}(\mathbf{G}))} n(\mathbf{H}) = n.$$

(Here $n(\mathbf{H})$ is the number of marked points.)

It will be useful to label the edges of $\mathrm{Bl}(\mathbf{G})$ as follows. Since $\mathrm{Bl}(\mathbf{G})$ is a tree, deleting any edge $\epsilon = (v, B)$ divides $\mathrm{Bl}(\mathbf{G})$ into two connected components. Let S be the set of vertices in the part containing $B \in V(\mathrm{Bl}(\mathbf{G}))$; then we label the edge ϵ

$$(g(v, B), n(v, B)) := \left(\sum_{\mathbf{H} \in S} g(\mathbf{H}), \sum_{\mathbf{H} \in S} n(\mathbf{H}) \right).$$

A property of this labeling that we record for later use is that for every $v \in \mathcal{A}$,

$$\sum_{B \ni v} g((v, B)) + w(v) = g, \quad \text{and} \quad \sum_{B \ni v} n((v, B)) + |m^{-1}(v)| = n. \quad (9)$$

Example 4.23   Let $g > 0$ and $n \geq 0$ with $(g, n) \neq (1, 0), (1, 1)$. Suppose $\mathbf{G} = (G, m, w)$ has a single vertex v and h loops. Then v has weight $g - h$ and n markings, and there are h blocks of **G**, each a single unweighted, unmarked loop based at v. There is a single articulation point v, equipped with weight $g - h$ and all n markings. The block graph $\mathrm{Bl}(\mathbf{G})$ is a star tree with h edges from v, each labeled $(1, 0)$.

We make the following observations.

Lemma 4.24   *Let* $\mathbf{G} \in \Gamma_{g,n}$.

1. *If* $e \in E(G)$ *is a bridge then its image vertex v in G/e is an articulation point.*
2. *Let v be an articulation point of* **G**, *with weight $u \geq 0$ and markings $m^{-1}(v) = M$, and with edges of* $\mathrm{Bl}(\mathbf{G})$ *at v labeled $(g_1, n_1), \ldots, (g_s, n_s)$. Then v may be expanded into a bridge, with the result a stable marked,*

weighted graph, in any of the following ways. Choose a partition of the edges of $\mathrm{Bl}(G)$ *at* v *into two parts* P_1 *and* P_2; *choose a partition of the set* M *into sets* M_1 *and* M_2; *and choose integers* $w_1, w_2 \geq 0$ *with* $w_1 + w_2 = u$, *such that for* $j = 1, 2$

$$\sum_{(v, B) \in P_j} \mathrm{val}_B(v) + |M_j| + 2w_j \geq 2.$$

Here and below, $\mathrm{val}_B(v)$ *denotes the number of half-edges at* v *lying in* B; *it does not count any marked points. By dividing the blocks, markings, and weight accordingly,* v *may be expanded into a bridge of type*

$$\left(\sum_{(v, B) \in P_1} g((v, B)) + w_1, \sum_{(v, B) \in P_1} n((v, B)) + |M_1| \right) \tag{10}$$

such that the result is stable; and no other stable expansions of v *into bridges are possible.*

3. *If* $\mathrm{Bl}(G)$ *has an edge* $\epsilon = (v, B)$ *labeled* (g', n'), *then* $G \in P^*_{g', n'}$.
4. *Suppose* $g' \geq 1$ *and* $w(v) = 0$ *for all* $v \in V(G)$, *and suppose every label* (g'', n'') *on* $E(\mathrm{Bl}(G))$ *satisfies either* $g'' > g'$, *or* $g'' = g'$ *and* $n'' > n'$. *Then* $G \notin P^*_{g', n'}$.

Proof Statements (1) and (2) are easy to check. Statement (4) then follows: if $G \in P^*_{g', n'}$ then (1) and (2) imply that some articulation point v may be expanded into a bridge of type (g', n'), with

$$(g', n') = \left(\sum_{(v, B) \in P_1} g((v, B)) + w_1, \sum_{(v, B) \in P_1} n((v, B)) + |M_1| \right)$$

for some choice of partition $P_1 \sqcup P_2$ of the blocks at v. Since $g' > 0$ and $w_1 = w(v) = 0$ we must have $P_1 \neq \emptyset$, but then the expression in (10) exceeds (g', n') in lexicographic order.

For statement (3), suppose $\epsilon = (v, B)$ is labeled (g', n'). If B itself is a (g', n')-bridge we are done. Otherwise $\mathrm{val}_B(v) \geq 2$. Write $B_1, \ldots, B_s$ for the remaining blocks at v. If $\sum_{j=1}^s \mathrm{val}_{B_j}(v) + |m^{-1}(v)| + 2w(v) \geq 2$ then v can be expanded into a (g', n')-bridge by (2). So assume

$$\sum_{j=1}^s \mathrm{val}_{B_j}(v) + |m^{-1}(v)| + 2w(v) \leq 1.$$

The only possibility consistent with v being an articulation point is $s = 1$, $\mathrm{val}_{B_1}(v) = 1$, $|m^{-1}(v)| = 0$, and $w(v) = 0$. Thus B_1 is a bridge, and the identities (9) show that B_1 is a $(g - g', n - n')$-bridge, which is the same as a (g', n')-bridge. $\qquad \square$

Now we turn to the proof of Theorem 4.19.

Proof of Theorem 4.19 Fix $g > 0$ and $n \geq 0$. If $n \geq 2$, let $P_1, P_2, \ldots$ be the sequence of properties

$$P_{0,n}, \ldots, P_{0,2}, P_{1,0}, \ldots, P_{1,n}, P_{2,0}, \ldots, P_{2,n}, \ldots$$

If $n = 0$ or 1 and $g \neq 1$, let $P_1, P_2, \ldots$ be the sequence of properties

$$P_{1,0}, \ldots, P_{1,n}, P_{2,0}, \ldots, P_{2,n}, \ldots$$

We need to check:

(i) for each $i = 2, 3, \ldots$ the properties $P = P_1 \cup \cdots \cup P_{i-1}$ and $Q = P_i$ satisfy the two conditions of Proposition 4.11, and every strictly Q-simplex is in P^*.

(ii) the symmetric orbits of X admit canonical co-Q-maximal faces.

Item (ii) above is exactly the statement that the properties $P = \emptyset$ and $Q = P_1$ satisfy the second condition of Proposition 4.11.

Condition (2) of Proposition 4.11. For each $i = 1, 2, \ldots$, let $P = P_1 \cup \cdots \cup P_{i-1}$ and $Q = P_i$. Let us check that condition (2) of Proposition 4.11 holds. Let $Q = P_{g',n'}$. Suppose $\mathbf{G} \in \Gamma_{g,n}$ is not in P^*. We need to show that $\mathbf{G}$ admits a maximal uncontraction $\alpha \colon \widetilde{\mathbf{G}} \to \mathbf{G}$ by (g', n')-bridges, which is canonical in the sense that for any $\alpha' \colon \widetilde{\mathbf{G}} \to \mathbf{G}$, there exists an automorphism $\theta \colon \widetilde{\mathbf{G}} \to \widetilde{\mathbf{G}}$ such that $\alpha'\theta = \alpha$. Informally speaking, we are saying that $\widetilde{\mathbf{G}}$ may be described in a way that is intrinsic to $\mathbf{G}$. We treat three cases:

- $Q = P_{g',n'}$ with $g' \geq 1$ and $(g', n') \neq (1, 0)$;
- $Q = P_{1,0}$; and
- $Q = P_{0,n'}$ for some n'.

The case $Q = P_{0,n'}$ is only needed when $n \geq 2$.

First, assume $Q = P_{g',n'}$ with $g' \geq 1$ and $(g', n') \neq (1, 0)$. Let v be any articulation point. Now either $n \leq 1$, or $n \geq 2$ and $\mathbf{G}$ is assumed not to be a $(0, 2)$-contraction since $P_{0,2} \subset P$. Therefore $\mathbf{G}$ has no repeated markings. Since $P_{1,0} \subset P$ and $\mathbf{G}$ is assumed not to be in $P^*(X)$, $\mathbf{G}$ has no vertex weights. Let $B \in \mathrm{Bl}(v)$, and let (g'', n'') be the label of $\epsilon = (v, B) \in E(\mathrm{Bl}(\mathbf{G}))$. Then by Lemma 4.24(3), $\mathbf{G} \in P^*_{g'',n''}$. Referring to the chosen ordering of properties, it follows that $g'' \geq g'$, and if $g'' = g'$ then $n'' > n'$. Now using the criterion of Lemma 4.24(2), we conclude that the only (g', n')-bridge expansions admitted at v are along the pairs (v, B) labeled exactly (g', n') where B is not itself a bridge, and there is a unique maximal such expansion which is canonical in the previously described sense.

Second, assume $Q = P_{1,0}$. Then by Lemma 4.24(2), the maximal $(1,0)$-bridge expansion of **G** is obtained by replacing, for any vertex v with $\text{val}(v) + 2w(v) > 3$, every loop based at v with a bridge from v to a loop; adding $w(v)$ bridges to vertices of weight 1, and setting $w(v) = 0$. Moreover this expansion is canonical.

Finally, assume $Q = P_{0,n'}$; this case is only needed when $n \geq 2$. Consider an articulation point v, and let $B_1, \ldots, B_k$ be the blocks at v labeled $(0, n_1), \ldots, (0, n_k)$ for some n_i. We are assuming that **G** is not in P^*; in this case the chosen ordering of properties implies that $P_{0,n''} \subset P$ for each $n'' > n'$. Therefore, by Lemma 4.24(2), $\sum n_i + |m^{-1}(v)| \leq n'$. Furthermore, v can be expanded into a $(0, n')$-bridge if and only if equality holds, so long as it is not the case that $k = 1$ and B_1 is itself a $(0, n')$-bridge. This analysis, performed at all articulation points, produces the unique maximal $(0, n')$-bridge expansion of **G**, and this expansion is canonical. This verifies condition (2) of Proposition 4.11.

Condition (1) of Proposition 4.11. Again, let $i = 1, 2, \ldots$, let $P = P_1 \cup \cdots \cup P_{i-1}$ and $Q = P_i$; we now check that condition (1) of Proposition 4.11 holds. Suppose $Q = P_{g',n'}$. We want to show that if **G** $\in \Gamma_{g,n}$ is not in P^* and **G**$'$ is obtained by contracting (g', n')-bridges, then **G**$'$ is also not in P^*. We consider the same three cases.

First, assume $g' = 0$, that is, $Q = P_{0,n'}$; we only need this case if $n \geq 2$. The assumption **G** $\notin P^*$ means that **G** $\notin P^*_{0,n''}$ for any $n'' > n'$. Let us describe what these assumptions imply. First, let C denote the core of **G**, as defined in §2.1. Then $G \setminus E(C)$ is a disjoint union of trees $\{Y_v\}_{v \in V(C)}$. Say that a core vertex $v \in V(C)$ *supports* a marked point $\alpha \in \{1, \ldots, n\}$ if $m(\alpha) \in Y_v$. Then observe that for any **G** $\in \Gamma_{g,n}$, the following are equivalent:

(1) **G** $\notin P^*_{0,n''}$ for any $n'' > n'$;
(2) every core vertex of **G** supports at most n' markings.

Now, we are assuming that **G** satisfies (1), so it satisfies (2). Moreover (2) is evidently preserved by contracting $(0, n')$-bridges, since those operations never increase the number of markings supported by a core vertex. So (1) is also preserved by contracting $(0, n')$-bridges, which is what we wanted to show.

Second, assume $Q = P_{1,0}$. If $n \leq 1$ then $P = \emptyset$ and we are done. Otherwise, $P = P_{0,n} \cup \cdots \cup P_{0,2}$, and a graph **G** is in P^* if and only if **G** has repeated markings. The property P is evidently preserved by uncontracting $(1,0)$-bridges, so we are done.

Third, assume $Q = P_{g',n'}$ with $g' \geq 1$ and $(g',n') \neq (1,0)$. Let $e \in E(G)$ be a (g',n')-bridge; we assume that $\mathbf{G} \not\subseteq P^*$ and we wish to show that $\mathbf{G}/e \not\subseteq P^*$.

First, in the case $n \geq 2$, we need to show that $\mathbf{G}/e \not\subseteq P^*_{0,n''}$ for any n'', i.e., $\mathbf{G}/e$ has no repeated markings. Since $\mathbf{G}$ has no repeated markings, it suffices to show that not both ends of $e = v_1 v_2$ are marked. We may assume that the edge (v_1, e) in $\mathrm{Bl}(\mathbf{G})$ was labeled $(g - g', n - n')$; we will show v_1 is unmarked. Since $\mathbf{G} \not\subseteq P^*_{0,n''}$ for any n'' and $\mathbf{G} \not\subseteq P^*_{1,0}$, we have $w(v_1) = 0$ and v_1 is at most once-marked. Therefore, there is at least one other block $B \neq e$ at v_1 and $(v_1, B) \in E(\mathrm{Bl}(\mathbf{G}))$ is labeled $(> g', *)$ or $(g', \geq n')$. In light of Equation (9), the only possibility is that there is only one such block B, (v_1, B) is labeled (g', n'), and v_1 is unmarked. Therefore $\mathbf{G}/e$ has no repeated markings.

Next, by Lemma 4.24(3), every label (g'', n'') on $E(\mathrm{Bl}(\mathbf{G}))$ satisfies either $g'' > g$, or $g'' = g'$ and $n'' \geq n'$. Furthermore, the labels on $E(\mathrm{Bl}(\mathbf{G}/e))$ are a subset of those on $E(\mathrm{Bl}(\mathbf{G}))$. Therefore by Lemma 4.24(4), $\mathbf{G}/e \not\subseteq P^*_{g'',n''}$ for any $g'' < g'$ or $g'' = g$ and $n'' < n'$, as long as $g'' \geq 1$. We have verified that condition (1) of Proposition 4.11 holds in the required cases.

To treat the last condition, regarding strictly co-Q faces being in P^*, we assume all edges of $\mathbf{G} \in \Gamma_{g,n}$ are (g',n')-bridges. Then G must be a tree with a single non-leaf vertex v, while every other vertex v' has $w(v') = g'$ and $|m^{-1}(v)| = n'$. Now we treat the following cases.

Suppose $Q = P_{g',n'}$ with $g' \geq 1$ and $(g',n') \neq (1,0)$. If $\mathbf{G}$ has only (g',n')-edges then $\mathbf{G} \in P^*_{1,0}$, since $\mathbf{G}$ has positive weights. Therefore $\mathbf{G} \in P^*$.

Next, suppose $Q = P_{1,0}$. If $\mathbf{G}$ has only $(1,0)$-edges, then either $n \leq 1$ and there is nothing to check, or $n \geq 2$ and so v supports $n \geq 2$ markings. Then $\mathbf{G} \in P^*_{0,2}$, so $\mathbf{G} \in P^*$.

Finally, suppose $n \geq 2$ and $Q = P_{0,n'}$ for $n' < n$. If $\mathbf{G}$ has only $(0,n')$-edges for some $n' < n$, note that $w(v) = g$ and so v may be expanded into a $(g,0)$-bridge, equivalently a $(0,n)$-bridge. So $\mathbf{G} \in P^*_{0,n}$, and hence $\mathbf{G} \in P^*$, as required. □

To close this section, we record a related contractibility result that will be useful for future applications. Let $\Delta^{\mathrm{w}}_{g,n}$ denote the subcomplex of $\Delta_{g,n}$ parametrizing tropical curves that have at least one positive vertex weight.

Lemma 4.25 *For all $g > 0$, $\Delta^{\mathrm{w}}_{g,n} \cup \Delta^{\mathrm{rep}}_{g,n}$ is contractible, unless $(g,n) = (1,1)$ in which case it is empty.*

Proof Regard $X = \Delta^{\mathrm{w}}_{g,n} \cup \Delta^{\mathrm{rep}}_{g,n}$ as a symmetric Δ-complex. In the above notation, the properties

$$P_{0,n}, \ldots, P_{0,2}, P_{1,0}$$

satisfy the hypotheses of Corollary 4.18. The conclusion is that X admits a strong deformation retraction to the point in X corresponding to the 1-edge graph $\mathbf{B}(g, 0)$. □

5 Calculations on $\Delta_{g,n}$

In §5.1, we apply the contractibility of $\Delta_{g,n}^{\mathrm{rep}}$ to calculate the S_n-equivariant homotopy type of $\Delta_{1,n}$ and prove Theorem 1.2 from the introduction. In §5.2 we prove Theorem 1.4. In §5.3 we construct a transfer homomorphism from the cellular chain complex of Δ_g to that of $\Delta_{g,1}$. Calculations of the rational homology of $\Delta_{g,n}$ in a range of cases for $g \geq 2$ are in Appendix A.

5.1 The Case $g = 1$

We now restate and prove Theorem 1.2, showing that contracting $\Delta_{1,n}^{\mathrm{rep}}$ produces a bouquet of $(n - 1)!/2$ spheres indexed by cyclic orderings of the set $\{1, \ldots, n\}$, up to order reversal, and then computing the representation of S_n on the reduced homology of this bouquet of spheres induced by permuting the marked points.

Theorem 1.2 *For $n \geq 3$, the space $\Delta_{1,n}$ is homotopy equivalent to a wedge sum of $(n - 1)!/2$ spheres of dimension $n - 1$. The representation of S_n on $H_{n-1}(\Delta_{1,n}; \mathbb{Q})$ induced by permuting marked points is*

$$\mathrm{Ind}_{D_n, \phi}^{S_n} \mathrm{Res}_{D_n, \psi}^{S_n} \mathrm{sgn}.$$

Here, $\phi: D_n \to S_n$ is the dihedral group of order $2n$ acting on the vertices of an n-gon, $\psi: D_n \to S_n$ is the action of the dihedral group on the edges of the n-gon, and sgn denotes the sign representation of S_n.

Proof Recall that the core of a weighted, marked graph is the smallest connected subgraph containing all cycles and all vertices of positive weight. The core of a genus 1 tropical curve is either a single vertex of weight 1 or a cycle. If $\Gamma \in \Delta_{1,n} \setminus \Delta_{1,n}^{\mathrm{rep}}$ then the core of Γ cannot be a vertex of weight 1, since then the underlying graph would be a tree, whose leaves would support repeated markings. Therefore the core of Γ is a cycle with all vertices of weight zero. Since $\Gamma \notin \Delta_{1,n}^{\mathrm{rep}}$, each vertex supports at most one marked point. The stability condition then ensures that each vertex supports exactly one marked point. In other words, the combinatorial types of tropical curves that appear outside the repeated marking locus consist of an n-cycle with the markings $\{1, \ldots, n\}$ appearing around that cycle in a specified order. There are $(n-1)!/2$

Figure 5 $\Delta_{1,2}$, shown with two maximal symmetric orbits of simplices, retracts onto the subcomplex $\Delta_{1,2}^{\text{rep}}$, shown in bold.

possible orders τ of $\{1, \ldots, n\}$ up to symmetry, so we have $(n-1)!/2$ such combinatorial types $\mathbf{G}_\tau$.

For $n \geq 3$, each $\mathbf{G}_\tau$ has no nontrivial automorphisms, so the image of the interior of $\sigma^1(\mathbf{G}_\tau)$ in $\Delta_{1,n}$ is an $(n-1)$-disc whose boundary is in $\Delta_{1,n}^{\text{rep}}$. Now it follows from Theorem 1.1 that $\Delta_{1,n}$ has the homotopy type of a wedge of $(n-1)!/2$ spheres of dimension $n-1$.

It remains to identify the representation of S_n on

$$V = H_{n-1}(\Delta_{1,n}/\Delta_{1,n}^{\text{rep}}; \mathbb{Q})$$

obtained by permuting the marked points. We have already shown that V has a basis given by the homology classes of the $(n-1)$-spheres in the wedge $\Delta_{1,n}/\Delta_{1,n}^{\text{rep}}$, which are in bijection with the $(n-1)!/2$ unoriented cyclic orderings of $\{1, \ldots, n\}$. Let $\phi: D_n \to S_n$ be the embedding of the dihedral group as a subgroup of the permutations of the vertices $\{1, \ldots, n\}$ of an n-cycle. Choose left coset representatives $\sigma_1, \ldots, \sigma_k$, where $k = (n-1)!/2$, and write $[\sigma_i]$ for the corresponding basis elements of V. For any $\pi \in S_n$, we have $\pi \sigma_i = \sigma_j \pi'$ for some $\pi' \in D_n$. Then $\pi \cdot [\sigma_i] = \pm[\sigma_j]$, where the sign depends exactly on the sign of the permutation on the *edges* of the n-cycle induced by π'. This is because the ordering of the edges determines the orientation of the corresponding sphere in $\Delta_{1,n}/\Delta_{1,n}^{\text{rep}}$. Therefore the representation of S_n on V is exactly $\text{Ind}_{D_n, \phi}^{S_n} \text{Res}_{D_n, \psi}^{S_n} \text{sgn}$, where the restriction is according to the embedding of $\psi: D_n \to S_n$ into the group of permutations of edges of the n-cycle. $\square$

Remark 5.1 We remark that $\Delta_{1,1}$ and $\Delta_{1,2}$ are contractible. Indeed, $\Delta_{1,1}$ is a point. And the unique cell of $\Delta_{1,2}$ not in $\Delta_{1,2}^{\text{rep}}$ consists of two vertices and two edges between them. Exchanging the edges gives a nontrivial $\mathbb{Z}/2\mathbb{Z}$ automorphism on this cell, which then retracts to $\Delta_{1,2}^{\text{rep}}$. So $\Delta_{1,2}$ is contractible by Theorem 1.1. See Figure 5.

5.2 Proof of Theorem 1.4

The results of §4 allow us to prove Theorem 1.4 from the introduction, restated below.

Theorem 1.4 *For $g \geq 1$ and $2g - 2 + n \geq 2$, there is a natural surjection of chain complexes*

$$C_*(\Delta_{g,n}; \mathbb{Q}) \to G^{(g,n)},$$

decreasing degrees by $2g - 1$, inducing isomorphisms on homology

$$\widetilde{H}_{k+2g-1}(\Delta_{g,n}; \mathbb{Q}) \xrightarrow{\cong} H_k(G^{(g,n)})$$

for all k.

Proof Consider the cellular chain complex $C_*(\Delta_{g,n}, \mathbb{Q})$. It is generated in degree p by $[\mathbf{G}, \omega]$ where $\mathbf{G} \in \Gamma_{g,n}$ and $\omega: E(\mathbf{G}) \to [p] = \{0, 1, \ldots, p\}$ is a bijection. These generators are subject to the relations $[\mathbf{G}, \omega] = \mathrm{sgn}(\sigma)[\mathbf{G}', \omega']$ if there is an isomorphism $\mathbf{G} \to \mathbf{G}'$ inducing the permutation σ of the set $[p]$.

Let $B^{(g,n)}$ be the subcomplex of $C_*(\Delta_{g,n}, \mathbb{Q})$ spanned by the generators $[\mathbf{G}, \omega]$ with at least one nonzero vertex weight. Note that $B^{(g,n)}$ is in fact a subcomplex, since one-edge contractions of graphs with positive vertex weights have positive vertex weights.

Define $A^{(g,n)}$ by the short exact sequence

$$0 \to B^{(g,n)} \to C_*(\Delta_{g,n}, \mathbb{Q}) \to A^{(g,n)} \to 0.$$

Then $A^{(g,n)}$ is isomorphic to the marked graph complex $G^{(g,n)}$, up to shifting degrees by $2g - 1$: a graph with e edges is in degree $e - 1$ in $A^{(g,n)}$ and in degree $e - 2g$ in $G^{(g,n)}$. And $B^{(g,n)}$ is the cellular chain complex associated to $\Delta_{g,n}^{w}$, which is contractible whenever it is nonempty by Theorem 1.1. Therefore, when $\Delta_{g,n}^{w}$ is nonempty then $B^{(g,n)}$ is an acyclic complex, and the theorem follows. $\square$

Remark 5.2 The $n = 0$ case is proved in [14, §4]. The proof here is analogous, but carried out on the level of spaces.

The proof of Theorem 1.4 relied on the contractibility of $\Delta_{g,n}^{w}$. Using other natural contractible subcomplexes in place of $\Delta_{g,n}^{w}$ would produce analogous results. We pause to record a particular version which will be useful for applications in [10].

Let $K^{(g,n)}$ denote the following variant the marked graph complex $G^{(g,n)}$ from §2.4. As a graded vector space, it has generators $[\Gamma, \omega, m]$ for each connected graph Γ of genus g (Euler characteristic $1 - g$) with or without loops, equipped with a total order ω on its set of edges and an *injective* marking function $m: \{1, \ldots, n\} \to V(G)$, such that the valence of each vertex plus the size of its preimage under m is at least 3. These generators are subject to the relations

$$[\Gamma, \omega, m] = \text{sgn}(\sigma)[\Gamma', \omega', m']$$

if there exists an isomorphism of graphs $\Gamma \cong \Gamma'$ that identifies m with m', and under which the edge orderings ω and ω' are related by the permutation σ. The homological degree of $[\Gamma, \omega, m]$ is $e - 2g$. The differential on $K^{(g,n)}$ of $[\Gamma, \omega, m]$ is defined as before (2), with the added convention that if e_i is a loop edge of Γ then we interpret $[\Gamma/e, \omega|_{E(\Gamma) \smallsetminus \{e\}}, \pi_i \circ m]$ as 0.

Proposition 5.3 *Fix $g > 0$ and $n \geq 0$ with $2g - 2 + n > 0$, excluding $(g, n) = (1, 1)$. For all k, we have isomorphisms on homology*

$$H_k(K^{(g,n)}) \xrightarrow{\cong} \widetilde{H}_{k+2g-1}(\Delta_{g,n}; \mathbb{Q}).$$

Proof The complex $K^{(g,n)}$ is isomorphic, after shifting degrees by $2g - 1$, to the relative cellular chain complex $C_*(\Delta_{g,n}, \Delta_{g,n}^{\text{w}} \cup \Delta_{g,n}^{\text{rep}}; \mathbb{Q})$ of the pair of symmetric Δ-complexes $\Delta_{g,n}^{\text{w}} \cup \Delta_{g,n}^{\text{rep}} \subset \Delta_{g,n}$, as defined in §3.2. But $\Delta_{g,n}^{\text{w}} \cup \Delta_{g,n}^{\text{rep}}$ is contractible by Lemma 4.25, so we have identifications

$$H_k(K^{(g,n)}) \cong H_{k+2g-1}(C_*(\Delta_{g,n}, \Delta_{g,n}^{\text{w}} \cup \Delta_{g,n}^{\text{rep}}; \mathbb{Q})) \cong \widetilde{H}_{k+2g-1}(\Delta_{g,n}; \mathbb{Q}).$$

$\square$

5.3 A cellular transfer map

In [14], we showed that $\bigoplus_g H_*(\Delta_g; \mathbb{Q})$ is large and has a rich structure; its dual contains the Grothendieck-Teichmüller Lie algebra $\mathfrak{grt}_1$, and $\dim_{\mathbb{Q}} H_*(\Delta_g; \mathbb{Q})$ grows at least exponentially with g. Here we restate and prove Theorem 1.7, showing that nontrivial homology classes on Δ_g give rise to nontrivial classes with a marked point.

Theorem 1.7 *For $g \geq 2$, there is a natural homomorphism of cellular chain complexes*

$$t: C_*(\Delta_g; \mathbb{Q}) \to C_*(\Delta_{g,1}; \mathbb{Q})$$

which descends to a homomorphism $G^{(g)} \to G^{(g,1)}$ and induces injections $\widetilde{H}_k(\Delta_g; \mathbb{Q}) \hookrightarrow \widetilde{H}_k(\Delta_{g,1}; \mathbb{Q})$ and $H_k(G^{(g)}) \hookrightarrow H_k(G^{(g,1)})$, for all k.

Proof We begin by defining the map $t: C_*(\Delta_g; \mathbb{Q}) \to C_*(\Delta_{g,1}; \mathbb{Q})$. For each vertex v in a stable, vertex weighted graph $\mathbf{G} \in \Gamma_g$, let $\chi(v) = 2w(v) - 2 + \text{val}(v)$. Note that, for a vertex weighted graph $\mathbf{G}$ of genus g, we have $\sum_{v \in V(\mathbf{G})} \chi(v) = 2g - 2$.

Now, consider an element $[\mathbf{G}, \omega]$ of $\Delta_g([p])$, i.e., the isomorphism class of a pair $(\mathbf{G}, \omega)$, where $\mathbf{G}$ is a stable graph of genus g and ω is an ordering of its

$p + 1$ edges. For each vertex v, let $[\mathbf{G}_v, \omega] \in \Delta_{g,1}([p])$ be the stable marked graph with ordered edges obtained by marking the vertex v. The linear map $\mathbb{Q}\Delta_g([p]) \to \mathbb{Q}\Delta_{g,1}([p])$ given by

$$[\mathbf{G}, \omega] \mapsto \sum_v \chi(v)[\mathbf{G}_v, \omega]$$

commutes with the action of S_{p+1}, and hence, after tensoring with $\mathbb{Q}^{\mathrm{sgn}}$ and taking coinvariants, induces a map $t_p \colon C_p(\Delta_g; \mathbb{Q}) \to C_p(\Delta_{g,1}; \mathbb{Q})$.

We claim that $t = \bigoplus_p t_p$ commutes with the differentials on $C_*(\Delta_g; \mathbb{Q})$ and $C_*(\Delta_{g,1}; \mathbb{Q})$. Recall that each differential is obtained as a signed sum over contractions of edges. The claim then follows from the observation that, if v is the vertex obtained by contracting an edge with endpoints v' and v'', then $\chi(v) = \chi(v') + \chi(v'')$. This shows that t is a map of chain complexes. Furthermore, by construction, t maps graphs without loops or vertices of positive weight to marked graphs without loops or vertices of positive weight, and hence takes the subcomplex $G^{(g)}$ into $G^{(g,1)}$.

It remains to show that these maps of chain complexes induce injections $\widetilde{H}_k(\Delta_g; \mathbb{Q}) \hookrightarrow \widetilde{H}_k(\Delta_{g,1}; \mathbb{Q})$ and $H_k(G^{(g)}) \hookrightarrow H_k(G^{(g,1)})$, for all k. To see this, we construct a map $\pi \colon C_*(\Delta_{g,1}; \mathbb{Q}) \to C_*(\Delta_g; \mathbb{Q})$ such that $\pi \circ t$ is multiplication by $2g - 2$.

Let $\widetilde{\pi} \colon \mathbb{Q}\Delta_{g,1}([p]) \to \mathbb{Q}\Delta_g([p])$ be the linear map obtained by forgetting the marked point. More precisely, if forgetting the marked point on $\mathbf{G} \in \Gamma_{g,1}$ yields a stable graph $\mathbf{G}_0 \in \Gamma_g$, then $\widetilde{\pi}$ maps $[\mathbf{G}, \omega]$ to $[\mathbf{G}_0, \omega]$. If forgetting the marked point on $\mathbf{G}$ yields an unstable graph, then $\widetilde{\pi}$ maps $[\mathbf{G}, \omega]$ to 0. (Forgetting the marked point yields an unstable graph exactly when the marking is carried by a weight zero vertex incident to exactly two half-edges.) The resulting linear map $\widetilde{\pi}$ commutes with the action of S_{p+1}, so tensoring with $\mathbb{Q}^{\mathrm{sgn}}$ and taking coinvariants gives $\pi_p \colon C_p(\Delta_{g,1}; \mathbb{Q}) \to C_p(\Delta_g; \mathbb{Q})$. Let $\pi = \bigoplus_p \pi_p$. One then checks directly that $\pi \circ t$ is multiplication by $2g - 2$, and that π commutes with the differentials.

The only subtlety to check is as follows. Suppose $[\mathbf{G}, \omega] \in \Delta_{g,1}([p])$ is such that forgetting the marked point results in an unstable graph. Then the vertex supporting the marked point is incident to exactly two edges e, e'. Then in the expression

$$\partial[\mathbf{G}, \omega] = \sum_{i=0}^{p} (-1)^i [\mathbf{G}/e_i, \omega/e_i]$$

in all but exactly two terms $(-1)^i [\mathbf{G}/e_i, \omega/e_i]$, forgetting the marked point results in an unstable graph. The two exceptional terms correspond to the two edges e, e', and these cancel under π. $\qquad\square$

Corollary 5.4 *We have*

$$\dim H_{2g-1}(\Delta_{g,1}; \mathbb{Q}) > \beta^g + constant$$

for any $\beta < \beta_0$, *where* $\beta_0 \approx 1.32\ldots$ *is the real root of* $t^3 - t - 1 = 0$.

Proof The analogous result for Δ_g is proved in [14]; now combine with Theorem 1.7.   □

Remark 5.5 As mentioned in the introduction, the splitting on the level of cohomology was constructed earlier in [52]. The splitting they construct is induced by Lie bracket with a graph L which is a single edge between two vertices, tracing through definitions, this is (at least up to signs and grading conventions) dual to the restriction of our t to a chain map $G^{(g)} \to G^{(g,1)}$.

Remark 5.6 For all n, there is a natural map $M_{g,n+1}^{\mathrm{trop}} \to M_{g,n}^{\mathrm{trop}}$ obtained by forgetting the marked point and stabilizing [2]. When $n = 0$, the preimage of $\bullet_g$ is $\bullet_{g,1}$, so there is an induced map on the link $\Delta_{g,1} \to \Delta_g$. This continuous map of topological spaces does not come from a map of symmetric Δ-complexes (because some cells of $\Delta_{g,1}$ are mapped to cells of lower dimension in Δ_g), but one can check that the pushforward on rational homology is induced by π. When $n > 1$, the preimage of $\bullet_{g,n}$ includes graphs other than $\bullet_{g,n+1}$, and there is no induced map from $\Delta_{g,n+1}$ to $\Delta_{g,n}$.

6 Applications to, and from, $\mathcal{M}_{g,n}$

6.1 The Boundary Complex of $\mathcal{M}_{g,n}$

We recall that the dual complex $\Delta(D)$ of a normal crossings divisor D in a smooth, separated Deligne–Mumford (DM) stack X is naturally defined as a symmetric Δ-complex [14, §5.2]. Over $\mathbb{C}$, for each $p \geq -1$, $\Delta(D)_p$ is the set of equivalence classes of pairs (x, σ), where x is a point in a stratum of codimension p in D and σ is an ordering of the $p + 1$ analytic branches of D that meet at x. The equivalence relation is generated by paths within strata: if there is a path from x to x' within the codimension p stratum and a continuous assignment of orderings of branches along the path, starting at (x, σ) and ending at (x', σ'), then we set $(x, \sigma) \sim (x', \sigma')$.

This dual complex can equivalently be defined (and, more generally, over fields other than $\mathbb{C}$), using normalization and iterated fiber product. Let $\tilde{D} \to D$ be the normalization of D and write

$$\widetilde{D}^{[p]} = \widetilde{D} \times_X \cdots \times_X \widetilde{D}$$

for the $(p + 1)$-fold iterated fiber product. Define $\widetilde{D}([p]) \subset \widetilde{D}^{[p]}$ as the open subvariety consisting of $(p + 1)$-tuples of pairwise distinct points in $\widetilde{D}^{[p]}$ that all lie over the same point of D. We can then define $\Delta(D)_p$ to be the set of irreducible components of $\widetilde{D}([p])$. (Note that, over $\mathbb{C}$, a point of $\widetilde{D}([p])$ encodes exactly the same data as a point in the codimension p stratum together with an ordering of the $p + 1$ analytic branches of D at that point.) For further details, see [14, §5].

If X is proper then the simple homotopy type of this dual complex depends only on the open complement $X \smallsetminus D$ [29, 46], and its reduced rational homology is naturally identified with the top weight cohomology of $X \smallsetminus D$. More precisely, if X has pure dimension d, then

$$\widetilde{H}_{k-1}(\Delta(D); \mathbb{Q}) \cong \mathrm{Gr}_{2d}^W H^{2d-k}(X \smallsetminus D; \mathbb{Q}). \tag{11}$$

See [14, Theorem 5.8].

Most important for our purposes is the special case where $X = \overline{\mathcal{M}}_{g,n}$ is the Deligne–Mumford stable curves compactification of $\mathcal{M}_{g,n}$ and $D = \overline{\mathcal{M}}_{g,n} \smallsetminus \mathcal{M}_{g,n}$ is the boundary divisor.

Theorem 6.1 *The dual complex of the boundary divisor in the moduli space of stable curves with marked points* $\Delta(\overline{\mathcal{M}}_{g,n} \smallsetminus \mathcal{M}_{g,n})$ *is* $\Delta_{g,n}$.

Proof Modulo the translation between smooth generalized cone complexes and symmetric Δ-complexes, this is one of the main results of [2], to which we refer for a thorough treatment. The details of the construction for $n = 0$ are also explained in [14, Corollaries 5.6 and 5.7], and the general case is similar. □

As an immediate consequence of Theorem 6.1 and (11), the reduced rational homology of $\Delta_{g,n}$ agrees with the top weight cohomology of $\mathcal{M}_{g,n}$.

Corollary 6.2 *There is a natural isomorphism*

$$\mathrm{Gr}_{6g-6+2n}^W H^{6g-6+2n-k}(\mathcal{M}_{g,n}; \mathbb{Q}) \xrightarrow{\sim} \widetilde{H}_{k-1}(\Delta_{g,n}; \mathbb{Q}),$$

identifying the reduced rational homology of $\Delta_{g,n}$ *with the top graded piece of the weight filtration on the cohomology of* $\mathcal{M}_{g,n}$.

In the case $g = 1$, we have a complete understanding of the rational homology of $\Delta_{1,n}$, from Theorem 1.2. Thus we immediately deduce a similarly complete understanding of the top weight cohomology of $\mathcal{M}_{1,n}$, stated as Corollary 1.3 in the introduction.

Corollary 1.3   *The top weight cohomology of* $\mathcal{M}_{1,n}$ *is supported in degree n, with rank* $(n-1)!/2$, *for* $n \geq 3$. *Moreover, the representation of* S_n *on* $\mathrm{Gr}_{2n}^W H_n(\mathcal{M}_{1,n}; \mathbb{Q})$ *induced by permuting marked points is*

$$\mathrm{Ind}_{D_n, \phi}^{S_n} \mathrm{Res}_{D_n, \psi}^{S_n} \mathrm{sgn}.$$

Remark 6.3   The fact that the top weight cohomology of $\mathcal{M}_{1,n}$ is supported in degree n can also be seen without tropical methods, as follows. The rational cohomology of a smooth Deligne–Mumford stack agrees with that of its coarse moduli space, and the coarse space $M_{1,n}$ is affine. To see this, note that $M_{1,1}$ is affine, and the forgetful map $M_{g,n+1} \to M_{g,n}$ is an affine morphism for $n \geq 1$. It follows that $M_{1,n}$ has the homotopy type of an n-dimensional CW-complex, by [3, 36], and hence $H^*(\mathcal{M}_{1,n}; \mathbb{Q})$ is supported in degrees less than or equal to n. The weights on H^k are always between 0 and $2k$, so the top weight $2n$ can appear only in degree n.

Remark 6.4   Getzler has calculated an expression for the S_n-equivariant Serre characteristic of $\mathcal{M}_{1,n}$ [25, (5.6)]. Since the top weight cohomology is supported in a single degree, it is determined as a representation by this equivariant Serre characteristic. We do not know how to deduce Corollary 1.3 directly from Getzler's formula. However, C. Faber has shown a formula for $\mathrm{Gr}_{2n}^W H_n(\mathcal{M}_{1,n}; \mathbb{Q})$, as an S_n-representation, that is derived from [24, Theorem 2.5]. See [10, Theorem 1.5].

Remark 6.5   Petersen explains that it is possible to adapt the methods from [47] to recover the fact that the top weight cohomology of $\mathcal{M}_{1,n}$ has rank $(n-1)!/2$, using the Leray spectral sequence for $\mathcal{M}_{1,n} \to \mathcal{M}_{1,1}$ and the Eichler–Shimura isomorphism [48].

Using Corollary 6.2 and the transfer homomorphism from §5.3, we also deduce an exponential growth result for top-weight cohomology of $\mathcal{M}_{g,1}$, stated as Corollary 1.8 in the introduction.

Corollary 1.8   *We have*

$$\dim \mathrm{Gr}_{6g-4}^W H^{4g-4}(\mathcal{M}_{g,1}; \mathbb{Q}) > \beta^g + constant$$

for any $\beta < \beta_0$, *where* $\beta_0 \approx 1.32\ldots$ *is the real root of* $t^3 - t - 1 = 0$.

Proof   This follows from Corollary 5.4 and Corollary 6.2.   □

Remark 6.6   Corollary 1.8 above, and the existence of a natural injection $H_k(\Delta_g; \mathbb{Q}) \to H_k(\Delta_{g,1}; \mathbb{Q})$, may also be deduced purely algebro-geometrically. Indeed, pulling back along the forgetful map $\mathcal{M}_{g,1} \to \mathcal{M}_g$ and composing with cup product with the Euler class is injective on rational singular cohomology. This is because further composing with the Gysin map (proper push-forward) induces multiplication by $2g - 2$ on $H^*(\mathcal{M}_g; \mathbb{Q})$. Furthermore, this injection maps top weight cohomology into top weight cohomology, because cup product with the Euler class increases weight by 2. Identifying top weight cohomology of $\mathcal{M}_{g,n}$ with rational homology of $\Delta_{g,n}$ for $n = 0$ and $n = 1$ then gives a natural injection $H_k(\Delta_g; \mathbb{Q}) \to H_k(\Delta_{g,1}; \mathbb{Q})$, as claimed. Presumably, this map agrees with the one defined in §5.3 up to some normalization constant.

The following is a strengthening of Corollary 1.9.

Corollary 6.7 *Let* Mod_g^1 *denote the mapping class group of a connected oriented 2-manifold of genus g, with one marked point, and let G be any group fitting into an extension*

$$\mathbb{Z} \to G \to \mathrm{Mod}_g^1,$$

Then

$$\dim H^{4g-3}(G; \mathbb{Q}) > \beta^g + constant$$

for any $\beta < \beta_0$, *where* $\beta_0 \approx 1.32\ldots$ *is the real root of* $t^3 - t - 1 = 0$.

In particular, $G = \mathrm{Mod}_{g,1}$, *the mapping class group with one parametrized boundary component, satisfies this dimension bound on its cohomology.*

Proof Let us write $B\mathrm{Mod}_g^1$ for the classifying space of the discrete group Mod_g^1, i.e., a $K(\pi, 1)$ for this group. Its homology is the group homology of Mod_g^1, and its rational cohomology is canonically isomorphic to $H^*(\mathcal{M}_{g,1}; \mathbb{Q})$. In particular $\dim H^{4g-4}(B\mathrm{Mod}_g^1; \mathbb{Q})$ is bounded below by Corollary 1.8.

The extension is classified by a class $e \in H^2(\mathrm{Mod}_g^1; \mathbb{Z})$, which is the first Chern class of some principal $U(1)$-bundle $\pi \colon P \to B\mathrm{Mod}_g^1$, and we have $P \simeq BG$. Part of the Gysin sequence for π looks like

$$\cdots \xrightarrow{e\cup} H^{4g-3}(B\mathrm{Mod}_g^1) \xrightarrow{\pi^*} H^{4g-3}(BG) \xrightarrow{\pi_*} H^{4g-4}(B\mathrm{Mod}_g^1)$$
$$\xrightarrow{e\cup} H^{4g-2}(B\mathrm{Mod}_g^1) \xrightarrow{\pi^*} \cdots,$$

and by Harer's theorem [28] that Mod_g^1 is a virtual duality group of virtual cohomological dimension $4g - 3$ we get $H^{4g-2}(B\mathrm{Mod}_g^1; \mathbb{Q}) = 0$, and hence a surjection $\pi_* \colon H^{4g-3}(BG) \to H^{4g-4}(B\mathrm{Mod}_g^1)$.

It is well-known that $G = \mathrm{Mod}_{g,1}$ fits into such an extension, where the $\mathbb{Z}$ is generated by Dehn twist along a boundary-parallel curve (see e.g., [22, Proposition 3.19]). $\qquad\square$

Remark 6.8   By methods similar to those outlined in [14, §6], the dimension bounds in the Corollaries above may be upgraded to explicit injections of graded vector spaces, from the Grothendieck–Teichmüller Lie algebra to $\prod_{g\geq 3} H_{4g-4}(\mathcal{M}_{g,1};\mathbb{Q})$ and $\prod_{g\geq 3} H_{4g-3}(\mathrm{Mod}_{g,1};\mathbb{Q})$, respectively.

6.2 Support of the Rational Homology of $\Delta_{g,n}$

In [14], we observed that known vanishing results for the cohomology of $\mathcal{M}_g$ imply that the reduced rational homology of Δ_g is supported in the top $g-2$ degrees, and that the homology of $G^{(g)}$ vanishes in negative degrees. Here we prove the analogous result with marked points, stated as Theorem 1.6 in the introduction, using Harer's computation of the virtual cohomological dimension of $\mathcal{M}_{g,n}$ from [28].

Theorem 1.6   *The marked graph homology $H_k(G^{(g,n)})$ vanishes for $k <$* $\max\{0, n-2\}$. *Equivalently, $\widetilde{H}_k(\Delta_{g,n};\mathbb{Q})$ vanishes for $k < \max\{2g-1, 2g-$* $3+n\}$.

Proof   The case $n = 0$ is proved in [14] and the case $g = 1$ follows from Theorem 1.3.

Suppose $g \geq 2$ and $n \geq 1$. By [28], the virtual cohomological dimension of $\mathcal{M}_{g,n}$ is $4g-4+n$. Furthermore, when $n = 1$, we have $H^{4g-3}(\mathcal{M}_{g,1};\mathbb{Q}) = 0$, by [11]. Therefore, the top weight cohomology of $\mathcal{M}_{g,n}$ is supported in degrees less than $4g-3+n-\delta_{1,n}$, where $\delta_{i,j}$ is the Kronecker δ-function. By Corollary 6.2, it follows that $\widetilde{H}_k(\Delta_{g,n};\mathbb{Q})$ is supported in degrees less than $\max\{2g-1, 2g+n-3\}$, as required. $\square$

It would be interesting to have a proof of this vanishing result using the combinatorial topology of $\Delta_{g,n}$.

7 Remarks on Stability

It is natural to ask whether the homology of $\Delta_{g,n}$ can be related to known instances of *homological stability* for the complex moduli space of curves $\mathcal{M}_{g,n}$ and for the free group F_g. Here, we comment briefly on the reasons that the tropical moduli space $\Delta_{g,n}$ relates to both $\mathcal{M}_{g,n}$ and F_g.

Homological stability has been an important point of view in the understanding of $\mathcal{M}_{g,n}$; we are referring to the fact that the cohomology group $H^k(\mathcal{M}_{g,n};\mathbb{Q})$ is independent of g as long as $g \geq 3k/2 + 1$ [27, 34, 7]. The structure of the rational cohomology in this stable range was famously conjectured by Mumford, for $n = 0$, and proved by Madsen and Weiss [45]; see [41, Proposition 2.1] for the extension to $n > 0$. There are certain *tautological classes* $\kappa_i \in H^{2i}(\mathcal{M}_{g,n})$ and $\psi_j \in H^2(\mathcal{M}_{g,n})$, and the induced map

$$\mathbb{Q}[\kappa_1, \kappa_2, \ldots] \otimes \mathbb{Q}[\psi_1, \ldots, \psi_n] \to H^*(\mathcal{M}_{g,n}; \mathbb{Q})$$

is an isomorphism in the "stable range" of degrees up to $2(g - 1)/3$.

A very similar homological stability phenomenon happens for *automorphisms of free groups*. If F_g denotes the free group on g generators and $\mathrm{Aut}(F_g)$ is its automorphism group, then Hatcher and Vogtmann [31] proved that the group cohomology of $H^k(\mathrm{Aut}(F_g))$ is independent of g as long as $g \gg k$. In [23] it was proved that an analogue of the Madsen–Weiss theorem holds for these groups: the rational cohomology $H^k(B\mathrm{Aut}(F_g); \mathbb{Q})$ vanishes for $g \gg k \geq 1$.

The tropical moduli space $\Delta_{g,n}$ is closely related to *both* of these objects. On the one hand its reduced rational homology is identified with the top weight cohomology of $\mathcal{M}_{g,n}$. On the other hand it is also closely related to $\mathrm{Aut}(F_g)$, as we shall now briefly explain.

7.1 Relationship with Automorphism Groups of Free Groups

Let us follow the terminology of [8] and call a tropical curve *pure* if all its vertices have zero weights. Isomorphism classes of pure tropical curves are parametrized by an open subset

$$\Delta_{g,n}^{\mathrm{pure}} = |\Delta_{g,n}| \setminus |\Delta_{g,n}^w| \subset |\Delta_{g,n}|.$$

Its points are isometry classes of triples (G, m, w, ℓ) with $(G, m, w) \in \Gamma_{g,n}$, such that $w = 0$ and $\ell(e) > 0$ for all $e \in E(G)$. These spaces are related to Culler and Vogtmann's "outer space" [20] and its versions with marked points, e.g., [31]. Indeed, for $n = 1$ for example, outer space $X_{g,1}$ can be regarded as the space of isometry classes of triples (G, m, w, ℓ, h) where $(G, m, w) \in \Gamma_{g,1}$ are as before, with $w = 0$ and $\ell(e) > 0$ for all e, and $h: F_g \to \pi_1(G, m(1))$ is a specified isomorphism between the free group F_g on g generators and the fundamental group of G at the point $m(1) \in V(G)$. The group $\mathrm{Aut}(F_g)$ acts on $X_{g,1}$ by changing h, and the forgetful map $(G, m, w, \ell, h) \mapsto (G, m, w, \ell)$ factors over a homeomorphism

$$X_{g,1} / \mathrm{Aut}(F_g) \xrightarrow{\approx} \Delta_{g,1}^{\mathrm{pure}}.$$

Since $X_{g,1}$ is contractible ([20, 31]) and the stabilizer of any point in $X_{g,1}$ is finite, there is a map

$$B\mathrm{Aut}(F_g) \to \Delta_{g,1}^{\mathrm{pure}}$$

which induces an isomorphism in rational cohomology. (Recall $B\mathrm{Aut}(F_g)$ denotes the *classifying space* of the discrete group $\mathrm{Aut}(F_g)$. It is a $K(\pi, 1)$ space whose singular cohomology is isomorphic to the group cohomology of $\mathrm{Aut}(F_g)$.) More generally, there are groups $\Gamma_{g,n}$ defined up to isomorphism by $\Gamma_{g,0} \cong \mathrm{Out}(F_g)$, and $\Gamma_{g,n} = \mathrm{Aut}(F_g) \ltimes F_g^{n-1}$ for $n > 0$, [30, 32]. The groups $\Gamma_{g,n}$ are also isomorphic to the groups denoted $A_{g,n}$ in [33].

By a similar argument as above, which ultimately again rests on contractibility of outer space, the space $\Delta_{g,n}^{\mathrm{pure}}$ is a rational model for the group $\Gamma_{g,n}$, in the sense that there is a map

$$B\Gamma_{g,n} \to \Delta_{g,n}^{\mathrm{pure}}$$

inducing an isomorphism in rational homology. A similar rational model for $B\Gamma_{g,n}$ was considered in [17, §6], and may in fact be identified with a deformation retract of $\Delta_{g,n}^{\mathrm{pure}}$. (We shall not need this last fact, but let us nevertheless point out that the subspace $Q_{g,n} \subset \Delta_{g,n}^{\mathrm{pure}}$ defined as parametrizing stable tropical curves with zero vertex weights in which the n marked points are on the *core*, as defined in §2.1, is a strong deformation retract of $\Delta_{g,n}^{\mathrm{pure}}$. The deformation retraction is given by uniformly shrinking the non-core edges and lengthening the core edges, where the rate of lengthening of each core edge is proportional to its length. This $Q_{g,n}$ is homeomorphic to the space considered in [17, §6] under the same notation.)

Therefore, the inclusion $\iota \colon \Delta_{g,n}^{\mathrm{pure}} \subset \Delta_{g,n}$ induces a map in rational homology

$$\widetilde{H}_*(\Gamma_{g,n}; \mathbb{Q}) \cong \widetilde{H}_*(\Delta_{g,n}^{\mathrm{pure}}; \mathbb{Q}) \xrightarrow{\iota_*} \widetilde{H}_*(\Delta_{g,n}; \mathbb{Q})$$

$$\cong \mathrm{Gr}_{6g-6+2n}^W H^{6g-7+2n-*}(\mathcal{M}_{g,n}; \mathbb{Q}). \tag{12}$$

By compactness of $\Delta_{g,n}$ and, for $g > 0$ and $2g+n > 3$, contractibility of $\Delta_{g,n}^W$, this map may equivalently be described as the linear dual of the canonical map from compactly supported cohomology to cohomology of $\Delta_{g,n}^{\mathrm{pure}}$. For $n = 0$ the map ι_* in particular gives a map

$$\widetilde{H}_*(\mathrm{Out}(F_g); \mathbb{Q}) \to \mathrm{Gr}_{6g-6}^W H^{6g-7-*}(\mathcal{M}_g; \mathbb{Q}).$$

It is intriguing to note, as emphasized to us by a referee, that group homology of $\Gamma_{g,n}$ is also calculated by a kind of graph complex, although different from $G^{(g,n)}$. For $n = 0$ for instance, this is the "Lie graph complex" calculating group cohomology of $\mathrm{Out}(F_r)$ (see [38, 39] and [21, Proposition 21, Theorem 2]). In that complex, vertices of valence m are labeled by elements of $\mathrm{Lie}(m-1)$, operations of arity $m-1$ in the Lie operad. The complex $G^{(g)}$ and the Lie graph complex both have boundary homomorphism involving contraction of edges, but the Lie graph complex calculates cohomology $H^*(\mathrm{Out}(F_g); \mathbb{Q}) = H^*(\Delta_g^{\mathrm{pure}}; \mathbb{Q})$. The dual of the Lie graph complex then calculates homology $H_*(\mathrm{Out}(F_g); \mathbb{Q}) = H_*(\Delta_g^{\mathrm{pure}}; \mathbb{Q})$, but the differential on this dual involves expanding vertices, instead of collapsing edges. The homomorphism ι_* therefore seems a bit mysterious from this point of view, going from homology of a graph chain complex to cohomology of a graph cochain complex. It would be interesting to understand this better on a chain/cochain level, but at the moment we have nothing substantial to say about it.

7.2 (Non-)triviality of ι_*

Known properties of $\mathcal{M}_{g,n}$ and $\Delta_{g,n}$ severely limit the possible degrees in which (12) may be non-trivial. Indeed, by Theorem 1.6, the reduced homology of $\Delta_{g,n}$ vanishes in degrees below $\max\{2g-1, 2g-3+n\}$. On the other hand, $H_*(\Gamma_{g,n}; \mathbb{Q})$ is supported in degrees at most $2g-3+n$ by [17, Remark 4.2]. It follows that ι_* vanishes in all degrees except possibly $* = 2g-3+n$, for $n > 0$, where it gives a homomorphism

$$H_{2g-3+n}(\Delta_{g,n}^{\mathrm{pure}}; \mathbb{Q}) \to H_{2g-3+n}(\Delta_{g,n}; \mathbb{Q}).$$

In this degree, the homomorphism is *not* always trivial. Indeed, for $g = 1$, the domain $H_{n-1}(\Delta_{1,n}^{\mathrm{pure}}; \mathbb{Q})$ is one-dimensional and the map into $H_{n-1}(\Delta_{1,n}; \mathbb{Q}) \cong \mathbb{Q}^{(n-1)!/2}$ is injective.

Proposition 7.1 *For $n \geq 3$ odd, the map*

$$H_{n-1}(\Delta_{1,n}^{\mathrm{pure}}; \mathbb{Q}) \to H_{n-1}(\Delta_{1,n}; \mathbb{Q})$$

is nontrivial.

Proof sketch The subspace $\Delta_{1,n}^{\mathrm{pure}} \subset \Delta_{1,n}$ is homotopy equivalent to the space $Q_{1,n}$, which is the orbit space $((S^1)^n)/O(2)$, where $O(2)$ acts by rotating and reflecting all S^1 coordinates. Moreover, $\Delta_{1,n}$ is homotopy equivalent to the orbit space $((S^1)^n/R)/O(2)$, where $R \subset (S^1)^n$ is the "fat diagonal" consisting of points where two coordinates agree. $(S^1)^n/R$ denotes the quotient space obtained by collapsing R, and the homotopy equivalence $\Delta_{1,n} \simeq ((S^1)^n/R)/O(2)$ follows from the contractibility of the bridge locus.

In this description, the inclusion of $\Delta_{1,n}^{\mathrm{pure}} \hookrightarrow \Delta_{1,n}$ is modeled by the obvious quotient map collapsing R to a point. We have the homeomorphism

$(S^1)^n/O(2) = (S^1)^{n-1}/O(1)$ and the map $H_{n-1}(\Delta_{1,n}^{\mathrm{pure}}; \mathbb{Q}) \to H_{n-1}(\Delta_{1,n}; \mathbb{Q})$ becomes identified with the $O(1)$ coinvariants of the map $\mathbb{Q} \cong H_{n-1}((S^1)^{n-1}; \mathbb{Q}) \to H_{n-1}((S^1)^n/R; \mathbb{Q}) \cong \mathbb{Q}^{(n-1)!}$, which sends the fundamental class of $(S^1)^{n-1}$ to the "diagonal" class, i.e., the sum of all fundamental classes of S^{n-1} in our description of $\Delta_{1,n}$ as a wedge of (S^{n-1})'s. $\qquad\square$

7.3 Stable Homology

One of the initial motivations for this paper was to use the tropical moduli space to provide a direct link between moduli spaces of curves, automorphism groups of free groups, and their homological stability properties. In light of homological stability for $\Gamma_{g,n}$ and $\mathcal{M}_{g,n}$, it is natural to try to form some kind of direct limit of $\Delta_{g,n}$ as $g \to \infty$. For $n = 1$, there is indeed a map $\Delta_{g,1} \to \Delta_{g+1,1}$, which sends a tropical curve $G \in \Delta_{g,1}$ to "$G \vee S^1$". More precisely, the map adds a single loop to G at the marked point, and appropriately normalizes edge lengths (for example, multiply all edge lengths in G by $\frac{1}{2}$ and give the loop length $\frac{1}{2}$). This map fits with the stabilization map for $B\mathrm{Aut}(F_g)$ into a commutative diagram of spaces,

$$
\begin{array}{ccccc}
B\mathrm{Aut}(F_g) & \xrightarrow{\;\simeq_{\mathbb{Q}}\;} & \Delta_{g,1}^{\mathrm{pure}} & \longrightarrow & \Delta_{g,1} \\
\downarrow & & \downarrow & & \downarrow \\
B\mathrm{Aut}(F_{g+1}) & \xrightarrow[\simeq_{\mathbb{Q}}]{} & \Delta_{g+1,1}^{\mathrm{pure}} & \longrightarrow & \Delta_{g+1,1}.
\end{array}
\tag{13}
$$

The leftmost vertical arrow is studied in [31], where it is shown to induce an isomorphism in homology in degree up to $(g - 3)/2$.

For the outer automorphism group $\mathrm{Out}(F_g)$, there is a similar comparison diagram

$$
\begin{array}{ccccc}
B\mathrm{Aut}(F_g) & \xrightarrow{\;\simeq_{\mathbb{Q}}\;} & \Delta_{g,1}^{\mathrm{pure}} & \longrightarrow & \Delta_{g,1} \\
\downarrow & & \downarrow & & \downarrow \\
B\mathrm{Out}(F_g) & \xrightarrow[\simeq_{\mathbb{Q}}]{} & \Delta_{g,0}^{\mathrm{pure}} & \longrightarrow & \Delta_{g,0}.
\end{array}
$$

In light of this relationship between $\mathcal{M}_{g,n}$ and $\mathrm{Out}(F_g)$ and $\Delta_{g,n}$ and $\Delta_{g,n}^{\mathrm{pure}}$, and in light of [45] and [23], it is tempting to ask about a limiting cohomology of $\Delta_{g,1}$ as $g \to \infty$. However, this limit seems to be of a different nature from the corresponding limits for $B\mathrm{Aut}(F_g)$ and $\mathcal{M}_{g,n}$, as in the observation below.

Observation 7.2   *The stabilization maps* $\Delta_{g,1} \to \Delta_{g+1,1}$ *in* (13) *are nullhomotopic. Hence the limiting cohomology vanishes.*

Proof sketch   Recall that the map $\Delta_{g,1} \to \Delta_{g+1,1}$ sends $G \mapsto G \vee S^1$, with total edge length of $G \subset G \vee S^1$ being $\frac{1}{2}$ and the loop $S^1 \subset G \vee S^1$ also having length $\frac{1}{2}$. Continuously changing the length distribution from $(\frac{1}{2}, \frac{1}{2})$ to $(0, 1)$ defines a homotopy which starts at the stabilization map and ends at the constant map $\Delta_{g,1} \to \Delta_{g+1,1}$ sending any weighted tropical curve Γ to a loop of length 1 based at a vertex of weight g. ☐

Remark 7.3   It is natural to ask if the homology groups $H_*(\Delta_{g,n}; \mathbb{Q})$, viewed as S_n-representations, may fit into the framework of *representation stability* from [9]. First, if we fix both k and g, then $H_k(\Delta_{g,n}; \mathbb{Q})$ vanishes for $n \gg 0$. This follows from the contractibility of $\Delta_{g,n}^{\mathrm{br}}$, because there is a natural CW complex structure on $\Delta_{g,n}/\Delta_{g,n}^{\mathrm{br}}$ in which all positive dimensional cells have dimension at least $n - 5g + 5$. See [13, Theorem 1.3 and Claim 9.3]. This vanishing may be compared with the stabilization with respect to marked points for homology of the pure mapping class group, which says that, for fixed g and k, the sequence $H_k(\mathcal{M}_{g,n}; \mathbb{Q})$ is representation stable [35].

Now suppose we fix g and a small *codegree* k and study the sequence of S_n-representations $H_{3g-4+n-k}(\Delta_{g,n}; \mathbb{Q})$. We still do not expect representation stability to hold in general: as shown in [9], representation stability implies polynomial dimension growth, whereas already the dimension of $H_{n-1}(\Delta_{1,n}; \mathbb{Q})$ grows super exponentially with n.

Nevertheless, Wiltshire-Gordon points out that $H_{n-1}(\Delta_{1,n}; \mathbb{Q})$ admits a natural filtration whose graded pieces are representation stable. Contracting the repeated marking locus gives a homotopy equivalence between $\Delta_{1,n}$ and the one point compactification of a disjoint union of $(n-1)!/2$ open balls. These balls are the connected components of the configuration space of n distinct labeled points on a circle, up to rotation and reflection. There is then a natural identification of $H_{n-1}(\Delta_{1,n}; \mathbb{Q}) \otimes \mathrm{sgn}$ with H^0 of this configuration space. By [42, 44, 53], this H^0 carries a natural filtration, induced by localization on a larger configuration space with S^1-action whose graded pieces are finitely generated FI-modules.

References

[1] E. Arbarello, M. Cornalba, and P. Griffiths, *Geometry of algebraic curves. Volume II*, Grundlehren der Mathematischen Wissenschaften, vol. 268, Springer, Heidelberg, 2011.

[2] D. Abramovich, L. Caporaso, and S. Payne, *The tropicalization of the moduli space of curves*, Ann. Sci. Éc. Norm. Supér. (4) **48** (2015), no. 4, 765–809.

[3] A. Andreotti and T. Frankel, *The Lefschetz theorem on hyperplane sections*, Ann. of Math. (2) **69** (1959), 713–717.

[4] A. Andersson, T. Willwacher, and M. Živković, *Oriented hairy graphs and moduli spaces of curves*, preprint, arXiv:2005.00439v1, 2020.

[5] S. Brannetti, M. Melo, and F. Viviani, *On the tropical Torelli map*, Adv. Math. **226** (2011), no. 3, 2546–2586.

[6] D. Bar-Natan and B. McKay, *Graph cohomology – an overview and some computations*, available at www.math.toronto.edu/~drorbn/papers/GCOC/GCOC.ps.

[7] S. Boldsen, *Improved homological stability for the mapping class group with integral or twisted coefficients*, Math. Z. **270** (2012), 297–329.

[8] L. Caporaso, *Algebraic and tropical curves: comparing their moduli spaces*, Handbook of moduli. Vol. I, Adv. Lect. Math. (ALM), vol. 24, Int. Press, Somerville, MA, 2013, pp. 119–160.

[9] T. Church and B. Farb, *Representation theory and homological stability*, Adv. Math. **245** (2013), 250–314.

[10] M. Chan, C. Faber, S. Galatius, and S. Payne, *The S_n-equivariant top-weight Euler characteristic of $\mathcal{M}_{g,n}$*, preprint, arXiv:1904.06367, 2019.

[11] T. Church, B. Farb, and A. Putman, *The rational cohomology of the mapping class group vanishes in its virtual cohomological dimension*, Int. Math. Res. Not. IMRN (2012), no. 21, 5025–5030.

[12] _____, *A stability conjecture for the unstable cohomology of* $SL_n\mathbb{Z}$, *mapping class groups, and* Aut(F_n), Algebraic topology: applications and new directions, Contemp. Math., vol. 620, Amer. Math. Soc., Providence, RI, 2014, pp. 55–70.

[13] M. Chan, S. Galatius, and S. Payne, *Tropicalization of the moduli space of curves II: Topology and applications*, preprint, arXiv:1604.03176, 2016.

[14] _____, *Tropical curves, graph complexes, and top weight cohomology of* $\mathcal{M}_g$, J. Amer. Math. Soc. **34** (2021), no. 2, 565–594.

[15] J. Conant, F. Gerlits, and K. Vogtmann, *Cut vertices in commutative graphs*, Q. J. Math. **56** (2005), no. 3, 321–336.

[16] M. Chan, *Topology of the tropical moduli spaces* $M_{2,n}$, Beitr. Algebra Geom. (2021). https://doi.org/10.1007/s13366-021-00563-6.

[17] J. Conant, A. Hatcher, M. Kassabov, and K. Vogtmann, *Assembling homology classes in automorphism groups of free groups*, Comment. Math. Helv. **91** (2016), no. 4, 751–806.

[18] J. Conant, M. Kassabov, and K. Vogtmann, *Hairy graphs and the unstable homology of* Mod(g,s), Out(F_n) *and* Aut(F_n), J. Topol. **6** (2013), no. 1, 119–153.

[19] _____, *Higher hairy graph homology*, Geom. Dedicata **176** (2015), 345–374.

[20] M. Culler and K. Vogtmann, *Moduli of graphs and automorphisms of free groups*, Invent. Math. **84** (1986), no. 1, 91–119.

[21] J. Conant and K. Vogtmann, *On a theorem of Kontsevich*, Algebr. Geom. Topol. **3** (2003), 1167–1224.

[22] B. Farb and D. Margalit, *A primer on mapping class groups*, Princeton Mathematical Series, vol. 49, Princeton University Press, Princeton, NJ, 2012.

[23] S. Galatius, *Stable homology of automorphism groups of free groups*, Ann. of Math. (2) **173** (2011), no. 2, 705–768.

[24] E. Getzler, *The semi-classical approximation for modular operads*, Comm. Math. Phys. **194** (1998), no. 2, 481–492.

[25] ———, *Resolving mixed Hodge modules on configuration spaces*, Duke Math. J. **96** (1999), no. 1, 175–203.

[26] E. Getzler and M. Kapranov, *Modular operads*, Compositio Math. **110** (1998), no. 1, 65–126.

[27] J. Harer, *Stability of the homology of the mapping class groups of orientable surfaces*, Ann. of Math. (2) **121** (1985), no. 2, 215–249.

[28] ———, *The virtual cohomological dimension of the mapping class group of an orientable surface*, Invent. Math. **84** (1986), no. 1, 157–176.

[29] A. Harper, *Factorization for stacks and boundary complexes*, preprint, arXiv:1706.07999, 2017.

[30] A. Hatcher, *Homological stability for automorphism groups of free groups*, Comment. Math. Helv. **70** (1995), no. 1, 39–62.

[31] A. Hatcher and K. Vogtmann, *Cerf theory for graphs*, J. London Math. Soc. (2) **58** (1998), no. 3, 633–655.

[32] ———, *Homology stability for outer automorphism groups of free groups*, Algebr. Geom. Topol. **4** (2004), 1253–1272 (electronic).

[33] A. Hatcher and N. Wahl, *Stabilization for mapping class groups of 3-manifolds*, Duke Math. J. **155** (2010), no. 2, 205–269.

[34] N. Ivanov, *On the homology stability for Teichmüller modular groups: closed surfaces and twisted coefficients*, Mapping class groups and moduli spaces of Riemann surfaces (Göttingen, 1991/Seattle, WA, 1991), Contemp. Math., vol. 150, Amer. Math. Soc., Providence, RI, 1993, pp. 149–194.

[35] R. Jiménez Rolland, *Representation stability for the cohomology of the moduli space $\mathcal{M}_g^n$*, Algebr. Geom. Topol. **11** (2011), no. 5, 3011–3041.

[36] K. Karčjauskas, *A generalized Lefschetz theorem*, Funkcional. Anal. i Priložen. **11** (1977), no. 4, 80–81.

[37] M. Kontsevich and Yu. Manin, *Gromov-Witten classes, quantum cohomology, and enumerative geometry*, Comm. Math. Phys. **164** (1994), no. 3, 525–562.

[38] M. Kontsevich, *Formal (non)commutative symplectic geometry*, The Gel'fand Mathematical Seminars, 1990–1992, Birkhäuser Boston, Boston, MA, 1993, pp. 173–187.

[39] ———, *Feynman diagrams and low-dimensional topology*, First European Congress of Mathematics, Vol. II (Paris, 1992), Progr. Math., vol. 120, Birkhäuser, Basel, 1994, pp. 97–121.

[40] Anton Khoroshkin, Thomas Willwacher, and Marko Živković, *Differentials on graph complexes II: hairy graphs*, Lett. Math. Phys. **107** (2017), no. 10, 1781–1797.

[41] E. Looijenga, *Stable cohomology of the mapping class group with symplectic coefficients and of the universal Abel-Jacobi map*, J. Algebraic Geom. **5** (1996), no. 1, 135–150.

[42] D. Moseley, *Equivariant cohomology and the Varchenko-Gelfand filtration*, J. Algebra **472** (2017), 95–114.

[43] S. Maggiolo and N. Pagani, *Generating stable modular graphs*, J. Symbolic Comput. **46** (2011), no. 10, 1087–1097.

[44] D. Moseley, N. Proudfoot, and B. Young, *The Orlik-Terao algebra and the cohomology of configuration space*, Exp. Math. **26** (2017), no. 3, 373–380.

[45] I. Madsen and M. Weiss, *The stable moduli space of Riemann surfaces: Mumford's conjecture*, Ann. of Math. (2) **165** (2007), no. 3, 843–941.

[46] S. Payne, *Boundary complexes and weight filtrations*, Michigan Math. J. **62** (2013), 293–322.

[47] D. Petersen, *The structure of the tautological ring in genus one*, Duke Math. J. **163** (2014), no. 4, 777–793.

[48] ———, personal communication, October 2015.

[49] A. Robinson and S. Whitehouse, *The tree representation of Σ_{n+1}*, J. Pure Appl. Algebra **111** (1996), no. 1–3, 245–253.

[50] P. A. Songhafouo Tsopméné and V. Turchin, *Euler characteristics for spaces of string links and the modular envelope of $\mathcal{L}_\infty$*, Homology Homotopy Appl. **20** (2018), no. 2, 115–144.

[51] ———, *Rational homology and homotopy of high-dimensional string links*, Forum Math. **30** (2018), no. 5, 1209–1235.

[52] V. Turchin and T. Willwacher, *Commutative hairy graphs and representations of* Out(F_r), J. Topol. **10** (2017), no. 2, 386–411.

[53] A. Varchenko and I. Gel'fand, *Heaviside functions of a configuration of hyperplanes*, Funktsional. Anal. i Prilozhen. **21** (1987), no. 4, 1–18, 96.

[54] K. Vogtmann, *Local structure of some* Out(F_n)-*complexes*, Proc. Edinburgh Math. Soc. (2) **33** (1990), no. 3, 367–379.

[55] T. Willwacher, *M. Kontsevich's graph complex and the Grothendieck-Teichmüller Lie algebra*, Invent. Math. **200** (2015), no. 3, 671–760.

[56] Claudia Yun, *The S_n-equivariant rational homology of the tropical moduli spaces* $\Delta_{2,n}$, Exp. Math. (2021). https://doi.org/10.1080/10586458.2020.1870179.

Appendix A   Calculations for $g \geq 2$

We now present some calculations of $\widetilde{H}_*(\Delta_{g,n}; \mathbb{Q})$ for $g \geq 2$. Apart from some small cases, these were carried out by computer using the cellular homology theory for $\Delta_{g,n}$ as a symmetric Δ-complex. This is notably more efficient than other available methods, e.g., equipping $\Delta_{g,n}$ with a cell structure via barycentric subdivision. We further simplified the computer calculations via relative cellular homology and the contractibility of subcomplexes given by Theorem 1.1. We also used the program boundary [43] which efficiently enumerates symmetric orbits of boundary strata of $\overline{\mathcal{M}}_{g,n}$, and hence of cells in $\Delta_{g,n}$. By (1), these calculations detect top weight cohomology groups $\mathrm{Gr}^W_{6g-6+2n} H^*(\mathcal{M}_{g,n}; \mathbb{Q})$.

In the case $n = 0$, our calculations replicate those from earlier manuscripts of Bar-Natan and McKay [6], given the identification of $H_*(\Delta_g; \mathbb{Q})$ with $H_{*-2g+1}(G^{(g)})$. We refer to that manuscript for further remarks on homology computations for the basic graph complex. When $n > 0$ there is no reason the

Table A.1. *For each* (g, n) *shown, the dimensions of*
$\widetilde{H}_{i-1}(\Delta_{g,n}; \mathbb{Q})$ *for* $i = 1, \ldots, 3g - 3 + n$.

(g, n)	Reduced Betti numbers of $\Delta_{g,n}$ for $i = 0, \ldots, 3g - 4 + n$
$(2, 0)$	$(0, 0, 0)$
$(2, 1)$	$(0, 0, 0, 0)$
$(2, 2)$	$(0, 0, 0, 0, 1)$
$(2, 3)$	$(0, 0, 0, 0, 0, 0)$
$(2, 4)$	$(0, 0, 0, 0, 0, 1, 3)$
$(2, 5)$	$(0, 0, 0, 0, 0, 0, 5, 15)$
$(2, 6)$	$(0, 0, 0, 0, 0, 0, 0, 26, 86)$
$(2, 7)$	$(0, 0, 0, 0, 0, 0, 0, 0, 155, 575)$
$(2, 8)$	$(0, 0, 0, 0, 0, 0, 0, 0, 0, 1066, 4426)$
$(3, 0)$	$(0, 0, 0, 0, 0, 1)$
$(3, 1)$	$(0, 0, 0, 0, 0, 1, 0)$
$(3, 2)$	$(0, 0, 0, 0, 0, 0, 0, 0)$
$(3, 3)$	$(0, 0, 0, 0, 0, 0, 0, 0, 0, 1)$
$(3, 4)$	$(0, 0, 0, 0, 0, 0, 0, 0, 0, 3, 2)$
$(4, 0)$	$(0, 0, 0, 0, 0, 0, 0, 0, 0, 0)$
$(4, 1)$	$(0, 0, 0, 0, 0, 0, 0, 0, 0, 0, 0)$
$(4, 2)$	$(0, 0, 0, 0, 0, 0, 0, 0, 0, 0, 0, 0)$
$(4, 3)$	$(0, 0, 0, 0, 0, 0, 0, 0, 0, 0, 0, 2, 1)$
$(5, 0)$	$(0, 0, 0, 0, 0, 0, 0, 0, 0, 0, 1, 0, 0)$
$(5, 1)$	$(0, 0, 0, 0, 0, 0, 0, 0, 0, 0, 1, 0, 0, 0)$
$(6, 0)$	$(0, 0, 0, 0, 0, 0, 0, 0, 0, 0, 0, 0, 0, 0, 0, 1)$

computations of $H_*(\Delta_{g,n}; \mathbb{Q})$ could not have been performed earlier, but since we are currently unaware of an appropriate reference, we include them in Table A.1. Closely related computations that do appear in the literature, such as those in [40], involve graphs with unlabeled marked points. The computations in the case $g = 2$ were also given in [16], where it is also proved that $\widetilde{H}_*(\Delta_{2,n}; \mathbb{Z})$ is supported in the top two degrees. More recently, they were achieved S_n-equivariantly in [56].

Some of the homology classes displayed in Table A.1 have representatives with small enough support that it is feasible to describe them explicitly. For instance, for $(g, n) = (2, 2)$, it is easy to explicitly describe the unique nonzero homology class in $\Delta_{2,2}$; it is represented by the graph shown in Figure A.1. Every edge of the graph is contained in a triangle, so the graph-theoretic lemma below shows immediately that it is a cycle in homology. Moreover it is obviously nonzero since it is in top degree.

Figure A.1 The graph appearing in the unique nonzero reduced homology class in $\Delta_{2,2}$.

Figure A.2 The graph W_5.

Lemma A.1 *If* $G \in \Gamma_{g,n}$ *has the property that every edge is contained in a triangle, then* G *represents a rational cycle in* $\Delta_{g,n}$.

Proof The boundary of G in the cellular chain complex $C_*(\Delta_{g,n}; \mathbb{Q})$ is a sum, with appropriate signs, of 1-edge contractions of G. Each such contraction has parallel edges and hence a non-alternating automorphism, so is zero as a cellular chain. $\qquad\qquad\qquad\qquad\qquad\qquad\qquad\qquad\qquad\qquad\qquad\qquad\quad$ $\square$

For $(g,n) = (3,3)$ and $(6,0)$, the unique nonzero homology group is in top degree, so there is a unique nontrivial cycle, up to scaling. We have explicit descriptions of these cycles, as linear combinations of trivalent graphs, as shown in Figures A.3 and A.4.

For $(g,n) = (3,0), (3,1), (5,0)$, and $(5,1)$, the unique nonzero homology groups in Table A.1 are spanned by the classes of "wheel graphs." Given any g, let W_g be a genus g *wheel*: the graph obtained from a g-cycle C_g by adding a vertex w that is simply adjacent to each vertex of C_g. See Figure A.2.

We also regard W_g as an object of $\Gamma_{g,0}$ and let W_g' be the object of $\Gamma_{g,1}$ obtained by marking w, the central vertex. W_g and W_g' represent cells of degree $2g - 1$ in Δ_g and $\Delta_{g,1}$, respectively. When g is even W_g and W_g' have automorphisms that act by odd permutations on the edges, and hence are zero as cellular chains. When g is odd, these graphs do not have automorphisms that act by odd permutations on the edges, and Lemma A.1 implies immediately that W_g and W_g' represent rational cycles on Δ_g and $\Delta_{g,1}$ respectively. Moreover, they are nonzero.

Lemma A.2 *For* $g \geq 3$ *odd,* W_g *and* W_g' *represent nonzero homology classes in* $\widetilde{H}_{2g-1}(\Delta_g; \mathbb{Q})$ *and* $\widetilde{H}_{2g-1}(\Delta_{g,1}; \mathbb{Q})$ *respectively.*

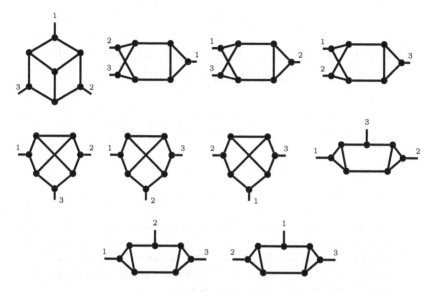

Figure A.3 The graphs appearing in the unique nonzero reduced homology class in $\Delta_{3,3}$, with unsigned coefficients $1,1,1,1,1,1,1,2,2,2$.

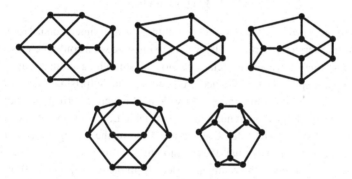

Figure A.4 The graphs appearing in the unique nonzero reduced homology class in Δ_6, with unsigned coefficients $2,3,6,3,4$.

Proof The fact that W_g represents a nontrivial class is established by [55]; see [14, Theorem 2.6]. As for W_g', we apply the chain map $\pi: C_*(\Delta_{g,1};\mathbb{Q}) \to C_*(\Delta_g;\mathbb{Q})$ from the proof of Theorem 1.7. It sends the cycle W_g' to $\pm W_g$, so the homology class represented by W_g' is also nontrivial. □

5

Mirror Symmetry and Smoothing Gorenstein Toric Affine 3-folds

Alessio Corti[a], Matej Filip[b] and Andrea Petracci[c]

Abstract. We state two conjectures that together allow one to describe the set of smoothing components of a Gorenstein toric affine 3-fold in terms of a combinatorially defined and easily studied set of Laurent polynomials called 0-*mutable polynomials*. We explain the origin of the conjectures in mirror symmetry and present some of the evidence.

1 Introduction

We explore mirror symmetry for smoothings of a 3-dimensional Gorenstein toric affine variety V. Specifically, we try to imagine what consequences mirror symmetry may have for the classification of smoothing components of the deformation space Def V. Conjecture A makes the surprising statement that the set of smoothing components of Def V is in bijective correspondence with a set of easily defined and enumerated 2-variable Laurent polynomials, called 0-mutable polynomials. Our Conjecture B — in the strong form stated in Remark 4.2 — asserts that these smoothing components are themselves smooth, and computes their tangent spaces from the corresponding 0-mutable polynomials.

As is customary in toric geometry, V is associated to a strictly convex 3-dimensional rational polyhedral cone $\sigma \subseteq N_{\mathbb{R}}$, where N is a 3-dimensional lattice; the Gorenstein condition means that the integral generators of the rays of σ all lie on an integral affine hyperplane ($u = 1$) for some $u \in M := \mathrm{Hom}_{\mathbb{Z}}(N, \mathbb{Z})$. We denote by F the convex hull of the integral generators of the rays of σ, i.e.

$$F := \sigma \cap (u = 1);$$

[a] Imperial College London
[b] University of Ljubljana
[c] Università di Bologna

this is a lattice polygon (i.e. a lattice polytope of dimension 2) embedded in the affine 2-dimensional lattice ($u = 1$). The isomorphism class of the toric variety V depends only on the affine equivalence class of the polygon F.

If V has an isolated singularity, then $\operatorname{Def} V$ is finite dimensional and we know from the work of Altmann [4] that there is a 1-to-1 correspondence between the set of irreducible components of $\operatorname{Def} V$ and integral maximal Minkowski decompositions of F. Altmann also shows that, when taken with their *reduced* structure, these components are all themselves smooth.

We are interested in the case when V has non-isolated singularities. Very little is known at this level of generality, but examples show that the picture for non-isolated singularities is very different from the one just sketched for isolated singularities. Our main reason for wanting to work with non-isolated singularities is the Fanosearch project: we wish to prove a general criterion for smoothing a toric Fano variety, and Conjecture A here is just the local case.

Conjecture A characterizes *smoothing components* of $\operatorname{Def} V$ in terms of the combinatorics of the polygon F. Specifically, we define the set $\mathfrak{B}$ of 0-mutable Laurent polynomials with Newton polygon F, and the conjecture states that there is a canonical bijective correspondence $\kappa : \mathfrak{B} \to \mathfrak{A}$, where $\mathfrak{A}$ is the set of smoothing components of $\operatorname{Def} V$.

At first sight the formulation of the conjecture seems strange; however, the statement makes sense in the context of mirror symmetry, where (conjecturally) the 0-mutable polynomials are the mirrors of the corresponding smoothing components.

In Section 4, we state a new Conjecture B,[1] which implies the existence of a map $\kappa : \mathfrak{B} \to \mathfrak{A}$ — see Remark 4.1. In that section, we also explain how to (conjecturally) construct a deformation directly from a 0-mutable polynomial in the spirit of the intrinsic mirror symmetry of Gross–Siebert [20, 19] and work of Gross–Hacking–Keel [22].

The coefficients of the 0-mutable Laurent polynomials that appear in our conjecture ought themselves to enumerate certain holomorphic discs in the corresponding smoothing, and we would love to see a precise statement along these lines.

In our view, the conjectures together are nothing other than a statement of mirror symmetry as a one-to-one correspondence between two sets of objects, similar to the conjectures made in [2] in the context of orbifold del Pezzo surfaces, and the correspondence between Fano 3-folds and Minkowski polynomials discovered in [10].

---

[1] The statement of Conjecture B comes after Sec. 3 but does not logically depend on it: if you wish, you can skip directly from Sec. 2.3 to Sec. 4.1.

In Section 3 we give some equivalent characterisations of 0-mutable polynomials and begin to sketch some of their general properties. These properties make it very easy to enumerate the 0-mutable polynomials with given Newton polygon. The material here is rather sketchy — full details will appear elsewhere; it serves for context, but it is not logically necessary for the statement of the conjectures.

The suggestion that there is a simple structure to the set of smoothing components is surprising in a subject that — as all serious practitioners know — is marred by Murphy's law. In fact, there is a substantial body of direct and circumstantial evidence for the conjectures, some of which we present in Section 5.

In the final Section 6 we compute in detail the deformation space of the variety V_F associated to the polygon F of Example 2.12, giving evidence for the conjectures. Some of the reasons for choosing this particular example are:

(1) The variety V_F is of codimension 5 and hence it lies outside the — still rudimentary but very useful — structure theory of codimension-4 Gorenstein rings [30], see also [9, 8];

(2) For this reason, V_F is a good test of the technology of [14, 13] as a tool for possibly proving the conjectures;

(3) The polygon F appears as a facet of some of the 3-dimensional reflexive polytopes and hence it is immediately relevant for the Fanosearch project.

As things stand, we are some distance away from being able to prove the conjectures. We had a tough time even with the example of Section 6: while a treatment based on [14, 13] seems possible, the task became so tedious that we decided instead to rely on Ilten's Macaulay2 [18] package *Versal deformations and local Hilbert schemes* [26]. That package makes it possible to test the conjectures in many other examples in codimension ≥ 5.

Notation and Conventions

We work over $\mathbb{C}$, but everything holds over an algebraically closed field of characteristic zero. We refer the reader to [16] for an introduction to toric geometry. All the toric varieties we consider are normal. We use the following notation.

F a lattice polygon
V the Gorenstein toric affine 3-fold associated to the cone over F put at height 1
∂V the toric boundary of V

$\overline{X}$ the projective toric surface associated to the normal fan of F

$\overline{B}$ the toric boundary of $\overline{X}$

A the ample line bundle on $\overline{X}$ given by F

X the cluster surface associated to F (the non-toric blowup of $\overline{X}$ constructed in Section 3)

B the strict transform of $\overline{B}$ in X

$W^{\flat}$ the toric 3-fold constructed in Section 4

$\overline{W}$ the toric blowup of $W^{\flat}$ constructed in Section 4

W the mirror cluster variety (the non-toric blowup of $\overline{W}$ constructed in Section 4)

Thanks to Bill from AC

I was lucky to be a L.E. Dickson Instructor at the University of Chicago in the years 1993–1996, where I worked in the group led by Bill Fulton. I had studied deformations of singularities in a seminar held in 1988–89 at the University of Utah, but in Bill's seminar I learnt about Chow groups and quantum cohomology. It was a wonderful time in my work life thanks largely to Bill. This paper features many of the ideas that I learnt in Bill's seminar and it is very nice to see that they are so relevant in the study of deformations of singularities.

Acknowledgements

We thank Klaus Altmann, Tom Coates, Mark Gross, Paul Hacking, Al Kasprzyk, Giuseppe Pitton, and Thomas Prince for many helpful conversations.

We particularly thank Al Kasprzyk for sharing with us and allowing us to present some of his ideas on maximally mutable Laurent polynomials [11].

We discussed with Mark Gross some of our early experiments with 0-mutable polynomials.

We owe special thanks to Paul Hacking who read and corrected various mistakes in earlier versions of the paper — the responsibility for the mistakes that are left is of course ours.

Giuseppe Pitton ran computer calculations that provide indirect evidence for the conjectures.

The idea that mirror Laurent polynomials are characterised by their mutability goes back to Sergey Galkin [17].

It will be clear to all those familiar with the issues that this paper owes a very significant intellectual debt to the work of Gross, Hacking and Keel [22] and the intrinsic mirror symmetry of Gross and Siebert [20, 19].

Last but not least, it is a pleasure to thank the anonymous referee who read our manuscript very carefully, and told us about Ilten's Macaulay2 package.

2 Conjecture A

2.1 Gorenstein Toric Affine Varieties

Consider a rank-n lattice $N \simeq \mathbb{Z}^n$ (usually $n = 3$) and, as usual in toric geometry, its dual lattice $M = \mathrm{Hom}_{\mathbb{Z}}(N, \mathbb{Z})$. The n-dimensional torus

$$\mathbb{T} = \mathrm{Spec}\,\mathbb{C}[M]$$

is referred to simply as "the" torus.[2] Consider a strictly convex full-dimensional rational polyhedral cone $\sigma \subset N_{\mathbb{R}}$ and the corresponding affine toric variety

$$V = \mathrm{Spec}\,\mathbb{C}[\sigma^{\vee} \cap M].$$

This is a normal Cohen–Macaulay n-dimensional variety.

By definition V is Gorenstein if and only if the pre-dualising sheaf

$$\omega_V^0 = \mathcal{H}^{-n}(\omega_V^{\bullet})$$

is a line bundle. Since our V is Cohen–Macaulay, this is the same as insisting that all the local rings of V are local Gorenstein rings. It is known and not difficult to show that V is Gorenstein if and only if there is a vector $u \in M$ such that the integral generators $\rho_1, \ldots, \rho_m$ of the rays of the cone σ all lie on the affine lattice $\mathbb{L} = (u = 1) \subset N$. Such vector u is called the *Gorenstein degree*.

If V is Gorenstein, then the toric boundary of V is the following effective reduced Cartier divisor on V:

$$\partial V = \mathrm{Spec}\,\mathbb{C}[\sigma^{\vee} \cap M]/(x^u).$$

If V is Gorenstein, we set

$$F := \sigma \cap (u = 1);$$

---

[2] Sometimes we denote this torus by $\mathbb{T}_N$, that is, the commutative group scheme $N \otimes_{\mathbb{Z}} \mathbb{G}_{\mathrm{m}}$ such that for all rings R $\mathbb{T}_N(R) = N \otimes_{\mathbb{Z}} R^{\times}$.

this is an $(n-1)$-dimensional lattice polytope embedded in the affine lattice $\mathbb{L} = (u = 1)$ and it is the convex hull of the integral generators of the rays of the cone σ. One can prove that the isomorphism class of V depends only on the affine equivalence class of the lattice polytope F in the affine lattice $\mathbb{L}$. Therefore we will say that V is associated to the polytope F.

This establishes a 1-to-1 correspondence between isomorphism classes of Gorenstein toric affine n-folds without torus factors and $(n-1)$-dimensional lattice polytopes up to affine equivalence.

2.2 Statement of Conjecture A

In this section we explain everything that is needed to make sense of the following:

Conjecture A *Consider a lattice polygon F in a 2-dimensional affine lattice $\mathbb{L}$. Let V be the Gorenstein toric affine 3-fold associated to F.*

Then there is a canonical bijective function $\kappa : \mathfrak{B} \to \mathfrak{A}$ where:

- *$\mathfrak{A}$ is the set of smoothing components of the miniversal deformation space Def V,*
- *$\mathfrak{B}$ is the set of 0-mutable polynomials $f \in \mathbb{C}[\mathbb{L}]$ with Newton polygon F.*

Remark 2.1 It is absolutely crucial to appreciate that we are not assuming that V has an isolated singularity at the toric 0-stratum. If V does not have isolated singularities, Def V is infinite-dimensional. A few words are in order to clarify what kind of infinite dimensional space Def V is.

In full generality, there is some discussion of this issue in the literature on the analytic category, see for example [24, 25].[3] In the special situation of interest in this paper, we take a naïve approach, which we briefly explain, based on the following two key facts:

(i) If V is a Gorenstein toric affine 3-fold, then T_V^2 is finite dimensional. Indeed V has transverse A_*-singularities in codimension two, hence it is unobstructed in codimension two, hence T_V^2 is a finite length module supported on the toric 0-stratum. In fact, there is an explicit description of T_V^2 as a representation of the torus, see [6, Section 5], an example of which is in Lemma 6.4 below. This shows that Def V is cut out by finitely many equations.

---

[3] We thank Jan Stevens for pointing out these references to us. We are not aware of a similar discussion in the algebraic literature.

(ii) On the other hand, the known explicit description of T_V^1 as a representation of the torus [5, Theorem 4.4], together with the explicit description of T_V^2 just mentioned, easily implies that each of the equations can only use finitely many variables.[4]

Thus we can take $\mathrm{Def}\, V$ to be the Spec of a non-Noetherian ring, that is, the simplest kind of infinite-dimensional scheme.[5]

Remark 2.2 Conjecture A does not state what the function κ is, nor what makes it "canonical." The existence of a function $\kappa \colon \mathfrak{B} \to \mathfrak{A}$ is implied by Conjecture B stated in Section 4.1 below, see Remark 4.1.

Remark 2.3 Let V be a Gorenstein toric affine 3-fold and let ∂V be the toric boundary of V. Let $\mathrm{Def}(V, \partial V)$ be the deformation functor (or the base space of the miniversal deformation) of the pair $(V, \partial V)$. There is an obvious forgetful map $\mathrm{Def}(V, \partial V) \to \mathrm{Def}\, V$. In this case, since ∂V is an effective Cartier divisor in V and V is affine, this map is smooth of relative dimension equal to the dimension of $\mathrm{coker}\left(H^0(\theta_V) \to H^0(N_{\partial V/V})\right)$, where θ_V is the sheaf of derivations on V and $N_{\partial V/V} = \mathcal{O}_{\partial V}(\partial V)$ is the normal bundle of ∂V inside V. In particular, this implies that $\mathrm{Def}\, V$ and $\mathrm{Def}(V, \partial V)$ have exactly the same irreducible components. One can also see that a smoothing component in $\mathrm{Def}\, V$ is the image of a component of $\mathrm{Def}(V, \partial V)$ where also ∂V is smoothed.

In other words, we can equivalently work with deformations of V or deformations of the pair $(V, \partial V)$. The right thing to do in mirror symmetry is to work with deformations of the pair $(V, \partial V)$; however, the literature on deformations of singularities is all written in terms of V. In most cases it is not difficult to make the translation but this paper is not the right place for doing that. Thus when possible we work with deformations of V. The formulation of Conjecture B in Section 4 requires that we work with deformations of the pair $(V, \partial V)$.

2.3 The Definition of 0-mutable Laurent Polynomials

This subsection is occupied by the definition of 0-mutable Laurent polynomials. The simple key idea — and the explanation for the name "0-mutable" — is that an irreducible Laurent polynomial f is 0-mutable if and only if there is a sequence of mutations

---

[4] The key point is to show that for all fixed weights $\mathbf{m} \in M$, the equation $\mathbf{m} = \sum m_i \mathbf{v}_i$ has finitely many solutions for $m_i \in \mathbb{N} \setminus \{0\}$ and $\mathbf{v}_i \in M$ a weight that appears non-trivially in T_V^1. This type of consideration is used extensively in the detailed example discussed in Section 6.

[5] The situation is not so simple for the universal family $\mathscr{U} \to \mathrm{Def}\, V$. Indeed the equations of $\mathscr{U}$ naturally involve all the infinitely many coordinate functions on T_V^1. Thus, $\mathscr{U}$ is a bona fide ind-scheme. The language to deal with this exists but it is not our concern here.

$$f \mapsto f_1 \mapsto \cdots \mapsto f_p = 1$$

starting from f and ending with the constant monomial $\mathbf{1}$. The precise definition is given below after some preliminaries on mutations. The definition is appealing and it is meaningful in all dimensions, but it is not immediately useful if you want to study 0-mutable polynomials. Indeed, for example, to prove that a given polynomial f is 0-mutable one must produce a chain of mutations as above and it may not be obvious where to look. It is even less clear how to prove that f is not 0-mutable. In Section 3, Theorem 3.5, we prove two useful characterizations of 0-mutable polynomials in two variables. The first of these states that a polynomial is 0-mutable if and only if it is rigid maximally mutable. From this property it is easy to check that a given polynomial is 0-mutable, that it is not 0-mutable, and to enumerate 0-mutable polynomials with given Newton polytope. The second characterization states that a (normalized, see below) Laurent polynomial in two variables is 0-mutable if and only if the irreducible components of its vanishing locus are -2-curves on the cluster surface: see Section 3 for explanations and details.

Let $\mathbb{L}$ be an affine lattice and let $\mathbb{L}_0$ be its underlying lattice.[6]

In other words, $\mathbb{L}_0$ is a free abelian group of finite rank and $\mathbb{L}$ is a set together with a free and transitive $\mathbb{L}_0$-action. We denote by $\mathbb{C}[\mathbb{L}]$ the vector space over $\mathbb{C}$ whose basis is made up of the elements of $\mathbb{L}$. For every $l \in \mathbb{L}$ we denote by x^l the corresponding element in $\mathbb{C}[\mathbb{L}]$. Elements of $\mathbb{C}[\mathbb{L}]$ will be called (Laurent) polynomials. It is clear that $\mathbb{C}[\mathbb{L}]$ is a rank-1 free module over the $\mathbb{C}$-algebra $\mathbb{C}[\mathbb{L}_0]$. The choice of an origin in $\mathbb{L}$ specifies an isomorphism $\mathbb{C}[\mathbb{L}] \simeq \mathbb{C}[\mathbb{L}_0]$.

Definition 2.4 A Laurent polynomial $f \in \mathbb{C}[\mathbb{L}]$ is *normalized* if for all vertices $v \in \mathrm{Newt}\, f$ we have $a_v = 1$ where a_v is the coefficient of the monomial x^v as it appears in f.

In this paper all polynomials are assumed to be normalized unless explicitly stated otherwise.

If $f \in \mathbb{C}[\mathbb{L}]$ then we say that f *lives* on the smallest saturated affine sublattice $\mathbb{L}' \subseteq \mathbb{L}$ such that $f \in \mathbb{C}[\mathbb{L}']$. The property of being 0-mutable only depends on the lattice where f lives — here it is crucial that we only allow saturated sub-lattices. We say that f is an r-variable polynomial if f lives on a rank-r affine lattice.

Let us start by defining 0-mutable polynomials in 1 variable.

---

[6] In our setup $\mathbb{L} = (u = 1) \subset N$ does not have a canonical origin. We try to be pedantic and write $\mathbb{L}_0$ for the underlying lattice – in our example, $\mathbb{L}_0 = \mathrm{Ker}\, u$. In practice this distinction is not super-important and you are free to choose an origin anywhere you want.

Definition 2.5   Let $\mathbb{L}$ be an affine lattice of rank 1 and let $\mathbb{L}_0$ be its underlying lattice. Let $v \in \mathbb{L}_0$ be one of the two generators of $\mathbb{L}_0$. A polynomial $f \in \mathbb{C}[\mathbb{L}]$ is called 0-*mutable* if

$$f = (1 + x^v)^k x^l$$

for some $l \in \mathbb{L}$ and $k \in \mathbb{N}$.

(It is clear that the definition does not depend on the choice of v.)

If f is a Laurent polynomial in 1 variable and its Newton polytope is a segment of lattice length k, then f is 0-mutable if and only if the coefficients of f are the $k + 1$ binomial coefficients of weight k.

The definition of 0-mutable polynomials in more than 1 variable will be given recursively on the number of variables. Thus from now on we fix $r \geq 2$ and we assume to know already what it means for a polynomial of $< r$ variables to be 0-mutable. Before we can state what it means for a polynomial f of r variables to be 0-mutable, we need to explain how to mutate f.

If $\mathbb{L}$ is an affine lattice, we denote by $\mathrm{Aff}(\mathbb{L}, \mathbb{Z})$ the lattice of affine-linear functions $\varphi \colon \mathbb{L} \to \mathbb{Z}$. If $\mathbb{L}_0$ denotes the underlying lattice of $\mathbb{L}$, φ has a well-defined *linear part* which we denote by $\varphi_0 \colon \mathbb{L}_0 \to \mathbb{Z}$.

Definition 2.6   Let $r \geq 2$ and fix a rank-r affine lattice $\mathbb{L}$.

A *mutation datum* is a pair (φ, h) of a non-constant affine-linear function $\varphi \colon \mathbb{L} \to \mathbb{Z}$ and a 0-mutable polynomial $h \in \mathbb{C}[\mathrm{Ker}\, \varphi_0]$.

Given a mutation datum (φ, h) and $f \in \mathbb{C}[\mathbb{L}]$, write (uniquely)

$$f = \sum_{k \in \mathbb{Z}} f_k \qquad \text{where} \qquad f_k \in \mathbb{C}[(\varphi = k) \cap \mathbb{L}].$$

We say that f is (φ, h)-*mutable* if for all $k < 0$ h^{-k} divides f_k (equivalently, if for all $k \in \mathbb{Z}$ $h^k f_k \in \mathbb{C}[\mathbb{L}]$). If f is (φ, h)-mutable, then the *mutation* of f, with respect to the mutation datum (φ, h), is the polynomial:

$$\mathrm{mut}_{(\varphi, h)} f = \sum_{k \in \mathbb{Z}} h^k f_k.$$

Remark 2.7   The notion of mutation goes back (at least) to Fomin–Zelevinsky [15]. We first learned of mutations from the work of Galkin–Usnich [17] and Akhtar–Coates–Galkin–Kasprzyk [1]. This paper owes a significant intellectual debt to the interpretation of mutations developed in work by Gross, Hacking and Keel, for instance [23, 21, 22].[7]

---

[7] Many mathematicians work on mutations from different perspectives and we apologize for not even trying to quote all the relevant references here.

The following is a recursive definition. The base step is given by Definition 2.5.

Definition 2.8 Let $\mathbb{L}$ be an affine lattice of rank $r \geq 2$. We define the set of 0-mutable polynomials on $\mathbb{L}$ in the following recursive way.

(i) If $\mathbb{L}'$ is a saturated affine sub-lattice of $\mathbb{L}$ and $f \in \mathbb{C}[\mathbb{L}']$ is 0-mutable, then f is 0-mutable in $\mathbb{C}[\mathbb{L}]$.

(ii) If $f = f_1 f_2$ is reducible, then f is 0-mutable if both factors f_1, f_2 are 0-mutable.[8]

(iii) If f is irreducible, then f is 0-mutable if a mutation of f is 0-mutable.

Equivalently, the set of 0-mutable polynomials of $\leq r$ variables is the smallest subset of $\mathbb{C}[\mathbb{L}]$ that contains all 0-mutable polynomials of $< r$ variables and that is closed under the operations of taking products (in particular translation) and mutations of irreducible polynomials.

Remark 2.9 It follows easily from the definition that 0-mutable polynomials are normalized. Indeed:

(1) The polynomials in Definition 2.5 (the base case) are 0-mutable;
(2) The product of two normalized polynomials is normalized;
(3) The mutation of a normalized polynomial is normalized.

Example 2.10 Let $\mathbb{L}$ be an affine lattice and let v be a primitive vector in the underlying lattice $\mathbb{L}_0$. Definition 2.5 and Definition 2.8(i) imply that $(1+x^v)^k x^l$ is 0-mutable for all $l \in \mathbb{L}$ and $k \in \mathbb{N}$.

Example 2.11 Consider the triangle

$$F = \mathrm{conv}\,((0,0),(3,0),(3,2)) \tag{1}$$

in the lattice $\mathbb{L} = \mathbb{Z}^2$. Let us identify $\mathbb{C}[\mathbb{L}]$ with $\mathbb{C}[x^{\pm}, y^{\pm}]$. One can prove that there are exactly two 0-mutable polynomials with Newton polytope F, namely:

$$(1+x)^3 + 2(1+x)x^2 y + x^3 y,$$
$$(1+y)^2 x^3 + 3(1+y)x^2 + 3x + 1.$$

In Figure 1 we have written the coefficients of these two polynomials next to the lattice point of F associated to the corresponding monomial.

---

[8] In order to take the "product" $f_1 f_2$ one has to choose an origin in $\mathbb{L}$. This choice makes no difference to the definition.

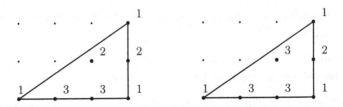

Figure 1 The two 0-mutable polynomials whose Newton polytope is the triangle F defined in (1).

It is shown in [9] that Def V has two components, and that they are both smoothing components, confirming our conjectures. The calculation there goes back to unpublished work by Jan Stevens, but see also [5].

Example 2.12   Consider the quadrilateral

$$F = \operatorname{conv}((-1, -1), (2, -1), (1, 1), (-1, 2)) \qquad (2)$$

in the lattice $\mathbb{L} = \mathbb{Z}^2$. Let us identify $\mathbb{C}[\mathbb{L}]$ with $\mathbb{C}[x^\pm, y^\pm]$. Consider the polynomial

$$g = \frac{(1 + x)^3 + (1 + y)^3 - 1 + x^2 y^2}{xy},$$

which is obtained by giving binomial coefficients to the lattice points of the boundary of F and by giving zero coefficient to the interior lattice points of F. By Lemma 3.1(2) every 0-mutable polynomial with Newton polytope F must coincide with g on the boundary lattice points of F. One can prove that there are exactly three 0-mutable polynomials with Newton polytope F, namely:

$$\alpha = g + 5 + 2x + 2y = \frac{(1 + x + 2y + y^2)(1 + 2x + x^2 + y)}{xy},$$

$$\beta = g + 6 + 3x + 4y = \frac{(1 + x)^3 + 3y(1 + x)^2 + y^2(1 + x)(3 + x) + y^3}{xy},$$

$$\gamma = g + 6 + 4x + 3y = \frac{(1 + y)^3 + 3x(1 + y)^2 + x^2(1 + y)(3 + y) + x^3}{xy}.$$

In Figure 2 we have written down the coefficients of these three polynomials. The polynomial α is reducible and it is easy to show that its factors are 0-mutable.

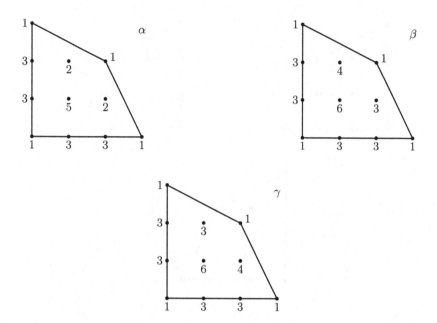

Figure 2 The three 0-mutable polynomials α, β and γ whose Newton polytope is the quadrilateral F defined in (2).

Let us consider the following affine-linear functions $\mathbb{L} = \mathbb{Z}^2 \to \mathbb{Z}$:

$$-m_{2,1}^1 : (a,b) \mapsto b - 1,$$
$$-m_{3,1}^1 : (a,b) \mapsto b - 2,$$
$$-m_{3,2}^1 : (a,b) \mapsto 2b - 1.$$

The level sets of these three affine-linear functions are depicted in Figure 3. We now consider the mutation data $(-m_{2,1}^1, 1 + x)$, $(-m_{3,2}^1, 1 + x)$ and $(-m_{3,1}^1, 1 + x)$. The polynomial α is mutable with respect to $(-m_{2,1}^1, 1 + x)$ and to $(-m_{3,2}^1, 1 + x)$ and

$$\mathrm{mut}_{(-m_{2,1}^1, 1+x)}\, \alpha = \frac{1 + x}{xy} + \frac{3 + 2x}{x} + \frac{y(3 + 2x + x^2)}{x} + \frac{y^2(1 + x)}{x},$$

$$\mathrm{mut}_{(-m_{3,2}^1, 1+x)}\, \alpha = \frac{1}{xy} + \frac{3 + 2x}{x} + \frac{y(3 + 5x + 3x^2 + x^3)}{x} + \frac{y^2(1 + x)^3}{x},$$

but α is not mutable with respect to $(-m_{3,1}^1, 1 + x)$. The polynomial β is mutable with respect to all three mutation data $(-m_{2,1}^1, 1 + x)$, $(-m_{3,2}^1, 1 + x)$ and $(-m_{3,1}^1, 1 + x)$, and the mutations are:

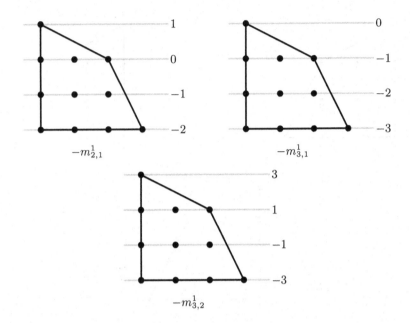

Figure 3 The level sets of the three affine-linear functions $\mathbb{L} = \mathbb{Z}^2 \to \mathbb{Z}$ considered in Example 2.12.

$$\text{mut}_{(-m_{2,1}^1, 1+x)} \beta = \frac{1+x}{xy} + \frac{3(1+x)}{x} + \frac{y(3+x)(1+x)}{x} + \frac{y^2(1+x)}{x},$$

$$\text{mut}_{(-m_{3,1}^1, 1+x)} \beta = \frac{1}{xy} + \frac{3}{x} + \frac{y(3+x)}{x} + \frac{y^2}{x} = \frac{(1+y)^3 + xy^2}{xy},$$

$$\text{mut}_{(-m_{3,2}^1, 1+x)} \beta = \frac{1}{xy} + \frac{3(1+x)}{x} + \frac{y(3+x)(1+x)^2}{x} + \frac{y^2(1+x)^3}{x}.$$

The polynomial γ is not mutable with respect to any of the mutation data $(-m_{2,1}^1, 1+x)$, $(-m_{3,2}^1, 1+x)$, $(-m_{3,1}^1, 1+x)$.

3 Properties of 0-mutable Polynomials

3.1 Some Easy Properties

If $\mathbb{L}$ is an affine lattice, $f \in \mathbb{C}[\mathbb{L}]$ and $F = \text{Newt } f$, then we write

$$f = \sum_{l \in F \cap \mathbb{L}} a_l x^l \quad \text{with} \quad a_l \in \mathbb{C}.$$

For every subset $A \subseteq \mathbb{L}_{\mathbb{R}}$, we write

$$f|_A = \sum_{l \in A \cap \mathbb{L}} a_l x^l.$$

Lemma 3.1 *(1) (Non-negativity and integrality) If f is 0-mutable, then every coefficient of f is a non-negative integer.*
(2) (Boundary terms) If $f \in \mathbb{C}[\mathbb{L}]$ is 0-mutable and $F \leq$ Newt f is a face, then $f|_F$ is 0-mutable.

Sketch of proof (1) is obvious due to the recursive definition of 0-mutable polynomials. Also (2) is easy because it is enough to observe that mutations and products behave well with respect to restriction to faces of the Newton polytope. ∎

Remark 3.2 A 0-mutable polynomial may have a zero coefficient at a lattice point of its Newton polytope, e.g. $(1 + x)(1 + xy^2)$.

3.2 Rigid Maximally Mutable Polynomials

From now on we focus on the two-variable case $r = 2$. In what follows, we give two equivalent characterizations of 0-mutable polynomials, one geometric in terms of the associated cluster variety and one combinatorial in terms of rigid maximally mutable polynomials.[9]

Let $\mathbb{L}$ be an affine lattice of rank 2. For a Laurent polynomial $f \in \mathbb{C}[\mathbb{L}]$, we set

$$\mathscr{S}(f) = \Big\{ \text{mutation data } s = (\varphi, h) \; \Big| \; f \text{ is } s\text{-mutable} \Big\}.$$

Conversely, if $\mathscr{S}$ is a set of mutation data, we denote by

$$L(\mathscr{S}) = \Big\{ f \in \mathbb{C}[\mathbb{L}] \; \Big| \; \forall s \in \mathscr{S}, f \text{ is } s\text{-mutable} \Big\}$$

the vector space of Laurent polynomials f that are s-mutable for all the mutation data $s \in \mathscr{S}$. For every polynomial $f \in \mathbb{C}[\mathbb{L}]$, it is clear that $f \in L(\mathscr{S}(f))$.

Definition 3.3 (Kasprzyk) Let $\mathbb{L}$ be an affine lattice of rank 2, and $f \in \mathbb{C}[\mathbb{L}]$.

(i) If $f = f_1 f_2$ is the product of normalized polynomials f_1, f_2, then f is rigid maximally mutable if both factors f_1, f_2 are rigid maximally mutable;

---

[9] The concept of rigid maximally mutable is due to Al Kasprzyk [11]. We thank him for allowing us to include his definition here.

(ii) If f is normalized and irreducible, then f is rigid maximally mutable if

$$L(\mathscr{S}(f)) = \{\lambda f \mid \lambda \in \mathbb{C}\}.$$

3.3 Cluster Varieties

Definition 3.4   A *Calabi–Yau (CY) pair* is a pair (Y, ω) of an n-dimensional quasiprojective normal variety Y and a degree n rational differential $\omega \in \Omega^n_{k(X)}$, such that

$$D = -\operatorname{div}_Y \omega \geq 0$$

is an effective reduced Cartier divisor on Y.[10]

A *torus chart* on (Y, ω) is an open embedding

$$j : (\mathbb{C}^\times)^n \hookrightarrow Y \setminus D \qquad \text{such that} \qquad j^\star(\omega) = \frac{1}{(2\pi i)^n} \frac{dx_1}{x_1} \wedge \cdots \wedge \frac{dx_n}{x_n}.$$

A *cluster variety* is an n-dimensional CY pair (Y, ω) that has a torus chart.

In our situation, the pair (Y, D) will always be log smooth.

3.4 The Cluster Surface

We construct a cluster surface from a lattice polygon.

Let $\mathbb{L}$ be an affine lattice of rank 2. There is a canonical bijection between the set of lattice polygons $F \subset \mathbb{L}_\mathbb{R}$ up to translation and the set of pairs $(\overline{X}, A)$ of a projective toric surface $\overline{X}$ and an ample line bundle A on $\overline{X}$: the torus in question is $\operatorname{Spec} \mathbb{C}[\mathbb{L}_0]$; the fan of the surface $\overline{X}$ is the normal fan of F, and it all works out such that there is a natural 1-to-1 correspondence between $F \cap \mathbb{L}$ and a basis of $H^0(\overline{X}, A)$.

Fix a lattice polygon F in $\mathbb{L}$ and consider the corresponding polarised toric surface $(\overline{X}, A)$. Denote by $\overline{B}$ the toric boundary of $\overline{X}$. For each edge $E \leq F$, let $\ell(E)$ be the lattice length of E and let $\overline{B}_E$ be the prime component of $\overline{B}$ corresponding to E; we have that $\overline{B}_E$ is isomorphic to $\mathbb{P}^1$ and the line bundle $A|_{\overline{B}_E}$ has degree $\ell(E)$. Denote by $x_E \in \overline{B}_E$ the point $[1 : -1] \in \mathbb{P}^1$.

For all edges $E \leq F$, blow up $\ell(E)$ times above x_E in the proper transform of $\overline{B}_E$ and denote by

$$p : (X, B) \longrightarrow (\overline{X}, \overline{B})$$

---

[10] Like most people, we mostly work with the pair (Y, D) and omit explicit reference to ω.

the resulting surface, where $B \subset X$ is the proper transform of the toric boundary $\overline{B} = \sum_{E \leq F} \overline{B}_E$. We call the pair (X, B) the *cluster surface* associated to the lattice polygon F.

Theorem 3.5 *Let* $\mathbb{L}$ *be a rank-2 lattice. The following are equivalent for a normalized Laurent polynomial* $f \in \mathbb{C}[\mathbb{L}]$:

(1) f *is 0-mutable;*
(2) f *is rigid maximally mutable;*
(3) *Let* $p \colon (X, B) \to (\overline{X}, \overline{B})$ *be the cluster surface associated to the polygon* $F = \mathrm{Newt}\, f$. *Denote by* $Z \subset \overline{X}$ *the divisor of zeros of* f *and by* $Z' \subset X$ *the proper transform of* Z. *Every irreducible component* $\Gamma \subset Z'$ *is a smooth rational curve with self-intersection* $\Gamma^2 = -2$. *(Necessarily then* $B \cdot \Gamma = 0$ *hence* Γ *is disjoint from the boundary* B.)

Remark 3.6 The support of Z' is not necessarily a normal crossing divisor. The irreducible components need not meet transversally, and ≥ 3 of them may meet at a point.

Sketch of proof In proving all equivalences we may and will assume that the polynomial f is irreducible hence Z is reduced and irreducible.

The proof uses the following ingredients, which we state without further discussion or proof:

(i) To give a torus chart $j \colon \mathbb{C}^{\times 2} \hookrightarrow X \setminus B$ in X is the same as to give a *toric model* of (X, B), that is a projective morphism $q \colon (X, B) \to (X', B')$ where (X', B') is a toric pair and q maps $j(\mathbb{C}^{\times 2})$ isomorphically to the torus $X' \setminus B'$;

(ii) The work of Blanc [7] implies that any two torus charts in X are connected by a sequence of mutations between torus charts in X;

(iii) A set $\mathscr{S}$ of mutation data specifies a line bundle $\mathcal{L}(\mathscr{S})$ on X such that $H^0(X, \mathcal{L}(\mathscr{S})) = L(\mathscr{S})$ and, conversely, every line bundle on X is isomorphic to a line bundle of the form $\mathcal{L}(\mathscr{S})$.

Let us show first that (1) implies (3). To say that an irreducible polynomial f is 0-mutable is to say that there exists a sequence of mutations that mutates f to the constant polynomial **1**. This sequence of mutations constructs a new torus chart $j_1 \colon \mathbb{T} = \mathbb{C}^{\times 2} \hookrightarrow X \setminus B$ such that the proper transform Z' – which is, by assumption, irreducible – is disjoint from $j_1(\mathbb{T})$. This new toric chart gives a new toric model $p_1 \colon (X, B) \to (X_1, B_1)$ that maps $j_1(\mathbb{T})$ isomorphically to the torus $X_1 \setminus B_1$ and hence contracts Z' to a boundary point. Z' is not a -1-curve, because those are all p-exceptional, hence Z' is a -2-curve.

To show that (3) implies (1), by Lemma 3.7 below, there is a new toric model $p_1 : (X, B) \rightarrow (X_1, B_1)$ that contracts Z' to a point in the boundary. The new toric model then gives a new torus chart $j_1 : \mathbb{T} \hookrightarrow X$ such that Z' is disjoint from $j_1(\mathbb{T})$. By Blanc the induced birational map of tori

$$j_1^{-1} j : \mathbb{T} \dashrightarrow \mathbb{T}$$

is a composition of mutations that mutates f to the constant polynomial.

Let us now show that (3) implies (2). By some tautology, Z' is the zero divisor of the section f of the line bundle $\mathcal{L}(\mathscr{S})$ on X specified by the set of mutation data $\mathscr{S} = \mathscr{S}(f)$, and $H^0(X, Z') = L(\mathscr{S})$. Since Z' is a -2-curve, $L(\mathscr{S})$ is 1-dimensional, which is to say that f is rigid maximally mutable.

Finally we show that (2) implies (3). Denote by $\mathcal{L} = \mathcal{L}(\mathscr{S})$ the line bundle on X specified by the set of mutations $\mathscr{S} = \mathscr{S}(f)$, so that Z' is the zero-locus of a section of $\mathcal{L}$. Note that:

$$h^2(X, \mathcal{L}) = h^0(X, K_X - Z') = h^0(Y, -\overline{B} - Z') = 0.$$

Riemann–Roch and the fact that f is rigid give:

$$1 = h^0(X, Z') = h^0(X, \mathcal{L}) \geq \chi(X, \mathcal{L}) = 1 + \frac{1}{2}(Z'^2 + Z'\overline{B})$$

and hence, because $Z'\overline{B} \geq 0$, we conclude that $Z'^2 \leq 0$ and:

$$2 p_a(Z') - 2 = \deg \omega_{Z'} = Z'^2 - Z'\overline{B} \leq 0$$

so either:

(i) $\deg \omega_{Z'} < 0$ and then Z' is a smooth rational curve, and then as above Z' is not a -1 curve therefore it is a -2-curve, or

(ii) $\omega_{Z'} = \mathcal{O}_{Z'}$ and $Z'^2 = Z'\overline{B} = 0$. It follows that Z' is actually disjoint from $\overline{B}$ and $\mathcal{O}_{Z'}(Z') = \mathcal{O}_{Z'}$. The homomorphism

$$H^0(X, Z') \rightarrow H^0\left(Z', \mathcal{O}_{Z'}(Z')\right) = \mathbb{C}$$

is surjective, hence actually $h^0(X, Z') = 2$, a contradiction.

This means that we must be in case (i) where Z' is a -2-curve.   □

Lemma 3.7 *Let (Y, D) be a cluster surface, and $Z' \subset Y$ an interior -2-curve. Then there is a toric model $q : (Y, D) \rightarrow (X', B')$ that contracts Z'.*

Sketch of proof First contract Z' to an interior node and then run a MMP. There is a small number of cases to discuss depending on how the MMP terminates.   □

4 Conjecture B

In this section we state a new conjecture — Conjecture B — which implies, see Remark 4.1, the existence of the map $\kappa : \mathfrak{B} \to \mathfrak{A}$ of Conjecture A. In the strong form stated in Remark 4.2, together with Conjecture A, Conjecture B asserts that the smoothing components of $\mathrm{Def}(V, \partial V)$ are themselves smooth, and computes their tangent space explicitly as a representation of the torus. We conclude by explaining how the two Conjectures A and B originate in mirror symmetry. This last discussion is central to how we arrived at the formulation of the conjectures, but it is not logically necessary for making sense of their statement. We work with the version of mirror symmetry put forward in [22] and [19, 20].[11]

For the remainder of this section fix a lattice polygon F and denote, as usual, by V the corresponding Gorenstein toric affine 3-fold with toric boundary ∂V.

In this section we always work with the space $\mathrm{Def}(V, \partial V)$. Also fix a 0-mutable polynomial $f \in \mathbb{C}[\mathbb{L}]$ with $\mathrm{Newt}\, f = F$. Conjecture B associates to f a $\mathbb{T}$-equivariant family $(\mathscr{U}_f, \mathscr{D}_f) \to \mathscr{M}_f$ of deformations of the pair $(V, \partial V)$.

In the last part of this section we construct from f a 3-dimensional cluster variety (W, D), conjecturally the *mirror* of $\mathscr{M}_f$, and hint at an explicit conjectural construction of the family $\mathscr{U}_f \to \mathscr{M}_f$ from the degree-0 quantum log cohomology of (W, D).

Denote by $\sigma \subset N_{\mathbb{R}}$ the cone over F at height 1. As usual, $u \in M = \mathrm{Hom}_{\mathbb{Z}}(N, \mathbb{Z})$ denotes the Gorenstein degree, so $F = \sigma \cap \mathbb{L}$ where $\mathbb{L} = (u = 1)$.

4.1 Statement of Conjecture B

As in Sec. 3.2, denote by $\mathscr{S}(f)$ the set of mutation data of f. Recall that an element of $\mathscr{S}(f)$ is a pair (φ, h) consisting of an affine function $\varphi \in \mathrm{Aff}(\mathbb{L}, \mathbb{Z})$ and a Laurent polynomial $h \in \mathbb{C}[\mathrm{Ker}\, \varphi_0]$. Using the restriction isomorphism $M \simeq \mathrm{Aff}(\mathbb{L}, \mathbb{Z})$, when it suits us we view a mutation datum (φ, h) as a pair of an element $\varphi \in M$ and a polynomial $h \in \mathbb{C}[\mathbb{L}_0 \cap \mathrm{Ker}\, \varphi] \subseteq \mathbb{C}[N]$.

The most useful mutation data are those where φ is strictly negative somewhere on F, and then the minimum of φ on F is achieved on an edge $E \leq F$.[12] In the present discussion we want to focus on these mutation data:

---

[11] We thank Paul Hacking for several helpful discussions on mirror symmetry and for correcting earlier drafts of this section. We are of course responsible for the mistakes that are left.
[12] Recall that in Definition 2.6 we explicitly assume that φ is not constant on F.

$$\mathscr{S}_-(f) = \left\{ (\varphi, h) \in \mathscr{S}(f) \;\middle|\; \begin{array}{l} \text{for some edge } E \leq F, \\ \varphi|_E \text{ is constant and } < 0 \end{array} \right\}.$$

We want to define a seed $\widetilde{\mathscr{F}}(f)$ on N, that is, a set of pairs (φ, h) of a character $\varphi \in M$ and a Laurent polynomial $h \in \mathbb{C}[\operatorname{Ker}\varphi]$. The seed we want is

$$\widetilde{\mathscr{F}}(f) = \mathscr{S}_-(f) \cup \left\{ (-ku, h) \;\middle|\; h \text{ is a prime factor of } f \text{ of multiplicity } k \right\}.$$

Conjecture B *In this statement, if U is a representation of the torus $\mathbb{T} = \operatorname{Spec}\mathbb{C}[M]$ and $m \in M$ is a character of $\mathbb{T}$, we denote by $U(m)$ the direct summand of U on which $\mathbb{T}$ acts with pure weight m.*

Let $F \subset \mathbb{L}$ be a lattice polygon, V the corresponding Gorenstein toric affine 3-fold, and $f \in \mathbb{C}[\mathbb{L}]$ a 0-mutable polynomial with Newt $f = F$. For every integer $k \geq 1$, denote by n_k the number of prime factors of f of multiplicity $\geq k$.

Then there is a $\mathbb{T}$-invariant submanifold $\mathscr{M}_f \subset \operatorname{Def}(V, \partial V)$ such that

$$\dim T_0 \mathscr{M}_f(m) = \begin{cases} 1 & \text{if } m \notin \langle -u \rangle_+ \text{ and there exists } (\varphi, h) \in \widetilde{\mathscr{F}}(f) \\ & \text{such that } m = \varphi, \\ n_k & \text{if } m = -ku \text{ for some integer } k \geq 1, \\ 0 & \text{otherwise,} \end{cases}$$

and the general fibre of the family over $\mathscr{M}_f$ is a pair consisting of a smooth variety and a smooth divisor.

Remark 4.1 If Conjecture B holds then by openness of versality a general point of $\mathscr{M}_f$ lies in a unique component of $\operatorname{Def}(V, \partial V)$ and this gives the map $\kappa \colon \mathfrak{B} \to \mathfrak{A}$ in the statement of Conjecture A (see also Remark 2.3).

Remark 4.2 (Strong form of Conjecture B) A strong form of Conjecture B states that the families $\mathscr{M}_f$ are precisely the smoothing components of $\operatorname{Def}(V, \partial V)$.

Example 4.3 (Example 2.12 continued) Consider the quadrilateral F in $\mathbb{L} = \mathbb{Z}^2$ defined in (2) and the three 0-mutable polynomials α, β and γ with Newton polytope F. Set $N = \mathbb{L} \oplus \mathbb{Z} = \mathbb{Z}^3$ and consider the following linear functionals in $M = \operatorname{Hom}_{\mathbb{Z}}(N, \mathbb{Z}) = \mathbb{Z}^3$:

$$-m_{2,1}^1 = (0, 1, -1) \qquad\qquad -m_{2,1}^2 = (1, 0, -1)$$
$$-m_{3,1}^1 = (0, 1, -2) \qquad\qquad -m_{3,1}^2 = (1, 0, -2)$$
$$-m_{3,2}^1 = (0, 2, -1) \qquad\qquad -m_{3,2}^2 = (2, 0, -1)$$

and $-u = (0, 0, -1)$. The names $-m_{2,1}^1$, $-m_{3,1}^1$ and $-m_{3,2}^1$ are compatible with the affine-linear functions considered in Example 2.12 via the restriction isomorphism $M \simeq \mathrm{Aff}(\mathbb{L}, \mathbb{Z})$. Set $x = x^{(1,0,0)} \in \mathbb{C}[N]$ and $y = x^{(0,1,0)} \in \mathbb{C}[N]$. Then we have:

$$\widetilde{\mathscr{F}}(\alpha) \supseteq \left\{ (-u, 1 + x + 2y + y^2), (-u, 1 + y + 2x + x^2) \right\},$$

$$\widetilde{\mathscr{F}}(\alpha) \supseteq \left\{ (-m_{2,1}^1, 1 + x), (-m_{3,2}^1, 1 + x), (-m_{2,1}^2, 1 + y), (-m_{3,2}^2, 1 + y) \right\},$$

$$\widetilde{\mathscr{F}}(\beta) \supseteq \left\{ (-u, \beta), (-m_{2,1}^1, 1 + x), (-m_{3,1}^1, 1 + x), (-m_{3,2}^1, 1 + x) \right\},$$

$$\widetilde{\mathscr{F}}(\gamma) \supseteq \left\{ (-u, \gamma), (-m_{2,1}^2, 1 + y), (-m_{3,1}^2, 1 + y), (-m_{3,2}^2, 1 + y) \right\}.$$

Let V be the Gorenstein toric affine 3-fold associated to F. Conjecture B states that there are three submanifolds $\mathscr{M}_\alpha$, $\mathscr{M}_\beta$ and $\mathscr{M}_\gamma$ of $\mathrm{Def}(V, \partial V)$ such that the dimensions of $T_0 \mathscr{M}_\alpha(m)$, $T_0 \mathscr{M}_\beta(m)$ and $T_0 \mathscr{M}_\gamma(m)$ for $m \in \{-u, -m_{2,1}^1, -m_{3,1}^1, -m_{3,2}^1, -m_{2,1}^2, -m_{3,1}^2, -m_{3,2}^2\}$ are written down in the table below.

	$-u$	$-m_{2,1}^1$	$-m_{3,1}^1$	$-m_{3,2}^1$	$-m_{2,1}^2$	$-m_{3,1}^2$	$-m_{3,2}^2$
$\dim T_0 \mathscr{M}_\alpha(m)$	2	1	0	1	1	0	1
$\dim T_0 \mathscr{M}_\beta(m)$	1	1	1	1	0	0	0
$\dim T_0 \mathscr{M}_\gamma(m)$	1	0	0	0	1	1	1

4.2 Mirror Symmetry Interpretation

Denote by $\mathscr{U}_f \to \mathscr{M}_f$ the deformation family of V induced by the composition $\mathscr{M}_f \hookrightarrow \mathrm{Def}(V, \partial V) \twoheadrightarrow \mathrm{Def}\, V$. We sketch a construction of $\mathscr{U}_f$ in the spirit of intrinsic mirror symmetry [19, 20].

Let $\sigma^\vee \subset M_\mathbb{R}$ be the dual cone, and let $s_1, \ldots, s_r$ be the primitive generators of the rays of $\sigma^\vee$. Denote by $W^\flat$ the toric variety — for the dual torus $\mathbb{T}_M = \mathrm{Spec}\,\mathbb{C}[N]$ — constructed from the fan consisting of the cones $\{0\}$, $\langle -u \rangle_+$, the $\langle s_j \rangle_+$, and the two-dimensional cones

$$\langle -u, s_j \rangle_+$$

(for $j = 1, \ldots, r$), and let $D^\flat \subset W^\flat$ be the toric boundary.

Now, the set of edges $E \leq F$ is in 1-to-1 correspondence with the set $\{s_1, \ldots, s_r\}$, where E corresponds to s_j if $s_j|_E = 0$. If $(\varphi, h) \in \widetilde{\mathscr{F}}(f)$ and $\varphi \notin \langle -u \rangle_+$, then there is a unique j such that φ is in the cone $\langle -u, s_j \rangle_+$ spanned by $-u$ and s_j.

Let

$$(\overline{W}, \overline{D}) \longrightarrow (W^\flat, D^\flat)$$

be the toric variety obtained by adding the rays $\langle \varphi \rangle_+ \subset M_{\mathbb{R}}$ whenever $(\varphi, h) \in \tilde{\mathscr{F}}(f)$, and (infinitely many) two-dimensional cones subdividing the cones $\langle -u, s_j \rangle_+$. Note that $\overline{W}$ is not quasi-compact and not proper.

Finally, we construct a projective morphism

$$(W, D) \longrightarrow (\overline{W}, \overline{D})$$

by a sequence of blowups.

In what follows for all $(\varphi, h) \in \tilde{\mathscr{F}}(f)$ we denote by $D_{\langle \varphi \rangle_+} \subset \overline{W}$ the corresponding boundary component, and set

$$Z_h = (h = 0) \subset D_{\langle \varphi \rangle_+}.$$

The following simple remarks will be helpful in describing the construction.

(a) For all positive integers k, $(k\varphi, h) \in \mathscr{S}_-(f)$ if and only if $(\varphi, h^k) \in \mathscr{S}_-(f)$.

(b) By construction, if $(\varphi, h) \in \mathscr{S}_-(f)$, then one has $h = (1 + x^e)^k$ (up to translation) for some positive integer k, where $e \in M$ is a primitive lattice vector along the edge $E \leq F$ where φ achieves its minimum.

Let $R \subset M_{\mathbb{R}}$ be a ray of the fan of $\overline{W}$ other than $\langle -u \rangle_+$, and let $\varphi \in M$ be the primitive generator of R. It follows from the remarks just made that there is a largest positive integer k_R such that $(k_R \varphi, 1 + x^e) \in \mathscr{S}_-(f)$.

Our mirror W is obtained from $\overline{W}$ by:

(1) First, as R runs through all the rays of the fan of $\overline{W}$ other than $\langle -u \rangle_+$ in some order, blow up k_R times above $(1 + x^e = 0)$ in the proper transform of D_R. It can be seen, and it is a nontrivial fact, that after doing all these blowups the Z_h in the proper transforms of the $D_{\langle -u \rangle_+}$ are smooth;

(2) Subsequently, if $f = \prod h^{k(h)}$ where the h are irreducible, blow up in any order $k(h)$ times above Z_h in the proper transform of $D_{\langle -u \rangle_+}$.

The resulting CY pair (W, D) is log smooth, because we have blown up a sequence of smooth centres. This (W, D) is the *mirror* of $\mathscr{M}_f$.

One can see that the Mori cone $\mathrm{NE}(W/\overline{W})$ is simplicial, and hence there is an identification:

$$\mathscr{M}_f = \mathrm{Spec}\, \mathbb{C}[\mathrm{NE}(W/\overline{W})].$$

Mirror symmetry suggests — modulo issues with infinite-dimensionality — that the ring $QH^0_{\log}(W, D; \mathbb{C}[\mathrm{NE}(W/\overline{W})])$ has a natural filtration and that one recovers the universal family of pairs $(\mathscr{U}_f, \mathscr{D}_f) \to \mathscr{M}_f$ from this ring out of the Rees construction.

Remark 4.4 In the context of Conjecture B, it would be very nice to work out an interpretation of the coefficients of the 0-mutable polynomial f as counting certain holomorphic disks on the general fibre of the family $\mathscr{U}_f \to \mathscr{M}_f$.

5 Evidence

We have already remarked that the variety V of Example 2.11 confirms Conjecture A. Here we collect some further evidence. Section 6 is a study of Def V where V is the variety of Example 2.12 and Example 4.3.

5.1 Isolated Singularities

Here we fix a lattice polygon F with unit edges, i.e. edges with lattice length 1. Let V be the Gorenstein toric affine 3-fold associated to F; we have that V has an isolated singularity.

Altmann [4] proved that there is a 1-to-1 correspondence between the irreducible components of Def V and the maximal Minkowski decompositions of F. This restricts to a 1-to-1 correspondence between the smoothing components of Def V and the Minkowski decompositions of F with summands that are either unit segments or standard triangles. Here a standard triangle is a lattice triangle that is $\mathbb{Z}^2 \rtimes \mathrm{GL}_2(\mathbb{Z})$-equivalent to $\mathrm{conv}\,((0,0),(1,0),(0,1))$.

On a polygon F with unit edges, the 0-mutable polynomials are exactly those that are associated to the Minkowski decompositions of F into unit segments and standard triangles. This confirms Conjecture A.

5.2 Local Complete Intersections

Nakajima [28] has characterised the affine toric varieties that are local complete intersection (lci for short). These are Gorenstein toric affine varieties associated to certain lattice polytopes called Nakajima polytopes. We refer the reader to [12, Lemma 2.7] for an inductive characterisation of Nakajima polytopes. From this characterisation it is very easy to see that every Nakajima polygon is affine equivalent to

$$F_{a,b,c} = \mathrm{conv}\,((0,0),(a,0),(0,b),(a,b+ac)) \qquad (3)$$

in the lattice $\mathbb{Z}^2$, for some non-negative integers a,b,c such that $a \geq 1$ and $b + c \geq 1$. It is easy to show that the Gorenstein toric affine 3-fold associated to the polygon $F_{a,b,c}$ is

$$V_{a,b,c} = \mathrm{Spec}\,\mathbb{C}[x_1,x_2,x_3,x_4,x_5]/(x_1x_2 - x_4^c x_5^b,\ x_3x_4 - x_5^a).$$

There is a unique 0-mutable polynomial on $F_{a,b,c}$: this is associated to the unique Minkowski decomposition of $F_{a,b,c}$ into a copies of the triangle $F_{1,0,c} = \text{conv}((0,0),(1,0),(0,c))$ and b copies of the segment $\text{conv}((0,0),(0,1))$. On the other hand, as $V_{a,b,c}$ is lci, we have that $V_{a,b,c}$ is unobstructed and smoothable, therefore there is a unique smoothing component in the miniversal deformation space of $V_{a,b,c}$. This confirms Conjecture A.

6 A Worked Example

We explicitly compute the smoothing components of the miniversal deformation space of the Gorenstein toric affine 3-fold V associated to the polygon F of Example 2.12 (continued in Example 4.3). We saw that there exist exactly three 0-mutable polynomials with Newton polytope F: α, β and γ. We explicitly compute the miniversal deformation space of V and see that it has three irreducible components, all of which are smoothing components. This confirms our conjectures.

6.1 The Equations of V

We consider the quadrilateral

$$F = \text{conv}\left(\begin{pmatrix} -1 \\ -1 \end{pmatrix}, \begin{pmatrix} 2 \\ -1 \end{pmatrix}, \begin{pmatrix} 1 \\ 1 \end{pmatrix}, \begin{pmatrix} -1 \\ 2 \end{pmatrix}\right) \tag{4}$$

in the lattice $\mathbb{L} = \mathbb{Z}^2$. We consider the cone σ obtained by placing F at height 1, i.e. σ is the cone generated by

$$a_1 = \begin{pmatrix} -1 \\ -1 \\ 1 \end{pmatrix} \qquad a_2 = \begin{pmatrix} 2 \\ -1 \\ 1 \end{pmatrix} \qquad a_3 = \begin{pmatrix} 1 \\ 1 \\ 1 \end{pmatrix} \qquad a_4 = \begin{pmatrix} -1 \\ 2 \\ 1 \end{pmatrix}$$

in the lattice $N = \mathbb{L} \oplus \mathbb{Z} = \mathbb{Z}^3$, and the corresponding Gorenstein toric affine 3-fold $V = \text{Spec}\,\mathbb{C}[\sigma^\vee \cap M]$, where $\sigma^\vee$ is the dual cone of σ in the dual lattice $M = \text{Hom}_{\mathbb{Z}}(N, \mathbb{Z}) \simeq \mathbb{Z}^3$.

The Gorenstein degree is

$$u = (0,0,1) \in M.$$

The primitive generators of the rays of the dual cone $\sigma^\vee \subseteq M_{\mathbb{R}}$ are the vectors

$$s_1 = (0,1,1), \quad s_2 = (1,0,1), \quad s_3 = (-1,2,3), \quad s_4 = (-2,-1,3)$$

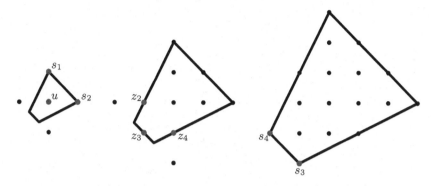

Figure 4 The intersections of the cone $\sigma^\vee$ with the planes $\mathbb{R}^2 \times \{1\}$, $\mathbb{R}^2 \times \{2\}$ and $\mathbb{R}^2 \times \{3\}$ in $M_\mathbb{R} = \mathbb{R}^3$.

which are orthogonal to the 4 edges of F. The Hilbert basis of the monoid $\sigma^\vee \cap M$ is the set of the vectors

$$u, \ s_1, \ z_2 = (-1, 0, 2), \ s_4, \ z_3 = (-1, -1, 2), \ s_3, \ z_4 = (0, -1, 2), \ s_2.$$

Notice that these are the Gorenstein degree u and certain lattice vectors on the boundary of $\sigma^\vee$. The elements of the Hilbert basis of $\sigma^\vee \cap M$ are depicted in Figure 4.

The elements of the Hilbert basis of $\sigma^\vee \cap M$ give a closed embedding of V inside $\mathbb{A}^8$ such that the ideal is generated by binomial equations. By using rolling factors formats (see [32] and [33, §12]), one can[13] see that these equations are:

$$\operatorname{rank} \begin{pmatrix} x_{s_1} & x_{z_2} & x_u & x_{s_2} & x_{z_4} \\ x_{z_2} & x_{s_4} & x_{z_3} & x_{z_4} & x_{s_3} \end{pmatrix} \leq 1,$$

$$x_{s_4} x_{s_3} - x_{z_3}^3 = 0 \qquad x_{z_2} x_{s_3} - x_{z_3}^2 x_u = 0,$$

$$x_{z_2} x_{z_4} - x_{z_3} x_u^2 = 0 \qquad x_{s_1} x_{z_4} - x_u^3 = 0.$$

The singular locus of V has two irreducible components of dimension 1: V has generically transverse A_2-singularities along each of these.

6.2 The Tangent Space

We consider the tangent space to the deformation functor of V, i.e. $T_V^1 = \operatorname{Ext}^1_{\mathcal{O}_V}(\Omega_V^1, \mathcal{O}_V)$. This is a $\mathbb{C}$-vector space with an M-grading. For every $m \in M$ we denote by $T_V^1(-m)$ the graded component of T_V^1 of degree $-m$.

---

[13] We are obliged to the referee for suggesting these equations to us.

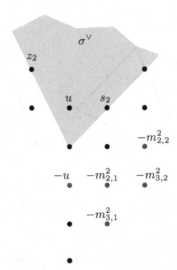

Figure 5 Some degrees of T_V^1 in the plane $\mathbb{R} \times \{0\} \times \mathbb{R} \subseteq M_{\mathbb{R}} = \mathbb{R}^3$.

Lemma 6.1 *We define* $J := \{(p,q) \in \mathbb{Z}^2 \mid 2 \le p \le 3,\ q \ge 1\}$. *For all* $p,q \in \mathbb{Z}$ *we set* $m_{p,q}^1 := pu - qs_1$ *and* $m_{p,q}^2 := pu - qs_2$.
  Then

$$
\dim T_V^1(-m) = \begin{cases}
1 & \text{if } m = u, \\
1 & \text{if } m = m_{p,q}^1 \text{ with } (p,q) \in J, \\
1 & \text{if } m = m_{p,q}^2 \text{ with } (p,q) \in J, \\
0 & \text{otherwise.}
\end{cases}
$$

Proof This is a direct consequence of [5, Theorem 4.4]. □

Some of the degrees of T_V^1 are depicted in Figure 5.

The base of the miniversal deformation of V is the formal completion (or germ) at the origin of a closed subscheme of the countable-dimensional affine space T_V^1. We denote by t_m the coordinate on the 1-dimensional $\mathbb{C}$-vector space $T_V^1(-m)$, when $m = u$ or $m \in \{m_{p,q}^1, m_{p,q}^2\}$ with $(p,q) \in J$. Since we want to understand the structure of $\mathrm{Def}\, V$, we want to analyse the equations of $\mathrm{Def}\, V \hookrightarrow T_V^1$ in the variables t_u, $t_{m_{p,q}^1}$ and $t_{m_{p,q}^2}$ for $(p,q) \in J$.

The first observation is that each homogeneous first order deformation of V is unobstructed as we see in the following two remarks.

Remark 6.2 The 1-dimensional $\mathbb{C}$-vector space $T_V^1(-u)$ gives a first order deformation of V, i.e. an infinitesimal deformation of V over $\mathbb{C}[t_u]/(t_u^2)$.

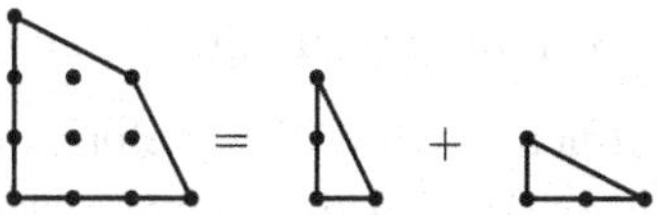

Figure 6 The Minkowski decomposition (5) of the quadrilateral F defined in (4).

This deformation can be extended to an algebraic deformation of V over $\mathbb{C}[t_u]$ as follows (see [3]).

Consider the unique non-trivial Minkowski decomposition of F (see Figure 6):

$$F = \operatorname{conv}\left(\begin{pmatrix}0\\0\end{pmatrix},\begin{pmatrix}1\\0\end{pmatrix},\begin{pmatrix}0\\2\end{pmatrix}\right) + \operatorname{conv}\left(\begin{pmatrix}-1\\-1\end{pmatrix},\begin{pmatrix}1\\-1\end{pmatrix},\begin{pmatrix}-1\\0\end{pmatrix}\right). \quad (5)$$

Let $\widetilde{F}$ be the Cayley polytope associated to this Minkowski sum; $\widetilde{F}$ is a 3-dimensional lattice polytope. Let $\tilde{\sigma}$ be the cone over $\widetilde{F}$ at height 1, i.e. $\tilde{\sigma}$ is the 4-dimensional cone generated by

$$\begin{pmatrix}0\\0\\1\\0\end{pmatrix},\begin{pmatrix}1\\0\\1\\0\end{pmatrix},\begin{pmatrix}0\\2\\1\\0\end{pmatrix},\begin{pmatrix}-1\\-1\\0\\1\end{pmatrix},\begin{pmatrix}1\\-1\\0\\1\end{pmatrix},\begin{pmatrix}-1\\0\\0\\1\end{pmatrix}.$$

Let $\widetilde{V}$ be the Gorenstein toric affine 4-fold $\widetilde{V}$ associated to $\tilde{\sigma}$. Consider the difference of the two regular functions on $\widetilde{V}$ associated to the characters $(0,0,1,0)$ and $(0,0,0,1)$; if we consider this regular function on $\widetilde{V}$ as a morphism $\widetilde{V} \to \mathbb{A}^1$, we obtain the following cartesian diagram

$$\begin{array}{ccc} V & \longrightarrow & \widetilde{V} \\ \downarrow & & \downarrow \\ \{0\} & \longrightarrow & \mathbb{A}^1 \end{array}$$

which gives the wanted 1-parameter deformation of V.

Remark 6.3 For every $m \in \{m^1_{p,q}, m^2_{p,q}\}$ with $(p,q) \in J$, the first order deformation of V corresponding to $T^1_V(-m) \simeq \mathbb{C}$ can be extended to an algebraic deformation of V over $\mathbb{C}[t_m]$ thanks to [5, Theorem 3.4] (see also [27, 29]).

6.3 The Obstruction Space

We now consider the obstruction space of the deformation functor of V: $T_V^2 =$ $\text{Ext}^2_{\mathcal{O}_V}(\Omega^1_V, \mathcal{O}_V)$. This is an M-graded $\mathbb{C}$-vector space. For all $m \in M$ we denote by $T_V^2(-m)$ the direct summand of degree $-m$.

Lemma 6.4 *If $m \in \{4u-s_1, 4u-s_2, 5u-s_1-s_2, 6u-s_1-s_2, 9u-2s_1-2s_2\}$, then* $\dim T_V^2(-m) = 1$. *Otherwise* $\dim T_V^2(-m) = 0$.

Proof This is a direct computation using formulae in [6, Section 5].   □

Remark 6.5 It immediately follows from the computation in [6, Section 5] that $T_V^2(-m) = 0$ if there exists a_i such that $\langle m, a_i \rangle \leq 0$.

6.4 Verifying the Conjectures

Since $\dim T_V^2 = 5$, the ideal of the closed embedding $\text{Def}\, V \hookrightarrow T_V^1$ has at most 5 generators. We have the following:

Proposition 6.6 *The equations of the closed embedding* $\text{Def}\, V \hookrightarrow T_V^1$ *are*

$$t_{m^1_{3,1}}\, t_u = 0,$$

$$t_{m^2_{3,1}}\, t_u = 0,$$

$$t_{m^1_{3,1}} t_{m^2_{2,1}} + t_{m^1_{2,1}} t_{m^2_{3,1}} = 0,$$

$$t_{m^1_{3,1}} t_{m^2_{3,1}} = 0,$$

$$t^2_{m^2_{3,1}} t_{m^1_{3,2}} - t^2_{m^1_{3,1}} t_{m^2_{3,2}} = 0.$$

Moreover, $\text{Def}\, V$ is non-reduced and has exactly 3 irreducible components; their equations inside T_V^1 are:

(1) $t_{m^1_{3,1}} = t_{m^2_{3,1}} = 0$,

(2) $t_u = t_{m^1_{3,1}} = t_{m^1_{2,1}} = t_{m^1_{3,2}} = 0$,

(3) $t_u = t_{m^2_{3,1}} = t_{m^2_{2,1}} = t_{m^2_{3,2}} = 0$.

Every irreducible component of $\text{Def}\, V$ is smooth and is a smoothing component.

Proof The proof of the equations of $\text{Def}\, V \hookrightarrow T_V^1$ is postponed to the next section and relies on some computer calculations performed with Macaulay2. We now assume to know these equations.

The fact that Def V is non-reduced and has 3 irreducible components C_1, C_2, C_3 with the equations given above can be checked by taking the primary decomposition of the ideal of Def $V \hookrightarrow T_V^1$. For each $i = 1, 2, 3$, from the equations of C_i it is obvious that C_i is smooth. We need to prove that C_i is a smoothing component, i.e. the general fibre over C_i is smooth.

The component C_1 contains the 1-parameter deformation constructed in Remark 6.2. The singular locus of the general fibre of this deformation has 2 connected components with everywhere transverse A_2-singularities; therefore the general fibre of this deformation is smoothable.

In order to prove that the general fibre over C_2 (resp. C_3) is smooth, we prove that the general fibre over the 2-parameter deformation of V with parameters $t_{m_{3,1}^2}$ and $t_{m_{2,2}^1}$ (resp. $t_{m_{3,1}^1}$ and $t_{m_{2,2}^2}$) is smooth. This can be done by applying the jacobian criterion to the output of the computer calculations that we will describe below. □

We now illustrate Conjecture A and Conjecture B in our example. Let C_1, C_2, C_3 be the 3 irreducible components of Def V, whose equations are given in Proposition 6.6. By Remark 2.3 Def$(V, \partial V)$ has 3 irreducible components, $\mathcal{M}_1$, $\mathcal{M}_2$, $\mathcal{M}_3$, each of which lie over exactly one of C_1, C_2, C_3. For each $i \in \{1, 2, 3\}$ the smooth morphism $\mathcal{M}_i \to C_i$ induces a surjective linear map $T_0\mathcal{M}_i \to T_0C_i$ of linear representations of the torus Spec $\mathbb{C}[M]$.

Let α, β and γ be the three 0-mutable polynomials with Newton polytope F (see Example 2.12). By comparing the degrees of T_0C_1, T_0C_2, T_0C_3 with the seeds $\tilde{\mathscr{S}}(\alpha)$, $\tilde{\mathscr{S}}(\beta)$, $\tilde{\mathscr{S}}(\gamma)$ in Example 4.3, we have that α (resp. β, resp. γ) corresponds to $\mathcal{M}_1$ (resp. $\mathcal{M}_2$, resp. $\mathcal{M}_3$).

6.5 Computer Computations

Here we present a proof of Proposition 6.6 which uses the software Macaulay2 [18], in particular the package `VersalDeformations` [26, 31].

By observing the degrees of T_V^1 (Lemma 6.1) and the degrees of T_V^2 (Lemma 6.4) it is immediate to see that each of the 5 equations of Def $V \hookrightarrow T_V^1$ can only involve the following 9 variables:

$$t_u \quad t_{m_{3,1}^1} \quad t_{m_{2,1}^1} \quad t_{m_{3,2}^1} \quad t_{m_{2,2}^1} \quad t_{m_{3,1}^2} \quad t_{m_{2,1}^2} \quad t_{m_{3,2}^2} \quad t_{m_{2,2}^2}.$$

We call the corresponding 9 degrees of T_V^1 the 'interesting' degrees of T_V^1. This implies that there is a smooth morphism Def $V \to G$, where G is a finite dimensional germ with embedding dimension 9. We now want to use the computer to determine G.

We consider the vector

$$\begin{pmatrix} 3 \\ 4 \\ 5 \end{pmatrix} \in N = \mathbb{Z}^3.$$

This gives a homomorphism $M \to \mathbb{Z}$ and a $\mathbb{Z}$-grading on the algebra $\mathbb{C}[\sigma^\vee \cap M] = H^0(V, \mathcal{O}_V)$, on T_V^1, and on T_V^2. We have chosen this particular $\mathbb{Z}$-grading because the corresponding linear projection is injective on the set $\{u, m_{3,1}^1, m_{2,1}^1, m_{3,2}^1, m_{2,2}^1, m_{3,1}^2, m_{2,1}^2, m_{3,2}^2, m_{2,2}^2\}$, which will allow us to identify our 9 variables above with the corresponding output of Macaulay2 below. In the following tables we write down the degrees in $\mathbb{Z}$ of the Hilbert basis of $\sigma^\vee \cap M$, of the interesting degrees of T_V^1, and of the degrees of T_V^2.

s_1	z_2	s_4	z_3	s_3	z_4	s_2	u
9	7	5	3	4	6	8	5

$-u$	$-m_{3,1}^1$	$-m_{2,1}^1$	$-m_{3,2}^1$	$-m_{2,2}^1$	$-m_{3,1}^2$	$-m_{2,1}^2$	$-m_{3,2}^2$	$-m_{2,2}^2$
-5	-6	-1	3	8	-7	-2	1	6

$4u - s_1$	$4u - s_2$	$5u - s_1 - s_2$	$6u - s_1 - s_2$	$9u - 2s_1 - 2s_2$
-11	-12	-8	-13	-11

One can see that all non-interesting summands of T_V^1 have degree ≥ 9. Therefore we are interested in the summands of T_V^1 with degree between -7 and 8. Now we run the following Macaulay2 code, which was suggested to us by the referee.

```
S = QQ[s1,z2,s4,z3,s3,z4,s2,u,Degrees=>{9,7,5,3,4,6,8,5}];
M = matrix {{s1,z2,u,s2,z4},{z2,s4,z3,z4,s3}};
I = minors(2,M) +
    ideal(s4*s3-z3^3,z2*s3-z3^2*u,z2*z4-z3*u^2,s1*z4-u^3);
needsPackage "VersalDeformations"
T1 = cotangentCohomology1(-7,8,I)
T2 = cotangentCohomology2(I)
(F,R,G,C) = versalDeformation(gens(I),T1,T2);
G
```

The output T1 describes how the equations of $V \hookrightarrow \mathbb{A}^8$ are perturbed, at the first order, by the coordinates $t_1, \ldots, t_9$ of the interesting part of T_V^1. From these perturbations one can compute the degrees of these coordinates and discover the following conversion table between our notation and the output of Macaulay2.

t_1	t_2	t_3	t_4	t_5	t_6	t_7	t_8	t_9
$t_{m_{3,1}^2}$	$t_{m_{3,2}^2}$	$t_{m_{2,2}^2}$	$t_{m_{2,1}^2}$	$t_{m_{3,1}^1}$	$t_{m_{3,2}^1}$	$t_{m_{2,2}^1}$	$t_{m_{2,1}^1}$	t_u

The output G describes the miniversal deformation space of V with degrees between -7 and 8, i.e. the germ G we wanted to study. This implies that G is the germ at the origin of the closed subscheme of $\mathbb{A}^9$ defined by the following equations:

$$t_5 t_9 = 0,$$
$$t_1 t_9 = 0,$$
$$t_4 t_5 + t_1 t_8 = 0,$$
$$t_1 t_5 = 0,$$
$$t_2 t_5^2 - t_1^2 t_6 = 0.$$

These equations are those in Proposition 6.6. The output F gives the equations of the deformation of V over the germ G.

Remark 6.7 The equations of the germ G are only well defined up to a homogeneous change of coordinates whose jacobian is the identity. In particular, the quadratic terms of these equations are well defined and can be computed by analysing the cup product $T_V^1 \otimes T_V^1 \to T_V^2$: this can be done via toric methods [14, 13].

References

[1] Akhtar, Mohammad, Coates, Tom, Galkin, Sergey, and Kasprzyk, Alexander M. 2012. Minkowski polynomials and mutations. *SIGMA Symmetry Integrability Geom. Methods Appl.*, **8**, Paper 094, 17.

[2] Akhtar, Mohammad, Coates, Tom, Corti, Alessio, Heuberger, Liana, Kasprzyk, Alexander, Oneto, Alessandro, Petracci, Andrea, Prince, Thomas, and Tveiten, Ketil. 2016. Mirror symmetry and the classification of orbifold del Pezzo surfaces. *Proc. Amer. Math. Soc.*, **144**(2), 513–527.

[3] Altmann, Klaus. 1995. Minkowski sums and homogeneous deformations of toric varieties. *Tohoku Math. J. (2)*, **47**(2), 151–184.

[4] Altmann, Klaus. 1997. The versal deformation of an isolated toric Gorenstein singularity. *Invent. Math.*, **128**(3), 443–479.

[5] Altmann, Klaus. 2000. One parameter families containing three-dimensional toric Gorenstein singularities. Pages 21–50 of: *Explicit birational geometry of 3-folds*. London Math. Soc. Lecture Note Ser., vol. 281. Cambridge Univ. Press, Cambridge.

[6] Altmann, Klaus, and Sletsjøe, Arne B. 1998. André-Quillen cohomology of monoid algebras. *J. Algebra*, **210**(2), 708–718.

[7] Blanc, Jérémy. 2013. Symplectic birational transformations of the plane. *Osaka J. Math.*, **50**(2), 573–590.

[8] Brown, Gavin, Kerber, Michael, and Reid, Miles. 2012. Fano 3-folds in codimension 4, Tom and Jerry. Part I. *Compos. Math.*, **148**(4), 1171–1194.

[9] Brown, Gavin, Reid, Miles, and Stevens, Jan. 2021. Tutorial on Tom and Jerry: the two smoothings of the anticanonical cone over $\mathbb{P}(1, 2, 3)$. *EMS Surv. Math. Sci.*, **8**, 25–38. https:/doi.org/10.4171/emss/43.

[10] Coates, Tom, Corti, Alessio, Galkin, Sergey, and Kasprzyk, Alexander. 2016. Quantum periods for 3-dimensional Fano manifolds. *Geom. Topol.*, **20**(1), 103–256.

[11] Coates, Tom, Kasprzyk, Alexander M., Pitton, Giuseppe, and Tveiten, Ketil. 2021. Maximally mutable Laurent polynomials. *Proc. Roy. Soc. A*, **477**(2254). https://doi.org/10.1098/rspa.2021.0584.

[12] Dais, Dimitrios I., Haase, Christian, and Ziegler, Günter M. 2001. All toric local complete intersection singularities admit projective crepant resolutions. *Tohoku Math. J. (2)*, **53**(1), 95–107.

[13] Filip, Matej. 2018. Hochschild cohomology and deformation quantization of affine toric varieties. *J. Algebra*, **506**, 188–214.

[14] Filip, Matej. 2021. The Gerstenhaber product $HH^2(A) \times HH^2(A) \to HH^3(A)$ of affine toric varieties. *Communications in Algebra*, **49**(3), 1146–1162. https://doi.org/10.1080/00927872.2020.1828906.

[15] Fomin, Sergey, and Zelevinsky, Andrei. 2002. Cluster algebras. I. Foundations. *J. Amer. Math. Soc.*, **15**(2), 497–529.

[16] Fulton, William. 1993. *Introduction to toric varieties*. Annals of Mathematics Studies, vol. 131. Princeton University Press, Princeton, NJ. The William H. Roever Lectures in Geometry.

[17] Galkin, Sergey, and Usnich, Alexandr. 2010. Mutations of potentials. *preprint IPMU*, 10–0100.

[18] Grayson, Daniel R., and Stillman, Michael E. *Macaulay2. A software system for research in algebraic geometry*. Available at http://www.math.uiuc.edu/Macaulay2/.

[19] Gross, Mark, and Siebert, Bernd. 2018. Intrinsic mirror symmetry and punctured Gromov–Witten invariants. Pages 199–230 of: *Algebraic geometry: Salt Lake City 2015*. Proc. Sympos. Pure Math., vol. 97. Amer. Math. Soc., Providence, RI.

[20] Gross, Mark, and Siebert, Bernd. 2019. *Intrinsic Mirror Symmetry*. arXiv:1909.07649.

[21] Gross, Mark, Hacking, Paul, and Keel, Sean. 2015a. Birational geometry of cluster algebras. *Algebr. Geom.*, **2**(2), 137–175.

[22] Gross, Mark, Hacking, Paul, and Keel, Sean. 2015b. Mirror symmetry for log Calabi–Yau surfaces I. *Publ. Math. Inst. Hautes Études Sci.*, **122**, 65–168.

[23] Gross, Mark, Hacking, Paul, and Keel, Sean. 2015c. Moduli of surfaces with an anti-canonical cycle. *Compos. Math.*, **151**(2), 265–291.

[24] Hauser, Herwig. 1983. An algorithm of construction of the semiuniversal deformation of an isolated singularity. Pages 567–573 of: *Singularities, Part 1 (Arcata, Calif., 1981)*. Proc. Sympos. Pure Math., vol. 40. Amer. Math. Soc., Providence, R.I.

[25] Hauser, Herwig. 1985. La construction de la déformation semi-universelle d'un germe de variété analytique complexe. *Ann. Sci. École Norm. Sup. (4)*, **18**(1), 1–56.

[26] Ilten, Nathan Owen. 2012. Versal deformations and local Hilbert schemes. *J. Softw. Algebra Geom.*, **4**, 12–16.

[27] Mavlyutov, Anvar R. 2009. *Deformations of toric varieties via Minkowski sum decompositions of polyhedral complexes.* arXiv:0902.0967.

[28] Nakajima, Haruhisa. 1986. Affine torus embeddings which are complete intersections. *Tohoku Math. J. (2)*, **38**(1), 85–98.

[29] Petracci, Andrea. 2021. Homogeneous deformations of toric pairs. *Manuscripta Mathematica*, **166**, 37–72. https://doi.org/10.1007/s00229-020-01219-w.

[30] Reid, Miles. 2015. Gorenstein in codimension 4: the general structure theory. Pages 201–227 of: *Algebraic geometry in east Asia—Taipei 2011.* Adv. Stud. Pure Math., vol. 65. Math. Soc. Japan, Tokyo.

[31] Stevens, Jan. 1995. Computing versal deformations. *Experiment. Math.*, **4**(2), 129–144.

[32] Stevens, Jan. 2001. Rolling factors deformations and extensions of canonical curves. *Doc. Math.*, **6**, 185–226.

[33] Stevens, Jan. 2003. *Deformations of singularities.* Lecture Notes in Mathematics, vol. 1811. Springer-Verlag, Berlin.

6

Vertex Algebras of CohFT-type

Chiara Damiolini[a], Angela Gibney[b] and Nicola Tarasca[c]

Dedicated to Bill Fulton on the occasion of his 80th birthday

Abstract. Representations of certain vertex algebras, here called of CohFT-type, can be used to construct vector bundles of coinvariants and conformal blocks on moduli spaces of stable curves [24]. We show that such bundles define semisimple cohomological field theories. As an application, we give an expression for their total Chern character in terms of the fusion rules, following the approach and computation in [70] for bundles given by integrable modules over affine Lie algebras. It follows that the Chern classes are tautological. Examples and open problems are discussed.

1 Introduction

Vertex algebras, fundamental in a number of areas of mathematics and mathematical physics, have recently been shown to be a source of new constructions for vector bundles on moduli of curves [46, 24]. In particular, given an n-tuple of modules M^i over a vertex algebra V satisfying certain natural hypotheses (stated in §3.1), one may construct the *vector bundle of coinvariants* $\mathbb{V}_g(V; M^{\bullet})$ on the moduli space $\overline{\mathcal{M}}_{g,n}$ of n-pointed stable curves of genus g [24]. The fiber at a pointed curve $(C, P_{\bullet})$ is the vector space of coinvariants, i.e., the largest quotient of $\otimes_{i=1}^{n} M^i$ by the action of a Lie algebra determined by $(C, P_{\bullet})$ and the vertex algebra V.

Such vector bundles generalize the classical coinvariants of integrable modules over affine Lie algebras [80, 81]. Bundles of coinvariants from vertex

[a] Rutgers University, New Jersey
[b] Rutgers University, New Jersey
[c] Virginia Commonwealth University

algebras have much in common with their classical counterparts. For instance, both support a projectively flat logarithmic connection [81, 25] and satisfy factorization [81, 24], a property that makes recursive arguments about ranks and Chern classes possible.

Following [70], bundles of coinvariants from integrable modules over affine Lie algebras give cohomological field theories (CohFTs for short). Here we show the same is true for their generalizations. We say that a vertex algebra V is of *CohFT-type* if V satisfies the hypotheses of §3.1.

We prove:

Theorem 1.1 *For a vertex algebra V of CohFT-type, the collection consisting of the Chern characters of all vector bundles of coinvariants from finitely-generated V-modules forms a semisimple CohFT.*

In particular, the ranks of the bundles of coinvariants form a *topological quantum field theory* (TQFT), namely, the degree zero part of the CohFT. As such, the ranks are recursively determined by the *fusion rules*, that is, the dimension of spaces of coinvariants on a three-pointed rational curve (Proposition 4.1). The fusion rules have been computed in the literature for many classes of vertex algebras of CohFT-type (see §6 for a few examples).

In fact, the CohFTs from Theorem 1.1 are determined by the fusion rules. Indeed, after work of Givental and Teleman [55, 56, 78], a semisimple CohFT is determined by its TQFT part together with some additional structure (see also [73]). As in [70], the explicit computation of the Atiyah algebra giving rise to the projectively flat logarithmic connection allows one to determine the recursion. As the Atiyah algebra in the case of bundles of coinvariants from vertex algebras was determined in [25], one is able to extend the reconstruction of the CohFTs of coinvariants from affine Lie algebras in [70] to the general case of vertex algebras.

Namely, following [70], there exists a polynomial $P_V(a_\bullet)$ with coefficients in $H^*\big(\overline{\mathcal{M}}_{g,n}\big)$, explicitly given in §5, such that the following holds:

Corollary 1.2 *For a vertex algebra V of CohFT-type and an n-tuple $M^\bullet$ of simple V-modules with M^i of conformal dimension a_i, the Chern character of the vector bundle of coinvariants $\mathbb{V}_g(V; M^\bullet)$ is*

$$\mathrm{ch}\big(\mathbb{V}_g(V; M^\bullet)\big) = P_V(a_\bullet) \quad in \ H^*\big(\overline{\mathcal{M}}_{g,n}\big).$$

By Corollary 1.2, Chern classes of bundles of coinvariants defined by vertex algebras of CohFT-type lie in the tautological ring of $\overline{\mathcal{M}}_{g,n}$. As an explicit example of the classes, the first Chern class in $\mathrm{Pic}_{\mathbb{Q}}(\overline{\mathcal{M}}_{g,n})$ is given by

Corollary 1.3, given next. As can be seen in the statement for the first Chern class, all Chern classes depend on the central charge of the vertex algebra and the conformal dimensions (or weights) of the modules. Since V is of CohFT-type, the central charge and the conformal dimensions of the modules are rational [39].

Corollary 1.3 *Let V be a vertex algebra of CohFT-type and central charge c, and let M^i be simple V-modules of conformal dimension a_i. Then*

$$c_1\big(\mathbb{V}_g(V;M^\bullet)\big) = \operatorname{rank}\mathbb{V}_g(V;M^\bullet)\left(\frac{c}{2}\lambda + \sum_{i=1}^{n} a_i\,\psi_i\right) - b_{\mathrm{irr}}\delta_{\mathrm{irr}} - \sum_{i,I} b_{i:I}\delta_{i:I}$$

with $\displaystyle b_{\mathrm{irr}} = \sum_{W\in\mathscr{W}} a_W \cdot \operatorname{rank}\mathbb{V}_{g-1}\big(V;M^\bullet \otimes W \otimes W'\big)$

and $\displaystyle b_{i:I} = \sum_{W\in\mathscr{W}} a_W \cdot \operatorname{rank}\mathbb{V}_i\left(V;M^I \otimes W\right) \cdot \operatorname{rank}\mathbb{V}_{g-i}\left(V;M^{I^c} \otimes W'\right).$

Here $\mathscr{W}$ is the set of finitely many simple V-modules; a_W is the conformal dimension of a simple V-module W (§2.3); for $I \subseteq [n] = \{1,\ldots,n\}$, we set $M^I := \otimes_{i\in I} M^i$; and the last sum is over i,I such that $i \in \{0,\ldots,g\}$ and $I \subseteq [n]$, modulo the relation $(i,I) \equiv (g-i,I^c)$.

Plan of paper: We start in §2 with some background on vertex algebras. In particular, there we describe the sheaf of coinvariants $\mathbb{V}_g(V;M^\bullet)$ and its dual, the sheaf of conformal blocks. In §3 we review a selection of results on vector bundles defined by representations of vertex algebras of CohFT-type, mainly from [25] and [24], which will be used to prove the statements above. Theorem 1.1 is proved in §4 and Corollary 1.2 in §5. In §6 we review the invariants necessary to compute the Chern classes in several examples, including the moonshine module vertex algebra $V^\natural$ and even lattice vertex algebras. We discuss the problem for commutant and orbifold vertex algebras, illustrating with parafermion vertex algebras, and orbifolds of lattice and parafermion vertex algebras.

From this and prior work, it is clear that the vector bundle of coinvariants from modules over vertex algebras of CohFT-type have a number of properties in common with their classical analogues, for which much has already been discovered. For instance, bundles of coinvariants defined from modules over affine Lie algebras are particularly interesting on $\overline{\mathcal{M}}_{0,n}$, where they are globally generated, and their sections define morphisms [43]. In particular, by studying their Chern classes one can learn about the maps they define. In §7 we discuss questions one might explore with this in mind.

2 Background

Here we briefly review vertex algebras, their modules, related Lie algebras, and the vector spaces of coinvariants they define. We refer the reader to [50, 67, 46, 25, 24] for details. We use notation as in [24], where further information and references on these topics can be found.

2.1 The Virasoro Algebra

The *Witt (Lie) algebra* Der $\mathcal{K}$ is the Lie algebra $\mathbb{C}((z))\partial_z$ generated by $L_p :=$ $-z^{p+1}\partial_z$, for $p \in \mathbb{Z}$, with Lie bracket given by $[L_p, L_q] = (p - q)L_{p+q}$.

The *Virasoro (Lie) algebra* Vir is a central extension of Der $\mathcal{K}$ which is generated by a formal vector K and the elements L_p, for $p \in \mathbb{Z}$, with Lie bracket given by

$$[K, L_p] = 0, \qquad [L_p, L_q] = (p - q)L_{p+q} + \frac{1}{12}(p^3 - p)\delta_{p+q,0}\,K.$$

A representation of Vir has *central charge* $c \in \mathbb{C}$ if $K \in$ Vir acts as $c \cdot$ id.

2.2 Vertex Operator Algebras

A *vertex operator algebra* is a four-tuple

$$\left(V, \mathbf{1}^V, \omega, Y(\cdot, z)\right)$$

with: $V = \oplus_{i \geq 0} V_i$ a $\mathbb{Z}_{\geq 0}$-graded $\mathbb{C}$-vector space with dim $V_i < \infty$; two elements $\mathbf{1}^V \in V_0$ (the *vacuum vector*) and $\omega \in V_2$ (the *conformal vector*); a linear map $Y(\cdot, z) \colon V \to \text{End}(V) \llbracket z, z^{-1} \rrbracket$ that assigns to $A \in V$ the *vertex operator* $Y(A, z) := \sum_{i \in \mathbb{Z}} A_{(i)} z^{-i-1}$. These data are required to satisfy suitable axioms, see e.g., [24, §1.1]. We review below some of the consequences which will be used in what follows. When no confusion arises, we refer to the four-tuple as V.

The Fourier coefficients of the fields $Y(\cdot, z)$ endow V with a series of products indexed by $\mathbb{Z}$, that is, $A *_i B := A_{(i)}B$, for $A, B \in V$. These products are *weakly* commutative and *weakly* associative.

The *conformal structure* of V realizes the Fourier coefficients $\omega_{(i)}$ as a representation of the Virasoro algebra on V via the identifications $L_p = \omega_{(p+1)}$ and $K = c \cdot \text{id}_V$ for a constant $c \in \mathbb{C}$ called the *central charge* of V. Moreover, L_0 is required to act as a degree operator on V, i.e., $L_0|_{V_i} = i \cdot \text{id}_{V_i}$, and L_{-1} (the *translation operator*) is given by $L_{-1}A = A_{(-2)}\mathbf{1}^V$, for $A \in V$.

Consequently, one has $A_{(i)}V_k \subseteq V_{k+\deg(A)-i-1}$ for homogeneous $A \in V$ [83], hence the *degree* of the operator $A_{(i)}$ is defined as $\deg A_{(i)} := \deg(A) - i - 1$.

2.3 Modules of Vertex Operator Algebras

There are a number of ways to define a module over a vertex operator algebra V. We take a V-module M to be a module over the universal enveloping algebra $\mathcal{U}(V)$ of V (defined by I. Frenkel and Zhu [48], see also [46, §5.1.5]) satisfying three finiteness properties. Namely, we assume that: (i) M is a finitely generated $\mathcal{U}(V)$-module; (ii) $F^0 \mathcal{U}(V)v$ is finite-dimensional, for every v in M; and (iii) for every v in M, there exists a positive integer k such that $F^k \mathcal{U}(V)v = 0$. These conditions are as in [72, Def. 2.3.1]. Here, $F^k \mathcal{U}(V) \subset \mathcal{U}(V)$ is the vector subspace topologically generated by compositions of operators with total degree less than or equal to $-k$.

E. Frenkel and Ben-Zvi [46, Thm 5.1.6] showed that there is an equivalence of categories between $\mathcal{U}(V)$-modules satisfying property (iii) and the so-called *weak* V-modules, which a priori are not graded. However, with the additional assumptions (i) and (ii), one can show the modules have a grading by the natural numbers. Such a V-module consists of a pair $(M, Y^M(\cdot, z))$, where $M = \oplus_{i \geq 0} M_i$ is a $\mathbb{Z}_{\geq 0}$-graded $\mathbb{C}$-vector space, and $Y^M(\cdot, z): V \to \mathrm{End}(M) [\![z, z^{-1}]\!]$ is a linear function that assigns to $A \in V$ an $\mathrm{End}(M)$-valued vertex operator $Y^M(A, z) := \sum_{i \in \mathbb{Z}} A_{(i)}^M z^{-i-1}$. Moreover, by condition (i), if $A \in V$ is homogeneous, then $A_{(i)}^M M_k \subseteq M_{k+\deg(A)-i-1}$.

The V-modules we work with are also known in the literature as finitely generated *admissible* V-modules (see for instance, [1] for the definitions of weak and admissible V-modules).

As for V, one has that M is also naturally equipped with an action of the Virasoro algebra with central charge c, induced by the identification of $\omega_{(p+1)}^M$ with L_p. When M is a simple V-module, there exists $a_M \in \mathbb{C}$, called the *conformal dimension* (or *conformal weight*) of M, such that $L_0(v) = (\deg(v) + a_M)v$, for every homogeneous v in M [83].

The vertex algebra V is a module over itself, sometimes referred to as the *adjoint module* [67, §4.1] or the *trivial module*. In what follows the set of simple modules over V is denoted $\mathcal{W}$.

2.4 Contragredient Modules

Contragredient modules provide a notion of duality for V-modules. We recall their definition following [50, §5.2]. Given a vertex algebra V and a V-module $(M = \oplus_{i \geq 0} M_i, Y^M(-, z))$, its *contragredient module* is

$$\left(M', Y^{M'}(-,z)\right),$$

where M' is the graded dual of M, that is, $M' := \oplus_{i \geq 0} M_i^\vee$, with $M_i^\vee := \mathrm{Hom}_{\mathbb{C}}(M_i, \mathbb{C})$, and

$$Y^{M'}(-,z) \colon V \to \mathrm{End}\left(M'\right)\left[\!\left[z, z^{-1}\right]\!\right]$$

is the unique linear map determined by

$$\left\langle Y^{M'}(A,z)\psi, m\right\rangle = \left\langle \psi, Y^M\left(e^{zL_1}(-z^{-2})^{L_0}A, z^{-1}\right)m\right\rangle$$

for $A \in V$, $\psi \in M'$, and $m \in M$. Here $\langle \cdot, \cdot \rangle$ is the natural pairing between a vector space and its graded dual.

2.5 The Lie Algebra Ancillary to V

The *Lie algebra ancillary to V* is defined as the quotient

$$\mathfrak{L}(V) = \mathfrak{L}_t(V) := \left(V \otimes \mathbb{C}((t))\right)\big/\mathrm{Im}\,\partial,$$

where t is a formal variable and $\partial := L_{-1} \otimes \mathrm{id}_{\mathbb{C}((t))} + \mathrm{id}_V \otimes \partial_t$. The image of $A \otimes t^i \in V \otimes \mathbb{C}((t))$ in $\mathfrak{L}(V)$ is denoted by $A_{[i]}$. Observe that $\mathfrak{L}(V)$ is spanned by series of the form $\sum_{i \geq i_0} c_i A_{[i]}$, for $A \in V$, $c_i \in \mathbb{C}$, and $i_0 \in \mathbb{Z}$. The Lie bracket is induced by

$$\left[A_{[i]}, B_{[j]}\right] := \sum_{k \geq 0} \binom{i}{k}\left(A_{(k)} \cdot B\right)_{[i+j-k]}.$$

There is a canonical Lie algebra isomorphism between $\mathfrak{L}(V)$ and the current Lie algebra in [72]. In what follows, the formal variable t is interpreted as a formal coordinate at a point P on an algebraic curve. A coordinate-free description of $\mathfrak{L}(V)$ is provided in §2.6.

For a V-module M, the Lie algebra homomorphism $\mathfrak{L}(V) \to \mathrm{End}(M)$ defined by

$$\sum_{i \geq i_0} c_i A_{[i]} \mapsto \mathrm{Res}_{z=0}\, Y^M(A,z) \sum_{i \geq i_0} c_i z^i dz$$

induces an action of $\mathfrak{L}(V)$ on M. For instance, $A_{[i]}$ acts as the Fourier coefficient $A_{(i)}$ of the vertex operator $Y^M(A,z)$.

2.6 The Vertex Algebra Bundle and Chiral Lie Algebra

Let $(C, P_\bullet)$ be a stable n-pointed curve. As illustrated in [46] for smooth curves and in [24] for stable curves, one can construct a vector bundle $\mathcal{V}_C$

(the *vertex algebra bundle*) on C whose fiber at each point of C is (non-canonically) isomorphic to V. For a smooth open subset $U \subset C$ admitting a global coordinate (e.g., if there exists an étale map $U \to \mathbb{A}^1$), the choice of a global coordinate on U gives a trivialization $\mathcal{V}_C|_U \cong V \times U$. When C is smooth the vertex algebra bundle $\mathcal{V}_C$ is constructed via descent along the torsor of formal coordinates at points in C. We refer to [24] for the description of $\mathcal{V}_C$ in the nodal case. The bundle $\mathcal{V}_C$ is naturally equipped with a flat connection $\nabla \colon \mathcal{V}_C \to \mathcal{V}_C \otimes \omega_C$ such that, up to the choice of a formal coordinate t_i at P_i, one can identify

$$H^0\left(D_{P_i}^\times, \mathcal{V}_C \otimes \omega_C / \mathrm{Im}\nabla\right) \cong \mathfrak{L}_{t_i}(V). \tag{1}$$

Here $D_{P_i}^\times$ is the punctured formal disk about the marked point $P_i \in C$. As shown in [46, §§19.4.14, 6.6.9], the isomorphism (1) induces the structure of a Lie algebra independent of coordinates on the left-hand side.

The *chiral Lie algebra* is defined as

$$\mathscr{L}_{C \setminus P_\bullet}(V) := H^0(C \setminus P_\bullet, \mathcal{V}_C \otimes \omega_C / \mathrm{Im}\nabla).$$

This space has indeed the structure of a Lie algebra after [46, §19.4.14].

2.7 The Action of the Chiral Lie Algebra on V-modules

Consider the linear map φ given by restriction of sections from $C \setminus P_\bullet$ to the n punctured formal disks $D_{P_i}^\times$ using the formal coordinates t_i at P_i:

$$\varphi \colon \mathscr{L}_{C \setminus P_\bullet}(V) \to \oplus_{i=1}^n \mathfrak{L}_{t_i}(V).$$

After [46, §19.4.14], φ is a homomorphism of Lie algebras. The map φ thus induces an action of $\mathscr{L}_{C \setminus P_\bullet}(V)$ on $\mathfrak{L}(V)^{\oplus n}$-modules which is used to construct coinvariants. See also [24, Proposition 3.3.2].

2.8 Sheaves of Coinvariants and Conformal Blocks

We briefly recall how to construct sheaves of coinvariants on $\overline{\mathcal{M}}_{g,n}$, and refer to [25, §5] for a detailed exposition. To a stable n-pointed curve $(C, P_\bullet)$ of genus g such that $C \setminus P_\bullet$ is affine, and to V-modules $M^1, \dots, M^n$, we associate the space of coinvariants

$$\mathbb{V}(V; M^\bullet)_{(C, P_\bullet)} := M^\bullet_{\mathscr{L}_{C \setminus P_\bullet}(V)} = M^\bullet / \mathscr{L}_{C \setminus P_\bullet}(V) \cdot M^\bullet,$$

where $M^\bullet = \otimes_{i=1}^n M^i$. Thanks to the *propagation of vacua*, it is possible to define these spaces also when $C \setminus P_\bullet$ is not affine via a direct limit. Carrying out the construction relatively over $\overline{\mathcal{M}}_{g,n}$, one defines the quasi-coherent *sheaf of* coinvariants $\mathbb{V}_g(V; M^\bullet)$ on $\overline{\mathcal{M}}_{g,n}$ assigned to $M^\bullet$. The dual sheaf $\mathbb{V}_g(V; M^\bullet)^\dagger$ is the *sheaf of conformal blocks* assigned to $M^\bullet$.

A brief history of coinvariants and conformal blocks and of the work on their properties can be found in [24, §§0.1 and 0.2].

3 Vector Bundles of Coinvariants

Here we review a number of results about vector bundles of coinvariants defined from representations of vertex algebras satisfying certain natural hypotheses. Motivated by the new results proved here, we name vertex algebras satisfying such hypotheses as *vertex algebras of CohFT-type*.

3.1 Vertex Algebras of CohFT-type

We define a vertex algebra V to be *of CohFT-type* if V is a simple, self-contragredient vertex operator algebra such that:

(I) $V = \oplus_{i \in \mathbb{Z}_{\geq 0}} V_i$ with $V_0 \cong \mathbb{C}$;
(II) V is *rational*, that is, every finitely generated V-module is a direct sum of simple V-modules; and
(III) V is C_2-*cofinite*, that is, the subspace

$$C_2(V) := \mathrm{span}_{\mathbb{C}} \left\{ A_{(-2)} B \ : \ A, B \in V \right\}$$

has finite codimension in V.

The set $\mathscr{W}$ of simple modules over a rational vertex algebra is finite, and a simple module $M = \oplus_{i \geq 0} M_i$ over a rational vertex algebra satisfies $\dim M_i < \infty$ [37].

The assumptions (I)-(III) on the vertex algebra have been found in [24] to imply that the sheaves of coinvariants are in fact vector bundles:

Theorem 3.1 *[24, VB Corollary] For a vertex algebra V of CohFT-type, the sheaf of coinvariants $\mathbb{V}_g(V; M^\bullet)$ assigned to finitely generated admissible V-modules $M^1, \ldots, M^n$ is a vector bundle of finite rank on the moduli space $\overline{\mathcal{M}}_{g,n}$.*

Since V is rational and C_2-cofinite, the central charge of V and the conformal dimension of every simple V-module are rational numbers [39]. When V is of CohFT-type, the Chern character of the bundles of coinvariants form a cohomological field theory, as we verify below, hence the name.

3.2 The Connection

Following [25], the restriction of the vector bundles $\mathbb{V}_g(V; M^\bullet)$ to $\mathcal{M}_{g,n}$ support a projectively flat connection. We can explicitly describe this using the language of Atiyah algebras [8]. Given a line bundle L on a variety, the Atiyah algebra $\mathcal{A}_L$ is the sheaf of first order differential operators acting on L. An analoguous construction holds for a virtual line bundle L^x, where $x \in \mathbb{C}$, yielding the Atiyah algebra $x\mathcal{A}_L$ [79]. With this terminology, the connection on $\mathbb{V}_g(V; M^\bullet)$ is explicitly described as follows:

Theorem 3.2 ([25])   *For n simple modules M^i of conformal dimension a_i over a vertex algebra V of CohFT-type and central charge c, the Atiyah algebra $\frac{c}{2}\mathcal{A}_\Lambda + \sum_{i=1}^n a_i \mathcal{A}_{\Psi_i}$ acts on the restriction of $\mathbb{V}_g(V; M^\bullet)$ to $\mathcal{M}_{g,n}$, specifying a twisted $\mathcal{D}$-module structure.*

Here Λ is the Hodge line bundle on $\mathcal{M}_{g,n}$ and Ψ_i is the cotangent line bundle at the i-th marked point on $\mathcal{M}_{g,n}$. Theorem 3.2 generalizes the analogous statement for bundles of coinvariants of integrable representations at a fixed level over affine Lie algebras [79]. This is proved more generally for *quasi-coherent* sheaves of coinvariants in [25, §7].

3.3 Chern Classes on $\mathcal{M}_{g,n}$

The explicit description of the connection determines the Chern character of the restriction of $\mathbb{V}_g(V; M^\bullet)$ on $\mathcal{M}_{g,n}$:

Corollary 3.3 ([25, Corollary 2])   *Let V be a vertex algebra of CohFT-type and central charge c, and let $M^1, \ldots, M^n$ be n simple V-modules of conformal dimension a_i. Then*

$$\mathrm{ch}\left(\mathbb{V}_g(V; M^\bullet)\right) = \mathrm{rank}\,\mathbb{V}_g(V; M^\bullet) \cdot \exp\left(\frac{c}{2}\lambda + \sum_{i=1}^n a_i \psi_i\right) \in H^*\left(\mathcal{M}_{g,n}\right).$$

Here $\lambda = c_1(\Lambda)$ and $\psi_i = c_1(\Psi_i)$. The corollary follows from: (a) a vector bundle E over a smooth base S with an action of the Atiyah algebra $\mathcal{A}_{L^{\otimes a}}$, for $L \to S$ a line bundle and $a \in \mathbb{Q}$, satisfies $c_1(E) = (\mathrm{rank}\,E)\,aL$ [69, Lemma 5]; and (b) the projectively flat connection implies that $\mathrm{ch}(E) = (\mathrm{rank}\,E)\exp(c_1(E)/\mathrm{rank}\,E)$ [62, (2.3.3)].

From Corollary 3.3, the total Chern class is

$$c\left(\mathbb{V}_g(V; M^\bullet)\right) = \left(1 + \frac{c}{2}\lambda + \sum_{i=1}^n a_i \psi_i\right)^{\mathrm{rank}\,\mathbb{V}_g(V; M^\bullet)} \in H^*\left(\mathcal{M}_{g,n}\right).$$

3.4 The Factorization Property

After [24, Factorization Theorem], for V of CohFT-type, the bundles $\mathbb{V}_g(V;M^\bullet)$ satisfy the *factorization property*. Assume that the curve C has one nodal point Q, let $\widetilde{C}$ be the partial normalization of C at Q, and let $Q_\bullet = (Q_+, Q_-)$ be the pair of preimages of Q in $\widetilde{C}$.

Theorem 3.4 ([24, Factorization Theorem]) *Let V be a vertex algebra of CohFT-type. Then*

$$\mathbb{V}\left(V;M^\bullet\right)_{(C,P_\bullet)} \cong \bigoplus_{W \in \mathscr{W}} \mathbb{V}\left(V;M^\bullet \otimes W \otimes W'\right)_{(\widetilde{C},P_\bullet \sqcup Q_\bullet)}.$$

When $\widetilde{C} = C_+ \sqcup C_-$ is disconnected, with $Q_\pm \in C_\pm$, one has:

$$\mathbb{V}\left(V;M^\bullet \otimes W \otimes W'\right)_{(\widetilde{C},P_\bullet \sqcup Q_\bullet)} \cong \mathbb{V}\left(V;M^\bullet_+ \otimes W\right)_{X_+} \otimes \mathbb{V}\left(V;M^\bullet_- \otimes W'\right)_{X_-}$$

where $X_\pm = \left(C_\pm, P_\bullet|_{C_\pm} \sqcup Q_\pm\right)$, and $M^\bullet_\pm$ are the modules at the $P_\bullet$ on $C_\pm$. The factorization property extends in families of nodal curves. More generally, the *sewing property* holds, extending the factorization property over the formal neighborhood of families of nodal curves [24].

3.5 Finding Ranks Through Recursions

As a consequence of the factorization property, the rank of $\mathbb{V}_g(V;M^\bullet)$, equal to the dimension of the vector space of coinvariants $\mathbb{V}(V;M^\bullet)_{(C,P_\bullet)}$ at any pointed curve $(C,P_\bullet)$, can be computed when $(C,P_\bullet)$ is maximally degenerate. Using propagation of vacua [46, §10.3.1] and inserting points with the adjoint module V if necessary, one may reduce to the case when all irreducible components of C are rational curves with three special points, and thus the rank may be expressed as sum of products of dimensions of vector spaces of coinvariants on three-pointed rational curves.

As the following two examples show, formulas for the ranks can be readily identified in some simple cases. A third such recursive calculation is carried out for bundles defined from modules over even lattice vertex algebras in Example 6.6.

Example 3.5 The rank of $\mathbb{V}_1(V;V)$ on $\overline{\mathcal{M}}_{1,1}$ equals the cardinality of the set of simple V-modules $\mathscr{W}$. This follows from factorization and the equality (2) in the next section.

Example 3.6 Let V be a vertex algebra of CohFT-type with no nontrivial modules. Bundles of coinvariants of modules over V on $\overline{\mathcal{M}}_{g,n}$ have rank 1.

In §4.2 we use the formalism of semisimple TQFTs to reconstruct the ranks from the fusion rules. Rank computations for bundles defined from modules over even lattice vertex algebras are done from this perspective in Examples 6.5 and 6.7.

4 Proof of Theorem 1.1 and TQFT Computations

In this section, using the results from §3, we prove Theorem 1.1, showing bundles of coinvariants defined by representations of vertex algebras of CohFT-type give rise to cohomological field theories (for more on CohFTs, see e.g., [70, §2]). Theorem 1.1 is proved as in the case of coinvariants from affine Lie algebras treated in [70, §3], and we follow their approach.

4.1 The CohFT of Chern Characters of Coinvariants

Let us start by defining the data of the CohFT. Let V be a vertex algebra of CohFT-type. As V is rational, the set $\mathscr{W}$ of simple V-modules is finite. The Hilbert space of the CohFT is $\mathcal{F}(V) := \mathbb{Q}^{\mathscr{W}} = \oplus_{W \in \mathscr{W}} \mathbb{Q} h_W$. The CohFT is defined by the classes

$$\mathrm{ch}\left(\mathbb{V}_g(V; M^\bullet)\right) \in H^*\left(\overline{\mathcal{M}}_{g,n}\right),$$

for $2g - 2 + n > 0$ and V-modules $M^1, \ldots, M^n$ viewed as elements of $\mathcal{F}(V)$ (hence necessarily finitely-generated), and extending by linearity. The vector space has a pairing η defined by $\eta(h_M, h_N) = \delta_{M, N'}$, where N' is the contragredient of N. In addition, $\mathcal{F}(V)$ has a commutative, associative product $*$ defined by

$$h_M * h_N := \sum_{W \in \mathscr{W}} \dim \mathbb{V}_0\left(V; M, N, W'\right) h_W,$$

for $M, N \in \mathscr{W}$ and extending by linearity, with unit h_V corresponding to the adjoint module V. This product extends linearly to the *fusion algebra* $\mathcal{F}(V)_{\mathbb{C}} := \mathcal{F}(V) \otimes_{\mathbb{Q}} \mathbb{C}$. This is a commutative, associative Frobenius algebra with unit. In particular, one has $\eta(a * b, c) = \eta(a, b * c)$, for $a, b, c \in \mathcal{F}(V)_{\mathbb{C}}$.

Proof of Theorem 1.1 The axioms necessary to form a CohFT are verified thanks to the factorization property in families [24, Thm 8.2.2], propagation of vacua in families ([46, §10.3.1] for a single smooth pointed curve, and [20, Prop. 3.6] for families of stable curves, see also [25, Thm 5.1]), and the fact that given simple V-modules M and N, one has

$$\text{rank } \mathbb{V}_0(V; M, N, V) = \delta_{M,N'}.$$ (2)

This follows from the identification of the three-point, genus zero conformal blocks for simple V-modules M, N, and W with the vector space $\text{Hom}_{A(V)}(A(W) \otimes_{A(V)} M_0, N_0^\vee)$ [48, 68]. Here $A(V)$ is Zhu's semisimple associative algebra, and $A(W)$ is an $A(V)$-bimodule generalizing the algebra $A(V)$ for a V-module W [48]. For $W = V$, the space of conformal blocks is thus isomorphic to $\text{Hom}_{A(V)}(M_0, N_0^\vee)$. The assumption that M and N are simple V-modules gives that M_0 and $N_0^\vee$ are simple $A(V)$-modules, and thus (2) follows from Schur's lemma. We note that statements asserting (2) can be found in case $V \cong V'$ or for even lattice vertex algebras in [58, 50, 32], and (2) is implicitly assumed elsewhere in the literature.

Finally, the CohFT is semisimple, or equivalently, the Frobenius algebra $\mathcal{F}(V)_\mathbb{C}$ is semisimple. This follows from the general argument in [6, §5 and Prop. 6.1] using contragredient duality in genus zero, non-negativity of ranks, and the factorization property. Contragredient duality, that is,

$$\text{rank } \mathbb{V}_0 \left(V; M^1, \dots, M^n \right) = \text{rank } \mathbb{V}_0 \left(V; (M^1)', \dots, (M^n)' \right),$$ (3)

is obtained in the affine Lie algebra case as a consequence of the stronger statement [6, Prop. 2.8]. In our case, we proceed as follows: By propagation of vacua, we can assume that enough of the modules M^i are equal to V, and by the factorization property, we can reduce to compute the rank over a totally degenerate stable rational curve, such that each component has at least one marked point where the adjoint module V is assigned. Then (3) follows from (2). $\qquad\qquad\square$

4.2 Computing the TQFT

As a consequence of Theorem 1.1, the ranks of vector bundles of coinvariants of vertex algebras of CohFT-type form a semisimple TQFT. Hence all ranks are determined by the dimensions of vector spaces of coinvariants on a rational curve with three marked points. We describe here the reconstruction of the TQFT of the ranks from the fusion rules following results on semisimple TQFTs [78, 66].

Let $\mathcal{F}(V)_\mathbb{C}$ be the fusion ring given from the semisimple TQFT determined by a vertex algebra V of CohFT-type, as in §3.1. Let $\{e_i\}_i$ be a *semisimple basis* of $\mathcal{F}(V)$, that is, $\eta(e_i, e_j) = \delta_{i,j}$ and $e_i * e_j = \delta_{i,j} \lambda_i e_i$, for some $\lambda_i \in \mathbb{C}$. The values λ_i are known as the *semisimple values* of the TQFT. Let $\{e^i\}_i$ be the dual basis to $\{e_i\}_i$.

Proposition 4.1   *The ranks of the vector bundles of coinvariants on* $\overline{\mathcal{M}}_{g,n}$ *assigned to* n *finitely-generated modules over a vertex algebra* V *of CohFT-type is given by the following linear functional on* $\mathcal{F}(V)^{\otimes n}$:

$$\sum_i \lambda_i^{2(g-1)+n}\ \underbrace{e^i \otimes \cdots \otimes e^i}_{n \text{ times}}.$$

The statement is a special case of [66, Prop. 4.1] applied to our TQFT. Examples are given in §6. In the case of coinvariants from modules over affine Lie algebras, the above formula reproduces the classical *Verlinde numbers* after some algebraic manipulations (see e.g., [6, 57]).

5 Chern Classes on $\overline{\mathcal{M}}_{g,n}$

In this section we prove Corollary 1.2 following [70]. We work with a vertex algebra V of CohFT-type. In particular the set $\mathscr{W}$ of simple V-modules is finite. The Chern characters of bundles of coinvariants are given by the polynomial $P_V(a_\bullet)$ defined as a sum over stable graphs. We start by reviewing stable graphs below, and then define the contributions to $P_V(a_\bullet)$ corresponding to vertices, edges, and legs.

5.1 Stable Graphs and Module Assignments

A *stable graph* is the dual graph of a stable curve. We only recall the basic features; for more details see, e.g., [74]. A stable graph Γ comes with a vertex set $V(\Gamma)$, an edge set $E(\Gamma)$, a half-edge set $H(\Gamma)$, and a leg set $L(\Gamma)$. Each leg has a label $i \in \{1, \ldots, n\}$, and this gives an isomorphism $L(\Gamma) \cong \{1, \ldots, n\}$. Each edge $e \in E(\Gamma)$ is the union of two half-edges $e = \{h, h'\}$, with $h, h' \in H(\Gamma)$. Each vertex $v \in V(\Gamma)$ has a genus label $g_v \in \mathbb{Z}_{\geq 0}$ and a valence n_v counting the number of half-edges and legs incident to v. The genus of the graph Γ is defined as $\sum_{v \in V(\Gamma)} g_v + h^1(\Gamma)$, where $h^1(\Gamma)$ is the first Betti number of Γ.

A stable graph Γ of genus g with n legs identifies a locally closed stratum in $\overline{\mathcal{M}}_{g,n}$ equal to the image of the glueing map of degree $|\mathrm{Aut}(\Gamma)|$:

$$\xi_\Gamma \colon \prod_{v \in V(\Gamma)} \overline{\mathcal{M}}_{g_v, n_v} =: \overline{\mathcal{M}}_\Gamma \to \overline{\mathcal{M}}_{g,n}.$$

Given a stable graph Γ, a *module assignment* is a function of type

$$\mu \colon H(\Gamma) \longrightarrow \mathscr{W}$$

such that for $(h, h') \in E(\Gamma)$, one has $\mu(h') = \mu(h)'$, that is, $\mu(h')$ is the contragredient module of $\mu(h)$ (§2.4).

5.2 Vertex Contributions

Fix a stable graph Γ and a module assignment $\mu: H(\Gamma) \to \mathcal{W}$. To each vertex v of Γ is assigned a collection of simple V-modules $M^{i_1}, \dots, M^{i_{n_v}}$, one for each leg or half-edge incident to v: for each leg i incident to v, the module M^i is assigned to v; and for each half-edge h incident to v, the module $\mu(h)$ is assigned to v. The vertex contribution is defined as

$$\mathrm{Cont}_\mu(v) := \mathrm{rank}\, \mathbb{V}_{g_v}\left(V; M^{i_1}, \dots, M^{i_{n_v}}\right).$$

5.3 Edge Contributions

Fix a stable graph Γ and a module assignment $\mu: H(\Gamma) \to \mathcal{W}$. For each edge $e = \{h, h'\}$, let $a_{\mu(h)}$ be the conformal dimension of $\mu(h)$. The edge contribution is defined as

$$\mathrm{Cont}_\mu(e) := \frac{1 - e^{a_{\mu(h)}(\psi_h + \psi_{h'})}}{\psi_h + \psi_{h'}}.$$

This is well defined since $\mu(h)$ and $\mu(h')$ have equal conformal dimension.

5.4 The Polynomial $P_V(a_\bullet)$

Following the computations of [70], consider the following polynomial with coefficients in $H^*\left(\overline{\mathcal{M}}_{g,n}\right)$:

$$P_V(a_\bullet) := e^{\frac{c}{2}\lambda} \sum_{\Gamma,\, \mu} \frac{1}{|\mathrm{Aut}(\Gamma)|} (\xi_\Gamma)_* \left(\prod_{i=1}^{n} e^{a_i \psi_i} \prod_{v \in V(\Gamma)} \mathrm{Cont}_\mu(v) \prod_{e \in E(\Gamma)} \mathrm{Cont}_\mu(e) \right),$$

where c is equal to the central charge of V. The sum in the formula is over all isomorphism classes of stable graphs Γ of genus g with n legs, and over all module assignments μ. For degree reasons, the exponentials are finite sums, and $P_V(a_\bullet)$ is indeed a polynomial.

For the vector bundle of coinvariants of modules over an affine vertex algebra, [70] shows that $\mathrm{ch}\,(\mathbb{V}(V; M^\bullet)) = P_V(a_\bullet)$, where $V = L_\ell(\mathfrak{g})$ is the simple affine vertex algebra, and M^i is a simple $L_\ell(\mathfrak{g})$-module of conformal dimension a_i, for each i. We extend this result to prove Corollary 1.2:

Proof of Corollary 1.2   From [70, Lemma 2.2], a semisimple CohFT is uniquely determined by the restriction of the classes to $\mathcal{M}_{g,n}$. More precisely, the results in [70] imply that any semisimple CohFT whose restriction to $\mathcal{M}_{g,n}$ is as in Corollary 3.3 for some $c, a_i \in \mathbb{Q}$ is given by an expression as in the statement.   □

6 Examples and Projects

6.1 Vertex Algebras with No Nontrivial Modules

We start with coinvariants constructed from a *holomorphic* vertex algebra V of CohFT-type, that is, a vertex algebra of CohFT-type such that V is the unique simple V-module. Any bundle of coinvariants of modules over V on $\overline{\mathcal{M}}_{g,n}$ has rank 1 and first Chern class equal to $\frac{c}{2}\lambda$, where c is the central charge of V. The rank assertion is in Example 3.6. The first Chern class follows from Theorem 1.2, as the conformal dimension of the adjoint module is zero.

Example 6.1   There are 71 holomorphic vertex algebras of CohFT-type with conformal dimension 24. This very special class includes the moonshine module vertex algebra $V^\natural$ (whose automorphism group is the monster group), and the vertex algebra given by the Leech lattice [64]. For such $V = \oplus_{i=0}^\infty V_i$, the weight one Lie algebra V_1 is either semi-simple, abelian of rank 24, or 0. If V_1 is abelian of rank 24, then V is isomorphic to the Leech lattice vertex algebra. If $V_1 = 0$, it is conjectured that $V \cong V^\natural$. Vertex algebras with the 69 other possible Lie algebras V_1 have been constructed in [64]. Each gives a vector bundle of coinvariants of rank 1 and first Chern class 12λ.

6.2 Even Lattice Vertex Algebras

As we illustrate below, even lattice vertex algebras are of CohFT-type, hence following Theorem 3.1, their simple modules define vector bundles on $\overline{\mathcal{M}}_{g,n}$.

For the definitions of lattice vertex algebras we recommend [18, 49, 28, 67]. We briefly review the notation. Let L be a positive-definite even lattice. That is, L is a free abelian group of finite rank d together with a positive-definite bilinear form $(\cdot, \cdot)$ such that $(\alpha, \alpha) \in 2\mathbb{Z}$ for all $\alpha \in L$. One assigns to L the *even lattice vertex algebra* V_L. This has finitely many simple modules $\{V_{L+\lambda} \mid \lambda \in L'/L\}$, where $L' := \{\lambda \in L \otimes_\mathbb{Z} \mathbb{Q} \mid (\lambda, \mu) \in \mathbb{Z}, \text{ for all } \mu \in L\}$ is the dual lattice [28]. Contragredient modules are determined by $V'_{L+\lambda} = V_{L-\lambda}$. The following statement follows from results in the literature.

Proposition 6.2 ([18, 47, 28, 32, 39]) *For a positive-definite even lattice L of rank d, one has:*

(i) *The lattice vertex algebra V_L is of CohFT-type with central charge $c = d$;*

(ii) *The conformal dimension of the module $V_{L+\lambda}$ is $\min_{\alpha \in L} \frac{(\lambda+\alpha,\lambda+\alpha)}{2}$;*

(iii) *The product in the fusion algebra $\mathcal{F}(V_L)_{\mathbb{C}} = \bigoplus_{\lambda \in L'/L} \mathbb{C}h_\lambda$ is given by*

$$h_{\lambda_1} * h_{\lambda_2} = h_{\lambda_1 + \lambda_2}.$$

Proof By [18, 47] V_L satisfies property (I), by [28] V_L is rational, and hence satisfies property (II), and by [39, Proposition 12.5] V_L is C_2-cofinite, so satisfies property (III). The central charge is computed in [47, Theorem 8.10.2], and the conformal dimension is deduced implicitly in [28, page 260]. The fusion rules are described in [32, Chapter 12]. $\qquad\square$

Proposition 6.2 contains all ingredients needed to compute ranks of bundles of coinvariants from modules over V_L applying Proposition 4.1 and their Chern characters applying Corollary 1.2. Let us discuss some examples.

Remark 6.3 There are a number of lattices L one may use to construct the vertex algebras V_L, and it is straightforward to cook up a lattice of almost any rank, whose discriminant group L'/L has arbitrary order. The order of the discriminant is the determinant of the Gram matrix for a basis of the lattice. For instance, to obtain $L'/L \cong \mathbb{Z}/2k\mathbb{Z}$, for $k \in \mathbb{N}$, one can take a one-dimensional lattice with basis vector e such that $(e, e) = 2k$.

For any root system (see e.g., [61, pgs 352–355]), there is a root lattice Λ, and for those of type A, D, E, F, and G, the lattice is even, and gives rise to a vertex algebra V_Λ. Every irreducible root system Λ corresponds to a simple Lie algebra $\mathfrak{g}_\Lambda$. If one normalizes the associated bilinear form (encoded by the Dynkin diagram), then in these cases one has $V_\Lambda \cong L_1(\mathfrak{g}_\Lambda)$, the simple affine vertex algebra at level 1 (see [47] and [67, Rmk 6.5.8] for details). The weight lattice gives the dual lattice Λ'. For instance, for $\Lambda = A_{m-1}$ the root lattice has rank $m - 1$, so that $V_\Lambda \cong L_1(\mathfrak{sl}_m)$ has conformal dimension $m - 1$.

One may also construct vertex algebras V_L by taking L to be the direct sum of lattices described above, getting quotient lattices L'/L that are of the form $\mathbb{Z}/m_1\mathbb{Z} \oplus \cdots \oplus \mathbb{Z}/m_k\mathbb{Z}$ for arbitrary $m_1, \ldots, m_k$. Such lattices may be interpreted as Mordel-Weil lattices (see e.g., [77, 76]).

The root lattices are very special, and one may easily construct more general lattices with discriminant groups isomorphic to $\mathbb{Z}/m\mathbb{Z}$. For instance if m is prime and congruent to 0 or 3 mod 4, this can be done with a rank 2 even lattice (the order of the discriminant group for an even lattice of rank 2 is always congruent to 0 or 3 mod 4). As the diversity of quadratic forms cannot be overstated (see e.g., [22, 21, 16]), there are many potentially interesting classes from lattice vertex algebras.

Example 6.4   Let V_L be a vertex algebra given by an even unimodular lattice of rank d. Because the lattice is unimodular, V_L is self-contragredient and it has no nontrivial modules. In particular, any bundle of coinvariants from modules over V_L has rank one and first Chern class $\frac{d}{2}\lambda$.

Example 6.5   Consider an even lattice L of rank d with $L'/L \cong \mathbb{Z}/2\mathbb{Z}$. The vertex algebra V_L has two simple modules $V = V_L$ and W. From Proposition 6.2, the product in $\mathcal{F}(V_L)_\mathbb{C} = \mathbb{C}h_V \oplus \mathbb{C}h_W$ is given by

$$h_V * h_V = h_V, \qquad h_V * h_W = h_W, \qquad h_W * h_W = h_V.$$

With terminology as in §4.2, a semisimple basis for $\mathcal{F}(V_L)_\mathbb{C}$ is

$$e_1 := \frac{1}{\sqrt{2}}(h_V + h_W), \qquad e_2 := \frac{1}{\sqrt{2}}(h_V - h_W),$$

with semisimple values both equal to $\sqrt{2}$. One has $h_V = \frac{1}{\sqrt{2}}(e_1 + e_2)$ and $h_W = \frac{1}{\sqrt{2}}(e_1 - e_2)$. Applying Proposition 4.1, the rank of the bundle $\mathbb{V}_g\left(V; V^{\otimes p} \otimes W^{\otimes q}\right)$ on $\overline{\mathcal{M}}_{g,n}$ for $p + q = n$ is

$$\operatorname{rank} \mathbb{V}_g\left(V; V^{\otimes p} \otimes W^{\otimes q}\right) = \sqrt{2}^{\,2g-2+n}\left(\frac{1}{\sqrt{2}^n} + \frac{(-1)^q}{\sqrt{2}^n}\right) = 2^g\,\delta_{q,\text{even}}.$$

In particular, the rank vanishes when q is odd. Applying Corollary 1.2, when $q = 2r$, the Chern character is

$$\operatorname{ch} \mathbb{V}_g\left(V; V^{\otimes p} \otimes W^{\otimes 2r}\right)$$

$$= e^{\frac{d}{2}\lambda} \sum_\Gamma \frac{2^{g-h^1(\Gamma)}}{|\operatorname{Aut}(\Gamma)|}\,(\xi_\Gamma)_* \left(\prod_{i=1}^{2r} e^{a\psi_i} \prod_{e\in E(\Gamma)} \frac{1 - e^{a(\psi_h + \psi_{h'})}}{\psi_h + \psi_{h'}}\right).$$

Here a is the conformal dimension of W. The sum in the formula is over those isomorphism classes of stable graphs Γ of genus g with n legs such that for each vertex, the number of assigned W at the incident legs is even. Note that the only module assignment μ contributing nontrivially to the polynomial P_V in this case is $\mu \colon h \mapsto W$, for all half-edges h.

Example 6.6   Let L be an even lattice such that $L'/L \cong \mathbb{Z}/m\mathbb{Z}$, for $m \geq 2$. Let $\mathcal{W} = \{V = W_0, \dots, W_{m-1}\}$ be the set of simple V_L-modules. The fusion rules from [32] give

$$\text{rank}\, \mathbb{V}_0 \left(V_L; W_i \otimes W_j \otimes W_k \right) = \delta_{i+j+k \,\equiv_m\, 0}.$$

This implies that

$$\text{rank}\, \mathbb{V}_0 \left(V_L; W_0^{\otimes n_0} \otimes \cdots \otimes W_{m-1}^{\otimes n_{m-1}} \right) = \delta_{\sum_{j=0}^{m-1} jn_j \,\equiv_m\, 0}$$

and by induction on the genus and the factorization property, we can further deduce that

$$\text{rank}\, \mathbb{V}_g \left(V_L; W_0^{\otimes n_0} \otimes \cdots \otimes W_{m-1}^{\otimes n_{m-1}} \right) = m^g \, \delta_{\sum_{j=0}^{m-1} jn_j \,\equiv_m\, 0}. \qquad (4)$$

Example 6.7 One can also obtain the rank found in Example 6.6 using Proposition 4.1. Namely, for an even lattice L such that $L'/L \cong \mathbb{Z}/m\mathbb{Z}$, a semisimple basis for the fusion ring $\mathcal{F}(V_L)_{\mathbb{C}} = \oplus_{i=0}^{m-1} \mathbb{C}h_i$ is

$$e_i := \frac{1}{\sqrt{m}} \sum_{j=0}^{m-1} \rho^{ij} h_j, \qquad \text{for } i = 0, \ldots, m-1,$$

where $\rho \in \mathbb{C}$ is a primitive m-th root of unity. One checks $e_i * e_i = \sqrt{m}\, e_i$, hence the semisimple values are all $\sqrt{m}$. As in Example 6.5, one applies Proposition 4.1 to recover (4).

Remark 6.8 Examples 6.5–6.7 show that ranks of bundles of coinvariants from modules over an even lattice L such that $L'/L \cong \mathbb{Z}/m\mathbb{Z}$ coincide with ranks of bundles of coinvariants from modules over the affine Lie algebra $\widehat{\mathfrak{sl}}_m$ at level one. However, while the ranks depend only on L'/L, the Chern characters depend additionally on the quadratic form of L, responsible for the conformal dimension of the irreducible V_L-modules. It is reasonable to expect that classes from lattice vertex algebras could give a collection of CohFTs larger than the one obtained from the affine Lie algebra case.

6.3 Commutants and the Parafermion Vertex Algebras

For a vertex algebra V and a vertex subalgebra U of V, one may construct the *commutant*, or *coset*, vertex algebra $\text{Com}_V(U)$ of U in V [48]. It would be interesting to study the Chern classes for bundles of coinvariants of modules over $\text{Com}_V(U)$ for pairs $U \subset V$ such that $\text{Com}_V(U)$ is of CohFT-type.

Conjecturally, if U and V are both of CohFT-type, then $\text{Com}_V(U)$ is also of CohFT-type. However, U and V need not be of CohFT-type: one such example is given by the well-studied family of cosets of the Heisenberg vertex algebra in the affine vertex algebra $L_k(\mathfrak{g})$ for a finite-dimensional simple Lie algebra $\mathfrak{g}$ at level $k \in \mathbb{Z}$. The Heisenberg vertex algebra is not rational, nor C_2-cofinite (see e.g., [46] for a discussion of the Heisenberg vertex algebra).

Nevertheless, the *parafermion vertex algebras* are known to be of CohFT-type [33] (see also [4, 41, 40, 34, 35, 36]). These are related to W-algebras [4]. The necessary invariants for expressing the Chern classes of bundles of coinvariants of modules over parafermion vertex algebras are known from [42, 33, 2], and one could proceed as in §6.2.

6.4 Orbifold Vertex Algebras

Let G be a subgroup of the group of automorphisms of a vertex algebra V. The *orbifold vertex algebra* V^G consists of the fixed points of G in V. In case V is of CohFT-type, its full group of automorphisms G is a finite-dimensional algebraic group [30]. If G is also solvable, then V^G will also be of CohFT-type [71, 19]. Conjecturally, V^G is always of CohFT-type. We note that even if V is holomorphic, and therefore has no non-trivial modules, the vertex algebra V^G will not generally be holomorphic [51, 27, 26].

One could for instance consider orbifold vertex algebras created from parafermion vertex algebras. In some cases, the simple modules and the fusion rules are known in the literature [59, 60, 60].

Similarly, one could construct orbifold vertex algebras starting from lattice vertex algebras. In [5], simple modules for orbifolds V_L^G, where G is generated by an isometry of order two, are classified and their fusion rules are given. Explicit examples with root lattices and Dynkin diagram automorphisms are given in [5, §4].

7 Questions

Summarizing, bundles of coinvariants defined by modules over vertex algebras of CohFT-type share three important properties with their classical counterparts: They (i) support a projectively flat logarithmic connection [81, 25]; (ii) satisfy the *factorization property*, a reflection of their underlying combinatorial structure [81, 24]; and (iii) give rise to cohomological field theories, as we show here.

As in Remark 6.3, lattice vertex algebras are generalizations of those given by affine Lie algebras at level one. It is natural to expect that other known properties of the classical case extend to the vertex algebra case, and a number of questions come to mind.

Question 1

Given a simple, simply connected algebraic group G with associated Lie algebra $\mathrm{Lie}(G) = \mathfrak{g}$ and $\ell \in \mathbb{Z}_{>0}$, the simple vertex algebra $V = L_\ell(\mathfrak{g})$ is of CohFT-type [48, 39]. By [7, 44, 63], for a smooth algebraic curve C, there is a natural line bundle D on the moduli stack $\mathrm{Bun}_G(C)$ of G-bundles on C such that, for any point P in C, there is a canonical isomorphism

$$\mathbb{V}\left(L_\ell(\mathfrak{g}); L_\ell(\mathfrak{g})\right)^{\dagger}_{(C,P)} \cong H^0\left(\mathrm{Bun}_G(C), D^\ell\right),$$

where $L_\ell(\mathfrak{g})$ is the adjoint module over itself. By [75, 65], given V-modules $M^\bullet$, there is a line bundle L on the moduli stack of quasi-parabolic G-bundles $\mathrm{ParBun}_G(C, P_\bullet)$, for which $\mathbb{V}(V; M^\bullet)^{\dagger}_{(C,P_\bullet)}$ is isomorphic to the global sections of L on $\mathrm{ParBun}_G(C, P_\bullet)$. This geometric picture holds for stable curves with singularities [13] (see also [10]).

The automorphism group $\mathrm{Aut}(V)$ of a vertex algebra V of CohFT-type is a finite-dimensional algebraic group [30]. The connected component $\mathrm{Aut}(V)^0$ of $\mathrm{Aut}(V)$ containing the identity has been described in a number of cases [38, 29, 31, 30]. For instance for $V = L_\ell(\mathfrak{g})$, one has $\mathrm{Lie}\left(\mathrm{Aut}(V)^0\right) \cong V_1 \cong \mathfrak{g}$ [30]. For V of CohFT-type, can one find a geometric realization for conformal blocks defined from modules over V, for instance involving algebraic structures on curves related to $\mathrm{Aut}(V)^0$? Ideas in this direction have been considered in [82] and [15].

Question 2

Gromov–Witten invariants of smooth projective homogeneous spaces define base-point-free classes on $\overline{\mathcal{M}}_{0,n}$; divisors defined from Gromov–Witten invariants of $\mathbb{P}^r = \mathbb{G}r(1, r+1)$ are equivalent to first Chern classes of bundles from integrable modules at level one over $\mathfrak{sl}_{r+1}$ [9, Props 1.4 and 3.1]. Numerical evidence suggests a more general connection between classes of bundles at level ℓ with Gromov–Witten divisors for Grassmannians $\mathbb{G}r(\ell, r + \ell)$ [9]. By Witten's Dictionary, the quantum cohomology of Grassmannians can be used to compute ranks of conformal blocks bundles in type A for any level [12]. Are there connections between other Gromov–Witten theories and the more general bundles of coinvariants studied here?

Question 3

Vector bundles defined by representations of affine Lie algebras are globally generated in genus zero, and so Chern classes have valuable positivity properties. For instance, first Chern classes are base-point-free, giving rise to morphisms [43]. Can one give sufficient conditions so that the vector bundles of coinvariants are globally generated? Chern classes of bundles from certain Virasoro vertex algebras are not nef, so further assumptions must be made. See [23] for initial results along these lines.

Question 4

Bundles of coinvariants from affine Lie algebras give rise to morphisms from $\overline{\mathcal{M}}_{0,n}$ to Grassmannian varieties [43]. In the case where the Lie algebra is of type A and modules are at level one, we know the image varieties parametrize configurations of weighted points on rational normal curves in projective spaces [54, 52, 53]. If Chern classes given by representations of particular types of vertex algebras are base-point-free, can one give modular interpretations for the images of their associated maps?

Question 5

Chern classes of bundles of coinvariants on $\overline{\mathcal{M}}_{0,n}$ defined by $V = L_\ell(\mathfrak{g})$ satisfy scaling and level-rank identities, and are zero above a critical level, allowing one to give sufficient conditions for when they lie on extremal faces of the nef cone [3, 14, 11]. Do Chern classes studied here satisfy similar identities? Are there criteria to ensure they lie on extremal faces of cones of nef cycles? Do Chern classes of particular such bundles generate extremal rays? Exotic lattices may be relevant.

Question 6

In [17, Theorem 5.1], using factorization, Verlinde bundles constructed from level one integrable modules over $\mathfrak{sl}_{r+1}$ are shown to be isomorphic to both GIT bundles [17], and to the r-th tensor power of cyclic bundles studied in [45]. Are there other such identifications, for instance involving the line bundles of coinvariants on $\overline{\mathcal{M}}_{0,n}$ constructed from even lattice theories, as discussed in Example 6.6?

Acknowledgements

The authors are grateful to Yi-Zhi Huang for helpful discussions, and to Marian, Oprea, Pandharipande, Pixton, and Zvonkine, for their work in [70]. Conversations with Aaron Bertram from some years ago helped with TQFTs computations. The authors also thank Daniel Krashen and Bin Gui for discussions about lattices. We thank the referee for their valuable comments. Gibney was supported by NSF DMS–1902237.

References

[1] Abe, Toshiyuki, Buhl, Geoffrey, and Dong, Chongying. 2004. Rationality, regularity, and C_2-cofiniteness. *Trans. Amer. Math. Soc.*, **356**(8), 3391–3402.

[2] Ai, Chunrui, Dong, Chongying, Jiao, Xiangyu, and Ren, Li. 2018. The irreducible modules and fusion rules for the parafermion vertex operator algebras. *Trans. Amer. Math. Soc.*, **370**(8), 5963–5981.

[3] Alexeev, Valery, Gibney, Angela, and Swinarski, David. 2014. Higher-level $\mathfrak{sl}_2$ conformal blocks divisors on $\overline{\mathcal{M}}_{0,n}$. *Proc. Edinb. Math. Soc. (2)*, **57**(1), 7–30.

[4] Arakawa, Tomoyuki, Lam, Ching Hung, and Yamada, Hiromichi. 2014. Zhu's algebra, C_2-algebra and C_2-cofiniteness of parafermion vertex operator algebras. *Adv. Math.*, **264**, 261–295.

[5] Bakalov, Bojko, and Elsinger, Jason. 2015. Orbifolds of lattice vertex algebras under an isometry of order two. *J. Algebra*, **441**, 57–83.

[6] Beauville, Arnaud. 1996. Conformal blocks, fusion rules and the Verlinde formula. Pages 75–96 of: *Proceedings of the Hirzebruch 65 Conference on Algebraic Geometry (Ramat Gan, 1993)*. Israel Math. Conf. Proc., vol. 9. Bar-Ilan Univ., Ramat Gan.

[7] Beauville, Arnaud, and Laszlo, Yves. 1994. Conformal blocks and generalized theta functions. *Comm. Math. Phys.*, **164**(2), 385–419.

[8] Beilinson, A. A., and Schechtman, V. V. 1988. Determinant bundles and Virasoro algebras. *Comm. Math. Phys.*, **118**(4), 651–701.

[9] Belkale, P, and Gibney, A. 2019a. Basepoint Free Cycles on $\overline{M}_{0,n}$ from Gromov–Witten Theory. *International Mathematics Research Notices*, 09. rnz184.

[10] Belkale, P., and Gibney, A. 2019b. On finite generation of the section ring of the determinant of cohomology line bundle. *Trans. Amer. Math. Soc.*, **371**(10), 7199–7242.

[11] Belkale, P., Gibney, A., and Mukhopadhyay, S. 2016. Nonvanishing of conformal blocks divisors on $\overline{M}_{0,n}$. *Transform. Groups*, **21**(2), 329–353.

[12] Belkale, Prakash. 2010. The tangent space to an enumerative problem. Pages 405–426 of: *Proceedings of the International Congress of Mathematicians. Volume II*. Hindustan Book Agency, New Delhi.

[13] Belkale, Prakash, and Fakhruddin, Najmuddin. 2019. Triviality properties of principal bundles on singular curves. *Algebr. Geom.*, **6**(2), 234–259.

[14] Belkale, Prakash, Gibney, Angela, and Mukhopadhyay, Swarnava. 2015. Vanishing and identities of conformal blocks divisors. *Algebr. Geom.*, **2**(1), 62–90.

[15] Ben-Zvi, David, and Frenkel, Edward. 2004. Geometric realization of the Segal-Sugawara construction. Pages 46–97 of: *Topology, geometry and quantum field theory*. London Math. Soc. Lecture Note Ser., vol. 308. Cambridge Univ. Press, Cambridge.

[16] Bhargava, Manjul, and Hanke, John. 2005. Universal quadratic forms and the 290-Theorem. Preprint, www.wordpress.jonhanke.com/wp-content/uploads/2011/09/290-Theorem-preprint.pdf

[17] Bolognesi, Michele, and Giansiracusa, Noah. 2015. Factorization of point configurations, cyclic covers, and conformal blocks. *J. Eur. Math. Soc. (JEMS)*, **17**(10), 2453–2471.

[18] Borcherds, Richard E. 1986. Vertex algebras, Kac-Moody algebras, and the Monster. *Proc. Nat. Acad. Sci. U.S.A.*, **83**(10), 3068–3071.

[19] Carnahan, Scott, and Miyamoto, Masahiko. 2016. Regularity of fixed-point vertex operator subalgebras. Preprint, *arXiv:1603.05645*.

[20] Codogni, Giulio. 2019. Vertex algebras and Teichmüller modular forms. Preprint, *arXiv:1901.03079*, 1–31.

[21] Conway, J. H. 2000. Universal quadratic forms and the fifteen theorem. Pages 23–26 of: *Quadratic forms and their applications (Dublin, 1999)*. Contemp. Math., vol. 272. Amer. Math. Soc., Providence, RI.

[22] Conway, John H. 1997. *The sensual (quadratic) form*. Carus Mathematical Monographs, vol. 26. Mathematical Association of America, Washington, DC. With the assistance of Francis Y. C. Fung.

[23] Damiolini, Chiara, and Gibney, Angela. 2021. On global generation of vector bundles on the moduli space of curves from representations of vertex operator algebras. Preprint, arXiv:2107.06923.

[24] Damiolini, Chiara, Gibney, Angela, and Tarasca, Nicola. 2019. On factorization and vector bundles of conformal blocks from vertex algebras. *Submitted, arXiv:1909.04683*.

[25] Damiolini, Chiara, Gibney, Angela, and Tarasca, Nicola. 2021. Conformal blocks from vertex algebras and their connections on $\overline{\mathcal{M}}_{g,n}$. Geom. Topol., **25**(5), 2235–2286. https://doi.org/10.2140/gt.2021.25.2235.

[26] Dijkgraaf, R., Pasquier, V., and Roche, P. 1990. Quasi Hopf algebras, group cohomology and orbifold models. vol. 18B. Recent advances in field theory (Annecy-le-Vieux, 1990).

[27] Dijkgraaf, Robbert, Vafa, Cumrun, Verlinde, Erik, and Verlinde, Herman. 1989. The operator algebra of orbifold models. *Comm. Math. Phys.*, **123**(3), 485–526.

[28] Dong, Chongying. 1993. Vertex algebras associated with even lattices. *J. Algebra*, **161**(1), 245–265.

[29] Dong, Chongying, and Griess, Jr., Robert L. 1998. Rank one lattice type vertex operator algebras and their automorphism groups. *J. Algebra*, **208**(1), 262–275.

[30] Dong, Chongying, and Griess, Jr., Robert L. 2002. Automorphism groups and derivation algebras of finitely generated vertex operator algebras. *Michigan Math. J.*, **50**(2), 227–239.

[31] Dong, Chongying, and Griess, Jr., Robert L. 2005. The rank-2 lattice-type vertex operator algebras V_L^+ and their automorphism groups. *Michigan Math. J.*, **53**(3), 691–715.

[32] Dong, Chongying, and Lepowsky, James. 1993. *Generalized vertex algebras and relative vertex operators*. Progress in Maths, vol. 112. Birkhäuser Boston, Inc., Boston, MA.

[33] Dong, Chongying, and Ren, Li. 2017. Representations of the parafermion vertex operator algebras. *Adv. Math.*, **315**, 88–101.

[34] Dong, Chongying, and Wang, Qing. 2010. The structure of parafermion vertex operator algebras: general case. *Comm. Math. Phys.*, **299**(3), 783–792.

[35] Dong, Chongying, and Wang, Qing. 2011a. On C_2-cofiniteness of parafermion vertex operator algebras. *J. Algebra*, **328**, 420–431.

[36] Dong, Chongying, and Wang, Qing. 2011b. Parafermion vertex operator algebras. *Front. Math. China*, **6**(4), 567–579.

[37] Dong, Chongying, Li, Haisheng, and Mason, Geoffrey. 1998. Twisted representations of vertex operator algebras. *Math. Ann.*, **310**(3), 571–600.

[38] Dong, Chongying, Griess, Jr., Robert L., and Ryba, Alex. 1999. Rank one lattice type vertex operator algebras and their automorphism groups. II. E-series. *J. Algebra*, **217**(2), 701–710.

[39] Dong, Chongying, Li, Haisheng, and Mason, Geoffrey. 2000. Modular-invariance of trace functions in orbifold theory and generalized Moonshine. *Comm. Math. Phys.*, **214**(1), 1–56.

[40] Dong, Chongying, Lam, Ching Hung, and Yamada, Hiromichi. 2009. W-algebras related to parafermion algebras. *J. Algebra*, **322**(7), 2366–2403.

[41] Dong, Chongying, Lam, Ching Hung, Wang, Qing, and Yamada, Hiromichi. 2010. The structure of parafermion vertex operator algebras. *J. Algebra*, **323**(2), 371–381.

[42] Dong, Chongying, Kac, Victor, and Ren, Li. 2019. Trace functions of the parafermion vertex operator algebras. *Adv. Math.*, **348**, 1–17.

[43] Fakhruddin, Najmuddin. 2012. Chern classes of conformal blocks. Pages 145–176 of: *Compact moduli spaces and vector bundles*. Contemp. Math., vol. 564. Amer. Math. Soc., Providence, RI.

[44] Faltings, Gerd. 1994. A proof for the Verlinde formula. *J. Algebraic Geom.*, **3**(2), 347–374.

[45] Fedorchuk, Maksym. 2011. Cyclic covering morphisms on $\overline{\mathcal{M}}_{0,n}$. Preprint, *arXiv:1105.0655*.

[46] Frenkel, Edward, and Ben-Zvi, David. 2004. *Vertex algebras and algebraic curves*. Second edn. Mathematical Surveys and Monographs, vol. 88. Am Mathem Soc, Providence, RI.

[47] Frenkel, Igor, Lepowsky, James, and Meurman, Arne. 1988. *Vertex operator algebras and the Monster*. Pure and Applied Mathematics, vol. 134. Academic Press, Inc., Boston, MA.

[48] Frenkel, Igor B., and Zhu, Yongchang. 1992. Vertex operator algebras associated to representations of affine and Virasoro algebras. *Duke Math. J.*, **66**(1), 123–168.

[49] Frenkel, Igor B, Lepowsky, James, and Meurman, Arne. 1984. A natural representation of the Fischer-Griess Monster with the modular function J as character. *Proceedings of the National Academy of Sciences*, **81**(10), 3256–3260.

[50] Frenkel, Igor B., Huang, Yi-Zhi, and Lepowsky, James. 1993. On axiomatic approaches to vertex operator algebras and modules. *Mem. Amer. Math. Soc.*, **104**(494), viii+64.

[51] Gemünden, Thomas, and Keller, Christoph. 2018. Orbifolds of lattice vertex operator algebras at $d = 48$ and $d = 72$. Preprint, *arXiv:1802.10581*, Feb.

[52] Giansiracusa, Noah. 2013. Conformal blocks and rational normal curves. *J. Algebraic Geom.*, **22**(4), 773–793.

[53] Giansiracusa, Noah, and Gibney, Angela. 2012. The cone of type A, level 1, conformal blocks divisors. *Adv. Math.*, **231**(2), 798–814.

[54] Giansiracusa, Noah, and Simpson, Matthew. 2011. GIT compactifications of $\mathcal{M}_{0,n}$ from conics. *Int. Math. Res. Not. IMRN*, 3315–3334.

[55] Givental, Aleksandr Borisovich. 2001a. Gromov–Witten invariants and quantization of quadratic Hamiltonians. *Moscow Mathematical Journal*, **1**(4), 551–568.

[56] Givental, Alexander B. 2001b. Semisimple Frobenius structures at higher genus. *International mathematics research notices*, **2001**(23), 1265–1286.

[57] Goller, Thomas. 2017. A weighted topological quantum field theory for Quot schemes on curves. *Mathematische Zeitschrift*, 1–36.

[58] Huang, Yi-Zhi. 2005. Vertex operator algebras, the Verlinde conjecture, and modular tensor categories. *Proc. Natl. Acad. Sci. USA*, **102**(15), 5352–5356.

[59] Jiang, Cuipo, and Wang, Qing. 2017. Representations of $\mathbb{Z}_2$-orbifold of the parafermion vertex operator algebra $K(\mathfrak{sl}_2, k)$. Preprint, *arXiv:1904.01798v1*.

[60] Jiang, Cuipo, and Wang, Qing. 2019. Fusion rules for $\mathbb{Z}_2$-orbifolds of affine and parafermion vertex operator algebras. Preprint, *arXiv:1904.01798v1*.

[61] Knus, Max-Albert, Merkurjev, Alexander, Rost, Markus, and Tignol, Jean-Pierre. 1998. *The book of involutions*. American Mathematical Society Colloquium Publications, vol. 44. American Mathematical Society, Providence, RI. With a preface in French by J. Tits.

[62] Kobayashi, Shoshichi. 1987. *Differential geometry of complex vector bundles*. Publications of the Mathematical Society of Japan, vol. 15. Princeton University Press, Princeton, NJ. Kanô Memorial Lectures, 5.

[63] Kumar, Shrawan, Narasimhan, M. S., and Ramanathan, A. 1994. Infinite Grassmannians and moduli spaces of G-bundles. *Math. Ann.*, **300**(1), 41–75.

[64] Lam, Ching Hung, and Shimakura, Hiroki. 2019. 71 holomorphic vertex operator algebras of central charge 24. *Bull. Inst. Math. Acad. Sin. (N.S.)*, **14**(1), 87–118.

[65] Laszlo, Yves, and Sorger, Christoph. 1997. The line bundles on the moduli of parabolic G-bundles over curves and their sections. *Ann. Sci. École Norm. Sup. (4)*, **30**(4), 499–525.

[66] Lee, Y.-P., and Vakil, R. 2009. Algebraic structures on the topology of moduli spaces of curves and maps. Pages 197–219 of: *Surveys in differential geometry. Vol. XIV. Geometry of Riemann surfaces and their moduli spaces.* Surv. Differ. Geom., vol. 14. Int. Press, Somerville, MA.

[67] Lepowsky, James, and Li, Haisheng. 2004. *Introduction to vertex operator algebras and their representations.* Progress in Maths, vol. 227. Birkhäuser Boston, Inc., Boston, MA.

[68] Li, Haisheng. 1999. Determining fusion rules by $A(V)$-modules and bimodules. *J. Algebra*, **212**(2), 515–556.

[69] Marian, Alina, Oprea, Dragos, and Pandharipande, Rahul. 2015. The first Chern class of the Verlinde bundles. Pages 87–111 of: *String-Math 2012.* Proc. Sympos. Pure Math., vol. 90. Amer. Math. Soc., Providence, RI.

[70] Marian, Alina, Oprea, Dragos, Pandharipande, Rahul, Pixton, Aaron, and Zvonkine, Dimitri. 2017. The Chern character of the Verlinde bundle over $\overline{\mathcal{M}}_{g,n}$. *J. Reine Angew. Math.*, **732**, 147–163.

[71] Miyamoto, Masahiko. 2015. C_2-cofiniteness of cyclic-orbifold models. *Comm. Math. Phys.*, **335**(3), 1279–1286.

[72] Nagatomo, Kiyokazu, and Tsuchiya, Akihiro. 2005. Conformal field theories associated to regular chiral vertex operator algebras. I. Theories over the projective line. *Duke Math. J.*, **128**(3), 393–471.

[73] Pandharipande, Rahul. 2017. Cohomological field theory calculations. Preprint, *arXiv:1712.02528*.

[74] Pandharipande, Rahul, Pixton, Aaron, and Zvonkine, Dimitri. 2015. Relations on $\overline{\mathcal{M}}_{g,n}$ via 3-spin structures. *J. Amer. Math. Soc.*, **28**(1), 279–309.

[75] Pauly, Christian. 1996. Espaces de modules de fibrés paraboliques et blocs conformes. *Duke Math. J.*, **84**(1), 217–235.

[76] Schütt, Matthias, and Shioda, Tetsuji. 2010. Elliptic surfaces. Pages 51–160 of: *Algebraic geometry in East Asia—Seoul 2008*. Adv. Stud. Pure Math., vol. 60. Math. Soc. Japan, Tokyo.

[77] Shioda, Tetsuji. 1999. Mordell-Weil lattices for higher genus fibration over a curve. Pages 359–373 of: *New trends in algebraic geometry (Warwick, 1996)*. London Math. Soc. Lecture Note Ser., vol. 264. Cambridge Univ. Press, Cambridge.

[78] Teleman, Constantin. 2012. The structure of 2D semi-simple field theories. *Inventiones mathematicae*, **188**(3), 525–588.

[79] Tsuchimoto, Yoshifumi. 1993. On the coordinate-free description of the conformal blocks. *J. Math. Kyoto Univ.*, **33**(1), 29–49.

[80] Tsuchiya, Akihiro, and Kanie, Yukihiro. 1987. Vertex operators in the conformal field theory on $\mathbf{P}^1$ and monodromy representations of the braid group. *Lett. Math. Phys.*, **13**(4), 303–312.

[81] Tsuchiya, Akihiro, Ueno, Kenji, and Yamada, Yasuhiko. 1989. Conformal field theory on universal family of stable curves with gauge symmetries. 459–566 of: *Integrable systems in quantum field theory and statistical mechanics*. Adv. Stud. Pure Math., vol. 19. Academic Press, Boston, MA.

[82] Ueno, Kenji. 1995. On conformal field theory. Pages 283–345 of: *Vector bundles in algebraic geometry (Durham, 1993)*. London Math. Soc. Lecture Note Ser., vol. 208. Cambridge Univ. Press, Cambridge.

[83] Zhu, Yongchang. 1996. Modular invariance of characters of vertex operator algebras. *J. Amer. Math. Soc.*, **9**(1), 237–302.

7

The Cone Theorem and the Vanishing of Chow Cohomology

Dan Edidin[a] and Ryan Richey[b]

Abstract. We show that a cone theorem for $\mathbb{A}^1$-homotopy invariant contravariant functors implies the vanishing of the positive degree part of the operational Chow cohomology rings of a large class of affine varieties. We also discuss how this vanishing relates to a number of questions about representing Chow cohomology classes of GIT quotients in terms of equivariant cycles.

1 Introduction

In [11, 10], Fulton and MacPherson define for any scheme X a graded cohomology ring $A^*_{\mathrm{op}}(X)$ which equals the classical intersection ring when X is non-singular. An element of $A^*_{\mathrm{op}}(X)$ is a collection of operations on Chow groups of X-schemes compatible with basic operations in intersection theory. The product structure is given by composition. By construction there is a pullback of operational rings $A^*_{\mathrm{op}}(Y) \to A^*_{\mathrm{op}}(X)$ for any morphism $X \to Y$.

For an arbitrary singular scheme elements of $A^*_{\mathrm{op}}(X)$ do not have natural interpretations in terms of algebraic cycles, although Kimura [13] showed how $A^*_{\mathrm{op}}(X)$ can be related to the intersection ring of a resolution of singularities of X. Despite the formal structure, there is a class of singular varieties where the operational Chow groups are readily computable. Specifically, if X is a complete linear variety then Totaro [19] proved that the pairing

$$A^k_{\mathrm{op}}(X) \times A_k(X) \to A_0(X) = \mathbb{Z}, (c, \alpha) \mapsto c \cap \alpha$$

is perfect so $A^k_{\mathrm{op}}(X) = \mathrm{Hom}(A_k(X), \mathbb{Z})$, where $A_k(X)$ denotes the classical Chow group of k-dimensional cycles. When X is a complete toric variety

[a] University of Missouri, Columbia
[b] University of Missouri, Columbia

Fulton and Sturmfels [12] gave an explicit description of the operational product in terms of Minkowski weights. This was generalized by Payne [17] who showed that the T-equivariant operational Chow ring of a toric variety X can be identified with the ring of integral piecewise polynomial functions on the fan of X.

The Cox construction expresses any toric variety as a good quotient $(\mathbb{A}^n \setminus B)/G$ where G is a diagonalizable group. In [9] the first author and Satriano showed that if $X = Z/G$ is the good quotient of a smooth variety by a linearly reductive group then the rational operational Chow ring $A_{\mathrm{op}}^*(X)_{\mathbb{Q}}$ naturally embeds in the equivariant Chow ring $A_G^*(Z)_{\mathbb{Q}}$. Moreover, the image of an element in $A_{\mathrm{op}}^k(X)$ is represented by a class of the form $\sum c_i[Z_i]$ where the $Z_i \subset Z$ are codimension-k G-invariant subvarieties of Z which are saturated with respect to the quotient map $\pi : Z \to X$.

By the étale slice theorem [15] the local model for the good quotient of a smooth variety at a closed orbit $Gx \subset Z$ is the quotient $V \to V/G_x$ where $V = T_{x,Z}/T_{x,Gx}$ is the normal space to the orbit Gx at x, and G_x is the stabilizer of x. Therefore, a natural problem is to compute the operational Chow rings of good quotients V/G where V is a representation of a linearly reductive group G. Examples of such quotients are affine toric varieties associated to maximal dimensional strongly convex polyhedral cones. In [18, 6] it is shown that $A^*(X) = \mathbb{Z}$ for any affine toric variety X, and likewise that op $K^0(X) = \mathbb{Z}$ where op K^0 is the operational K-theory defined by Anderson and Payne [3].

The purpose of this paper is to show that stronger results hold. The contravariant functors A_{op}^* and op K^0 are both $\mathbb{A}^1$-homotopy invariants. The cone theorem (Theorem 3.2) implies the vanishing of such functors on a large class of naturally occurring varieties including affine toric varieties and quotients of representations of reductive groups.

Since quotients of the form V/G are also the local models for good moduli spaces of Artin stacks [2] we conclude with a discussion of how the vanishing of $A_{\mathrm{op}}^*(V/G)$ relates to questions about the image of $A_{\mathrm{op}}^*(X)$ in $A^*\mathcal{X}$ when X is the good moduli space of a smooth algberaic stack $\mathcal{X}$.

Dedication. It is a pleasure to dedicate this work to William Fulton on the occasion of his 80th birthday.

2 Homotopy Invariant Functors

Fix a ground field k and let $\mathcal{S}ch/k$ denote the category of k-schemes of finite type.

Definition 2.1 A homotopy invariant functor is a contravariant functor $H: Sch/k \to Ab$ such that the pullback $H(X) \xrightarrow{\pi^*} H(X \times \mathbb{A}^1)$ is an isomorphism for all X in Sch/k. Likewise if G is an algebraic group then a G-homotopy invariant functor is a contravariant functor $H^G: Sch^G/k \to Ab$ such that for any G-scheme X the pullback $H^G(X) \xrightarrow{\pi^*} H^G(X \times \mathbb{A}^1)$ is an isomorphism, where the action of G on $\mathbb{A}^1$ is trivial.

In this paper we focus on several homotopy invariant functors – the operational Chow cohomology ring defined in [11] and [10, Chapter 17] as well as its equivariant counterpart defined in [5] and the (equivariant) operational K-theory defined by Anderson and Payne in [3].

2.1 Chow Cohomology

Let X be a scheme. Following [10], let $A_k(X)$ denote the group of dimension k cycle classes modulo rational equivalence, and if X is equidimensional of dimension n we let $A^k(X)$ denote the group of $(n - k)$-dimensional cycle classes modulo rational equivalence. If X is smooth and equidimensional, then the intersection product on $A_k(X)$ as constructed in [10, Chapter 6.1] makes $A^*(X)$ into a commutative, graded ring.

For general schemes [10, Chapter 17] defines a graded operational Chow cohomology ring $A^*_{op}(X) := \bigoplus_{k \geq 0} A^k_{op}(X)$: an element $c \in A^k_{op}(X)$ is a collection of homomorphisms of groups:

$$c_g^{(k)}: A_p X' \to A_{p-k} X'$$

for every morphism $g : X' \to X$ which are compatible with respect to proper pushforward, and pullbacks along flat morphisms and regular embeddings (see [10, Definitions 17.1 and 17.3]). The product is given by composition and turns $A^*_{op}(X)$ into a graded ring called the **Chow cohomology ring** of X. Moreover, if X has a resolution of singularities (e.g. if the characteristic of the ground field is zero or if X is a toric variety) then $A^*_{op}(X)$ is known to be commutative. If X is smooth, then [10, Corollary 17.4] proves that the Poincaré duality map $A^k_{op}(X) \to A^k(X) = A_{n-k}(X)$ is an isomorphism of rings where the intersection product agrees with the product given by composition.

Remark 2.2 In [10, Chapter 17] the Chow cohomology ring is also denoted $A^*(X)$ without the inclusion of the subscript 'op'.

Proposition 2.3 Chow cohomology is an $\mathbb{A}^1$-homotopy invariant functor.

Proof Consider the pullback $\pi^* : A^*_{op}(X) \to A^*_{op}(X \times \mathbb{A}^1)$ where $\pi : X \times \mathbb{A}^1 \to X$ is the projection. First note that injectivity of π^* is a formal consequence of the functoriality; the composition $X \xrightarrow{\iota} X \times \mathbb{A}^1 \xrightarrow{\pi} X$ is the identity, where $\iota(x) = (x, 0)$.

Suppose that $c \in A^k_{\mathrm{op}}(X \times \mathbb{A}^1)$. We wish to show that $c = \pi^* d$ for some $d \in A^k_{\mathrm{op}}(X)$. Given a morphism $Y \xrightarrow{f} X$ let $g = f \times \mathrm{id} \colon Y \times \mathbb{A}^1 \to X \times \mathbb{A}^1$. Since the flat pullback $\pi^* \colon A_*(Y) \to A_*(Y \times \mathbb{A}^1)$ is an isomorphism and this isomorphism is compatible with other operations on Chow homology we can define a class $d \in A^k_{\mathrm{op}}(X)$ such that $\pi^* d = c$ by the formula $d_f(\alpha) = (\pi^*)^{-1} c_g(\pi^* \alpha)$. $\qquad\square$

2.1.1 Equivariant Chow Cohomology

An equivariant version of operational Chow cohomology was defined in [5]. An element $c \in A^k_{\mathrm{op},G}(X)$ is a collection operations on equivariant Chow groups $c_f \colon A^G_*(X') \to A^G_{*-k}G(X')$ for every equivariant morphism $X' \xrightarrow{f} X$ compatible with equivariant proper pushforward and equivariant flat maps and equivariant regular embeddings maps. [5, Corollary 2] states that if X admits a resolution of singularities then $A^k_{\mathrm{op},G}(X)$ can be identified with the operational Chow group $A^k_{\mathrm{op}}(X_G)$ where X_G is an algebraic space of the form $X \times^G U$. Here U is an open set in a representation V of G on which G acts freely and $\mathrm{codim}(V \setminus U) > k$. It follows from this identification that the equivariant Chow cohomology groups $A^k_{\mathrm{op},G}(X)$ enjoy all of the formal properties of ordinary operational Chow cohomology. In particular, the functor $A^*_{\mathrm{op},G}$ is a homotopy invariant functor on the category of schemes or algebraic spaces with a G-action.

2.2 Operational K-theory

Following [3], if X is a scheme we denote by $K_0(X)$ the Grothendieck group of coherent sheaves, and $K^0(X)$ the Grothendieck group of perfect complexes. If X has an ample family of line bundles, then $K^0(X)$ is the same as the naive Grothendieck group of vector bundles.

For any scheme X, Anderson and Payne define the **operational K-theory** op $K^0(X)$ of X as follows. An element $c \in$ op $K^0(X)$ is a collection of operators $c_f \colon K_0(X') \to K_0(X')$ indexed by morphisms $X' \xrightarrow{f} X$ compatible with proper pushforward, flat pullback and pullback along regular embeddings.

For any scheme X, there is a canonical map op $K^0(X) \to K_0(X)$ given by $c \mapsto c_{\mathrm{id}_X}(\mathcal{O}_X)$. If X is smooth, then [3, Corollary 4.5] states that this map is an isomorphism.

Theorem 2.4 [3, Theorem 1.1] op K^0 is a homotopy invariant functor.

2.2.1 Equivariant Operational K-theory

Anderson and Payne also define the equivariant operational K-theory ring as the ring of operations on the equivariant Grothendieck group of coherent sheaves, $K_0^G(X)$. Since the equivariant Grothendieck group is an $\mathbb{A}^1$-homotopy invariant the proof of [3, Theorem 1.1] goes through and we conclude that op K_G^0 is a homotopy invariant functor. If $G = T$ is a torus and X is smooth, then Anderson and Payne also prove that op $K_T^0(X)$ can be identified with $K_0^T(X)$.

3 The Cone Theorem

Fix a base scheme X of finite type defined over a field k.

Definition 3.1 An X-cone is a scheme of the form $C = \operatorname{Spec}_X S$ where $S = \oplus_{n=0}^{\infty} S_i$ is a finitely generated graded $\mathcal{O}_X$-algebra such that $S_0 = \mathcal{O}_X$. (Note that we do not require that S be locally generated in degree one.)

More generally, if G is an algebraic group and X is a G-scheme then we say that $C = \operatorname{Spec} S$ is a G-cone if the S_i are sheaves of G-$\mathcal{O}_X$ modules and multiplication of local sections is G-equivariant.

The inclusion $S_0 \to S$ defines a projection $\rho : C \to X$ and the identification of $S_0 = S/S^+$ defines an inclusion $\iota : X \to C$. Clearly, $\rho \circ \iota = \operatorname{id}_X$.

The key property of homotopy invariant functors is the following cone theorem.

Theorem 3.2 [20, cf. Exercise IV.11.5] The pullbacks ρ^* and ι^* are inverses. In particular $H(X) = H(C)$. Likewise if H^G is a G-homotopy invariant functor and $C = \operatorname{Spec} S$ is a G-cone then $H^G(X) = H^G(Y)$.

Proof We give the proof in the non-equivariant case as the proof in the equivariant case is identical.

Since $\rho \circ \iota = \operatorname{id}_X$, we know that $\iota^* \circ \rho^* : X \to X$ is the identity. In particular, ρ^* is injective. Thus it suffices to prove that $\rho^* \circ \iota^* : C \to C$ is an isomorphism.

Since C is a cone over $X = \operatorname{Spec} S_0$, the map of graded rings $S \to S[t]$, sending S_i to $t^i S_i$. defines an $\mathbb{A}^1$ action $\sigma : C \times \mathbb{A}^1 \to C$ with fixed scheme $X = \operatorname{Spec} S_0$.

Let $s_t : C \to C$ be the map $x \mapsto tx$. For $t \neq 0$, s_t is an isomorphism with inverse $s_{t^{-1}}$ and $s_0 : C \to C$ is the composition $\iota \circ \rho$. The map s_t is itself a composition

$$C \overset{i_t}{\to} C \times \mathbb{A}^1 \overset{\sigma}{\to} C$$

where $i_t : X \hookrightarrow X \times \mathbb{A}^1$ is the inclusion $x \mapsto (x, t)$.

Now if $\pi : X \times \mathbb{A}^1 \to X$ is the projection, then for any t, $\pi \circ i_t = \mathrm{id}$. Since π^* is assumed to be an isomorphism, i_t^* must also be an isomorphism for any t. Since, for $t \neq 0$ the composite $s_t = \sigma \circ i_t$ is an isomorphism we see that σ^* must also be an isomorphism. Hence, $s_0^* = (\sigma \circ i_0)^*$ is an isomorphism. But $s_0 = (\iota \circ \rho)$, so $(\iota \circ \rho)^*$ is an isomorphism as claimed. $\square$

Example 3.3   Let $X \subset Y$ be a closed subscheme and let $C_X Y$ be the normal cone of X in Y. Theorem 3.2 implies that the pullback $A_{\mathrm{op}}^*(X) \to A_{\mathrm{op}}^*(C_X Y)$ is an isomorphism. In particular if X is smooth then $A_{\mathrm{op}}^*(C_X Y)$ is identified with the Chow ring of X. If the closed embedding $X \hookrightarrow Y$ is a regular embedding (for example if X and Y are both smooth) then $C_X Y$ is the normal bundle to X in Y and this identification follows from the usual homotopy invariance of operational Chow rings.

3.1 Cone Theorem for Bivariant Groups

The operational Chow and K-theory rings defined by Fulton–MacPherson and Anderson–Payne are part of a more general construction associated to the covariant (for proper morphisms) functors K_0 and A_*. Given a morphism of schemes $Y \to X$ the bivariant Chow group $A_{\mathrm{op}}^k(Y \to X)$ is the graded abelian group consisting of a collection of operators $c_f : A_*(Y') \to A_{*-k} * (X')$ for each morphism $X' \xrightarrow{f} X$ compatible with proper pushforward, flat pullback, and pullback along regular embeddings, where $Y' = Y \times_X X'$. The group $\mathrm{op}\, K^0(Y \to X)$ is defined analogously.

The groups $A_{\mathrm{op}}^*(Y \to X)$ and $\mathrm{op}\, K^0(Y \to X)$ are contravariant functors on the category whose objects are morphisms of schemes $Y \to X$ and whose morphisms are cartesian diagrams

$$
\begin{array}{ccc}
Y' & \to & X' \\
\downarrow & & \downarrow \\
Y & \to & X
\end{array} \; .
$$

It is easy to show that pullback along the diagram

$$
\begin{array}{ccc}
Y \times \mathbb{A}^1 & \to & X \times \mathbb{A}^1 \\
\downarrow & & \downarrow \\
Y & \to & X
\end{array}
$$

induces isomorphisms $A_{\mathrm{op}}^*(Y \to X) \to A_{\mathrm{op}}^*(Y \times \mathbb{A}^1 \to X \times \mathbb{A}^1)$, $\mathrm{op}\, K^0(Y \to X) \to \mathrm{op}\, K^0(Y \times \mathbb{A}^1 \to X \times \mathbb{A}^1)$. As a corollary we obtain a cone isomorphism theorem for these bivariant groups.

Corollary 3.4   Given a morphism, $Y \to X$ and cone $C \to X$ let $C_Y \to Y$ be the cone obtained by base change. Then the pullbacks $A^*_{\text{op}}(Y \to X) \to A^*_{\text{op}}(C_Y \to C)$ and op $K^0(Y \to X) \to$ op $K^0(C_Y \to C)$ are isomorphisms.

4 Affine Toric Varieties

Theorem 4.1   If $X = X(\sigma)$ is an affine toric variety defined by a strongly convex rational cone σ in a lattice N, then for any homotopy invariant functor H on $\mathcal{S}ch/k$, $H(X) = H(T_0)$ where T_0 is an algebraic torus.

Proof   Since $X(\sigma)$ is an affine toric variety the proof of [4, Proposition 3.3.9] shows that we can decompose $X = X(\overline{\sigma}) \times T_0$ where T_0 is a torus and $\overline{\sigma}$ is a full dimensional cone. Since $\overline{\sigma}$ is full-dimensional the semi-group $S_{\overline{\sigma}}$ is generated in positive degree, so $R = k[X(\overline{\sigma})]$ is a positively graded ring with $R_0 = k$.

Hence $S = k[X(\sigma)] = k[T_0] \otimes_k R$ is a positively graded ring with $S_0 = k[T_0]$. Hence by the cone theorem, $H(X) = H(T_0)$ □

Corollary 4.2   If X is an affine toric variety then $A^0(X) = \mathbb{Z}$ and $A^k_{\text{op}}(X) = 0$ for $k > 0$. Likewise, op $K^0(X) = \mathbb{Z}$.

Proof   By the theorem we know that $A^*_{\text{op}}(X) = A^*_{\text{op}}(T_0)$ and op $K^0(X) =$ op $K^0(T_0)$. Since a torus is an open subset of $\mathbb{A}^n$, $A^k_{\text{op}}(T_0) = A^k(T_0) = 0$ if $k > 0$ and $A^0(X) = \mathbb{Z}$ for any X. Likewise, op $K^0(X) =$ op $K^0(T_0) = K_0(T_0) = \mathbb{Z}$.   □

Example 4.3   When X is a complete toric variety then we know that $A^k_{\text{op}}(X) = \text{Hom}(A_k(X), \mathbb{Z})$. However, for non-complete toric variety this result fails. For example if $X = \mathbb{A}^1$ then $A_0(X) = 0$ but $A^0(X) = \mathbb{Z}$. Another example is to let σ denote the cone generated by $\{(1, 0, 1), (0, -1, 1), (-1, 0, 1), (0, 1, 1)\}$ in $\mathbb{R}^3$. One can compute that $A_2(X(\sigma)) = \mathbb{Z}/2 \oplus \mathbb{Z}$. Thus, $\text{Hom}(A_2(X(\sigma)), \mathbb{Z}) = \mathbb{Z}$ but by Corollary 4.2, $A^2_{\text{op}}(X(\sigma)) = 0$.

Example 4.4   In the equivariant case we have an analogous result for the T-equivariant operational Chow ring and K-theory. To simplify the notation we assume that the cone σ is full dimensional.

Corollary 4.5   If X is an affine toric variety associated to a full dimensional cone σ then $A^*_{\text{op}, T}(X) = \text{Sym}(X(T))$ and op $K^0_T(X) = R(T)$. Here $\text{Sym}(X(T))$ is the polynomial algebra generated by the character group of T and $R(T)$ is the representation ring of T.

Proof In this case $S_0 = \operatorname{Spec} k$, so $A^*_{\text{op},T}(X) = A^*_T(\text{pt}) = \operatorname{Sym}(X(T))$ by [5, Section 3.2] and op $K^0_T(X) = K^0_T(\text{pt}) = R(T)$. □

4.1 An Alternative Proof of the Vanishing of Chow Cohomology and Operational K-theory on Affine Toric Varieties

In [18, 6] a more involved proof that $A^*_{\text{op}}(X) = \operatorname{op} K^0(X) = \mathbb{Z}$ when X is an affine toric variety is given. The proof of both of these statements rests on the fact that both $A^*_{\text{op}}(X)$ and op $K^0(X)$ satisfy the following descent property for proper surjective morphisms.

If $X' \to X$ is a proper surjective morphism and if H denotes either functor A^*_{op} or op K^0 then the sequence

$$0 \to H(X) \otimes \mathbb{Q} \to H(X') \otimes \mathbb{Q} \xrightarrow{p_1^* - p_2^*} H(X' \times_X X') \otimes \mathbb{Q}$$

is exact where p_1, p_2 are the two projections $X' \times_X X' \to X'$.

This descent property does not hold for arbitrary homotopy invariant functors – for example it need not hold for the functor op K^0_G when G is not a torus. However when the descent property holds for a functor H, it can be used as a tool to calculate H on singular schemes [13, 3].

5 Operational Chow Rings of Good Moduli Spaces

The goal of this section is to explain how the cone theorem for homotopy invariant functors can shed light on questions about the structure of the operational Chow ring for quotients of smooth varieties and, more generally, good moduli spaces of smooth Artin stacks.

5.1 Strong Cycles on Good Moduli Spaces of Artin Stacks

Let G be a linearly algebraic group acting on a scheme X. We say that a scheme Y equipped with a G-invariant morphism $p: X \to Y$ is a *good quotient* if p is affine and $(p_* \mathcal{O}_X)^G = \mathcal{O}_Y$. The basic example is the quotient $X^{ss} \to X^{ss}/G$ where X^{ss} is the set of semi-stable points (with respect to a choice of linearization) for the action of a linearly reductive group on a projective variety X. This definition was extended to Artin stacks by Alper.

Definition 5.1 ([1, Definition 4.1])   Let $\mathcal{X}$ be an Artin stack and let X be an algebraic space. We say that X is a *good moduli space of* $\mathcal{X}$ if there is a morphism $\pi: \mathcal{X} \to X$ such that

1. π is *cohomologically affine* meaning that the pushforward functor π_* on the category of quasi-coherent $\mathcal{O}_{\mathcal{X}}$-modules is exact.
2. π is *Stein* meaning that the natural map $\mathcal{O}_X \to \pi_* \mathcal{O}_{\mathcal{X}}$ is an isomorphism.

Remark 5.2   If $\mathcal{X} = [Z/G]$ where G is a linearly reductive algebraic group then the statement that X is a good moduli space for $\mathcal{X}$ is equivalent to the statement that X is the good quotient of Z by G.

Definition 5.3 ([7])   Let $\mathcal{X}$ be an Artin stack with good moduli space X and let $\pi: \mathcal{X} \to X$ be the good moduli space morphism. We say that a closed point x of $\mathcal{X}$ is *stable* if $\pi^{-1}(\pi(x)) = x$ under the induced map of topological spaces $|\mathcal{X}| \to |X|$. A closed point x of $\mathcal{X}$ is *properly stable* if it is stable and the stabilizer of x is finite.

We say $\mathcal{X}$ is stable (resp. properly stable) if there is a good moduli space $\pi: \mathcal{X} \to X$ and the set of stable (resp. properly stable) points is non-empty. Likewise we say that π is a stable (resp. properly stable) good moduli space morphism.

Remark 5.4   Again this definition is modeled on GIT. If G is a linearly reductive group and X^{ss} is the set of semistable points for a linearization of the action of G on a projective variety X then a (properly) stable point of $[X^{ss}/G]$ corresponds to a (properly) stable orbit in the sense of GIT. The stack $[X^{ss}/G]$ is stable if and only if $X^s \neq \emptyset$. Likewise $[X^{ss}/G]$ is properly stable if and only if $X^{ps} \neq \emptyset$. As is the case for GIT quotients, the set of stable (resp. properly stable points) is open [7].

Definition 5.5   [8, 9] Let $\mathcal{X}$ be an irreducible Artin stack with stable good moduli space $\pi: \mathcal{X} \to X$. A closed integral substack $\mathcal{Z} \subseteq \mathcal{X}$ is *strong* if $\mathrm{codim}_{\mathcal{X}} \mathcal{Z} = \mathrm{codim}_X \pi(\mathcal{Z})$ and $\mathcal{Z}$ is saturated with respect to π, i.e. $\pi^{-1}(\pi(\mathcal{Z})) = \mathcal{Z}$ as stacks. We say $\mathcal{Z}$ is *topologically strong* if $\mathrm{codim}_{\mathcal{X}} \mathcal{Z} = \mathrm{codim}_X \pi(\mathcal{Z})$ and $\pi^{-1}(\pi(\mathcal{Z}))_{red} = \mathcal{Z}$.

Let $A^*_{\mathrm{tst}}(\mathcal{X}/X)$ be the subgroup of $A^*(\mathcal{X})$ generated by topologically strong cycles and $A^*_{\mathrm{st}}(\mathcal{X}/X)$ be the subgroup generated by by strong cycles. (Here the Chow group $A^*(\mathcal{X})$ is the Chow group defined by Kresch [14]. When $\mathcal{X} = [Z/G]$ it can be identified with the equivariant Chow group $A^*_G(Z)$ of [5].)

The main result of [9] is the following theorem which says that if X is smooth then any operational class $c \in A^k(X)$ can be represented by a codimension k-cycle on the stack $\mathcal{X}$ which is saturated with respect to the good moduli space morphism $\mathcal{X} \to X$.

Theorem 5.6 [9, Theorem 1.1] Let $\mathcal{X}$ be a properly stable smooth Artin stack with good moduli space $\mathcal{X} \xrightarrow{\pi} X$. Then there is a pullback $\pi : A^*_{\mathrm{op}}(X)_\mathbb{Q} \to A^*(\mathcal{X})_\mathbb{Q}$ which is injective and factors through the subgroup $A^*_{\mathrm{tst}}(\mathcal{X})_\mathbb{Q}$.

[9, Example 3.24] shows that not every topologically strong cycle is in the image of $A^*_{\mathrm{op}}(X)$. However, [9, Theorem 1.7c] states that any strong lci cycle on $\mathcal{X}$ is in the image of $A^*_{\mathrm{op}}(X)_\mathbb{Q}$. (A cycle $\sum_i a_i [\mathcal{Z}_i]$ is lci if the $\mathcal{Z}_i$ are closed substacks of $\mathcal{X}$ such that the inclusion $\mathcal{Z}_i \hookrightarrow \mathcal{X}$ is an lci morphism.) This leads to a number of successively weaker questions about the operational Chow rings of good moduli spaces of smooth Artin stacks. They can be viewed as analogues for quotients of smooth varieties by reductive groups of Conjectures 2 and 3 of [16].

Question 5.7 Is the image of $A^*_{\mathrm{op}}(X)_\mathbb{Q}$ contained in the subgroup of $A^*(\mathcal{X})_\mathbb{Q}$ generated by strong lci cycles?

Question 5.8 Is the image of $A^*_{\mathrm{op}}(X)_\mathbb{Q}$ equal to the subring of $A^*(\mathcal{X})_\mathbb{Q}$ generated by strong lci cycles?

Question 5.9 Is $A^*_{\mathrm{op}}(X)_\mathbb{Q}$ generated by Chern classes of perfect complexes on X?

Remark 5.10 Note that Question 5.9 is an analogue of the question raised by Anderson and Payne about the surjectivity of the map $K^0(X) \to \mathrm{op}\, K^0(X)$ where $K^0(X)$ is the Grothendieck group of perfect complexes. Anderson and Payne prove that for 3-dimensional complete toric varieties this map is in fact surjective. By comparison in [9] the authors prove that $A^*(X)_\mathbb{Q} = A^*_{\mathrm{st}}(\mathcal{X}/X)_\mathbb{Q}$ and in the case of 3-dimensional toric varieties [18] shows that $A^*(X)_\mathbb{Q}$ is generated by strong lci cycles.

Given the relation between the operational Chow ring and strong cycles leads to the following additional question.

Question 5.11 Is $A^*_{\mathrm{st}}(\mathcal{X}/X)$ (resp. $A^*_{tst}(\mathcal{X}/X)$) a subring of $A^*(\mathcal{X})$; i.e., is the product of strong (resp. topologically strong) cycles strong?

5.2 The Cone Theorem and Local Models for Good Moduli

The étale slice theorem of Alper, Hall and Rydh [2] states if $\mathcal{X}$ is a smooth stack then at a closed point x of $\mathcal{X}$ the good moduli morphism $\mathcal{X} \to X$ is étale locally isomorphic to the quotient $[V/G_x] \to \mathrm{Spec}\, k[V]^{G_x}$ where G_x is the inertia group of x in $\mathcal{X}$ and V is a representation of G_x. Thus we may view stacks of the form $[V/G]$ with their good moduli spaces $\operatorname{Spec} k[V]^G$ as the "affine models" of smooth stacks with good moduli spaces.

The cone theorem has the following corollary.

Corollary 5.12 Let V be a representation of a reductive group G and let $X = \operatorname{Spec} k[V]^G$ be the quotient. Then for any homotopy invariant functor $H(X) = H(\operatorname{Spec} k)$. In particular, $A^*_{\mathrm{op}}(X) = \mathbb{Z}$ and op $K^0(X) = \mathbb{Z}$.

Proof Since G acts linearly on V the action of G on $k[V]$ preserves the natural grading. Hence the invariant ring $k[V]^G$ is also graded. Thus by Theorem 3.2 $H(X) = H(\operatorname{Spec} k)$ since k is the 0-th graded piece of $k[V]^G$. □

Combining Corollary 5.12 with [9, Theorem 1.7c] yields the following result.

Corollary 5.13 If $\mathcal{Z} \subset \mathcal{X} = [V/G]$ is a proper closed substack which is strongly regularly embedded then $[\mathcal{Z}]$ is torsion in $A^*(\mathcal{X}) = A^*_G(\mathrm{pt})$. In particular if $A^*_G(\mathrm{pt})$ is torsion free (for example if G is torus or GL_n) then $[\mathcal{Z}] = 0$.

Note that any integral strong divisor $\mathcal{D}$ on $\mathcal{X} = [V/G]$ is necessarily defined by a single G-invariant equation. In this case $\mathcal{O}(\mathcal{D})$ is an equivariantly trivial line bundle so $[\mathcal{D}] = c_1(\mathcal{O}(\mathcal{D})) = 0$ in the equivariant Chow ring $A^*_G(V)$. More generally, if an integral substack $\mathcal{Z} \subset [V/G]$ is a complete intersection of strong divisors, then its class in $A^*_G(V)$ is also 0. In particular it shows that in the "local case" the image of $A^*_{\mathrm{op}}(X)$ is contained in the subgroup generated by strong global complete intersections.

This leads to the following question.

Question 5.14 Are there examples of strong integral substacks $\mathcal{Z} \subset [V/G]$ such that $[\mathcal{Z}] \neq 0$ in $A^*([V/G])_\mathbb{Q}$?

5.3 Acknowledgments

The authors are grateful to Sam Payne and Angelo Vistoli for helpful comments. The first author was supported by Simons Collaboration Grant 315460.

References

[1] Jarod Alper, *Good moduli spaces for Artin stacks*, Ann. Inst. Fourier (Grenoble) **63** (2013), no. 6, 2349–2402.

[2] Jarod Alper, Jack Hall, and David Rydh, *A Luna étale slice theorem for algebraic stacks*, arXiv:1504.06467.

[3] Dave Anderson and Sam Payne, *Operational K-theory*, Doc. Math. **20** (2015), 357–399.

[4] David A. Cox, John B. Little, and Henry K. Schenck, *Toric varieties*, Graduate Studies in Mathematics, vol. 124, American Mathematical Society, Providence, RI, 2011.

[5] Dan Edidin and William Graham, *Equivariant intersection theory*, Invent. Math. **131** (1998), no. 3, 595–634.

[6] Dan Edidin and Ryan Richey, *The Chow cohomology of affine toric varieties*, Math Research Letters **27** (2020), no. 6, 1645–1667.

[7] Dan Edidin and David Rydh, *Canonical reduction of stabilizers for Artin stacks with good moduli spaces*, Duke Math J. **170** (2021), no. 5, 827–880.

[8] Dan Edidin and Matthew Satriano, *Strong cycles and intersection products on good moduli spaces*, K-Theory—Proceedings of the International Colloquium, Mumbai, 2016, Hindustan Book Agency, New Delhi, 2018, pp. 223–238.

[9] Dan Edidin and Matthew Satriano, *Towards an intersection Chow cohomology on GIT quotients*, Transform Groups **25** (2020), no. 4, 1103–1129.

[10] William Fulton, *Intersection theory*, Springer-Verlag, Berlin, 1984.

[11] William Fulton and Robert MacPherson, *Categorical framework for the study of singular spaces*, Mem. Amer. Math. Soc. **31** (1981), no. 243, vi+165.

[12] William Fulton and Bernd Sturmfels, *Intersection theory on toric varieties*, Topology **36** (1997), no. 2, 335–353.

[13] Shun-ichi Kimura, *Fractional intersection and bivariant theory*, Comm. Algebra **20** (1992), no. 1, 285–302.

[14] Andrew Kresch, *Cycle groups for Artin stacks*, Invent. Math. **138** (1999), no. 3, 495–536.

[15] Domingo Luna, *Slices étales*, Sur les groupes algébriques, Soc. Math. France, Paris, 1973, pp. 81–105. Bull. Soc. Math. France, Paris, Mémoire 33.

[16] Marco Maggesi and Gabriele Vezzosi, *Some elementary remarks on lci algebraic cycles*, arXiv:1612.04570.

[17] Sam Payne, *Equivariant Chow cohomology of toric varieties*, Math. Res. Lett. **13** (2006), no. 1, 29–41.

[18] Ryan Richey, Ph.D. thesis, University of Missouri, May 2019.

[19] Burt Totaro, *Chow groups, Chow cohomology, and linear varieties*, Forum Math. Sigma **2** (2014), e17, 25pages.

[20] Charles A. Weibel, *The K-book*, Graduate Studies in Mathematics, vol. 145, American Mathematical Society, Providence, RI, 2013, An introduction to algebraic K-theory.

8

Cayley–Bacharach Theorems with Excess Vanishing

Lawrence Ein[a] and Robert Lazarsfeld[b]

Dedicated to Bill Fulton on the occasion of his 80th birthday.

1 Introduction

A classical theorem of Cayley and Bacharach asserts that if $D_1, D_2 \subseteq \mathbf{P}^2$ are curves of degrees d_1 and d_2 meeting transversely, then any curve of degree $d_1 + d_2 - 3$ passing through all but one of the $d_1 d_2$ points of $D_1 \cap D_2$ must also contain the remaining point. Generalizations of this statement have been a source of fascination for decades. Algebraically, the essential point is that complete intersection quotients of a polynomial ring are Gorenstein. We refer the reader to [6, Part I] for a detailed overview.

The most natural geometric setting for results of this sort was introduced in the paper [7] of Griffiths and Harris. Specifically, let X be a smooth complex projective variety of dimension n, let E be a vector bundle on X of rank n, and set $L = \det E$. Suppose given a section $s \in \Gamma(X, E)$ that vanishes simply along a finite set $Z \subseteq X$. Griffiths and Harris prove that if

$$h \in \Gamma\big(X, \mathcal{O}_X(K_X + L)\big)$$

vanishes at all but one of the points of Z, then it vanishes at the remaining point as well. This of course implies statements for hypersurfaces in projective space by taking E to be a direct sum of line bundles.

The starting point of the present note was the paper [10] of Mu-Lin Li, who proposed an extension allowing for excessive vanishing. With X, E and L as above, suppose that $s \in \Gamma(X, E)$ is a section that vanishes scheme-theoretically

[a] University of Illinois at Chicago
[b] Stony Brook University, New York
L. E. was partially supported by NSF grant #1801870. R. L. was partially supported by NSF grant #1739285.

along a smooth subvariety $W \subseteq X$ of dimension $w \geq 0$ in addition to a non-empty reduced finite set $Z \subseteq X$:

$$\text{Zeroes}(s) = W \sqcup Z.$$

For example, one might imagine three surfaces in $\mathbf{P}^3$ cutting out the union of a smooth curve and a finite set. Assuming for simplicity that W is irreducible, its normal bundle $N_{W/X}$ sits naturally as a sub-bundle of the restriction $E \mid W$, giving rise to an exact sequence

$$0 \longrightarrow N_{W/X} \longrightarrow E \mid W \longrightarrow V \longrightarrow 0, \qquad (*)$$

where V is a vector bundle of rank w on W. Li's result is the following:

Theorem 1.1 ([10], Corollary 1.3) *Assume that the exact sequence* $(*)$ *splits. Then any section of* $\mathcal{O}_X(K_X + L)$ *vanishing on* W *and at all but one of the points of* Z *vanishes also on the remaining point of* Z.

His argument is analytic in nature, using what he calls "virtual residues."

It is natural to ask whether the statement remains true without assuming the splitting of $(*)$. The following example shows that this is not the case.

Example 1.2 Let $C \subseteq \mathbf{P}^3$ be a rational normal cubic curve, and fix general surfaces

$$Q_1, Q_2, F \supseteq C$$

containing C, with $\deg Q_1 = \deg Q_2 = 2$ and $\deg F = d > 2$. Recall that $Q_1 \cap Q_2 = C \cup L$ where L is a line meeting C at two points $a_1, a_2 \in L$. Therefore

$$Q_1 \cap Q_2 \cap F = C \sqcup Z,$$

where Z consists of the $(d - 2)$ additional points of intersection of F with L. (See Figure 1.) The conclusion of Li's theorem would be that any surface H of degree $2 + 2 + d - 4 = d$ passing through C and all but one of the points of Z passes through the remaining one. However this need not happen: for instance, one can take H to be the union of a general cubic through C and $(d - 3)$ planes each passing through exactly one of the points of Z. $\qquad \square$

On the other hand, staying in the setting of the Example, suppose that $H \subseteq \mathbf{P}^3$ is a surface of degree d that passes *doubly* through C and in addition contains all but one of the points of Z. Then H meets L twice at a_1 and a_2 as well as at $(d - 3)$ other points, and therefore $H \supseteq L$. In other words, in this case the conclusion of Li's theorem does hold if one looks at surfaces that have multiplicity ≥ 2 along C. This is an illustration of our first general result.

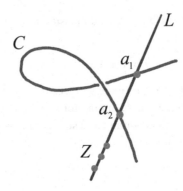

Figure 1 Twisted cubic and secant line: $C \cup L = Q_1 \cap Q_2$.

Theorem 1.3 *With X and E and $L = \det E$ as above, consider a section $s \in \Gamma(X, E)$ with*

$$\text{Zeroes}(s) =_{\text{scheme-theoretically}} W \sqcup Z,$$

where W is smooth of dimension w and Z is a non-empty reduced finite set. Suppose that

$$h \in \Gamma\left(X, \mathcal{O}_X(K_X + L) \otimes I_W^{w+1}\right)$$

is a section of $K_X + L$ vanishing to order $(w + 1)$ along W, as well as at all but one of the points of Z. Then h vanishes also at the remaining point of Z.

Note that if $\dim W = 0$, this reduces to the classical result. See also Example 3.6 for an application to statements closer to the spirit of [10].

Theorem 1.3 is a special case of a more general result involving multiplier ideals. Continuing to keep X, E and L as before, consider a section $s \in \Gamma(X, E)$ that vanishes simply along a non-empty finite set $Z \subseteq X$ and arbitrarily along a scheme disjoint from Z defined by an ideal $\mathfrak{b} \subseteq \mathcal{O}_X$. In other words, we ask that that the image of the map

$$E^* \longrightarrow \mathcal{O}_X$$

defined by s be the ideal $\mathfrak{b} \cdot I_Z$, with $\mathfrak{b} + I_Z = \mathcal{O}_X$. One can associate to $\mathfrak{b}$ and its powers *multiplier ideals* $\mathcal{J}(\mathfrak{b}^m) = \mathcal{J}(X, \mathfrak{b}^m) \subseteq \mathcal{O}_X$ that measure in a somewhat delicate way the singularities of elements of $\mathfrak{b}$. We prove:

Theorem 1.4 *Let*

$$h \in \Gamma\left(\mathcal{O}_X(K_X + L) \otimes \mathcal{J}(\mathfrak{b}^n)\right)$$

be a section of $\mathcal{O}_X(K_X + L)$ vanishing along the multiplier ideal $\mathcal{J}(\mathfrak{b}^n)$, and suppose that h vanishes at all but one of the points of Z. Then it vanishes also at the remaining point.

If $\mathfrak{b} = I_W$ is the ideal sheaf of a smooth subvariety of dimension w, then $\mathcal{J}(\mathfrak{b}^n) = I_W^{w+1}$, yielding Theorem 1.3. We remark that it is not essential that s vanish simply along the finite set Z, but then one has to reformulate (in a well-understood manner) what it means for h to vanish at all but one of the points of Z: see Remarks 2.5 and 3.4.

Theorem 1.4 follows almost immediately from the classical statement, but at the risk of making the result seem more subtle than it is let us explain conceptually why one expects multiplier ideals to enter the picture. When Zeroes$(s) = Z$ is a finite set, one can think of Cayley–Bacharach as arising via duality from the exactness of the Koszul complex

$$0 \longrightarrow \Lambda^n E^* \longrightarrow \Lambda^{n-1} E^* \longrightarrow \cdots \longrightarrow \Lambda^2 E^* \longrightarrow E^* \longrightarrow \mathcal{I}_Z \longrightarrow 0$$
$$\text{(Kos)}$$

determined by s. (See §1 for a review of the argument, which is due to Griffiths–Harris.) If s vanishes excessively this complex is no longer exact, which is why – as in the example above – the most naive analogue of Cayley–Bacharach fails. However (Kos) contains a subcomplex involving multiplier ideals that always is exact:

$$0 \longrightarrow \Lambda^n E^* \longrightarrow \Lambda^{n-1} E^* \otimes \mathcal{J}(\mathfrak{b}) \longrightarrow \cdots \longrightarrow E^* \otimes \mathcal{J}(\mathfrak{b}^{n-1})$$
$$\longrightarrow \mathcal{J}(\mathfrak{b}^n) \cdot \mathcal{I}_Z \longrightarrow 0. \quad \text{(Skod)}$$

(This is essentially the *Skoda complex* introduced in [5]: see Example 3.5 below.) One can view Theorem 1.4 as coming from (Skod) in much the same way that the classical result arises from (Koz).

There are variants of the Griffiths–Harris theorem that also extend to the setting of excess vanishing. As above, let $s \in \Gamma(X, E)$ be a section that vanishes simply on a finite set Z. Tan and Viehweg [13] in effect prove the following:

Theorem 1.5 *Fix an arbitrary line bundle A on X, and write $Z = Z_1 \sqcup Z_2$ as the union of two disjoint non-empty subsets. Set*

$$v_1 = \dim \operatorname{coker}\left(H^0(A) \longrightarrow H^0(A \otimes \mathcal{O}_{Z_1}) \right)$$

$$v_2 = \dim \operatorname{coker}\left(H^0(I_Z(K_X + L - A)) \hookrightarrow H^0(I_{Z_2}(K_X + L - A)) \right),$$

so that v_1 measures the failure of Z_1 to impose independent conditions on $H^0(A)$, while v_2 counts the number of sections of $\mathcal{O}_X(K_X + L - A)$ vanishing on Z_2 but not on Z_1. Then

$$v_2 \leq v_1. \tag{1}$$

So for example, if $A = \mathcal{O}_X$ and Z_1 consists of single point $x \in Z$, then $v_1 = 0$ and this reduces to the classical statement. Similarly, if we choose Z_1 in such a way that it imposes independent conditions on $H^0(A)$ then the assertion is that any section of $\mathcal{O}_X(K_X + L - A)$ vanishing on $Z_2 = Z - Z_1$ also vanishes on Z_1. The theorem of Tan–Viehweg generalizes analogous statements for hypersurfaces in projective space ([1], [4], [6]).[1]

We prove that in the case of possibly excessive vanishing, the analogous statement remains true taking into account multiplier ideal corrections.

Theorem 1.6 *Suppose as above that s defines the ideal* $\mathfrak{b} \cdot I_Z \subseteq \mathcal{O}_X$ *and that* $\mathfrak{b} + \mathcal{I}_Z = \mathcal{O}_X$. *Then the inequality* (1) *continues to hold provided that one takes*

$$v_2 = \dim \operatorname{coker} \Big(H^0\big(I_Z(K_X + L - A) \otimes \mathcal{J}(\mathfrak{b}^n)\big)$$
$$\hookrightarrow H^0\big(I_{Z_2}(K_X + L - A) \otimes \mathcal{J}(\mathfrak{b}^n)\big)\Big).$$

Again this follows quite directly from the classical statement.

Concerning the organization of this note, we start in §1 with a review of the theorem of Griffiths–Harris, and the extension in the spirit of Tan and Viehweg. As an application of the latter, we give at the end of the section a somewhat simplified and strengthened account of some results of Sun [11] concerning finite determinantal loci: see Theorem 2.6.[2] In §2 we derive the results involving multiplier ideals by applying the classical theorems on a log resolution of the base ideal. Since our primary interests lie on excess vanishing, we make the simplifying assumption throughout the main exposition that the finite zero-locus Z is reduced. The well-understood modifications needed in the general case are discussed in Remarks 2.5, 2.10 and 3.4. We work throughout over the complex numbers.

2 A Review of Cayley–Bacharach with Proper Vanishing

In this section we review the classical theorem of Cayley–Bacharach from the viewpoint of Griffiths–Harris and its extension in the spirit of Tan and Viehweg. As an application, we give at the end of the section some results of Cayley–Bacharach type for degeneracy loci.

---

[1] In the classical case, vanishings for the cohomology of line bundles on projective space yield the stronger assertion that $v_2 = v_1$.
[2] In an earlier version of this paper, we overlooked the work of Sun. We apologize for this omission.

Suppose then that X is a smooth complex projective variety of dimension n, and let E be a vector bundle of rank n on X, with $\det E = L$. We assume given a section $s \in \Gamma(X, E)$ vanishing simply on a non-empty finite set $Z \subseteq X$. Thus

$$\#Z = \int_X c_n(E).$$

In this setting, the basic result is due to Griffiths and Harris [7]:

Theorem 2.1 *Consider a section*

$$h \in \Gamma(X, \mathcal{O}_X(K_X + L))$$

vanishing at all but one of the points of Z. Then h vanishes on the remaining one as well.

We outline the argument of Griffiths and Harris from [7]. The starting point of the proof is to form the Koszul complex determined by s:

$$0 \longrightarrow \Lambda^n E^* \longrightarrow \Lambda^{n-1} E^* \longrightarrow \cdots \longrightarrow E^* \longrightarrow \mathcal{O}_X \longrightarrow \mathcal{O}_Z \longrightarrow 0.$$

Because X is smooth and s vanishes in the expected codimension n, this is exact. Now twist through by $\mathcal{O}_X(K_X + L)$. Recalling that $L = \Lambda^n E$ this gives a long exact sequence:

$$0 \longrightarrow \mathcal{O}_X(K_X) \longrightarrow \Lambda^{n-1} E^* \otimes \mathcal{O}_X(K_+L) \longrightarrow \cdots$$
$$\cdots \longrightarrow E^* \otimes \mathcal{O}_X(K_X + L) \longrightarrow \mathcal{O}_X(K_X + L) \longrightarrow \mathcal{O}_Z(K_K + L) \longrightarrow 0.$$
$$(2.1)$$

Splitting this into short exact sequences and taking cohomology, one arrives at maps

$$H^0\big(X, \mathcal{O}_X(K_X + L)\big) \longrightarrow H^0\big(Z, \mathcal{O}_Z(K_X + L)\big) \overset{\delta}{\longrightarrow} H^n\big(X, \mathcal{O}_X(K_X)\big)$$
$$(2.2)$$

whose composition is zero. (Absent additional vanishings, (2.2) might not be exact.)

On the other hand, note that $L \otimes \mathcal{O}_Z = \det(N_{Z/X})$ and hence there is a natural identification

$$\mathcal{O}_Z(K_X + L) = \omega_Z.$$

Therefore duality canonically identifies δ with a homomorphism

$$\delta' : H^0\big(Z, \mathcal{O}_Z\big)^* \longrightarrow H^0\big(X, \mathcal{O}_X\big)^*. \qquad (*)$$

Not surprisingly, one has the:

Lemma 2.2 *The mapping δ' in* (∗) *is the dual of the canonical restriction*

$$H^0(X, \mathcal{O}_X) \longrightarrow H^0(Z, \mathcal{O}_Z).$$

As Griffiths and Harris observe, this is ultimately a consequence of the functoriality of duality. We give the proof of a more general result – Lemma 2.4 below – in Appendix A.

Granting the Lemma, Theorem 2.1 follows at once. In fact, in terms of the natural basis for $H^0(Z, \mathcal{O}_Z)$ and its dual, the Lemma shows that δ' is given by the matrix $(1, \ldots, 1)$. Therefore if $h \in H^0\big(X, \mathcal{O}_X(K_X + L)\big)$ were to vanish at all but one of the points of Z but not at the remaining one, then $h|Z \notin \ker(\delta)$, contradicting the fact that the composition in (2.2) is the zero mapping.

We turn now to a result in the spirit of Tan and Viehweg [13].[3] Keeping assumptions and notation as above, write $Z = Z_1 \sqcup Z_2$ as the union of two non-empty subsets, and fix an arbitrary line bundle A. Recall the statement:

Theorem 2.3 *Define*

$$V_1 = \operatorname{coker}\Big(H^0(A) \longrightarrow H^0\big(A \otimes \mathcal{O}_{Z_1}\big)\Big),$$

$$V_2 = \operatorname{coker}\Big(H^0\big(I_Z(K_X + L - A)\big) \hookrightarrow H^0\big(I_{Z_2}(K_X + L - A)\big)\Big).$$

Then $\dim V_2 \leq \dim V_1$.

As noted in the Introduction, this implies Theorem 2.1 (at least when $\#Z \geq 2$).

For the proof, one starts by tensoring (2.1) by $\mathcal{O}_X(-A)$. Taking cohomology as before, one arrives at a complex

$$H^0\big(\mathcal{O}_X(K_X + L - A)\big) \xrightarrow{\text{res}} H^0\big(\mathcal{O}_Z(K_X + L - A)\big) \xrightarrow{\delta} H^n\big(\mathcal{O}_X(K_X - A)\big).$$
$$(2.3)$$

Moreover, via the decomposition

$$H^0\big(\mathcal{O}_Z(K_X + L - A)\big) = H^0\big(\mathcal{O}_{Z_1}(K_X + L - A)\big) \oplus H^0\big(\mathcal{O}_{Z_2}(K_X + L - A)\big),$$

this restricts to a subcomplex

$$H^0\big(\mathcal{O}_X(K_X + L - A) \otimes I_{Z_2}\big) \xrightarrow{\text{res}_1} H^0\big(\mathcal{O}_{Z_1}(K_X + L - A)\big)$$

$$\xrightarrow{\delta_1} H^n\big(\mathcal{O}_X(K_X - A)\big). \qquad (2.4)$$

Note next that

$$\ker(\text{res}_1) = H^0\big((K_X + L - A) \otimes I_Z\big),$$

---

[3] The actual statement and proof in [13] are rather more complicated, but Theorem 2.3 is essentially what is established there. In [12], Tan relates these statements to the Fujita conjecture.

and hence $V_2 = \text{Im}(\text{res}_1)$. On the other hand, since (2.4) is a complex, one has

$$\dim \text{Im}(\text{res}_1) \leq \dim \ker(\delta_1).$$

It is therefore sufficient to show that

$$\dim \ker(\delta_1) = \dim \text{coker}\Big(H^0(A) \longrightarrow H^0\big(A \otimes \mathcal{O}_{Z_1}\big)\Big). \qquad (*)$$

For this we again apply duality, which identifies δ and δ_1 with a diagram of maps

$$
\begin{array}{c}
H^0\big(Z, A \otimes \mathcal{O}_Z\big)^* \\
\Big\uparrow \qquad \qquad \searrow^{\delta'} \\
\qquad\qquad\qquad H^0\big(X, A\big)^*. \\
H^0\big(Z_1, A \otimes \mathcal{O}_{Z_1}\big)^* \quad {}^{\delta'_1}\nearrow
\end{array}
$$

As above the crucial point is to verify:

Lemma 2.4 *The mappings δ' and δ'_1 are dual to the natural restriction morphisms*

$$H^0\big(X, A\big) \longrightarrow H^0\big(Z, A \otimes \mathcal{O}_Z\big) \,, \;\; H^0\big(X, A\big) \longrightarrow H^0\big(Z_1, A \otimes \mathcal{O}_{Z_1}\big).$$

A proof of the Lemma appears in Appendix Appendix A. The Lemma implies that in fact

$$\ker \delta_1 = V_1^*,$$

and Theorem 2.3 is proved.

Remark 2.5 (Non-reduced zero schemes)   It is not necessary to assume that the finite scheme Z be reduced. In fact, since Z is Gorenstein, one can associate to any subscheme $Z_1 \subseteq Z$ a residual scheme $Z_2 \subseteq Z$ having various natural properties: see [6, p. 311 ff] for a nice discussion. The hypothesis in Theorem 2.1 should then be that h vanishes on the scheme residual to a point $x \in Z$. In Theorem 2.3 one works with a residual pair $Z_1, Z_2 \subseteq Z$. In this more general setting, one no longer has the embedding $\mathcal{O}_{Z_1} \subseteq \mathcal{O}_Z$ used in the proof. Instead, one replaces this with the canonical inclusion

$$\omega_{Z_1} \otimes \omega_Z^* \hookrightarrow \mathcal{O}_Z,$$

and then the argument with duality goes through. We leave details to the interested reader. □

We conclude this section by sketching an application of Theorem 2.3 to statements, essentially due to Sun [11], of Cayley–Bacharach type for determinantal loci. The present approach is somewhat different than that of [11], which uses Eagon–Northcott complexes.

We start with the set-up. Let X be a smooth projective variety of dimension n, and let E be a vector bundle on X of rank $n + e$ for some $e \geq 0$. Suppose given sections

$$s_0, \ldots, s_e \in \Gamma(X, E)$$

that drop rank simply along a reduced finite set Z, so that once again $\#Z = \int c_n(E)$. Denote by $W \subseteq H^0(E)$ the $(e + 1)$-dimensional subspace spanned by the s_i, and write $V = W^*$ for the dual of W. The s_i determine a natural vector bundle map $w \colon W_X \longrightarrow E$, where $W_X = W \otimes_{\mathbf{C}} \mathcal{O}_X$ is the trivial vector bundle with fibre W. By assumption w has rank exactly e at each point of Z, and hence its dual determines an exact sequence

$$E^* \xrightarrow{\ u\ } V_X \longrightarrow B_Z \longrightarrow 0. \tag{2.5}$$

where $V_X = V \otimes_{\mathbf{C}} \mathcal{O}_X$ and B_Z is a line bundle on Z. In particular, there is a natural mapping

$$\phi = \phi_u \colon Z \longrightarrow \mathbf{P}(V) = \mathbf{P}^e.$$

More concretely, ϕ sends each point $z \in Z$ to the one-dimensional quotient $\operatorname{coker}(u(z))$ of V. In particular, for any subset $Z' \subseteq Z$, and any $k \geq 0$, one gets a homomorphism

$$\rho_{Z',k} \colon H^0(\mathcal{O}_{\mathbf{P}}(k)) \longrightarrow H^0(\phi_* \mathcal{O}_{Z'}(k)) = H^0(Z', B_Z^{\otimes k} \mid Z').$$

Equivalently, this is the mapping

$$S^k V \longrightarrow H^0(Z', B_Z^{\otimes k} \mid Z')$$

arising from (2.5).

Theorem 2.6 *In the situation just described, set $L = \det E$, and write $Z = Z_1 \sqcup Z_2$ as the disjoint union of two non-empty subsets. Define*

$$c_1 = \dim \operatorname{coker}\!\left(\rho_{Z_1, n-1} \colon H^0(\mathcal{O}_{\mathbf{P}}(n - 1)) \longrightarrow H^0(\phi_* \mathcal{O}_{Z_1}(n - 1))\right),$$
$$c_2 = \dim \operatorname{coker}\!\left(H^0(X, I_Z(K_X + L)) \hookrightarrow H^0(X, I_{Z_2}(K_X + L))\right). \tag{2.6}$$

Then $c_2 \leq c_1$.

Example 2.7 Let $C, D \subseteq \mathbf{P}^2$ denote respectively a cubic and a quartic curve meeting transversely at twelve points. Take $O \in C \cap D$, and set

$$Z = (C \cap D) - \{O\},$$

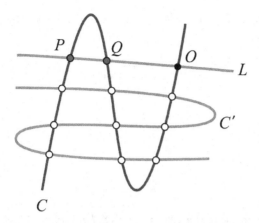

Figure 2 Eleven points in intersection of cubic and quartic.

so that Z consists of eleven of the twelve intersection points of C and D. Then Z is the degeneracy locus of a map

$$\mathcal{O}_{\mathbf{P}^2}^2 \xrightarrow{w} \mathcal{O}_{\mathbf{P}^2}(1) \oplus \mathcal{O}_{\mathbf{P}^2}(2) \oplus \mathcal{O}_{\mathbf{P}^2}(3),{}^4$$

whose dual (2.5) has the form

$$\mathcal{O}_{\mathbf{P}^2}(-1) \oplus \mathcal{O}_{\mathbf{P}^2}(-2) \oplus \mathcal{O}_{\mathbf{P}^2}(-3) \longrightarrow \mathcal{O}_{\mathbf{P}^2}^2.$$

Now pick two points $P, Q \in Z$ and take $Z_1 = \{P, Q\}$. If the line joining P and Q passes through O, then $c_1 = 1$, otherwise $c_1 = 0$. According to the Theorem, in the former case there may be an additional cubic passing through the remaining nine points of Z, but in the latter case there is none. If we take D to be the union of a cubic C' and a line L, and if $P, Q, O \in L$, then in fact $c_2 = 1$. (See Figure 2.) $\qquad\square$

Proof of Theorem 2.6 Set $Y = \mathbf{P}(V_X) = X \times \mathbf{P}(V)$, with projections

$$\mathrm{pr}_1 \colon Y \longrightarrow X \ , \ \mathrm{pr}_2 \colon Y \longrightarrow \mathbf{P}(V).$$

The plan is to use a well-known construction to realize Z as the zero-locus of a section of a vector bundle on Y, and then apply Theorem 2.3. Specifically, the presentation (2.5) determines an embedding $Z \subseteq Y$ under

---

[4] If O is defined by linear forms L_1 and L_2, then the equations defining C and D are expressed as

$$A_1 L_1 - A_2 L_2 \ , \ B_1 L_1 - B_2 L_2,$$

where $\deg A_i = 2$, $\deg B_i = 3$. The map w is then given by the matrix $\begin{pmatrix} L_1 & L_2 \\ A_2 & A_1 \\ B_2 & B_1 \end{pmatrix}$.

which $B_Z = \mathrm{pr}_2^*(\mathcal{O}_{\mathbf{P}(V)}(1))|Z$. Moreover, Z is defined in Y by the vanishing of the composition

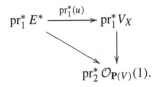

In other words, writing

$$F = \mathrm{pr}_1^* E \otimes \mathrm{pr}_2^* \mathcal{O}_{\mathbf{P}(V)}(1),$$

$Z \subseteq Y$ is the zero-locus of a section $s \in \Gamma(Y, F)$. Note next that

$$K_Y + \det F = \mathrm{pr}_1^*\big(\mathcal{O}_X(K_X + \det E)\big) \otimes \mathrm{pr}_2^* \mathcal{O}_{\mathbf{P}(V)}(n-1).$$

We now apply Theorem 2.3 with $A = \mathrm{pr}_2^* \mathcal{O}_{\mathbf{P}(V)}(n-1)$. Then on the one hand, for any $Z' \subseteq Z$, the restriction

$$H^0(Y, A) \longrightarrow H^0(Z', A|Z')$$

is identified with the map $\rho_{Z', n-1}$ appearing in (2.6). On the other hand, since

$$K_Y + \det F - A = \mathrm{pr}_1^*(K_X + \det E),$$

for any $Z' \subseteq Z$ one has

$$H^0(Y, \mathcal{O}_Y(K_Y + \det F - A) \otimes I_{Z'/Y}) = H^0(X, \mathcal{O}_X(K_X + \det E) \otimes I_{Z'/X}).$$

The Theorem follows.   □

Example 2.8   When Z_1 consists of a single point in Z, we find that if $s_0, \ldots, s_e \in \Gamma(X, E)$ drop rank along Z, then any section of $\mathcal{O}_X(K_X + L)$ vanishing at all but one of the points of Z vanishes at the remaining one. (This is a special case of [11, Theorem 4.1]. It also can be deduced directly from the theorem of Griffiths-Harris.)

Remark 2.9 (More general degeneracy loci)   An analogous statement – with essentially the same proof – holds for more general determinantal loci. Specifically, consider vector bundles V and E of ranks $e+1$ and $n+e$, and suppose that $w \colon V^* \longrightarrow E$ is a homomorphism that drops rank simply on a reduced finite set $Z \subseteq X$. As above, this gives rise to a surjection

$$E^* \longrightarrow V \longrightarrow B \otimes \mathcal{O}_Z \longrightarrow 0.$$

Then the statement of Theorem 2.6 remains valid provided that one takes $L = \det E + \det V$ and

$$c_1 = \dim \operatorname{coker}\left(H^0\left(X, S^{n-1}V\right) \longrightarrow H^0\left(X, B^{n-1} \otimes \mathcal{O}_Z\right)\right). \qquad \square$$

Remark 2.10 (Non-reduced degeneracy loci)   One can remove the hypothesis that the finite set Z be reduced by assuming instead that the map w (and hence also v) drops rank by exactly one at every point $x \in Z$. In this case $\operatorname{coker}(u)$ is still a line bundle on the degeneracy scheme Z defined by the vanishing of the maximal minors of u, as one sees by locally pulling back u from the space of all matrices. Then Z again embeds in Y, where it is the zero-locus of a section of a vector bundle. In particular Z is a local complete intersection scheme, and therefore Gorenstein, and one can proceed as in Remark 2.5. $\qquad \square$

3 Excess Vanishing

This section is devoted to the proofs of Theorems 1.4 and 1.6 from the Introduction.

We begin with a quick review of the basic facts about multiplier ideals, referring to [8, Chapter 9] or [9] for details. Let X be a smooth complex variety of dimension n, and let $\mathfrak{b} \subseteq \mathcal{O}_X$ be a coherent sheaf of ideals on X. One associates to $\mathfrak{b}$ and its powers a *multiplier ideal sheaf*

$$\mathcal{J}\left(\mathfrak{b}^m\right) = \mathcal{J}\left(X, \mathfrak{b}^m\right) \subseteq \mathcal{O}_X,$$

as follows. Start by forming a log resolution $\mu: X' \longrightarrow X$ of $\mathfrak{b}$, i.e. a proper birational map, with X' smooth, such that

$$\mathfrak{b} \cdot \mathcal{O}_{X'} = \mathcal{O}_{X'}(-B)$$

where B is an effective divisor on X' such that $B + K_{X'/X}$ has simple normal crossing support. One then takes

$$\mathcal{J}\left(\mathfrak{b}^m\right) = \mu_*\left(\mathcal{O}_X\left(K_{X'/X} - mB\right)\right). \qquad (3.1)$$[5]

One shows that the definition is independent of the choice of log-resolution. The intuition is that these multiplier ideals measure the singularities of functions $f \in \mathfrak{b}$, with "deeper" ideals corresponding to "greater singularities."

---

[5] More generally one can define $\mathcal{J}\left(\mathfrak{b}^c\right)$ for any rational number $c > 0$, but we will not requiren this.

Multiplier ideals satisfy many pleasant properties. We mention two here by way of orienting the reader. First, keeping notation as in (3.1), one has:

$$R^j \mu_* \left(\mathcal{O}_X \left(K_{X'/X} - mB \right) \right) = 0 \text{ for } j > 0. \tag{3.2}$$

This is known as the local vanishing theorem for multiplier ideals, and it guarantees in effect that the $\mathcal{J}(\mathfrak{b}^m)$ will be particularly well-behaved. Secondly, Skoda's theorem states that

$$\mathcal{J}(\mathfrak{b}^n) = \mathfrak{b} \cdot \mathcal{J}(\mathfrak{b}^{n-1}),$$

where as always $n = \dim X$. In particular, $\mathcal{J}(\mathfrak{b}^n) \subseteq \mathfrak{b}$, meaning that the multiplier ideals appearing in Theorems 3.2 and 3.3 below are at least as deep as $\mathfrak{b}$ itself.

Example 3.1 (Multiplier ideals of smooth subvarieties)   Suppose that $W \subseteq X$ is a smooth subvariety of dimension w. Let us show that

$$\mathcal{J}\left(X, I_W^m\right) = I_W^d,$$

where $d = \max\{0, w + 1 + (m - n)\}$. In fact, the blowing up $\mu : X' = \mathrm{Bl}_W(X) \longrightarrow X$ of W is a log resolution, with

$$K_{X'/X} = (n - w - 1)E \ , \ I_W \cdot \mathcal{O}_{X'} = \mathcal{O}_{X'}(-E),$$

$E \subseteq \mathcal{O}_{X'}$ being the exceptional divisor. Therefore

$$\mathcal{J}\left(I_W^m\right) = \mu_* \left(\mathcal{O}_X \left((n - w - 1 - m) E \right) \right) = I_W^d,$$

as claimed.

We now come to our main results. Let X be smooth complex projective variety of dimension n, and E a rank n vector bundle on X with $\det E = L$.

Theorem 3.2   *Let* $s \in \Gamma\left(X, E\right)$ *be a section whose zero-scheme is defined by the ideal*

$$\mathfrak{b} \cdot \mathcal{I}_Z \subseteq \mathcal{O}_X,$$

where $Z \subseteq X$ *is a non-empty reduced finite set, and* $\mathfrak{b} \subseteq \mathcal{O}_X$ *is an arbitrary ideal whose zero-locus is disjoint from* Z. *Suppose that*

$$h \in \Gamma\left(X, \mathcal{O}_X(K_X + L) \otimes \mathcal{J}(\mathfrak{b}^n)\right)$$

is a section vanishing at all but one of the points of Z. *Then* h *vanishes on the remaining point as well.*

Theorem 3.3 *In the setting of Theorem 3.2, write $Z = Z_1 \sqcup Z_2$ as the disjoint union of two non-empty subsets, and fix an arbitrary line bundle A on X. Write*

$$v_1 = \dim \operatorname{coker}\left(H^0(A) \longrightarrow H^0\left(A \otimes \mathcal{O}_{Z_1}\right) \right)$$

$$v_2 = \dim \operatorname{coker}\left(H^0\left(I_Z(K_X + L - A) \otimes \mathcal{J}(\mathfrak{b}^n)\right) \right.$$

$$\left. \hookrightarrow H^0\left(I_{Z_2}(K_X + L - A) \otimes \mathcal{J}(\mathfrak{b}^n)\right) \right).$$

Then $v_2 \le v_1$.

Observe that Theorem 1.3 from the Introduction follows from 3.2 together with Example 3.1.

Remark 3.4 (Non-reduced zeros) Provided that one proceeds as in Remark 2.5, one can remove the assumption that the zero-scheme Z be reduced. □

Proof of Theorems 3.2 and 3.3 We will deduce both results from the corresponding statements in §1. Specifically, let $\mu \colon X' \longrightarrow X$ be a log resolution of $\mathfrak{b}$, with

$$\mathfrak{b} \cdot \mathcal{O}_{X'} = \mathcal{O}_{X'}(-B),$$

where B is an effective divisor on X' with SNC support. We may and do suppose that μ is constructed by a sequence of blowings up over $\operatorname{Zeroes}(\mathfrak{b})$, so that in particular μ is an isomorphism over a neighborhood of Z. Therefore Z embeds naturally as a subset $Z' \subseteq X'$.

By assumption the image of the natural mapping $E^* \longrightarrow \mathcal{O}_X$ determined by s is the ideal $\mathfrak{b} \cdot \mathcal{I}_{Z/X}$. This mapping pulls back to a surjection

$$\mu^* E^* \longrightarrow\!\!\!\!\!\rightarrow \mathcal{O}_{X'}(-B) \cdot \mathcal{I}_{Z'/X'}.$$

In particular, setting

$$E' = \mu^* E \otimes \mathcal{O}_{X'}(-B),$$

s gives rise to a section $s' \in \Gamma\left(X', E'\right)$ vanishing exactly on $Z' \subseteq X'$.

For 3.2, we apply Theorem 2.1 to this section. That result asserts that every section

$$h' \in \Gamma\left(X', \mathcal{O}_{X'}(K_{X'} + \det E')\right)$$

vanishing at all but one of the points of Z' also vanshes at the remaining one. But observe that

$$K_{X'} + \det E' \equiv_{\lin} (K_{X'/X} - nB) + \mu^*(K_X + \det E).$$

Therefore

$$\mu_*\big(\mathcal{O}_{X'}(K_{X'} + \det E')\big) = \mathcal{O}_X(K_X + \det E) \otimes \mathcal{J}\big(\mathfrak{b}^n\big),$$

and in particular

$$\Gamma\big(X, \mathcal{O}_X(K_X + \det E) \otimes \mathcal{J}\big(\mathfrak{b}^n\big)\big) = \Gamma\big(X', \mathcal{O}_{X'}(K_{X'} + \det E')\big).$$

Recalling that μ is an isomorphism over a neighborhood of Z, Theorem 3.2 follows.

Theorem 3.3 follows in a similar manner from Theorem 2.3. Staying in the same setting, write $Z' = Z'_1 \sqcup Z'_2$ for the decomposition of Z' determined by Z_1 and Z_2, and put $A' = \mu^* A$. Then

$$H^0\big(X', A'\big) = H^0\big(X, A\big),$$

and hence

$$v_1 = \dim \operatorname{coker}\Big(H^0(A') \longrightarrow H^0\big(A' \otimes \mathcal{O}_{Z'_1}\big)\Big).$$

Moreover

$$H^0\big(X, I_Z(K_X + L - A) \otimes \mathcal{J}\big(\mathfrak{b}^n\big)\big) = H^0\big(X', I_{Z'}(K_{X'} + \det E' - A')\big)$$
$$H^0\big(X, I_{Z_2}(K_X + L - A) \otimes \mathcal{J}\big(\mathfrak{b}^n\big)\big) = H^0\big(X', I_{Z'_2}(K_{X'} + \det E' - A')\big),$$

and hence 2.3 yields 3.3. $\qquad\square$

Example 3.5 (Skoda–Koszul complex) Let $s \in \Gamma\big(X, E\big)$ be as in the hypothesis of Theorems 3.2 and 3.3. If $\dim \operatorname{Zeroes}(s) \geq 1$, then the Koszul complex determined by s is not exact. However we assert that the Koszul complex determined by s contains an *exact* subcomplex

$$0 \longrightarrow \wedge^n E^* \longrightarrow \wedge^{n-1} E^* \otimes \mathcal{J}\big(\mathfrak{b}\big) \longrightarrow \cdots$$
$$\longrightarrow E^* \otimes \mathcal{J}\big(\mathfrak{b}^{n-1}\big) \longrightarrow \mathcal{J}\big(\mathfrak{b}^n\big) \cdot \mathcal{I}_Z \longrightarrow 0. \qquad (*)$$

In fact, consider the Koszul complex on X' arising from $s' \in \Gamma\big(X', E'\big)$:

$$0 \longrightarrow \wedge^n E'^* \longrightarrow \wedge^{n-1} E'^* \longrightarrow \cdots \longrightarrow E'^* \longrightarrow \mathcal{I}_{Z'} \longrightarrow 0.$$

It is exact since s' vanishes in codimension n. Twisting by $\mathcal{O}_{X'}(K_{X'/X} - nB)$, one arrives at an exact sequence on X' with terms of the form

$$\mu^*\big(\wedge^{n-i} E^*\big) \otimes \mathcal{O}_{X'}(K_{X'/X} - iB).$$

These have vanishing higher direct images thanks to (3.2), and it follows that the direct image of the twisted Kosul complex remains exact. But

$$\mu_*\Big(\mu^*\big(\wedge^{n-i} E^*\big) \otimes \mathcal{O}_{X'}(K_{X'/X} - iB)\Big) = \wedge^{n-i} E^* \otimes \mathcal{J}\big(\mathfrak{b}^i\big),$$

yielding the long exact sequence (*). It would be possible to prove our main results using in the spirit of §2 using the Skoda complex to replace the Koszul complex.   □

Example 3.6 (Statements of Li-type)   By adjusting the numerics, one can deduce from Theorem 1.3 statements closer to the spirit of [10]. For example, suppose that $W \subseteq \mathbf{P}^n$ is a smooth variety of dimension w that is cut out scheme-theoretically by hypersurfaces of degree e, and consider hypersurfaces $D_1, \ldots, D_n$ of degrees $d_1, \ldots, d_n$ such that

$$D_1 \cap \cdots \cap D_n =_{\text{scheme-theoretically}} W \sqcup Z,$$

where Z is a non-empty reduced finite set. Then any hypersurface H with

$$\deg H = \left(\sum d_i \right) - (n+1) - w \cdot e$$

passing through W and all but one of the points of Z must also pass through the remaining point of Z. For instance, in the example from the Introduction this applies if H is a surface of degree $d - 2$ passing through C and all but one of the $(d - 2)$ points of Z. (In fact, we may choose hypersurfaces $H_1, \ldots, H_w$ of degree e passing through W but missing every point of Z. The assertion then follows by applying Theorem 1.3 to the hypersurface $H + H_1 + \cdots + H_w$.)   □

Example 3.7 (Fibres of rational coverings of projective space)   Let X be a smooth projective variety of dimension n, let

$$f \colon X \dashrightarrow \mathbf{P}^n$$

be a generically finite rational mapping, and let $Z \subseteq X$ be a generic fibre of f. Denote by $D \subseteq X$ the proper transform of a hyperplane in $\mathbf{P}^n$, so that f is defined by a linear series $|V| \subseteq |D|$, with base ideal $\mathfrak{b} = \mathfrak{b}(|V|)$. Then Z is the isolated zero-locus of a section of $\mathcal{O}_X^n(D)$ vanishing also along $\mathfrak{b}$, so Theorem 3.2 implies:

(*)   Any section of $\mathcal{O}_X(K_X + nD) \otimes \mathcal{J}(\mathfrak{b}^n)$ vanishing at all but one of the points of Z also vanishes at the remaining one.

Observe that we can find a section $t \in \Gamma(X, \mathcal{O}_X(D) \otimes \mathfrak{b})$ not vanishing at any point of Z, and then multiplication by t^n determines an embedding

$$\mathcal{O}_X(K_X) \hookrightarrow \mathcal{O}_X(K_X + nD) \otimes \mathcal{J}(\mathfrak{b}^n)$$

that is an isomorphism along Z. So we recover the statement – proved using Mumford's trace in [2] – that Z satisfies the Cayley–Bacharach property with respect to $|K_X|$. However (*) is a priori stronger since $|\mathcal{O}_X(K_X + nD) \otimes \mathcal{J}(\mathfrak{b}^n)|$ is typically larger than $|K_X|$. The statements for K_X are used in [2] and [3] to study the *degree of irrationality* of X – ie the least degree of a covering $f : X \dashrightarrow \mathbf{P}^n$. It would be interesting to know if (*) can lead to any improvements. In a similar vein, given an arbitrary line bundle A, Theorem 3.3 leads to a statement involving the linear series $|K_X - A|$ whose formulation we leave to the reader. $\square$

Remark 3.8 (Degeneracy loci)   We do not know whether or how one can generalize Theorem 2.6 to the case of vector bundle maps with excess degeneracies. $\square$

Appendix A Proof of Lemma 2.4

We sketch here one way to verify the identification asserted in Lemma 2.4. For clarity we work in a slightly more general setting.

Consider then a finite subscheme $Z \subseteq X$ of a smooth projective variety of dimension n. Fix a locally free acyclic complex

$$0 \longrightarrow L_{-n} \longrightarrow L_{-n+1} \longrightarrow \cdots \longrightarrow L_{-1} \longrightarrow L_0 \longrightarrow 0, \qquad (L_\bullet)$$

with $L_0 = \mathcal{O}_X$ resolving $\mathcal{O}_Z$ as an $\mathcal{O}_X$-module, so that $\mathcal{O}_Z = \mathcal{H}^0(L_\bullet)$. Let A be an arbitrary line bundle on X, and consider the complex $M_\bullet = \mathbf{D}_X(L_\bullet \otimes A) = L_\bullet^\vee \otimes A^* \otimes \omega_X[n]$. This has the form

$$0 \longrightarrow M_{-n} \longrightarrow M_{-n+1} \longrightarrow \cdots \longrightarrow M_{-1} \longrightarrow M_0 \longrightarrow 0, \qquad (M_\bullet)$$

where $M_{-n+i} = L_i^* \otimes \omega_X \otimes A^*$, and $M_\bullet$ is an acyclic resolution of the sheaf $Ext^n(A \otimes \mathcal{O}_Z, \omega_X)$. Breaking the long exact sequence

$$0 \to L_0^* \otimes \omega_X \otimes A^* \to L_1^* \otimes \omega_X \otimes A^* \to \cdots$$
$$\to L_{-n}^* \otimes \omega_X \otimes A^* \to Ext^n(A \otimes \mathcal{O}_Z, \omega_X) \to 0$$

into short exact sequences and taking cohomology, and recalling that $L_0 = \mathcal{O}_X$, one arrives at a homomorphism:

$$\delta : H^0\big(X, Ext^n(A \otimes \mathcal{O}_Z, \omega_X)\big) \longrightarrow H^n\big(X, \omega_X \otimes A^*\big).$$

Now

$$H^0\big(X, Ext^n(A \otimes \mathcal{O}_Z, \omega_X)\big) = Ext^n\big(A \otimes \mathcal{O}_Z, \omega_X\big)$$

is Grothendieck dual to $H^0\big(X, A \otimes \mathcal{O}_Z\big)$, and $H^n\big(X, \omega_X \otimes A^*\big)$ is dual to $H^0\big(X, A\big)$, so δ is identified with a mapping

$$\delta' : H^0\big(X, A \otimes \mathcal{O}_Z\big)^* \longrightarrow H^0\big(X, A\big)^*.$$

Proposition A.1 *The homomorphism δ' is dual to the restriction mapping*

$$\rho \colon H^0(X, A) \longrightarrow H^0(X, A \otimes \mathcal{O}_Z).$$

The first statement in Lemma 2.4 follows from this together with the self-duality of the Koszul complex. We leave the second statement – involving the subscheme Z_1 of Z – to the reader.

Proof This follows from the functoriality of Grothendieck–Verdier duality. Specifically, we view $L_\bullet$ as representing $\mathcal{O}_Z$ in $D^b(X)$. Recalling again that $L_0 = \mathcal{O}_X$, the restriction $A \longrightarrow A \otimes \mathcal{O}_Z$ is represented by the natural map of complexes $u \colon A[0] \longrightarrow L_\bullet \otimes A$, and $\rho \colon H^0(A) \longrightarrow H^0(A \otimes \mathcal{O}_Z)$ is given by the resulting homomorphism

$$\mathbf{H}^0(u) \colon \mathbf{H}^0\big(X, A[0]\big) \longrightarrow \mathbf{H}^0\big(X, L_\bullet \otimes A\big)$$

on hypercohomology. Now the map $\mathbf{D}(u) \colon \mathbf{D}_X(L_\bullet \otimes A) \longrightarrow \mathbf{D}_X(A[0])$ determined by u is the evident morphism

$$M_\bullet \longrightarrow M_{-n}[n] = \omega_X \otimes A^*[n].$$

By Grothendieck–Verdier duality, $\mathbf{H}^0(u)$ is dual to the resulting homomorphism

$$\mathbf{H}^0\big(X, M_\bullet\big) \longrightarrow \mathbf{H}^0\big(X, \omega_X \otimes A^*[n]\big).$$

The Proposition is therefore a consequence of the following $\qquad\qquad \square$

Lemma A.2 *Consider an acyclic complex*

$$0 \longrightarrow M_{-n} \longrightarrow M_{-n+1} \longrightarrow \cdots \longrightarrow M_{-1} \longrightarrow M_0 \longrightarrow 0, \qquad (M_\bullet)$$

resolving a sheaf $\mathcal{F}$ on the n-dimensional projective variety X. Then the mapping

$$H^0(X, \mathcal{F}) \longrightarrow H^n(X, M_{-n})$$

determined by splitting $(M_\bullet)$ into short exact sequences and taking cohomology coincides with the induced map on hypercohomology

$$\mathbf{H}^0(X, M_\bullet) \longrightarrow \mathbf{H}^0(X, M_{-n}[n]).$$

Proof Denote by $(M_\bullet)_{\le -i}$ the naive truncation of $M_\bullet$, obtained by replacing M_j with the zero sheaf for $j > -i$. There is an evident map $(M_\bullet)_{\le -i} \longrightarrow (M_\bullet)_{\le -(i+1)}$ of complexes that induces on $\mathbf{H}^0$ the connecting homomorphism

$$H^i\Big(X, \mathcal{H}^{-i}\big((M_\bullet)_{\le -i}\big)\Big) \longrightarrow H^{i+1}\Big(X, \mathcal{H}^{-i-1}\big((M_\bullet)_{\le -(i+1)}\big)\Big)$$

determined by the short exact sequence of sheaves

$$0 \longrightarrow \mathcal{H}^{-i-1}\big((M_\bullet)_{\leq -(i+1)}\big) \longrightarrow M_{-i} \longrightarrow \mathcal{H}^{-i}\big((M_\bullet)_{\leq -i}\big) \longrightarrow 0.$$

Applying this remark successively starting with $i = 0$, the Lemma follows.

$\square$

References

[1] I. Bacharach, Uber den Cayley'schen Schnittpunktsatz, Math. Ann. **26** (1886), 275–299.

[2] Francesco Bastianelli, R. Cortini and Pietro De Poi., The gonality theorem of Noether for hypersurfaces, *J. Algebraic Geometry* **23** (2014), 313–339.

[3] Francesco Bastianelli, Pietro De Poi, Lawrence Ein, Robert Lazarsfeld and Brooke Ullery, Measures of irrationality for hypersurfaces of large degree, *Compos. Math.* **153** (2017), 2368–2393.

[4] E. Davis, A. Geramita and F. Orecchia, Gorenstein algebras and the Cayley–Bacharach theorem, Proc. AMS **93** (1985), 593–597.

[5] Lawrence Ein and Robert Lazarsfeld, A geometric effective Nullstellensatz, *Invent. Math.* **137** (1999), 427–448.

[6] David Eisenbud, Mark Green and Joseph Harris, Cayley–Bacharach theorems and conjectures, *Bull. AMS* **33** (1996), 295–324.

[7] Phillip Griffiths and Joseph Harris, Residues and zero-cycles on algebraic varieties, *Ann. Math.* **108** (1978), 461–505.

[8] Robert Lazarsfeld, *Positivity in Algebraic Geometry*, Ergebnisse Math. vols 48 & 49, Springer 2004.

[9] Robert Lazarsfeld, A short course on multiplier ideals, in *Analytic and Algebraic Geometry*. IAS/Park City Math. Ser. **17**, 2010, 451–494.

[10] Mu-Lin Li, Virtual residue and a generalized Cayley–Bacharach theorem, Proc. AMS (to appear), doi.org/10.1090/proc/14606

[11] Hao Sun, On the Cayley–Bacharach property and the construction of vector bundles, Sci. China Math. **54** (2011), 1891–1898.

[12] S.-L. Tan, Cayley–Bacharach property of an algebraic variety and Fujita's conjecture, *J. Algebraic Geometry* **9** (2000), 201–222.

[13] S.-L. Tan and Eckart Viehweg, A note on the Cayley–Bacharach property for vector bundles, in *Complex Analysis and Algebraic Geometry*, de Greyter, Berlin, 2000, 361–373.

9

Effective Divisors on Hurwitz Spaces

Gavril Farkas[a]

To Bill Fulton on the occasion of his 80th birthday.

Abstract. We prove the effectiveness of the canonical bundle of a partial compactification of several Hurwitz spaces $\overline{\mathcal{H}}_{g,k}$ of degree k admissible covers of $\mathbf{P}^1$ from curves of genus $14 \leq g \leq 19$. This suggests that these moduli spaces are of general type.

Following a principle due to Mumford, most moduli spaces that appear in algebraic geometry (classifying curves, abelian varieties, $K3$ surfaces) are of general type, with a finite number of exceptions, which are unirational, or at least uniruled. Understanding the transition from negative Kodaira dimension to being of general type is usually quite difficult. The aim of this paper is to investigate to some extent this principle in the case of a prominent parameter space of curves, namely the Hurwitz space $\mathcal{H}_{g,k}$ classifying degree k covers $C \to \mathbf{P}^1$ with only simple ramification from a smooth curve C of genus g. Hurwitz spaces provide an interesting bridge between the much more accessible moduli spaces of pointed rational curves and the moduli space of curves. The study of covers of the projective line emerges directly from Riemann's Existence Theorem and Hurwitz spaces viewed as classifying spaces for covers of curves have been classically used by Clebsch [9] and Hurwitz [29] to establish the irreducibility of the moduli space of curves. Fulton was the first to pick up the problem in the modern age and his influential paper [19] initiated the treatment of Hurwitz spaces in arbitrary characteristic. Harris and Mumford [26] compactified Hurwitz spaces in order to show that $\overline{\mathcal{M}}_g$ is of general type for large g, see also [25]. Their work led the way to the theory of admissible covers which has many current ramifications in

[a] Humboldt University of Berlin

Gromov–Witten theory. Later, Harris and Morrison [27] bound from below the slope of all effective divisors on $\overline{\mathcal{M}}_g$ using the geometry of compactified Hurwitz spaces.

We denote by $\overline{\mathcal{H}}_{g,k}$ the moduli space of admissible covers constructed by Harris and Mumford [26] and studied further in [1], or in the book [3]. It comes equipped with a finite *branch map*

$$\mathfrak{b}\colon \overline{\mathcal{H}}_{g,k} \to \overline{\mathcal{M}}_{0,2g+2k-2}/\mathfrak{S}_{2g+2k-2},$$

where the target is the moduli space of *unordered* $(2g+2k-2)$-pointed stable rational curves, as well as with a map

$$\sigma\colon \overline{\mathcal{H}}_{g,k} \to \overline{\mathcal{M}}_g,$$

obtained by assigning to an admissible cover the stable model of its source. The effective cone of divisors of the quotient $\overline{\mathcal{M}}_{0,2g+2k-2}/\mathfrak{S}_{2g+2k-2}$ is spanned by the divisor classes $\widetilde{B}_2, \ldots, \widetilde{B}_{g+k-1}$, where $\widetilde{B}_i$ denotes the closure of the locus of stable rational curves consisting of two components such that one of them contains precisely i of the $2g+2k-2$ marked points.

It is a fundamental question to describe the birational nature of $\overline{\mathcal{H}}_{g,k}$ and one knows much less than in the case of other moduli spaces like $\overline{\mathcal{M}}_g$ or $\overline{\mathcal{A}}_g$. Recall that in a series of landmark papers [26], [25], [12] published in the 1980s, Harris, Mumford and Eisenbud proved that $\overline{\mathcal{M}}_g$ is a variety of general type for $g > 23$. This contrasts with the classical result of Severi [37] that $\overline{\mathcal{M}}_g$ is unirational for $g \le 10$. Later it has been shown that the Kodaira dimension of $\overline{\mathcal{M}}_g$ is negative for $g \le 15$. Precisely, $\overline{\mathcal{M}}_g$ is unirational for $g \le 14$, see [5], [36], [39], whereas $\overline{\mathcal{M}}_{15}$ is rationally connected [4]. It has been recently established in [15] that both $\overline{\mathcal{M}}_{22}$ and $\overline{\mathcal{M}}_{23}$ are of general type.

From Brill–Noether theory it follows that when $\frac{g+2}{2} \le k \le g$, every curve of genus g can be represented as a k-sheeted cover of $\mathbf{P}^1$, that is, the map $\sigma\colon \overline{\mathcal{H}}_{g,k} \to \overline{\mathcal{M}}_g$ is dominant. Using [28] it is easy to see that the generic fibre of the map σ is of general type, thus in this range $\overline{\mathcal{H}}_{g,k}$ is of general type whenever $\overline{\mathcal{M}}_g$ is. When $k \ge g+1$, the Hurwitz space is obviously uniruled being a Grassmannian bundle. Classically it has been known that $\mathcal{H}_{g,k}$ is unirational for $k \le 5$, see [2] and references therein for a modern treatment (or [32] and [33] for an alternative treatment using the Buchsbaum–Eisenbud structure theorem of Gorenstein rings of codimension 3). Geiss [23] using liaison techniques showed that most Hurwitz spaces $\mathcal{H}_{g,6}$ with $g \le 45$ are unirational. Schreyer and Tanturri [35] put forward the hypothesis that there exist only finitely many pairs (g,k), with $k \ge 6$, such that $\overline{\mathcal{H}}_{g,k}$ is *not* unirational. At the other end, when $k = g$, the Hurwitz space $\overline{\mathcal{H}}_{g,g}$ is birationally isomorphic to the universal symmetric product of degree $g-2$ over

$\overline{\mathcal{M}}_g$ and it has been showed in [18] that $\overline{\mathcal{H}}_{g,g}$ is of general type for $g \geq 12$ and uniruled in all the other cases. Further results on the unirationality of some particular Hurwitz spaces are obtained in [30] and [10].

In this paper, we are particularly interested in the range $\frac{g+2}{2} \leq k \leq g - 1$ (when $\overline{\mathcal{H}}_{g,k}$ dominates $\overline{\mathcal{M}}_g$) and $14 \leq g \leq 19$, so that $\overline{\mathcal{M}}_g$ is either unirational/uniruled in the cases $g = 14, 15$, or its Kodaira dimension is unknown, when $g = 16, 17, 18, 19$. We summarize our results showing the positivity of the canonical bundle of $\overline{\mathcal{H}}_{g,k}$ and we begin with the cases when $\overline{\mathcal{M}}_g$ is known to be (or thought to be) uniruled.

Theorem 0.1 *Suppose* $\overline{\mathcal{H}}$ *is one of the spaces* $\overline{\mathcal{H}}_{14,9}$ *or* $\overline{\mathcal{H}}_{16,9}$. *Then there exists an effective* $\mathbb{Q}$-*divisor class* E *on* $\overline{\mathcal{H}}$ *supported on the union* $\sum_{i \geq 3} \mathfrak{b}^*(\widetilde{B}_i)$, *such that the twisted canonical class* $K_{\overline{\mathcal{H}}} + E$ *is effective.*

A few comments are in order. We expect that the boundary divisor E is empty. Carrying this out requires an extension of the determinantal structure defining the Koszul divisors considered in this paper to the entire space $\overline{\mathcal{H}}_{g,k}$. This technically cumbersome step will be carried out in future work. Secondly, Theorem 0.1 is optimal. In genus 14, Verra [39] established the unirationality of the Hurwitz space $\mathcal{H}_{14,8}$, which is a finite cover of $\mathcal{M}_{14}$, and concluded in this way that $\mathcal{M}_{14}$ itself is unirational. In genus 16, the map

$$\sigma : \overline{\mathcal{H}}_{16,9} \to \overline{\mathcal{M}}_{16}$$

is generically finite. Chang and Ran [7] claimed that $\overline{\mathcal{M}}_{16}$ is uniruled (more precisely, that $K_{\overline{\mathcal{M}}_{16}}$ is not pseudo-effective, which by now, one knows that it implies uniruledness). However, it has become recently clear [38] that their proof contains a fatal error, so the Kodaira dimension of $\overline{\mathcal{M}}_{16}$ is currently unknown.

Theorem 0.2 *Suppose* $\overline{\mathcal{H}}$ *is one of the spaces* $\overline{\mathcal{H}}_{17,11}$ *or* $\overline{\mathcal{H}}_{19,13}$. *Then there exists an effective* $\mathbb{Q}$-*divisor class* E *on* $\overline{\mathcal{H}}$ *supported on the union* $\sum_{i \geq 3} \mathfrak{b}^*(\widetilde{B}_i)$, *such that the twisted canonical class* $K_{\overline{\mathcal{H}}} + E$ *is big.*

We observe that the maps $\overline{\mathcal{H}}_{17,11} \to \overline{\mathcal{M}}_{17}$ and $\overline{\mathcal{H}}_{19,13} \to \overline{\mathcal{M}}_{19}$ have generically 3 and respectively 5-dimensional fibres. As in the case of Theorem 0.1, here too we expect the divisor E to be empty.

Both Theorems 0.1 and 0.2 are proven at the level of a partial compactification $\widetilde{\mathcal{G}}^1_{g,k}$ (described in detail in Section 2) and which incorporates only admissible covers with a source curve whose stable model is irreducible. In each relevant genus we produce an explicit effective divisor $\widetilde{\mathcal{D}}_g$ on $\widetilde{\mathcal{G}}^1_{g,k}$ such that the canonical class $K_{\widetilde{\mathcal{G}}^1_{g,k}}$ can be expressed as a positive combination of a multiple of $[\widetilde{\mathcal{D}}_g]$ and the pull-back under the map σ of an effective divisor on $\overline{\mathcal{M}}_g$. Theorems 0.1 and 0.2 offer a very strong indication that the moduli spaces $\overline{\mathcal{H}}_{14,9}, \overline{\mathcal{H}}_{16,9}, \overline{\mathcal{H}}_{17,11}$ and $\overline{\mathcal{H}}_{19,13}$ are all of general type. To complete such a proof one would need to show that the singularities of $\overline{\mathcal{H}}_{g,k}$ impose no *adjunction conditions*, that is, every pluricanonical form on $\overline{\mathcal{H}}_{g,k}$ lifts to a pluricanonical form of a *resolution of singularities* of $\overline{\mathcal{H}}_{g,k}$. Establishing such a result, which holds at the level of $\overline{\mathcal{M}}_g$, see [26], or at the level of the moduli space of ℓ-torsion points $\overline{\mathcal{R}}_{g,\ell}$ for small levels [8], remains a significant challenge.

We describe the construction of these divisors in two instances and refer to Section 2 for the remaining cases and further details. First we consider the space $\overline{\mathcal{H}}_{16,9}$ which is a generically finite cover of $\overline{\mathcal{M}}_{16}$.

Let us choose a general pair $[C, A] \in \mathcal{H}_{16,9}$. We set $L := K_C \otimes A^\vee \in W_{21}^7(C)$. Denoting by $I_{C,L}(k) := \mathrm{Ker}\{\mathrm{Sym}^k H^0(C, L) \to H^0(C, L^{\otimes k})\}$, by Riemann-Roch one computes that $\dim I_{C,L}(2) = 9$ and $\dim I_{C,L}(3) = 72$. Therefore the locus where there exists a non-trivial syzygy, that is, the map

$$\mu_{C,L} : I_{C,L}(2) \otimes H^0(C, L) \to I_{C,L}(3)$$

is not an isomorphism is expected to be a divisor on $\mathcal{H}_{16,9}$. This indeed is the case and we define the *syzygy divisor*

$$\mathcal{D}_{16} := \left\{ [C, A] \in \mathcal{H}_{16,9} : I_{C, K_C \otimes A^\vee}(2) \otimes H^0(C, K_C \otimes A^\vee) \xrightarrow{\not\cong} I_{C, K_C \otimes A^\vee}(3) \right\}.$$

The ramification divisor of the map $\sigma : \mathcal{H}_{16,9} \to \mathcal{M}_{16}$, viewed as the *Gieseker–Petri divisor*

$$\mathcal{GP} := \left\{ [C, A] \in \mathcal{H}_{16,9} : H^0(C, A) \otimes H^0(C, K_C \otimes A^\vee) \xrightarrow{\not\cong} H^0(C, K_C) \right\}$$

also plays an important role. After computing the class of the closure $\widetilde{\mathcal{D}}_{16}$ of the Koszul divisor $\mathcal{D}_{16}$ inside the partial compactification $\widetilde{\mathcal{G}}_{16,9}^1$, we find the following explicit representative for the canonical class

$$K_{\widetilde{\mathcal{G}}_{16,9}^1} = \frac{2}{5}[\widetilde{\mathcal{D}}_{16}] + \frac{3}{5}[\widetilde{\mathcal{GP}}] \in CH^1(\widetilde{\mathcal{G}}_{16,9}^1). \tag{1}$$

This proves Theorem 0.1 in the case $g = 16$. We may wonder whether (1) is the only effective representative of $K_{\widetilde{\mathcal{G}}_{16,9}^1}$, which would imply that the Kodaira dimension of $\overline{\mathcal{H}}_{16,9}$ is equal to zero. To summarize, since the smallest Hurwitz cover of $\overline{\mathcal{M}}_{16}$ has an effective canonical class, it seems unlikely that the method of establishing the uniruledness/unirationality of $\mathcal{M}_{14}$ and $\mathcal{M}_{15}$ by studying a Hurwitz space covering it can be extended to higher genera $g \geq 17$.

The last case we discuss in this introduction is $g = 17$ and $k = 11$. We choose a general pair $[C, A] \in \mathcal{H}_{17,11}$. The residual linear system $L := K_C \otimes A^\vee \in W_{21}^6(C)$ induces an embedding $C \subseteq \mathbf{P}^6$ of degree 21. The multiplication map

$$\phi_L : \mathrm{Sym}^2 H^0(C, L) \to H^0(C, L^{\otimes 2})$$

has a 2-dimensional kernel. We impose the condition that the pencil of quadrics containing the image curve $C \overset{|L|}{\hookrightarrow} \mathbf{P}^6$ be *degenerate*, that is, it intersects the discriminant divisor in $\mathbf{P}(\mathrm{Sym}^2 H^0(C, L))$ non-transversally. We thus define the locus

$$\mathcal{D}_{17} := \left\{ [C, A] \in \mathcal{H}_{17,11} : \mathbf{P}(\mathrm{Ker}(\phi_{K_C \otimes A^\vee})) \text{ is a degerate pencil} \right\}.$$

Using [16], we can compute the class $[\widetilde{\mathcal{D}}_{17}]$ of the closure of $\mathcal{D}_{17}$ inside $\widetilde{\mathcal{G}}_{17,11}$. Comparing this class to that of the canonical divisor, we obtain the relation

$$K_{\widetilde{\mathcal{G}}_{17,11}^1} = \frac{1}{5}[\widetilde{\mathcal{D}}_{17}] + \frac{3}{5}\sigma^*(7\lambda - \delta_0) \in CH^1(\widetilde{\mathcal{G}}_{17,11}^1). \tag{2}$$

Since the class $7\lambda - \delta_0$ can be easily shown to be big on $\overline{\mathcal{M}}_{17}$, the conclusion of Theorem 0.2 in the case $g = 17$ now follows.

1 The birational geometry of Hurwitz spaces

We denote by $\mathcal{H}_{g,k}^o$ the Hurwitz space classifying degree k covers $f : C \to \mathbf{P}^1$ with source being a smooth curve C of genus g and having simple ramifications. Note that we choose an *ordering* $(p_1, \ldots, p_{2g+2k-2})$ of the set of the branch points of f. An excellent reference for the algebro-geometric study of Hurwitz spaces is [19]. Let $\overline{\mathcal{H}}_{g,k}^o$ be the (projective) moduli space of admissible covers (with an ordering of the set of branch points). The geometry of $\overline{\mathcal{H}}_{g,k}^o$ has been described in detail by Harris and Mumford [26] and further clarified in [1]. Summarizing their results, the stack $\overline{H}_{g,k}^o$ of admissible covers (whose coarse moduli space is precisely $\overline{\mathcal{H}}_{g,k}^o$) is isomorphic to the stack of *twisted stable* maps into the classifying stack $B\mathfrak{S}_k$ of the symmetric group $\mathfrak{S}_k$, that is, there is a canonical identification

$$\overline{H}_{g,k}^o := \overline{M}_{0,2g+2k-2}(B\mathfrak{S}_k).$$

Points of $\overline{\mathcal{H}}_{g,k}^o$ can be thought of as admissible covers

$$[f : C \to R, p_1, \ldots, p_{2g+2k-2}],$$

where the source C is a nodal curve of arithmetic genus g, the target R is a tree of smooth rational curves, f is a finite map of degree k satisfying $f^{-1}(R_{\mathrm{sing}}) = C_{\mathrm{sing}}$, and $p_1, \ldots, p_{2g+2k-2} \in R_{\mathrm{reg}}$ denote the branch points of f. Furthermore, the ramification indices on the two branches of C at each ramification point of f at a node of C must coincide. One has a finite *branch* morphism

$$\mathfrak{b} \colon \overline{\mathcal{H}}^{\mathrm{o}}_{g,k} \to \overline{\mathcal{M}}_{0,2g+2k-2},$$

associating to a cover its (ordered) branch locus. The symmetric group $\mathfrak{S}_{2g+2k-2}$ operates on $\overline{\mathcal{H}}^{\mathrm{o}}_{g,k}$ by permuting the branch points of each admissible cover. Denoting by

$$\overline{\mathcal{H}}_{g,k} := \overline{\mathcal{H}}^{\mathrm{o}}_{g,k}/\mathfrak{S}_{2g+2k-2}$$

the quotient parametrizing admissible covers *without* an ordering of the branch points, we introduce the projection $q \colon \overline{\mathcal{H}}^{\mathrm{o}}_{g,k} \to \overline{\mathcal{H}}_{g,k}$. Finally, let

$$\sigma \colon \overline{\mathcal{H}}_{g,k} \to \overline{\mathcal{M}}_g$$

be the map assigning to an admissible degree k cover the stable model of its source curve, obtained by contracting unstable rational components.

1.1 Boundary divisors of $\overline{\mathcal{H}}_{g,k}$

We discuss the structure of the boundary divisors on the compactified Hurwitz space. For $i = 0, \ldots, g+k-1$, let B_i be the boundary divisor of $\overline{\mathcal{M}}_{0,2g+2k-2}$, defined as the closure of the locus of unions of two smooth rational curves meeting at one point, such that precisely i of the marked points lie on one component and $2g + 2k - 2 - i$ on the remaining one. To specify a boundary divisor of $\overline{\mathcal{H}}^{\mathrm{o}}_{g,k}$, one needs the following combinatorial information:

(i) A partition $I \sqcup J = \{1, \ldots, 2g + 2k - 2\}$, such that $|I| \geq 2$ and $|J| \geq 2$.

(ii) Transpositions $\{w_i\}_{i \in I}$ and $\{w_j\}_{j \in J}$ in $\mathfrak{S}_k$, satisfying

$$\prod_{i \in I} w_i = u, \quad \prod_{j \in J} w_j = u^{-1},$$

for some permutation $u \in \mathfrak{S}_k$.

To this data (which is determined up to conjugation with an element from $\mathfrak{S}_k$), we associate the locus of admissible covers of degree k with labeled branch points

$$[f \colon C \to R, \ p_1, \ldots, p_{2g+2k-2}] \in \overline{\mathcal{H}}^{\mathrm{o}}_{g,k},$$

where $[R = R_1 \cup_p R_2, p_1, \ldots, p_{2g+2k-2}] \in B_{|I|} \subseteq \overline{\mathcal{M}}_{0,2g+2k-2}$ is a pointed union of two smooth rational curves R_1 and R_2 meeting at the point p. The marked points indexed by I lie on R_1, those indexed by J lie on R_2. Let $\mu :=$ $(\mu_1, \ldots, \mu_\ell) \vdash k$ be the partition corresponding to the conjugacy class of $u \in$ $\mathfrak{S}_k$. We denote by $E_{i:\mu}$ the boundary divisor on $\overline{\mathcal{H}}_k^o$ classifying twisted stable maps with underlying admissible cover as above, with $f^{-1}(p)$ having partition type μ, and exactly i of the points $p_1, \ldots, p_{2g+2k-2}$ lying on the component R_1. Passing to the unordered Hurwitz space $\overline{\mathcal{H}}_{g,k}$, we denote by $D_{i:\mu}$ the image $E_{i:\mu}$ under the map q, with its reduced structure. In general, both $D_{i:\mu}$ and $E_{i:\mu}$ may well be reducible.

The effect of the map $\mathfrak{b}$ on boundary divisors is summarized in the following relation that holds for $i = 2, \ldots, g + k - 1$, see [26] page 62, or [20] Lemma 3.1:

$$\mathfrak{b}^*(B_i) = \sum_{\mu \vdash k} \mathrm{lcm}(\mu) E_{i:\mu}. \tag{3}$$

The Hodge class on the Hurwitz space is by definition pulled back from $\overline{\mathcal{M}}_g$. Its class $\lambda := (\sigma \circ q)^*(\lambda)$ on $\overline{\mathcal{H}}_{g,k}^o$ has been determined in [31], or see [20] Theorem 1.1 for an algebraic proof. Remarkably, unlike on $\overline{\mathcal{M}}_g$, the Hodge class is always a boundary class:

$$\lambda = \sum_{i=2}^{g+k-1} \sum_{\mu \vdash k} \mathrm{lcm}(\mu) \left(\frac{i(2g + 2k - 2 - i)}{8(2g + 2k - 3)} \right.$$
$$\left. - \frac{1}{12} \left(k - \sum_{j=1}^{\ell(\mu)} \frac{1}{\mu_j} \right) \right) [E_{i:\mu}] \in CH^1(\overline{\mathcal{H}}_k^o). \tag{4}$$

The sum (4) runs over partitions μ of k corresponding to conjugacy classes of permutations that can be written as products of i transpositions. In the formula (4) $\ell(\mu)$ denotes the length of the partition μ. The only negative coefficient in the expression of $K_{\overline{\mathcal{M}}_{0,2g+2k-2}}$ is that of the boundary divisor B_2. For this reason, the components of $\mathfrak{b}^*(B_2)$ play a special role, which we now discuss. We pick an admissible cover

$$[f : C = C_1 \cup C_2 \to R = R_1 \cup_p R_2, \ p_1, \ldots, p_{2g+2k-2}] \in \mathfrak{b}^*(B_2),$$

and set $C_1 := f^{-1}(R_1)$ and $C_2 := f^{-1}(R_2)$ respectively. Note that C_1 and C_2 may well be disconnected. Without loss of generality, we assume $I = \{1, \ldots, 2g + 2k - 4\}$, thus $p_1, \ldots, p_{2g+2k-4} \in R_1$ and $p_{2g+2k-3}, p_{2g+2k-2} \in R_2$.

Let $E_{2:(1^k)}$ be the closure in $\overline{\mathcal{H}}_{g,k}^o$ of the locus of admissible covers such that the transpositions $w_{2g+2k-3}$ and $w_{2g+2k-2}$ describing the local monodromy in a neighborhood of the branch points $p_{2g+2k-3}$ and $p_{2g+2k-2}$ respectively, are equal. Let E_0 further denote the subdivisor of $E_{2:(1^k)}$ consisting of those admissible cover for which the subcurve C_1 is connected. This is the case precisely when $\langle w_1, \ldots, w_{2g+2k-4} \rangle = \mathfrak{S}_k$. Note that $E_{2:(1^k)}$ has many components not contained in E_0, for instance when C_1 splits as the disjoint union of a smooth rational curve mapping isomorphically onto R_1 and a second component mapping with degree $k-1$ onto R_1.

When the permutations $w_{2g+2k-3}$ and $w_{2g+2k-2}$ are distinct but share one element in their orbit, then $\mu = (3, 1^{k-3}) \vdash k$ and the corresponding boundary divisor is denoted by $E_{2:(3,1^{k-3})}$. Let E_3 be the subdivisor of $E_{2:(3,1^{k-3})}$ corresponding to admissible covers with $\langle w_1, \ldots, w_{2g+2k-4} \rangle = \mathfrak{S}_k$, that is, C_1 is a connected curve. Finally, in the case when $w_{2g+2k-3}$ and $w_{2g+2k-2}$ are disjoint transpositions, we obtain the boundary divisor $E_{2:(2,2,1^{k-4})}$. Similarly to the previous case, we denote by E_2 the subdivisor of $E_{2:(2,2,1^{k-4})}$ consisting of admissible covers for which $\langle w_1, \ldots, w_{2g+2k-4} \rangle = \mathfrak{S}_k$.

We denote by D_0, D_2 and D_3 the push-forward of E_0, E_2 and E_3 respectively under the map q. Using a variation of Clebsch's original argument [9] for establishing the irreducibility of $\overline{\mathcal{H}}_{g,k}$, it is easy to establish that D_0, D_2 and D_3 are all irreducibile, see also [22] Theorem 6.1. The boundary divisors E_0, E_2 and E_3, when pulled-back under the quotient map $q: \overline{\mathcal{H}}^o_{g,k} \to \overline{\mathcal{H}}_{g,k}$, verify the following formulas

$$q^*(D_0) = 2E_0, \ q^*(D_2) = E_2 \ \text{and} \ q^*(D_3) = 2E_3,$$

which we now explain. The general point of both E_0 and E_3 has no automorphism that fixes all branch points, but admits an automorphism of order two that fixes C_1 and permutes the branch points $p_{2g+2k-3}$ and $p_{2g+2k-2}$. The general admissible cover in E_2 has an automorphism group $\mathbb{Z}_2 \times \mathbb{Z}_2$ (each of the two components of C_2 mapping $2:1$ onto R_2 has an automorphism of order 2). In the stack $\overline{H}^o_{g,k}$ we have two points lying over this admissible cover and each of them has an automorphism group of order 2. In particular the map $\overline{H}^o_{g,k} \to \overline{\mathcal{H}}^o_{g,k}$ from the stack to the coarse moduli space is ramified with ramification index 1 along the divisor E_2.

One applies now the Riemann-Hurwitz formula to the branch map $\mathfrak{b}: \overline{\mathcal{H}}^o_{g,k} \to \overline{\mathcal{M}}_{0,2g+2k-2}$. Recall also that the canonical bundle of the moduli space of pointed rational curves is given by the formula

$$K_{\overline{\mathcal{M}}_{0,2g+2k-2}} = \sum_{i=2}^{g+k-1} \left(\frac{i(2g+2k-2-i)}{2g+2k-3} - 2 \right) [B_i].$$

All in all, we obtain the following formula for the canonical class of the Hurwitz stack:

$$K_{\overline{\mathcal{H}}^{\circ}_{g,k}} = \mathfrak{b}^* K_{\overline{\mathcal{M}}_{0,2g+2k-2}} + \text{Ram}(\mathfrak{b}), \tag{5}$$

where $\text{Ram}(\mathfrak{b}) = \sum_{i,\mu \vdash k} (\text{lcm}(\mu) - 1)[E_{i:\mu}]$.

1.2 A partial compactification of $\mathcal{H}_{g,k}$

Like in the paper [16], we work on a partial compactification of $\mathcal{H}_{g,k}$, for the Koszul-theoretic calculations are difficult to extend over all the boundary divisors $D_{i:\mu}$. The partial compactification of the Hurwitz space we consider is defined in the same spirit as in the part of the paper devoted to divisors on $\overline{\mathcal{M}}_g$. We fix an integer $k \leq \frac{2g+4}{3}$. Then $\rho(g,2,k) < -2$ and using [12] it follows then that the locus of curves $[C] \in \mathcal{M}_g$ such that $W^2_k(C) \neq \emptyset$ has codimension at least 2 in $\mathcal{M}_g$. We denote by $\widetilde{\mathcal{G}}^1_{g,k}$ the space of pairs $[C,A]$, where C is an irreducible nodal curve of genus g satisfying $W^2_k(C) = \emptyset$ and A is a base point free locally free sheaf of rank 1 and degree k on C with $h^0(C,A) = 2$. The rational map

$$\overline{\mathcal{H}}_{g,k} \dashrightarrow \widetilde{\mathcal{G}}^1_{g,k}$$

is of course regular outside a subvariety of $\overline{\mathcal{H}}_{g,k}$ of codimension at least 2, but can be made explicit over the boundary divisors D_0, D_2 and D_3, which we now explain.

Retaining the previous notation, to the general point $[f : C_1 \cup C_2 \to R_1 \cup_p R_2]$ of D_3 (respectively D_2), we assign the pair $[C_1, A_1 := f^* \mathcal{O}_{R_1}(1)] \in \widetilde{\mathcal{G}}^1_k$. Note that C_1 is a smooth curve of genus g and A_1 is a pencil on C_1 having a triple point (respectively two ramification points in the fibre over p). The spaces $\mathcal{H}_{g,k} \cup D_0 \cup D_2 \cup D_3$ and $\widetilde{\mathcal{G}}^1_{g,k}$ differ outside a set of codimension at least 2 and for divisor class calculations they will be identified. Using this, we copy the formula (4) at the level of the parameter space $\widetilde{\mathcal{G}}^1_{g,k}$ and obtain:

$$\lambda = \frac{g+k-2}{4(2g+2k-3)}[D_0] - \frac{1}{4(2g+2k-3)}[D_2]$$
$$+ \frac{g+k-6}{12(2g+2k-3)}[D_3] \in CH^1(\widetilde{\mathcal{G}}^1_{g,k}). \tag{6}$$

We now observe that the canonical class of $\widetilde{\mathcal{G}}^1_k$ has a simple expression in terms of the Hodge class λ and the boundary divisors D_0 and D_3. Quite remarkably, this formula is independent of both g and k!

Theorem 1.1   *The canonical class of the partial compactification $\widetilde{\mathcal{G}}^1_{g,k}$ is given by*

$$K_{\widetilde{\mathcal{G}}^1_{g,k}} = 8\lambda + \frac{1}{6}[D_3] - \frac{3}{2}[D_0].$$

Proof   We combine the equation (5) with the Riemann-Hurwitz formula applied to the quotient $q : \overline{\mathcal{H}}^o_{g,k} \dashrightarrow \widetilde{\mathcal{G}}^1_{g,k}$ and write:

$$q^*(K_{\widetilde{\mathcal{G}}^1_{g,k}}) = K_{\overline{\mathcal{H}}^o_{g,k}} - [E_0] - [E_2] - [E_3]$$

$$= -\frac{4}{2g+2k-3}[D_2] - \frac{2g+2k-1}{2g+2k-3}[D_0] + \frac{2g+2k-9}{2g+2k-3}[D_3].$$

To justify this formula, observe that the divisors E_0 and E_3 lie in the ramification locus of q, hence they have to subtracted from $K_{\overline{\mathcal{H}}^o_{g,k}}$. The morphism $\overline{\mathcal{H}}^o_{g,k} \to \overline{\mathcal{H}}^o_{g,k}$ from the stack to the coarse moduli space is furthermore simply ramified along E_2, so this divisor has to be subtracted as well. We now use (6) to express $[D_2]$ in terms of λ, $[D_0]$ and $[D_3]$ and obtain that $q^*(K_{\widetilde{\mathcal{G}}^1_{g,k}}) = 8\lambda + \frac{1}{3}[E_3] - 3[E_0]$, which yields the claimed formula. $\square$

Let $f : \mathcal{C} \to \widetilde{\mathcal{G}}^1_{g,k}$ be the universal curve and we choose a degree k Poincaré line bundle $\mathcal{L}$ on $\mathcal{C}$ (or on an étale cover if necessary). Along the lines of [16] Section 2 (where only the case $g = 2k - 1$ has been treated, though the general situation is analogous), we introduce two tautological codimension one classes:

$$\mathfrak{a} := f_*\big(c_1^2(\mathcal{L})\big) \text{ and } \mathfrak{b} := f_*\big(c_1(\mathcal{L}) \cdot c_1(\omega_f)\big) \in CH^1(\widetilde{\mathcal{G}}^1_{g,k}).$$

The push-forward sheaf $\mathcal{V} := f_*\mathcal{L}$ is locally free of rank 2 on $\widetilde{\mathcal{G}}^1_{g,k}$ (outside a subvariety of codimension at least 2). Its fibre at a point $[C, A]$ is canonically identified with $H^0(C, A)$. Although $\mathcal{L}$ is not unique, an easy exercise involving first Chern classes, convinces us that the class

$$\gamma := \mathfrak{b} - \frac{g-1}{k}\mathfrak{a} \in CH^1(\widetilde{\mathcal{G}}^1_{g,k}) \tag{7}$$

does not depend of the choice of a Poincaré bundle.

Proposition 1.2   *We have that $\mathfrak{a} = kc_1(\mathcal{V}) \in CH^1(\widetilde{\mathcal{G}}^1_{g,k})$.*

Proof   Simple application of the Porteous formula in the spirit of Proposition 11.2 in [16]. $\square$

The following locally free sheaves on $\widetilde{\mathcal{G}}_{g,k}^1$ will play an important role in several Koszul-theoretic calculations:

$$\mathcal{E} := f_*(\omega_f \otimes \mathcal{L}^{-1}) \quad \text{and} \quad \mathcal{F}_\ell := f_*(\omega_f^\ell \otimes \mathcal{L}^{-\ell}),$$

where $\ell \geq 2$.

Proposition 1.3 *The following formulas hold*

$$c_1(\mathcal{E}) = \lambda - \frac{1}{2}\mathfrak{b} + \frac{k-2}{2k}\mathfrak{a} \quad \text{and}$$

$$c_1(\mathcal{F}_\ell) = \lambda + \frac{\ell^2}{2}\mathfrak{a} - \frac{\ell(2\ell-1)}{2}\mathfrak{b} + \binom{\ell}{2}(12\lambda - [D_0]).$$

Proof Use Grothendieck–Riemann–Roch applied to the universal curve f, coupled with Proposition 1.2 in order to evaluate the terms. Use that $R^1 f_*(\omega_f^\ell \otimes \mathcal{L}^{-\ell}) = 0$ for $\ell \geq 2$. Similar to Proposition 11.3 in [16], so we skip the details. $\qquad\square$

We summarize the relation between the class γ and the classes $[D_0], [D_2]$ and $[D_3]$ as follows. Again, we find it remarkable that this formula is independent of g and k.

Proposition 1.4 *One has the formula* $[D_3] = 6\gamma + 24\lambda - 3[D_0]$.

Proof We form the fibre product of the universal curve $f: \mathcal{C} \to \widetilde{\mathcal{G}}_{g,k}^1$ together with its projections:

$$\mathcal{C} \xleftarrow{\;\pi_1\;} \mathcal{C} \times_{\widetilde{\mathcal{G}}_{g,k}^1} \mathcal{C} \xrightarrow{\;\pi_2\;} \mathcal{C}.$$

For $\ell \geq 1$, we consider the jet bundle $J_f^\ell(\mathcal{L})$, which sits in an exact sequence:

$$0 \longrightarrow \omega_f^{\otimes \ell} \otimes \mathcal{L} \longrightarrow J_f^\ell(\mathcal{L}) \longrightarrow J_f^{\ell-1}(\mathcal{L}) \longrightarrow 0. \tag{8}$$

Here $J_f^\ell(\mathcal{L}) := \left(\mathcal{P}_f^\ell(\mathcal{L})\right)^{\vee\vee}$ is the *reflexive closure* of the sheaf of principal parts defined as $\mathcal{P}_f^\ell(\mathcal{L}) := (\pi_2)_*\left(\pi_1^*(\mathcal{L}) \otimes \mathcal{I}_{(\ell+1)\Delta}\right)$, where $\Delta \subseteq \mathcal{C} \times_{\widetilde{\mathcal{G}}_{g,k}^1} \mathcal{C}$ is the diagonal.

One has a sheaf morphism $\nu_2: f^*(\mathcal{V}) \to J_f^2(\mathcal{L})$, which we think of as the *second Taylor map* associating to a section its first two derivatives. For points $[C, A, p] \in \mathcal{C}$ such that $p \in C$ is a smooth point, this map is simply the evaluation $H^0(C, A) \to H^0(A \otimes \mathcal{O}_{3p})$. Let $Z \subseteq \mathcal{C}$ be the locus where ν_2 is not injective. Over the locus of smooth curves, D_3 is the set-theoretic image of Z. A local analysis that we shall present shows that ν_2 is degenerate with multiplicity 1 at a point $[C, A, p]$, where $p \in C_{\text{sing}}$. Thus, D_0 is to be found with multiplicity 1 in the degeneracy locus of v_2. The Porteous formula leads to:

$$[D_3] = f_* c_2 \left(\frac{J_f^2(\mathcal{L})}{f^*(\mathcal{V})} \right) - [D_0] \in CH^1(\widetilde{\mathcal{G}}_{g,k}^1).$$

As anticipated, we now show that D_0 appears with multiplicity 1 in the degeneracy locus of v_2. To that end, we choose a family $F : X \rightarrow B$ of genus g curves of genus over a smooth 1-dimensional base B, such that X is smooth, and there is a point $b_0 \in B$ with $X_b := F^{-1}(b)$ is smooth for $b \in B \setminus \{b_0\}$, whereas X_{b_0} has a unique node $u \in X$. Assume furthermore that $A \in \text{Pic}(X)$ is a line bundle such that $A_b := L_{|X_b} \in W_k^1(X_b)$, for each $b \in B$. We further choose a local parameter $t \in \mathcal{O}_{B,b_0}$ and $x, y \in \mathcal{O}_{X,u}$, such that $xy = t$ represents the local equation of X around the point u. Then ω_F is locally generated by the meromorphic differential τ that is given by $\frac{dx}{x}$ outside the divisor $x = 0$ and by $-\frac{dy}{y}$ outside the divisor $y = 0$. Let us pick sections $s_1, s_2 \in H^0(X, A)$, where $s_1(u) \neq 0$, whereas s_2 vanishes with order 1 at the node u of X_{b_0}, along both its branches. Passing to germs of functions at u, we have the relation $s_{2,u} = (x + y)s_{1,u}$. Then by direct calculation in local coordinates, the map $H^0(X_{b_0}, A_{b_0}) \rightarrow H^0(X_{b_0}, A_{b_0|3u})$ is given by the 2×2 minors of the following matrix:

$$\begin{pmatrix} 1 & 0 & 0 \\ x + y & x - y & x + y \end{pmatrix}.$$

We conclude that D_0 appears with multiplicity 1 in the degeneracy locus of v_2.

From the exact sequence (8) one computes $c_1(J_f^2(\mathcal{L})) = 3c_1(\mathcal{L}) + 3c_1(\omega_f)$ and

$$c_2(J_f^2(\mathcal{L})) = c_2(J_f^1(\mathcal{L})) + c_1(J_f^1(\mathcal{L})) \cdot c_1(\omega_f^{\otimes 2} \otimes \mathcal{L})$$

$$= 3c_1^2(\mathcal{L}) + 6c_1(\mathcal{L}) \cdot c_1(\omega_f) + 2c_1^2(\omega_f).$$

Substituting, we find after routine calculations that

$$f_* c_2 \left(\frac{J_f^2(\mathcal{L})}{f^*(\mathcal{V})} \right) = 6\gamma + 2\kappa_1,$$

where $\kappa_1 = f_*(c_1(\omega_f)^{\otimes 2})$. Using Mumford's formula $\kappa_1 = 12\lambda - [D_0] \in CH^1(\widetilde{\mathcal{G}}_{g,k}^1)$, see e.g. [26] top of page 50, we finish the proof. $\square$

2 Effective divisors on Hurwitz spaces when $14 \leq g \leq 19$

We now describe the construction of four effective divisors on particular Hurwitz spaces interesting in moduli theory. The divisors in question are of syzygetic nature and resemble somehow the divisors on $\overline{\mathcal{M}}_g$ used recently in [15] to show that both $\overline{\mathcal{M}}_{22}$ and $\overline{\mathcal{M}}_{23}$, as well as those used earlier in [13] and [14] to disprove the Slope Conjecture formulated in [27]. The divisors on $\overline{\mathcal{H}}_{g,k}$ that we consider are defined directly in terms of a general element $[C, A]$ of $\mathcal{H}_{g,k}$ (where $A \in W_k^1(C)$), without making reference to other attributes of the curve C. This simplifies both the task of computing their classes and showing that the respective codimension 1 conditions in moduli lead to genuine divisor on $\mathcal{H}_{g,k}$. Using the irreducibility of $\mathcal{H}_{g,k}$, this amounts to exhibiting *one* example of a point $[C, A] \in \mathcal{H}_{g,k}$ outside the divisor, which can be easily achieved with the use of *Macaulay*. This is in sharp contrast to the situation in [15], where establishing the transversality statement ensuring that the virtual divisor on the moduli spaces in question are actual divisors turns out to be a major challenge. We mention finally also the papers [21], [11], where one studies other interesting divisors in Hurwitz spaces using the splitting type of the $(k-1)$-dimensional scroll canonically associated to a degree k cover $C \to \mathbf{P}^1$.

2.1 The Hurwitz space $\mathcal{H}_{14,9}$

We consider the morphism $\sigma : \overline{\mathcal{H}}_{14,9} \to \overline{\mathcal{M}}_{14}$, whose general fibre is 2-dimensional. We choose a general element $[C, A] \in \mathcal{H}_{14,9}$ and set $L := K_C \otimes A^\vee \in W_{17}^5(C)$ to be the residual linear system. Furthermore, L is very ample, else there exist points $x, y \in C$ such that $A(x + y) \in W_{13}^2(C)$, which contradicts the Brill–Noether Theorem. Note that

$$h^0(C, L^{\otimes 2}) = \dim \operatorname{Sym}^2 H^0(C, L) = 21$$

and we set up the (a priori virtual) divisor

$$\widetilde{\mathcal{D}}_{14} := \Big\{ [C, A] \in \widetilde{\mathcal{G}}_{14,9}^1 : \operatorname{Sym}^2 H^0(C, L) \xrightarrow{\phi_L} H^0(C, L^{\otimes 2})$$
$$\text{is not an isomorphism} \Big\}.$$

Our next results shows that, remarkably, this locus is indeed a divisor and it gives rise to an effective representative of the canonical divisor of $\widetilde{\mathcal{G}}_{14,9}^1$.

Proposition 2.1 *The locus $\mathcal{D}_{14}$ is a divisor on $\mathcal{H}_{14,9}$ and one has the following formula*

$$[\widetilde{\mathcal{D}}_{14}] = 4\lambda + \frac{1}{12}[D_3] - \frac{3}{4}[D_0] = \frac{1}{2}K_{\widetilde{\mathcal{G}}^1_{14,9}} \in CH^1(\widetilde{\mathcal{G}}^1_{14,9}).$$

Proof The locus $\widetilde{\mathcal{D}}_{14}$ is the degeneracy locus of the vector bundle morphism

$$\phi \colon \mathrm{Sym}^2(\mathcal{E}) \to \mathcal{F}_2. \tag{9}$$

The Chern class of both vector bundles $\mathcal{E}$ and $\mathcal{F}_2$ are computed in Proposition 1.3 and we have the formulas $c_1(\mathcal{E}) = \lambda - \frac{1}{2}\mathfrak{b} + \frac{7}{18}\mathfrak{a}$ and $c_1(\mathcal{F}_2) = 13\lambda + 2\mathfrak{a} - 3\mathfrak{b} - [D_0]$. Taking into account that $\mathrm{rk}(\mathcal{E}) = 6$, we can write:

$$[\widetilde{\mathcal{D}}_{14}] = c_1\big(\mathcal{F}_2 - \mathrm{Sym}^2(\mathcal{E})\big) = c_1(\mathcal{F}_2) - 7c_1(\mathcal{E}) = 6\lambda - [D_0] + \frac{1}{2}\gamma.$$

We now substitute γ in the formula given by Proposition 1.4 involving also the divisor D_3 on $\widetilde{\mathcal{G}}^1_{14,9}$ of pairs $[C, A]$, such that $A \in W^1_9(C)$ has a triple ramification point. We obtain that $[\widetilde{\mathcal{D}}_{14}] = 4\lambda + \frac{1}{12}[D_3] - \frac{3}{4}[D_0]$. Comparing with Theorem 1.1, the fact that $[\widetilde{\mathcal{D}}_{14}]$ is a (half-) canonical representative on $\widetilde{\mathcal{G}}^1_{14,9}$ now follows.

It remains to show that $\widetilde{\mathcal{D}}_{14}$ is indeed a divisor, that is, the morphism ϕ given by (9) is generically non-degenerate. Since $\mathcal{H}_{14,9}$ is irreducible, it suffices to construct one example of a smooth curve $C \subseteq \mathbf{P}^5$ of genus 14 and degree 17 which does not lie on any quadrics. To that end, we consider the *White surface* $X \subseteq \mathbf{P}^5$, obtained by blowing-up $\mathbf{P}^2$ at 15 points $p_1, \ldots, p_{15}$ in general position and embedded into $\mathbf{P}^5$ by the linear system

$$|H| := |5h - E_{p_1} - \cdots - E_{p_{15}}|,$$

where E_{p_i} is the exceptional divisor at the point $p_i \in \mathbf{P}^2$ and $h \in |\mathcal{O}_{\mathbf{P}^2}(1)|$. The White surface is known to projectively Cohen–Macaulay, its ideal being generated by the 3×3-minors of a certain 3×5-matrix of linear forms, see [24] Proposition 1.1. In particular, the map

$$\mathrm{Sym}^2 H^0(X, \mathcal{O}_X(1)) \to H^0(X, \mathcal{O}_X(2))$$

is an isomorphism and $X \subseteq \mathbf{P}^5$ lies on no quadrics. We now let $C \subseteq X$ be a general element of the linear system

$$\left| 12h - 3(E_{p_1} + \cdots + E_{p_{13}}) - 2(E_{p_{14}} + E_{p_{15}}) \right|.$$

Note that $\dim |\mathcal{O}_X(C)| = 6$ and a general element is a smooth curve $C \subseteq \mathbf{P}^5$ of degree 17 and genus 14. Since $H^0\big(X, \mathcal{O}_X(2H - C)\big) = 0$, it follows that C lies on no quadrics, which finishes the proof. $\square$

2.2 The Hurwitz space $\mathcal{H}_{19,13}$

The case $g = 19$ is analogous to the situation in genus 14. We have a morphism $\sigma : \overline{\mathcal{H}}_{19,13} \to \overline{\mathcal{M}}_{19}$ with generically 5-dimensional fibres. The Kodaira dimension of both $\overline{\mathcal{M}}_{19}$ and of the Hurwitz spaces $\overline{\mathcal{H}}_{19,11}$ and $\overline{\mathcal{H}}_{19,12}$ is unknown. For a general element $[C, A] \in \mathcal{H}_{19,13}$, we set $L := K_C \otimes A^{\vee} \in W_{23}^6(C)$. Observe that $W_{23}^7(C) = \emptyset$, that is, L must be a complete linear series. The multiplication

$$\phi_L : \text{Sym}^2 H^0(C, L) \to H^0(C, L^{\otimes 2})$$

is a map between between two vector spaces of the same dimension 28. We introduce the degeneracy locus

$$\widetilde{\mathcal{D}}_{19} := \left\{ [C, A] \in \widetilde{\mathcal{G}}_{19,13}^1 : \text{Sym}^2 H^0(C, L) \xrightarrow{\phi_L} H^0(C, L^{\otimes 2}) \right.$$

$$\left. \text{is not an isomorphism} \right\}.$$

Proposition 2.2 *One has the following formula*

$$[\widetilde{\mathcal{D}}_{19}] = \lambda + \frac{1}{6}[D_3] - \frac{1}{2}[D_0] \in CH^1(\widetilde{\mathcal{G}}_{19,13}^1).$$

Proof The class calculation is very similar to the one in the proof of Proposition 2.1. It remains to produce an example of a pair (C, L), where C is a smooth curve and $L \in W_{23}^6(C)$, for which the map ϕ_L is an isomorphism.

We blow-up $\mathbf{P}^2$ at 21 general points which we denote by $p, p_1, \ldots, p_{12}$ and $q_1, \ldots, q_8$ respectively. Let $X \subseteq \mathbf{P}^{12}$ be the surface obtained by embedding $\text{Bl}_{21}(\mathbf{P}^2)$ via the linear system

$$|H| := |6h - E_p - E_{p_1} - \cdots - E_{p_{12}} - E_{q_1} - \cdots - E_{q_8}|.$$

Using again [24], we have that the multiplication map

$$\phi_{\mathcal{O}_X(1)} : \text{Sym}^2 H^0(X, \mathcal{O}_X(1)) \to H^0(X, \mathcal{O}_X(2))$$

is an isomorphism, therefore $X \subseteq \mathbf{P}^6$ lies on no quadrics. We consider C to be a general element of the linear system

$$C \in \left| 12h - E_p - 4\sum_{i=1}^{12} E_{p_i} - 3\sum_{j=1}^{8} E_{q_j} \right|.$$

Then C is a smooth curve of degree 23 and genus 19. Since $2H - C$ is not an effective divisor it follows that $\text{Ker}(\phi_{\mathcal{O}_X(1)}) \cong \text{Ker}(\phi_{\mathcal{O}_C(1)})$ and we conclude that the multiplication map $\phi_{\mathcal{O}_C(1)}$ is an isomorphism as well. This finishes the proof. $\square$

We can now prove the case $g = 19$ from Theorem 0.2. Indeed, combining Proposition 2.2 and Theorem 1.1, we find

$$K_{\widetilde{\mathcal{G}}^1_{19,13}} = [\widetilde{\mathcal{D}}_{19}] + \sigma^*(7\lambda - \delta_0).$$

Since the class $7\lambda - \delta_0$ is big an $\overline{\mathcal{M}}_{19}$, it follows that $K_{\widetilde{\mathcal{G}}^1_{19,13}}$ is big.

2.3 The Hurwitz space $\mathcal{H}_{17,11}$

The minimal Hurwitz cover of $\overline{\mathcal{M}}_{17}$ is $\overline{\mathcal{H}}_{17,10}$, but its Kodaira dimension is unknown. We consider the next case $\sigma \colon \overline{\mathcal{H}}_{17,11} \to \overline{\mathcal{M}}_{17}$. As described in the Introduction, a general curve $C \subseteq \mathbf{P}^6$ of genus 17 and degree 21 (whose residual linear system is a pencil $A = K_C \otimes L^\vee \in W^1_{11}(C)$) lies on a pencil of quadrics. The general element of this pencil has full rank 7 and we consider the intersection of the pencil with the discriminant, that is, we require we require that two of the rank 6 quadrics contained in Ker ϕ_L coalesce. We define $\mathcal{D}_{17}$ to be the locus of pairs $[C, A] \in \mathcal{H}_{17,11}$ such that this intersection is not reduced.

Theorem 2.3 *The locus $\mathcal{D}_{17}$ is a divisor and the class of its closure $\widetilde{\mathcal{D}}_{17}$ in $\widetilde{\mathcal{G}}^1_{17,11}$ is given by*

$$[\widetilde{\mathcal{D}}_{17}] = \frac{1}{6}\left(19\lambda - \frac{9}{2}[D_0] + \frac{5}{6}[D_3]\right) \in CH^1(\widetilde{\mathcal{G}}^1_{17,11}).$$

Proof We are in a position to apply [16] Theorem 1.2, which deals precisely with degeneracy loci of this type. We obtain

$$[\widetilde{\mathcal{D}}_{17}] = 6(7c_1(F) - 52c_1(\mathcal{E})) = 6(39\lambda - 7[D_0] + 5\gamma).$$

Using once more Proposition 1.4, we obtained the claimed formula.

We now establish that $\widetilde{\mathcal{D}}_{17}$ is a genuine divisor on $\widetilde{\mathcal{G}}^1_{17,11}$. To that end, it suffices to exhibit *one* smooth curve $C \subseteq \mathbf{P}^6$ of genus 17 and degree 21 such that if $L = \mathcal{O}_C(1)$, then Ker(ϕ_L) is a non-degenerate pencil of quadrics. We proceed by picking 16 points in $\mathbf{P}^2$ in general position. We denote these points by $p, p_1, \ldots, p_4$ and $q_1, \ldots, q_{11}$. Let $X \subseteq \mathbf{P}^6$ be the smooth rational surface obtained by embedding the blow-up of $\mathbf{P}^2$ at these 16 points by the very ample linear system

$$\left| 7h - 3E_p - 2\sum_{i=1}^{4} E_{p_i} - \sum_{j=1}^{11} E_{q_j} \right|,$$

where $E, E_{p_1}, \ldots, E_{p_4}, E_{q_1}, \ldots, E_{q_{11}}$ are the exceptional divisors at the corresponding points in $\mathbf{P}^2$.

Note that $h^0(X, \mathcal{O}_X(1)) = 7$ and $h^0(X, \mathcal{O}_X(2)) = 26$. Furthermore, X is projectively normal, therefore $\mathrm{Ker}(\phi_{\mathcal{O}_X(1)})$ is a pencil of quadrics. It can be directly checked with *Macaulay* that this pencil is non-degenerate (see [16] Theorem 1.10, including the accompanying *Macaulay* file for how to carry that out). We now take a general element

$$C \in \left| 12h - 3E_p - 4\sum_{i=1}^{4} E_{p_i} - 2\sum_{j=1}^{11} E_{q_j} \right|.$$

One checks that $C \subseteq X \subseteq \mathbf{P}^6$ is a smooth curve of genus 17 and degree 21. Furthermore, taking cohomology in the short exact sequence

$$0 \longrightarrow \mathcal{I}_{X/\mathbf{P}^6}(2) \longrightarrow \mathcal{I}_{C/\mathbf{P}^6}(2) \longrightarrow \mathcal{I}_{C/X}(2) \longrightarrow 0,$$

we conclude that $H^0\big(\mathbf{P}^6, \mathcal{I}_{X/\mathbf{P}^6}(2)\big) \cong H^0\big(\mathbf{P}^6, \mathcal{I}_{C/\mathbf{P}^6}(2)\big)$, that is, the pencil of quadrics $\mathrm{Ker}(\phi_{\mathcal{O}_C(1)})$ is non-degenerate, which finishes the proof. $\qquad\square$

Substituting the expression of $[\widetilde{\mathcal{D}}_{17}]$ in the formula of the canonical class of the Hurwitz space, we find

$$K_{\widetilde{\mathcal{G}}^1_{17,11}} = \frac{1}{5}[\widetilde{\mathcal{D}}_{17}] + \frac{3}{5}\sigma^*(7\lambda - \delta_0).$$

Just like in the previous case, since the class $7\lambda - \delta_0$ is big on $\overline{\mathcal{M}}_{17}$ and λ is ample of $\overline{\mathcal{H}}_{17,11}$, Theorem 0.2 follows for $g = 17$ as well.

2.4 The Hurwitz space $\mathcal{H}_{16,9}$

This is the most interesting case, for (i) we consider a *minimal* Hurwitz cover $\sigma : \overline{\mathcal{H}}_{16,9} \to \overline{\mathcal{M}}_{16}$ of the $\overline{\mathcal{M}}_{16}$ and (ii) 16 is the smallest genus for which the Kodaira dimension of the moduli space of curves of that genus is unknown, see [38].

We fix a general point $[C, A] \in \mathcal{H}_{16,9}$ and, set $L := K_C \otimes A^\vee \in W^6_{21}(C)$. It is proven in [14] Theorem 2.7 that the locus $\mathcal{D}_{16}$ classifying pairs $[C, A]$ such that the multiplication map

$$\mu : I_2(L) \otimes H^0(C, L) \to I_3(L)$$

is not an isomorphism, is a divisor on $\mathcal{H}_{16,9}$.

First we determine the class of the Gieseker–Petri divisor, already mentioned in the introduction.

Proposition 2.4 *One has* $[\widetilde{\mathcal{GP}}] = -\lambda + \gamma \in CH^1(\widetilde{\mathcal{G}}^1_{16,9})$.

Proof Recall that we have introduced the sheaves $\mathcal{V}$ and $\mathcal{E}$ on $\widetilde{\mathcal{G}}^1_{16,19}$ with fibres canonically isomorphic to $H^0(C, A)$ and $H^0(C, \omega_C \otimes A^\vee)$ over a point $[C, A] \in \widetilde{\mathcal{G}}^1_{16,9}$. We have a natural morphism $\mathcal{E} \otimes \mathcal{V} \to f_*(\omega_f)$ and $\widetilde{\mathcal{GP}}$ is the degeneracy locus of this map. Accordingly, we can write

$$[\widetilde{\mathcal{GP}}] = \lambda - 2c_1(\mathcal{E}) - 8c_1(\mathcal{V}) = -\lambda + \left(\mathfrak{b} - \frac{5}{3}\mathfrak{a}\right) = -\lambda + \gamma. \qquad \square$$

We can now compute the class of the divisor $\widetilde{\mathcal{D}}_{16}$.

Theorem 2.5 *The locus* $\widetilde{\mathcal{D}}_{16}$ *is an effective divisor on* $\widetilde{\mathcal{G}}^1_{16,9}$ *and its class is given by*

$$[\widetilde{\mathcal{D}}_{16}] = \frac{65}{2}\lambda - 5[D_0] + \frac{3}{2}[\widetilde{\mathcal{GP}}] \in CH^1(\widetilde{\mathcal{G}}^1_{16,9}).$$

Proof Recall the definition of the vector bundles $\mathcal{F}_2$ and $\mathcal{F}_3$ on $\widetilde{\mathcal{G}}^1_{16,9}$, as well as the expression of their first Chern classes provided by Proposition 1.3. We define two further vector bundles $\mathcal{I}_2$ and $\mathcal{I}_3$ on $\widetilde{\mathcal{G}}^1_{16,9}$, via the following exact sequences:

$$0 \longrightarrow \mathcal{I}_\ell \longrightarrow \mathrm{Sym}^\ell(\mathcal{E}) \longrightarrow \mathcal{F}_\ell \longrightarrow 0,$$

for $\ell = 2, 3$. Note that $\mathrm{rk}(\mathcal{I}_2) = 9$, whereas $\mathrm{rk}(\mathcal{I}_3) = 72$. To make sure that these sequences are exact on the left outside a set of codimension at least 2 inside $\widetilde{\mathcal{G}}^1_{16,9}$, we invoke [13], Propositions 3.9 and 3.10. The divisor $\widetilde{\mathcal{D}}_{16}$ is then the degeneracy locus of the morphism of vector bundles of the same rank

$$\mu \colon \mathcal{I}_2 \otimes \mathcal{E} \to \mathcal{I}_3,$$

which globalizes the multiplication maps $\mu_{C,L} \colon I_{C,L}(2) \otimes H^0(C, L) \to I_{C,L}(3)$, where $L = \omega_C \otimes A^\vee$ and $[C, A] \in \widetilde{\mathcal{G}}^1_{16,9}$.

Noting that $c_1(\mathrm{Sym}^3(\mathcal{E})) = 45c_1(\mathcal{E})$ and $c_1(\mathrm{Sym}^2(\mathcal{E})) = 9c_1(\mathcal{E})$, we compute

$$[\widetilde{\mathcal{D}}_{16}] = c_1(\mathcal{I}_3) - 8c_1(\mathcal{I}_2) - 9c_1(\mathcal{E}) = 31\lambda - 5[D_0] + \frac{3}{2}\gamma.$$

Substituting $\gamma = \lambda + [\widetilde{\mathcal{GP}}]$, we obtain the claimed formula.

It remains to observe that it has already been proved in [13] Theorem 2.7 that for a general pair $[C, L]$, where $L \in W^6_{21}(C)$, the multiplication map $\mu_{C,L}$ is an isomorphism. $\qquad \square$

The formula (1) mentioned in the introduction follows now by using Theorem 2.5 and the Riemann-Hurwitz formula for the map $\sigma \colon \widetilde{\mathcal{G}}^1_{16,9} \to \overline{\mathcal{M}}_{16}$. One writes

$$K_{\widetilde{\mathcal{G}}^1_{16,9}} = 13\lambda - 2[D_0] + \frac{3}{5}[\widetilde{\mathcal{GP}}] + \frac{2}{5}[\widetilde{\mathcal{GP}}] = \frac{2}{5}[\widetilde{\mathcal{D}}_{16}] + \frac{3}{5}[\widetilde{\mathcal{GP}}].$$

References

[1] D. Abramovich, A. Corti and A Vistoli, *Twisted bundles and admissible covers*, Comm. Algebra **31** (2003), 3547–3618.

[2] E. Arbarello and M. Cornalba, *Footnotes to a paper of Beniamino Segre*, Mathematische Annalen **256** (1981), 341–362.

[3] J. Bertin and M. Romagny, *Champs de Hurwitz*, Mémoires de la Soc. Math. de France 125–126 (2011).

[4] A. Bruno and A. Verra, $\mathcal{M}_{15}$ *is rationally connected*, in: Projective varieties with unexpected properties, 51–65, Walter de Gruyter 2005.

[5] M. C. Chang and Z. Ran, *Unirationality of the moduli space of curves of genus 11, 13 (and 12)*, Inventiones Math. **76** (1984), 41–54.

[6] M. C. Chang and Z. Ran, *The Kodaira dimension of the moduli space of curves of genus 15*, Journal of Differential Geometry **24** (1986), 205–220.

[7] M. C. Chang and Z. Ran, *On the slope and Kodaira dimension of $\overline{\mathcal{M}}_g$ for small g*, Journal of Differential Geometry **34** (1991), 267–274.

[8] A. Chiodo and G. Farkas, *Singularities of the moduli space of level curves* Journal of the European Mathematical Society **19** (2017), 603–658.

[9] A. Clebsch, *Zur Theorie der algebraischen Funktionen*, Mathematische Annalen **29** (1887), 171–186.

[10] H. Damadi and F.-O. Schreyer, *Unirationality of the Hurwitz space $\mathcal{H}_{9,8}$*, Archiv der Mathematik, **109** (2017), 511–519.

[11] A. Deopurkar and A. Patel, *Syzygy divisors on Hurwitz spaces*, Contemporary Mathematics Vol. 703 (2018), 209–222.

[12] D. Eisenbud and J. Harris, *The Kodaira dimension of the moduli space of curves of genus $\geq$ 23* Inventiones Math. **90** (1987), 359–387.

[13] G. Farkas, *Syzygies of curves and the effective cone of $\overline{\mathcal{M}}_g$*, Duke Mathematical Journal **135** (2006), 53–98.

[14] G. Farkas, *Koszul divisors on moduli spaces of curves*, American Journal of Mathematics **131** (2009), 819–869.

[15] G. Farkas, D. Jensen and S. Payne, *The Kodaira dimension of $\overline{\mathcal{M}}_{22}$ and $\overline{\mathcal{M}}_{23}$*, arXiv:2005.00622.

[16] G. Farkas and R. Rimányi, *Quadric rank loci on moduli of curves and K3 surfaces*, arXiv:1707.00756, to appear in Annales L'Ecole Normale Sup.

[17] G. Farkas and A. Verra, *The geometry of the moduli space of odd spin curves*, Annals of Mathematics **180** (2014), 927–970.

[18] G. Farkas and A. Verra, *The universal theta divisor over the moduli space of curves*, Journal Math. Pures Appl. **100** (2023), 591–605.

[19] W. Fulton, *Hurwitz schemes and irreducibility of moduli of algebraic curves*, Annals of Mathematics **90** (1969), 542–575.

[20] G. van der Geer and A. Kouvidakis, *The Hodge bundle on Hurwitz spaces*, Pure Appl. Math. Quarterly **7** (2011), 1297–1308.

[21] G. van der Geer and A. Kouvidakis, *The cycle classes of divisorial Maroni loci*, International Math. Research Notices **11** (2017), 3463–3509.

[22] G. van der Geer and A. Kouvidakis, *The class of a Hurwitz divisor on the moduli of curves of even genus*, Asian Journal of Mathematics **16** (2012), 786–806.

[23] F. Geiss, *The unirationality of Hurwitz spaces of 6-gonal curves of small genus*, Documenta Mathematica **17** (2012), 627–640.

[24] A. Gimigliano, *On Veronesean surfaces*, Indagationes Math. **92** (1989), 71–85.

[25] J. Harris, *On the Kodaira dimension of the moduli space of curves II: The even genus case*, Inventiones Math. **75** (1984), 437–466.

[26] J. Harris and D. Mumford, *On the Kodaira dimension of* $\overline{\mathcal{M}}_g$, Inventiones Math. **67** (1982), 23–88.

[27] J. Harris and I. Morrison, *Slopes of effective divisors on the moduli space of curves*, Inventiones Math. **99** (1990), 321–355.

[28] J. Harris and L. Tu, *Chern numbers of kernel and cokernel bundles*, Invent. Math. **75** (1984), 467–475.

[29] A. Hurwitz, *Über Riemann'sche Fächen mit gegebenen Verzweigungspunkten*, Mathematische Annalen **39** (1891), 1–61.

[30] H. Keneshlou and F. Tanturri, *The unirationality of the Hurwitz schemes* $\mathcal{H}_{10,8}$ *and* $\mathcal{H}_{13,7}$, Atti Accad. Naz. Lincei Rend. Lincei Mat. Appl. **30** (2019), 31–39.

[31] A. Kokotov, D. Korotkin and P. Zograf, *Isomonodromic tau function on the space of admissible covers*, Advances in Math. **227** (2011), 586–600.

[32] K. Petri, *Über die invariante Darstellung algebraischer Funktionen einer Veränderlichen*, Mathematische Annalen, **88** (1923), 242–289.

[33] F.-O. Schreyer, *Syzygies of canonical curves and special linear series*, Mathematische Annalen **275** (1986), 105–137.

[34] F.-O. Schreyer, *Matrix factorizations and families of curves of genus* 15, Algebraic Geometry **2** (2015), 489–507.

[35] F.-O. Schreyer and F. Tanturri, *Matrix factorization and curves in* $\boldsymbol{P}^4$, Documenta Mathematica **23** (2018), 1895–1924.

[36] E. Sernesi, *L'unirazionalitá della varietá dei moduli delle curve di genere* 12, Ann. Scuola Normale Sup. Pisa, **8** (1981), 405–439.

[37] F. Severi, *Sulla classificazione delle curve algebriche e sul teorema d'esistenza di Riemann*, Rendiconti della R. Accad. Naz. Lincei **24** (1915), 877–888.

[38] D. Tseng, *On the slope of the moduli space of genus 15 and 16 curves*, arXiv:1905.00449.

[39] A. Verra, *The unirationality of the moduli space of curves of genus* $\leq$ 14, Compositio Mathematica **141** (2005), 1425–1444.

10

Chow Quotients of Grassmannians by Diagonal Subtori

Noah Giansiracusa[a] and Xian Wu[b]

Abstract. The literature on maximal torus orbits in the Grassmannian is vast; in this paper we initiate a program to extend this to diagonal subtori. Our main focus is generalizing portions of Kapranov's seminal work on Chow quotient compactifications of these orbit spaces. This leads naturally to discrete polymatroids, generalizing the matroidal framework underlying Kapranov's results. By generalizing the Gelfand–MacPherson isomorphism, these Chow quotients are seen to compactify spaces of arrangements of parameterized linear subspaces, and a generalized Gale duality holds here. A special case is birational to the Chen–Gibney-Krashen moduli space of pointed trees of projective spaces, and we show that the question of whether this birational map is an isomorphism is a specific instance of a much more general question that hasn't previously appeared in the literature, namely, whether the geometric Borel transfer principle in non-reductive GIT extends to an isomorphism of Chow quotients.

1 Introduction

The literature on maximal torus orbits in the Grassmannian and the torus-equivariant geometry (cohomology, K-theory, etc.) of the Grassmannian is extensive; it is a rich field beautifully interweaving combinatorics, representation theory, and geometry, with many applications across these disciplines. One of the seminal works is Kapranov's paper on Chow quotients in which he compactifies the space of maximal torus orbit closures [14]. The goal of the present paper is to initiate a program of studying diagonal subtorus orbits in the Grassmannian; we focus here on extending portions of Kapranov's paper to this setting and explore some consequences.

[a] Bentley University. Massachusetts
[b] Jagiellonian University, Krakow

1.1 Setup and notation

Fix a base field k. By a *diagonal subtorus* S we mean that coordinates in the maximal torus $T = (k^\times)^n$ acting on $\mathrm{Gr}(d,n)$ are allowed to coincide; that is, $S = (k^\times)^m$ for $m \leq n$ and we have an inclusion map $S \hookrightarrow T$ given by a matrix whose rows are all standard basis vectors. Up to permutation, every such subtorus is of the form

$$S = \{(\underbrace{t_1, \ldots, t_1}_{r_1}, \underbrace{t_2, \ldots, t_2}_{r_2}, \ldots, \underbrace{t_m, \ldots, t_m}_{r_m}) \mid t_i \in k^\times\} \subseteq T,$$

where $\sum r_i = n$. Setting $r_i = 1$ for all i recovers Kapranov's case of the maximal torus. In essence, the combinatorics in Kapranov's paper (matroids, matroid subdivisions, etc.) are generalized by replacing the set $[n] = \{1, 2, \ldots, n\}$ with the multiset

$$[\vec{r}] := \{\underbrace{1, \ldots, 1}_{r_1}, \underbrace{2, \ldots, 2}_{r_2}, \ldots, \underbrace{m, \ldots, m}_{r_m}\}$$

where i has multiplicity r_i. (Matroids on multisets appear in the literature under the name discrete polymatroids [13].) The hypersimplex $\Delta(d,n) \subseteq \mathbb{R}^n$, a polytope playing a fundamental role in Kapranov's paper, is replaced with its projection under the linear map

$$\lambda_{\vec{r}} \colon \mathbb{R}^n \to \mathbb{R}^m$$

given by the matrix $|e_1 \cdots e_1 \, e_2 \cdots e_2 \cdots e_m \cdots e_m|$, the transpose of the matrix defining the inclusion $S \hookrightarrow T$. These vague assertions will be made precise in what follows.

1.2 Results

The T-orbit closure of any k-point of $\mathrm{Gr}(d,n) \subseteq \mathbb{P}^{\binom{n}{d}-1}$ is a polarized toric variety whose corresponding lattice polytope is a subpolytope of the hypersimplex

$$\Delta(d,n) = \{(a_1, \ldots, a_n) \mid a_i \in [0,1] \text{ and } \sum a_i = d\} \subseteq \mathbb{R}^n.$$

This subpolytope has its vertices and edges among those of $\Delta(d,n)$; subpolytopes with this property are called matroid polytopes and are known to be in bijection with rank d matroids on $[n]$, with matroids representable over k identified with the polytopes of T-orbit closures in $\mathrm{Gr}(d,n)$ [9]. This perspective of matroid polytopes is a relatively recent advance in matroid theory that has fruitfully brought the subject closer to algebraic geometry (cf. [2, §1]). Via diagonal subtori, this story extends seamlessly to discrete polymatroids:

Theorem 1.1 *Rank d discrete polymatroids on the multiset $[\vec{r}]$ are in bijection with subpolytopes of $\lambda_{\vec{r}}(\Delta(d,n)) \subseteq \mathbb{R}^m$ whose vertices and edges are among the images of the vertices and edges of $\Delta(d,n)$; moreover, this bijection identifies the discrete polymatroids representable over k with the lattice polytopes corresponding to S-orbit closures in $\mathrm{Gr}(d,n)$.*

Now let $k = \mathbb{C}$. Kapranov's idea for compactifying the space of maximal torus orbit closures in $\mathrm{Gr}(d,n)$ is to take a sufficiently small T-invariant Zariski open locus $U \subseteq \mathrm{Gr}(d,n)$ such that the T-action on U is free and there is an inclusion $U/T \hookrightarrow \mathrm{Chow}(\mathrm{Gr}(d,n))$ sending each torus orbit to its Zariski closure, viewed as an algebraic cycle on the Grassmannian. The closure of the image of this embedding in the Chow variety is by definition the Chow quotient $\mathrm{Gr}(d,n)/\!/_{Ch}T$ [14]. We can apply the same idea here and study the diagonal subtorus Chow quotient $\mathrm{Gr}(d,n)/\!/_{Ch}S$. We compute some explicit examples of this Chow quotient, together with its natural closed embedding in the toric Chow quotient $\mathbb{P}^{\binom{n}{d}-1}/\!/_{Ch}S$, in §3.

Kapranov shows [14, Theorem 1.6.6] that the rational maps sending a linear space to its intersection with, and projection onto, a coordinate hyperplane induce morphisms

$$\mathrm{Gr}(d,n)/\!/_{Ch}T \to \mathrm{Gr}(d-1,n-1)/\!/_{Ch}T'$$

and

$$\mathrm{Gr}(d,n)/\!/_{Ch}T \to \mathrm{Gr}(d,n-1)/\!/_{Ch}T',$$

respectively, where $T' = (k^\times)^{n-1}$ is the maximal torus acting on these smaller Grassmannians. We have the following extension of this to the subtorus setting:

Theorem 1.2 *Fix an index $1 \leq i \leq m$, let $I \subseteq [n]$ index the r_i coordinates of S corresponding to the ith $\mathbb{G}_m$-factor, and let S_i denote the rank $m-1$ torus given by projecting S onto the complement of the I-coordinates. Then intersection with, and projection onto, the codimension r_i coordinate linear space defined by $x_j = 0$ for all $j \in I$ induce morphisms*

$$\mathrm{Gr}(d,n)/\!/_{Ch}S \to \mathrm{Gr}(d-r_i,n-r_i)/\!/_{Ch}S_i$$

and

$$\mathrm{Gr}(d,n)/\!/_{Ch}S \to \mathrm{Gr}(d,n-r_i)/\!/_{Ch}S_i,$$

respectively.

Kapranov's proof directly analyzes Chow forms to demonstrate their polynomial dependence, whereas we use polytopal subdivisions to apply a valuative criterion for regularity; thus, we obtain in particular a new variant of Kapranov's proof in the case of the maximal torus.

The Gelfand–MacPherson correspondence identifies generic torus orbits in the Grassmannian with generic general linear group orbits in a product of projective spaces, and Kapranov shows [14, Theorem 2.2.4] that this extends to an isomorphism of Chow quotients

$$\mathrm{Gr}(d,n)\,/\!/_{Ch}\,T \cong (\mathbb{P}^{d-1})^n\,/\!/_{Ch}\,\mathrm{GL}_d\,.$$

Thus, his Grassmannian Chow quotient can be viewed as compactifying the space of configurations of points in projective space, up to projectivity, or dually, the space of hyperplane arrangements. This has been a fruitful perspective [12, 1] and it generalizes to our setting as follows:

Theorem 1.3   *There is an isomorphism*

$$\mathrm{Gr}(d,n)\,/\!/_{Ch}\,S \cong \left(\prod_{i=1}^{m}\mathbb{P}\,\mathrm{Hom}(k^{r_i},k^d)\right)\,/\!/_{Ch}\,\mathrm{GL}_d$$

where GL_d acts diagonally by left matrix multiplication.

To prove this, we adapt an argument of Thaddeus in [17] and so also obtain a new proof of Kapranov's original result as a special case. We can view the right side of the above isomorphism as compactifying the space of arrangements of "parameterized" linear subspaces: $(L_1, \alpha_1, \ldots, L_m, \alpha_m)$ where $L_i \subseteq \mathbb{P}^{d-1}$ is a linear subspace of dimension $r_i - 1$ and $\alpha_i \in \mathrm{Aut}(L_i) \cong \mathrm{PGL}_{r_i}$.

Since orthogonal complement yields a T-equivariant isomorphism $\mathrm{Gr}(d,n) \cong \mathrm{Gr}(n-d,n)$ and hence an isomorphism of Chow quotients $\mathrm{Gr}(d,n)\,/\!/_{Ch}\,S \cong \mathrm{Gr}(n-d,n)\,/\!/_{Ch}\,S$ for any diagonal subtorus $S \subseteq T$, our generalized Gelfand–MacPherson isomorphism implies the following generalized Gale duality:

Corollary 1.4   *There is a natural involutive isomorphism*

$$\left(\prod_{i=1}^{m}\mathbb{P}\,\mathrm{Hom}(k^{r_i},k^d)\right)\,/\!/_{Ch}\,\mathrm{GL}_d$$

$$\cong \left(\prod_{i=1}^{m}\mathbb{P}\,\mathrm{Hom}(k^{r_i},k^{(\sum_{i=1}^{m} r_i)-d})\right)\,/\!/_{Ch}\,\mathrm{GL}_{(\sum_{i=1}^{m} r_i)-d}$$

In geometric terms, arrangements (up to projectivity) of m generic parameterized linear subspaces $L_i \hookrightarrow \mathbb{P}^{d-1}$ and their Chow limits are in natural bijection with arrangements (up to projectivity) of m generic parameterized linear subspaces, of the same dimensions, in $\mathbb{P}^{m-d-1+\sum \dim(L_i)}$ and their Chow limits.

Kapranov showed [14, Theorem 4.1.8] that his Chow quotients generalize the ubiquitous Grothendieck–Knudsen moduli spaces of stable pointed rational curves, namely

$$Gr(2,n) /\!\!/_{Ch} GL_2 \cong \overline{M}_{0,n}.$$

Another generalization was constructed by Chen–Gibney-Krashen in [4], where a moduli space denoted $T_{d,n}$ compactifying the space of n distinct points and a disjoint parameterized hyperplane in $\mathbb{P}^d$ up to projectivity was introduced and studied and shown to satisfy $T_{1,n} \cong \overline{M}_{0,n+1}$. Essentially $T_{d,n}$ is the locus in the Fulton–MacPherson configuration space $X[n]$ [7] where all n points have come together at a single fixed smooth point on a d-dimensional variety X [4, §3.1]. The space $T_{d,n}$ is birational to $Gr(d+1, n+d) /\!\!/_{Ch} S$, where

$$S = \{(\underbrace{t_1, \ldots, t_1}_{d}, t_2, \ldots, t_{n+1}) \mid t_i \in k^\times\},$$

since both compactify the space of n distinct points and a disjoint parameterized hyperplane $\mathbb{P}^{d-1} \hookrightarrow \mathbb{P}^d$ up to projectivity. Krashen has asked, informally, whether this birational map is actually an isomorphism. While we have not been able to answer this question, we conclude this paper by showing that Krashen's question is a specific instance of a much more general question that appears not to have been asked previously in the literature—namely, whether the classical Borel transfer principle (relating non-reductive invariants to reductive invariants) extends from GIT quotients [5] to Chow quotients.

Acknowledgement   We thank Gary Gordon and Felipe Rincon for drawing our attention to discrete polymatroids, and we thank Valery Alexeev, Danny Krashen, and Angela Gibney for helpful conversations on this project. This paper is part of the second author's PhD dissertation at the University of Georgia, supervised by the first author. The first author was supported in part by NSF grant DMS-1802263, NSA grant H98230-16-1-0015, and Simons Collaboration Grant 346304.

2 Discrete polymatroids

For a non-negative integer vector $v = (v_1, \ldots, v_m) \in \mathbb{Z}_{\geq 0}^m$, the *modulus* is $|v| = \sum v_i$. A *discrete polymatroid* on the ground set $[m] = \{1, 2, \ldots, m\}$ can be defined as a nonempty finite subset $B \subseteq \mathbb{Z}_{\geq 0}^m$ of vectors all of the same modulus (called the *rank* of B) satisfying the following exchange property: if $u, v \in B$ with $u_i > v_i$ for some $1 \leq i \leq m$, then there exists $1 \leq j \leq m$ such that $u_j < v_j$ and $u - e_i + e_j \in B$ [13, Theorem 2.3]. This can be reformulated in terms of multisets as follows. Given a discrete polymatroid B, let

$$[\vec{r}] := \{\underbrace{1, \ldots, 1}_{r_1}, \underbrace{2, \ldots, 2}_{r_2}, \ldots, \underbrace{m, \ldots, m}_{r_m}\}$$

be the multiset where i has multiplicity $r_i := \max_{v \in B}\{v_i\}$. Each element of B can then be viewed as a sub-multiset of $[\vec{r}]$. If one considers the usual basis definition of a matroid except replacing the word "set" with "multiset" then the discrete polymatroid B is a matroid on the multiset $[\vec{r}]$, and conversely any matroid on a multiset is a discrete polymatroid on the ground set given by the set underlying the multiset. We will freely switch between the multiset perspective and the integer vector perspective of discrete polymatroids.

Proof of Theorem 1.1 This can either be proven by adapting the original arguments in [9], or it can be reduced to the results in [9] by using a multiset projection map; we present here the latter approach.

Fix an integer $d \geq 1$ and a multiset $[\vec{r}]$ with underlying set $[m] = \{1, 2, \ldots, m\}$ where i has multiplicity $r_i \geq 1$. Let $\pi_{\vec{r}}: [n] \to [m]$ be the "projection" map sending $1, 2, \ldots, r_1$ to 1, and $r_1 + 1, \ldots, r_1 + r_2$ to 2, etc. By a slight abuse of notation, for a subset $A = \{a_1, \ldots, a_\ell\} \subseteq [n]$ we denote by $\pi_{\vec{r}}(A)$ the multiset $\{\pi_{\vec{r}}(a_1), \ldots, \pi_{\vec{r}}(a_\ell)\}$, in other words the multiplicity of j is the cardinality of the fiber $\pi_{\vec{r}}^{-1}(j) \cap A$. Clearly $\pi_{\vec{r}}$ then sends a rank d matroid on $[n]$ to a rank d discrete polymatroid on $[m]$, and conversely if B is a rank d discrete polymatroid on $[m]$ then $\{A \subseteq [n] \mid \pi_{\vec{r}}(A) \in B\}$ is a rank d matroid on $[n]$; we denote the latter matroid by $\pi_{\vec{r}}^{-1}(B)$.

Given a rank d discrete polymatroid B on the multiset $[\vec{r}]$, the rank d matroid $\pi_{\vec{r}}^{-1}(B)$ on $[n]$ has basis polytope P given by the convex hull of the vectors $e_A := \sum_{i \in A} e_i$ for $A \in \pi_{\vec{r}}^{-1}(B)$, and by the classical results of [9] the vertices and edges of P are among the vertices and edges of the hypersimplex $\Delta(d, n)$. It then follows trivially that the linear projection $\lambda_{\vec{r}}(P)$ has its vertices and edges among the images under $\lambda_{\vec{r}}$ of the vertices and edges of $\Delta(d, n)$. Moreover, $\lambda_{\vec{r}}(P) \subseteq \mathbb{R}^m$ is the convex hull of the basis vectors of B (where now we view B as a set of vectors in $\mathbb{Z}_{\geq 0}^m$), and by [13, Theorem 3.4] we can recover B from this convex hull (specifically, the integral vectors in this convex hull are the independent sets in B). This faithfully embeds the set of rank d discrete polymatroids on $[\vec{r}]$ into the set of subpolytopes of $\lambda_{\vec{r}}(\Delta(d, n))$ whose vertices and edges are among the images under $\lambda_{\vec{r}}$ of those of $\Delta(d, n)$. This association is also surjective, since if $Q \subseteq \lambda_{\vec{r}}(\Delta(d, n))$ is a subpolytope whose vertices and edges are among the images of those of $\Delta(d,n)$, then the preimage of Q under $\lambda_{\vec{r}}$ is a subpolytope of $\Delta(d,n)$ whose vertices and edges are among those of this hypersimplex, i.e., $\lambda_{\vec{r}}^{-1}(Q)$ is a matroid polytope, and the multiset image under $\pi_{\vec{r}}$ of the corresponding rank d matroid on $[n]$ is a rank d discrete polymatroid with Q as its associated polytope.

We now turn to the assertion about representability. Given a k-point of the Grassmannian $L \in \mathrm{Gr}(d,n)(k)$, the lattice polytope $\Delta_{\overline{S \cdot L}}$ for the projective toric variety $\overline{S \cdot L} \subseteq \mathbb{P}^{\binom{n}{d}-1}$ is the image of this torus orbit closure under the moment map $\mu_S \colon \mathbb{P}^{\binom{n}{d}-1} \to \mathbb{R}^m$ for S. This moment map is the composition of the moment map $\mu_T \colon \mathbb{P}^{\binom{n}{d}-1} \to \mathbb{R}^n$ for the maximal torus T with the linear projection $\lambda_{\vec{r}} \colon \mathbb{R}^n \to \mathbb{R}^m$. Thus,

$$\Delta_{\overline{S \cdot L}} = \mu_S(\overline{S \cdot L}) = \lambda_{\vec{r}}\left(\mu_T(\overline{S \cdot L})\right) = \lambda_{\vec{r}}\left(\mu_T(\overline{T \cdot L})\right) = \lambda_{\vec{r}}(\Delta_{\overline{T \cdot L}}),$$

which is the polytope associated to the discrete polymatroid $\pi_{\vec{r}}(M(L))$, where $M(L)$ is the matroid represented by L. But $\pi_{\vec{r}}(M(L))$ is also the discrete polymatroid represented by L. $\square$

The linear projection $\lambda_{\vec{r}} \colon \mathbb{R}^n \to \mathbb{R}^m$ may send vertices of the hypersimplex $\Delta(d,n)$ to non-vertex points of the polytope $\lambda_{\vec{r}}(\Delta(d,n))$, and for the above theorem it is crucial that our subpolytopes are allowed to use such points rather than just the actual vertices of $\lambda_{\vec{r}}(\Delta(d,n))$, as the following example illustrates:

Example 2.1   Let $\vec{r} = (1,2,2)$, so $n = 5$ and $m = 3$; the projection function $\pi_{\vec{r}}$ is $1 \mapsto 1$, and $2,3 \mapsto 2$, and $4,5 \mapsto 3$; in coordinates, the linear projection $\lambda_{\vec{r}} \colon \mathbb{R}^5 \to \mathbb{R}^3$ is $(x_1, x_2 + x_3, x_4 + x_5)$. Consider rank 3 matroids. The hypersimplex $\Delta(3,5)$ has 10 vertices, the permutations of the vector $(1,1,1,0,0)$; the images of these 10 vertices are $(1,1,1)$ four times, $(1,2,0)$ once, $(1,0,2)$ once, $(0,1,2)$ twice, and $(0,2,1)$ twice. The polytope $\lambda_{\vec{r}}(\Delta(3,5))$ is a trapezoid, and the point $(1,1,1)$ is not a vertex of this trapezoid even though it is the image of vertices of the hypersimplex (see Figure 1). The segment from, say, $(1,1,1)$ to $(1,2,0)$ is a discrete polymatroid even though it has a vertex that is not a vertex of the trapezoid. On the other hand, the four vertices of the trapezoid $(1,2,0),(1,0,2),(0,1,2),(0,2,1)$ do not form a discrete polymatroid because the trapezoid edge from $(1,2,0)$ to $(1,0,2)$ is not an edge of the projected hypersimplex, it is a union of two such edges (and indeed the basis exchange axiom fails on these two without the presence of the midpoint $(1,1,1)$).

The interior of the Chow quotient $\mathrm{Gr}(d,n)/\!\!/_{Ch}S$ consists, by definition, of torus orbit closures $\overline{S \cdot L}$ (viewed as algebraic cycles) for generic linear

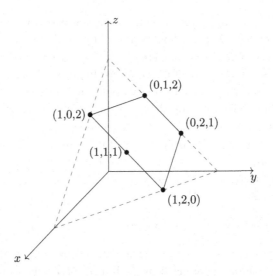

Figure 1 For the multiset $\{1,2,2,3,3\}$ the projected polytope $\lambda_{\vec{r}}(\Delta(3,5))$ is a trapezoid lying on a triangle. The point $(1,1,1)$ is not a vertex of the trapezoid even though it is the image under the linear projection $\lambda_{\vec{r}}\colon \mathbb{R}^5 \to \mathbb{R}^3$ of a vertex (in fact, four of them) of the hypersimplex $\Delta(3,5)$.

subspaces $L \in Gr(d,n)$; taking the closure of this interior locus in the Chow quotient adds limit points that are certain algebraic cycles

$$\sum_{i=1}^{\ell} m_i Z_i \in \text{Chow}\,(Gr(d,n)),$$

about which, following Kapranov, we can now say a bit more (cf. [14, Proposition 1.2.11]):

Proposition 2.2 *For each cycle $\sum_{i=1}^{\ell} m_i Z_i \in Gr(d,n)/\!\!/_{Ch}S$, the multiplicities m_i are all 1 and the irreducible cycles Z_i are single orbit closures $\overline{S \cdot L_i}$, $L_i \in Gr(d,n)$. The lattice polytopes $\Delta_{\overline{S\cdot L_i}}$, for $i = 1,\dots,\ell$, form a polyhedral decomposition of $\lambda_{\vec{r}}(\Delta(d,n))$.*

Proof The condition in Kapranov's [14, Theorem 0.3.1] is automatically satisfied here so each Z_i is a single orbit closure $\overline{S \cdot L_i}$. Kapranov's proof of [14, Proposition 1.2.15] shows that the index of the sub-lattice generated by the vertices of the representable matroid polytope $\Delta_{\overline{T\cdot L_i}}$ inside the lattice generated by the vertices of the hypersimplex $\Delta(d,n)$ is one. This index is preserved when applying the linear map $\lambda_{\vec{r}}$, and as we noted at the end of the proof of Theorem 1.1 we have $\Delta_{\overline{S \cdot L_i}} = \lambda_{\vec{r}}(\Delta_{\overline{T \cdot L_i}})$, so we have that the multiplicity m_i of our cycle $\overline{S \cdot L_i}$ is also one. The assertion about polyhedral decompositions follows the more general result [15, Proposition 3.6], since the torus equivariant Plücker embedding identifies each point of the Chow quotient $\mathrm{Gr}(d,n)/\!\!/_{Ch}S$ with a point of the toric Chow quotient $\mathbb{P}^{\binom{n}{d}-1}/\!\!/_{Ch}S$.   □

3 Examples of subtorus Chow quotients

In this section we describe some diagonal subtorus Chow quotients of $\mathrm{Gr}(2,4)$, starting with the case of the maximal torus that Kapranov worked out in [14, Example 1.2.12] so that we can present an explicit equational approach that generalizes to the other cases. First, let us recall the more general setup. The Plücker embedding $\mathrm{Gr}(d,n) \subseteq \mathbb{P}^{\binom{n}{d}-1}$ is maximal torus equivariant so induces a closed embedding of Chow quotients

$$\mathrm{Gr}(d,n)/\!\!/_{Ch}S \subseteq \mathbb{P}^{\binom{n}{d}-1}/\!\!/_{Ch}S$$

for any diagonal subtorus $S \subseteq T$. Since S acts here through the dense torus for $\mathbb{P}^{\binom{n}{d}-1}$, the Chow quotient $\mathbb{P}^{\binom{n}{d}-1}/\!\!/_{Ch}S$ is a projective toric variety; the lattice polytope for it is a secondary polytope that we now describe (see [15] and [14, §0.2]).

If we denote the coordinates on $\mathbb{P}^{\binom{n}{d}-1}$ by $x_I, I \in \binom{[n]}{d}$ then $t = (t_1, \ldots, t_n) \in T$ acts by $t \cdot x_I = \left(\prod_{i \in I} t_i\right) x_I$. These weights are encoded by the $\binom{n}{d}$ integer vectors $\sum_{i \in I} e_i \in \mathbb{Z}^n$. The weights for the rank m diagonal subtorus $S \subseteq T$ are then the images of these integer vectors under the linear map $\lambda_{\vec{r}}: \mathbb{R}^n \to \mathbb{R}^m$. Define the following cardinality $\binom{n}{d}$ multiset:

$$A := \left\{ \lambda_{\vec{r}} \left(\sum_{i \in I} e_i \right) \mid I \in \binom{[n]}{d} \right\}.$$

The lattice polytope for $\mathbb{P}^{\binom{n}{d}-1}/\!\!/_{Ch}S$ is the secondary polytope $\Sigma(A)$. Recall that this means $\Sigma(A)$ is the convex hull in $\mathbb{R}^A$ of the characteristic functions $\varphi_{\mathcal{T}}: A \to \mathbb{Z}$ where $\mathcal{T}$ is a triangulation of the pair $(\mathrm{Conv}(A), A)$—meaning a collection of simplices, intersecting only along common faces, whose union is $\mathrm{Conv}(A)$ and whose vertices lie in A—and where by definition the value of $\varphi_{\mathcal{T}}$ on $a \in A$ is the sum of the volumes of all simplices in $\mathcal{T}$ for which a is a vertex (with the volume form normalized by setting the volume of the smallest possible lattice simplex to be 1).

3.1 Gr(2, 4) with the maximal torus action

Here $\vec{r} = (1, 1, 1, 1)$ and $\lambda_{\vec{r}}$ is the identity on $\mathbb{R}^4$, so A consists of the six vertices of the octahedron $\Delta(2, 4)$, namely all permutations of the vector $(1, 1, 0, 0)$. There are three triangulations here: choose two of the three pairs of non-adjacent vertices and for each of these chosen pairs slice a plane through the remaining four vertices. The three characteristic functions are then the vectors $(4, 4, 2, 2, 2, 2)$, $(2, 2, 4, 4, 2, 2)$, and $(2, 2, 2, 2, 4, 4)$. These form an equilateral triangle whose lattice points, in addition to the three vertices, are the midpoints of the three edges, namely $(3, 3, 3, 3, 2, 2)$, $(3, 3, 2, 2, 3, 3)$, and $(2, 2, 3, 3, 3, 3)$. This lattice polytope defines the toric variety $\mathbb{P}^2$ polarized by the line bundle $\mathcal{O}(2)$; by labeling the lattice points, in the order listed above, we can view this as $\operatorname{Proj} k[x^2, y^2, z^2, xy, xz, yz]$.

The Grassmannian Gr(2, 4) is a hypersurface in $\mathbb{P}^5$, defined by a single Plücker relation, so the Chow quotient $\operatorname{Gr}(2, 4) /\!\!/_{Ch} S \subseteq \mathbb{P}^5 /\!\!/_{Ch} S \cong \mathbb{P}^2$ is also a hypersurface and our next task is finding the equation for it. If we write the coordinates for $\mathbb{P}^5$ as $(x_{12}, x_{34}, x_{13}, x_{24}, x_{14}, x_{23})$ then the monomials specified by the six lattice points described in the preceding paragraph, after dividing by the common factor that is the product of the squares of all the variables, are the following:

$$m_1 = x_{12}^2 x_{34}^2, \ m_2 = x_{13}^2 x_{24}^2, \ m_3 = x_{14}^2 x_{23}^2,$$

$$m_4 = x_{12} x_{34} x_{13} x_{24}, \ m_5 = x_{12} x_{34} x_{14} x_{23}, \ m_6 = x_{13} x_{24} x_{14} x_{23}.$$

Multiplying the Plücker relation

$$x_{12} x_{34} - x_{13} x_{24} + x_{14} x_{23} = 0$$

by $x_{12} x_{34}$ yields the relation $m_1 - m_4 + m_5$, and similarly multiplying by $x_{13} x_{24}$ yields $m_4 - m_2 + m_6 = 0$ and multiplying by $x_{14} x_{23}$ yields $m_5 - m_6 + m_3 = 0$. These are linear relations among the monomials m_i, so they are three quadratic relations among the variables x, y, z introduced at the end of the preceding paragraph, namely

$$x^2 - xy + xz = 0, \ xy - y^2 + yz = 0, \text{ and } xz - yz + z^2 = 0.$$

These quadratics generate a non-saturated ideal whose saturation is the principal ideal generated by $x - y + z = 0$; this linear relation is the defining equation for the Chow quotient $\operatorname{Gr}(2, 4) /\!\!/_{Ch} T \subseteq \mathbb{P}^2$ that we were seeking. Note that in [14, Example 1.2.12] Kapranov described this as a conic in the plane, whereas we see here more specifically it is a line in the plane together embedded by $\mathcal{O}(2)$ as a conic in the Veronese surface in $\mathbb{P}^5$.

3.2 Gr(2, 4) with a rank 3 diagonal subtorus

Now consider the rank 3 diagonal subtorus $S \subseteq T$ defined by $\vec{r} = (1,1,2)$, namely $S = \{(t_1,t_2,t_3,t_3) \mid t_i \in k^\times\}$, which acts on a subspace $L \in \mathrm{Gr}(2,4)$ represented by a 2×4 matrix by rescaling the first two columns independently and rescaling the last two columns together. Here $\mathrm{Gr}(2,4)/\!\!/_{Ch}S$ is a surface embedded in the toric threefold $\mathbb{P}^5/\!\!/_{Ch}S$. The linear map $\lambda_{\vec{r}} \colon \mathbb{R}^4 \to \mathbb{R}^3$ is $(x_1,x_2,x_3 + x_4)$. The multiset A, the image under this linear projection of the 6 vertices of the octahedron $\Delta(2,4)$, is $(1,1,0)$, $(1,0,1)$ with multiplicity 2, $(0,1,1)$ with multiplicity 2, and $(0,0,2)$. This is a square with a pair of non-adjacent vertices doubled, and there are eight triangulations: there are two ways of subdividing with a diagonal line segment, and for each of these there are four ways of choosing which of the doubled vertices to use in the resulting pair of triangles. The six characteristic functions, which we shall name $v_1, \ldots, v_4, w_1, \ldots, w_4$, then take the following form:

$$v_1 = (1,2,0,2,0,1), \quad v_2 = (1,2,0,0,2,1), \quad v_3 = (1,0,2,0,2,1), \quad v_4 = (1,0,2,2,0,1),$$

$$w_1 = (2,1,0,1,0,2), \quad w_2 = (2,1,0,0,1,2), \quad w_3 = (2,0,1,0,1,2), \quad w_4 = (2,0,1,1,0,2).$$

The convex hull of these is a 3-dimensional polytope. The convex hull of the v_i is a square and the convex hull of the w_i is a smaller square that is parallel to it, so altogether we have a truncated square pyramid. A square is the toric polytope description of $\mathbb{P}^1 \times \mathbb{P}^1$, extending this to a square pyramid corresponds to taking the projective cone over $\mathbb{P}^1 \times \mathbb{P}^1$, and truncating this pyramid corresponds to blowing up the torus-fixed cone point corresponding to the pyramid apex. In coordinates this can be written

$$\mathbb{P}^5/\!\!/_{Ch}S \cong \mathrm{Bl}\left(\mathrm{Proj}\, k[x_0,x_1,x_2,x_3,y]/(x_0x_3 - x_1x_2)\right),$$

and by computing lattice lengths one sees that the polarization is $\mathcal{O}(2H - E)$. To find the equations for the closed subvariety $\mathrm{Gr}(2,4)/\!\!/_{Ch}S$ inside here, we follow the approach in the previous example. Plugging the variables x_{ij} into the 8 vertices of our secondary polytope yields the following monomials:

$$m_1 = x_{12}x_{13}^2x_{23}^2x_{34}, \quad m_2 = x_{12}x_{13}^2x_{24}^2x_{34}, \quad m_3 = x_{12}x_{14}^2x_{24}^2x_{34}, \quad m_4 = x_{12}x_{14}^2x_{23}^2x_{34},$$

$$n_1 = x_{12}^2x_{13}x_{23}x_{34}^2, \quad n_2 = x_{12}^2x_{13}x_{24}x_{34}^2, \quad n_3 = x_{12}^2x_{14}x_{24}x_{34}^2, \quad n_4 = x_{12}^2x_{14}x_{23}x_{34}^2.$$

Multiplying the Plücker relation by $x_{12}x_{13}x_{24}x_{34}$ and by $x_{12}x_{14}x_{23}x_{34}$ yields the relations

$$n_2 - m_2 + \prod x_{ij} = 0 \quad \text{and} \quad n_4 - \prod x_{ij} + m_4 = 0$$

so our Grassmannian Chow quotient here is defined in the above toric Chow quotient by the single relation $m_2 - m_4 - n_2 - n_4 = 0$ in the polynomial Cox ring.

3.3 Gr(2, 4) with a balanced rank two diagonal subtorus

Next, consider the diagonal subtorus $\{(t_1, t_1, t_2, t_2) \mid t_i \in k^\times\}$ defined by $\vec{r} = (2, 2)$. The linear projection $\lambda_{\vec{r}} \colon \mathbb{R}^4 \to \mathbb{R}^2$ is $(x_1 + x_2, x_3 + x_4)$ which sends the vertices of $\Delta(2, 4)$ to $(2, 0)$, $(1, 1)$ four times, and $(0, 2)$. The result of course is an interval with a single interior lattice point that has been quadrupled. There are five triangulation, four from subdividing with the different midpoints and one from not subdividing at all; the characteristic functions are:

$$v_1 = (1, 2, 0, 0, 0, 1), v_2 = (1, 0, 2, 0, 0, 1), v_3 = (1, 0, 0, 2, 0, 1),$$
$$v_4 = (1, 0, 0, 0, 2, 1), v_5 = (2, 0, 0, 0, 0, 2).$$

The convex hull of $v_1, \ldots, v_4$ is a tetrahedron giving the polarized toric variety $(\mathbb{P}^3, \mathcal{O}(2))$, and
$P^5 /\!/_{Ch} S$ is the toric variety given by the convex cone over this tetrahedron with apex v_5. Plugging the variables x_{ij} into these five vertices yields

$$m_1 = x_{12}x_{13}^2 x_{34}, \ m_2 = x_{12}x_{14}^2 x_{34}, \ m_3 = x_{12}x_{23}^2 x_{34},$$
$$m_4 = x_{12}x_{24}^2 x_{34}, \ m_5 = x_{12}^2 x_{34}^2.$$

The Plücker relation can be expressed as

$$\sqrt{m_5} - \sqrt{\frac{m_1 m_4}{m_5}} + \sqrt{\frac{m_2 m_3}{m_5}} = 0,$$

which after some elementary algebra yields the relation

$$m_1^2 m_4^2 + m_2^2 m_3^2 + m_5^2 - 2m_1 m_2 m_3 m_4 - 2m_1 m_4 m_5^2 - 2m_2 m_3 m_5^2 = 0$$

defining $\mathrm{Gr}(2, 4) /\!/_{Ch} S$ in the Cox ring of our toric variety $\mathbb{P}^5 /\!/_{Ch} S$.

4 Maps between Chow quotients

Let us start here by generalizing Kapranov's [14, Theorem 1.6.6]; while one probably could have adapted Kapranov's proof nearly verbatim to our setting, we instead provide a slight variant that we feel brings out more prominently the elegant toric geometry underlying the result.

Proof of Theorem 1.2 Recall from the theorem statement that we have fixed an index i and denoted by I the index of the r_i columns acted upon nontrivially by the ith $\mathbb{G}_m$ factor of S and by S_i the projection of S onto the coordinates outside of I. So S has rank m and S_i has rank $m-1$. Let

$$a_i: \operatorname{Gr}(d,n)/\!\!/_{Ch}S \dashrightarrow \operatorname{Gr}(d-r_i,n-r_i)/\!\!/_{Ch}S_i$$

be the rational map sending a generic torus orbit closure $\overline{S \cdot L}$, $L \in \operatorname{Gr}(d,n)$, to the torus orbit closure $\overline{S_i \cdot (L \cap H_I)}$, where $H_I \subseteq k^n$ is the coordinate linear subspace defined by setting all coordinates in I equal to zero (and $\operatorname{Gr}(d-r_i, n-r_i)$ here parameterizes subspaces of $H_I \cong k^{n-r-i}$). Let

$$b_i: \operatorname{Gr}(d,n)/\!\!/_{Ch}S \dashrightarrow \operatorname{Gr}(d,n-r_i)/\!\!/_{Ch}S_i$$

be the rational map sending a generic $\overline{S \cdot L}$ to $\overline{S_i \cdot \pi_{I^c}(L)}$, where $\pi_{I^c}: k^n \to k^{n-r_i}$ projects away the I-coordinates.

To show that these rational maps extend to morphisms, we will use the valuative criterion provided in [10, Theorem 7.3], restated below in Lemma 4.1 for the reader's convenience (though in this proof we will use the analytic language of 1-parameter families, rather valuation rings, since we have restricted to the setting $k = \mathbb{C}$ anyway). This means we need to show that for any 1-parameter family of cycles Z_t, $t \in k^\times$, in the interior of $\operatorname{Gr}(d,n)/\!\!/_{Ch}S$, which necessarily maps to a 1-parameter family of cycles $a_i(Z_t)$ in the interior of $\operatorname{Gr}(d-r_i,n-r_i)/\!\!/_{Ch}S_i$, the limit cycle

$$\lim_{t \to 0} a_i(Z_t) \in \operatorname{Gr}(d-r_i,n-r_i)/\!\!/_{Ch}S_i \subseteq \operatorname{Chow}\left(\operatorname{Gr}(d-r_i,n-r_i)\right)$$

$$\subseteq \operatorname{Chow}\left(\mathbb{P}^{\binom{n-r_i}{d-r_i}-1}\right)$$

depends only on the limit cycle

$$Z_0 := \lim_{t \to 0} Z_t \in \operatorname{Gr}(d,n)/\!\!/_{Ch}S \subseteq \operatorname{Chow}\left(\operatorname{Gr}(d,n)\right) \subseteq \operatorname{Chow}\left(\mathbb{P}^{\binom{n}{d}-1}\right),$$

and similarly for b_i. We will do this by explicitly describing $\lim a_i(Z_0)$ and $\lim b_i(Z_t)$ in terms of Z_0.

Following Kapranov, let $G_j^+ \subseteq \operatorname{Gr}(d,n)$ be the locus of linear subspaces containing the jth coordinate axis, and let $G_j^- \subseteq \operatorname{Gr}(d,n)$ be the locus of linear subspaces contained in the hyperplane where the jth coordinate is zero. Then, as noted in [14, Proposition 1.6.10],

$$\operatorname{Gr}(d-1,n-1) \cong G_j^+ = \operatorname{Gr}(d,n) \cap \Pi_j^+$$

where $\Pi_j^+ \subseteq \mathbb{P}^{\binom{n}{d}-1}$ is the coordinate linear subspace defined by $x_J = 0$ for $J \not\ni j$, and

$$\mathrm{Gr}(d, n-1) \cong G_j^- = \mathrm{Gr}(d,n) \cap \Pi_j^-$$

where Π_j^- is the coordinate linear subspace defined by $x_J = 0$ for $J \ni j$. In our setting we shall need to consider certain intersections of these sub-Grassmannians, so let

$$\Pi_I^{\pm} := \bigcap_{j \in I} \Pi_j^{\pm} \quad \text{and} \quad G_I^{\pm} := \bigcap_{j \in I} G_j^{\pm} = Gr(d,n) \cap \Pi_I^{\pm}.$$

We claim that

$$\lim_{t \to 0} a_i(Z_t) = Z_0 \cap \Pi_I^+ \quad \text{and} \quad \lim_{t \to 0} b_i(Z_t) = Z_0 \cap \Pi_I^-.$$

Verifying this claim will establish the theorem, by the aforementioned valuative criterion.

The argument in Kapranov's [14, Lemma 1.6.13] applies equally well for diagonal subtori and shows that for $t \neq 0$ we have $a_i(Z_t) = Z_t \cap \Pi_I^+$ and $b_i(Z_t) = Z_t \cap \Pi_I^-$, and from this it immediately follows from elementary topology that

$$\lim_{t \to 0} a_i(Z_t) \subseteq Z_0 \cap \Pi_I^+ \quad \text{and} \quad \lim_{t \to 0} b_i(Z_t) \subseteq Z_0 \cap \Pi_I^-. \tag{1}$$

We claim that in both cases the intersection on the right has the same dimension as the limit on the left, namely $m - 2$ (the diagonal $\mathbb{G}_m$ where all torus coordinates are equal acts trivially so a full-dimensional orbit has dimension one less than the rank of the torus). To see this, first note that by Proposition 2.2 we can write $Z_0 = \sum_{j=1}^{\ell} \overline{S \cdot L_j}$ for linear subspaces L_j whose S-orbits have full dimension $m - 1$. Then

$$Z_0 \cap \Pi_I^{\pm} = \sum_{j=1}^{\ell} \left(\overline{S \cdot L_j} \cap \Pi_I^{\pm} \right).$$

If the dimension of this intersection were not equal to $m - 2$ it would have to be dimension $m - 1$, the dimension of Z_0, which means for at least one j we would have $L_j \subseteq \Pi_I^{\pm}$, But this would mean that the S-orbit of this L_j is not full-dimensional, contradicting our assumption on it. Indeed, if $L_j \subseteq \Pi_I^+$ then the rank one subtorus of S where all $\mathbb{G}_m$ factors except for the ith are trivial is in the stabilizer of L_j, since this $\mathbb{G}_m$ subtorus rescales equally by t^{r_i} the Plücker coordinates x_J where $J \supseteq I$ and by definition of Π_I^+ all remaining Plücker coordinates are zero; similarly, if $L_j \subseteq \Pi_I^-$ then this same

$\mathbb{G}_m$ factor is in the stabilizer of L_j, since here it acts trivially on the Plücker coordinates x_J where $J \cap I = \varnothing$ and by definition of Π_I^- all remaining Plücker coordinates are zero.

For each of the containments in Equation (1), since the dimensions of both sides are equal, to prove that the containment is an equality it suffices to prove that the degrees of both sides are equal. Now, $\lim_{t\to 0} a_i(Z_t)$ is a limit of generic S_i-orbit closures so it has the same degree as a generic orbit closure $\overline{S_i \cdot L}$, $L \in \mathrm{Gr}(d - r_i, n - r_i)^0$. But $\overline{S_i \cdot L}$ is a toric variety so its degree is the volume of the lattice polytope $\Delta_{\overline{S_i \cdot L}}$, and since L here is generic this lattice polytope is the full linearly projected hypersimplex $\lambda_{\pi_{[m]\setminus i}\vec{r}} (\Delta(d - r_i, n - r_i))$, where $\lambda_{\pi_{[m]\setminus i}\vec{r}} : \mathbb{R}^{n-r_i} \to \mathbb{R}^{m-1}$ is the linear projection map corresponding to the diagonal subtorus S_i of the maximal torus acting on $\mathrm{Gr}(d - r_i, n - r_i)$. On the other hand, by Proposition 2.2 for the limit cycle $Z_0 = \sum_{j=1}^{\ell} \overline{S \cdot L_j}$ the lattice polytopes $\Delta_{\overline{S \cdot L_1}}, \dots, \Delta_{\overline{S \cdot L_\ell}}$ form a polyhedral decomposition of $\lambda_{\vec{r}} (\Delta(d,n))$. Then the lattice polytopes $\Delta_{\overline{S \cdot L_1}} \cap \lambda_{\vec{r}}(\Gamma_I^+), \dots, \Delta_{\overline{S \cdot L_\ell}} \cap \lambda_{\vec{r}}(\Gamma_I^+)$ form a polyhedral decomposition of the face $\lambda_{\vec{r}}(\Gamma_I^+)$ of $\lambda_{\vec{r}}(\Delta(d,n))$, where $\Gamma_I^+ := \cap_{j\in I}\Gamma_j^+$ and Γ_j^+ is the face of $\Delta(d,n)$ that Kapranov identified in [14, Proposition 1.6.10] as the image under the moment map μ_T of $G_j^+ \subseteq \mathrm{Gr}(d,n)$. We claim

$$\deg(Z_0 \cap \Pi_I^+) \le \sum_{j=1}^{\ell} \deg\left(\overline{S \cdot L_j} \cap \Pi_I^+\right)$$

$$= \sum_{j=1}^{\ell} \mathrm{vol}\left(\Delta_{\overline{S \cdot L_j}} \cap \Gamma_I^+\right) = \mathrm{vol}\left(\lambda_{\vec{r}}(\Gamma_I^+)\right)$$

$$= \mathrm{vol}\left(\lambda_{\pi_{[m]\setminus i}\vec{r}} (\Delta(d - r_i, n - r_i))\right).$$

Indeed, the inequality here allows for the possibility that some of these intersected orbit closures are not full-dimensional, the first equality is Kapranov's observation in [14, Proposition 1.6.10] about the interplay between the moment map and the sub-Grassmannians G_j^+, the second equality is due to the above observation about having a polyhedral decomposition, and the final equality follows from the observation that the moment map μ_S restricted to the sub-Grassmannian $\Gamma_I^+ \cong \mathrm{Gr}(d - r_i, n - r_i)$ is identified by this isomorphism with the moment map $\mu_{S_i} = \lambda_{\pi_{[m]\setminus i}\vec{r}} \circ \mu_{T'}$ where T' is the maximal torus acting on $\mathrm{Gr}(d - r_i, n - r_i)$. This concludes the argument for a_i, and the volume calculation for b_i is entirely analogous. $\qquad\square$

Lemma 4.1 *[10, Theorem 7.3] Suppose X_1 and X_2 are proper schemes over a noetherian scheme S, with X_1 normal. Let $U \subseteq X_1$ be an open dense subset and $f : U \to X_2$ an S-morphism. Then f extends to an S-morphism $X_1 \to X_2$*

if and only if for any DVR with fraction field denoted K, and any morphism g: Spec $K \rightarrow U$, the point $\lim fg$ of X_2 is uniquely determined by the point $\lim g$ of X_1.

5 Generalized Gelfand–MacPherson correspondence and Gale duality

In [17] Thaddeus studies an interesting classical geometric situation related to the configuration spaces studied by Kapranov in [14], and while doing so he proves a handful of results that are in close analogy with results in Kapranov's paper—but in almost all cases, the proofs Thaddeus provides are new, not merely adaptations of Kapranov's. In particular, when studying Chow quotients Thaddeus avails himself of the functorial machinery developed by Kollár in [16], obviating the need to rely on the analytic methods for working with Chow varieties that were the only option for Kapranov at the time his paper was written. We adapt here one particular proof of Thaddeus (and a particularly clever one at that) which in our setting yields the generalized Gelfand–MacPherson isomorphism Theorem 1.3 stated in the introduction. Note that by specializing to the maximal torus this yields an explicit Thaddeus-esque proof of Kapranov's original Chow-theoretic Gelfand–MacPherson isomorphism [14, Theorem 2.2.4].

Proof of Theorem 1.3 The basic idea is, quite like the usual Gelfand–MacPherson correspondence, to observe that the GL_d-action on the affine space of $n \times d$ matrices (we have taken a transpose here to work with sub rather than quotient objects, but that is immaterial and just to ease notation) commutes with the torus action; taking the GL_d quotient first yields the Grassmannian $\mathrm{Gr}(d,n)$, whereas taking the S-quotient first projectivizes the size $r_i \times d$ matrix blocks, $i = 1, \ldots, m$, of this space of matrices resulting in a product of projective spaces. In fact, this already shows that the two sides of the claimed isomorphism are birational, so the work is to extend this birational map to an isomorphism. To do this, we follow and mildly adapt the argument of Thaddeus in his proof in [17, §6.3]. The main insight in Thaddeus' proof, translated to our situation, is that the two rational quotient maps

$$\mathbb{P}\operatorname{Hom}(k^d, k^n) \dashrightarrow \mathrm{Gr}(d,n) \qquad (2)$$

and

$$\mathbb{P}\operatorname{Hom}(k^d, k^n) \dashrightarrow \prod_{i=1}^{m} \mathbb{P}\operatorname{Hom}(k^d, k^{r_i}) \qquad (3)$$

have different base loci, and by resolving both it is easier to compare cycles by using pullback and pushforward properties of the Chow variety. We now go through these details in earnest.

The rational GL_d-quotient map in (2) sends an injective linear map $\varphi \colon k^d \hookrightarrow k^n$ to $[\varphi(k^d)] \in Gr(d,n)$, the point in the Grassmannian corresponding to the image of this linear map; the base locus is the set of linear maps $k^d \to k^n$ with nontrivial kernel. Let $\mathcal{S}_{d,n} \to Gr(d,n)$ denote the universal subbundle over the Grassmannian. Then the rational GL_d-quotient map is resolved by the space $\mathbb{P}\,\mathrm{Hom}(k^d, \mathcal{S}_{d,n})$:

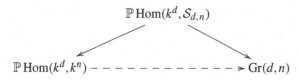

Indeed, the fiber over a point $\varphi \colon k^d \to k^n$ of $\mathbb{P}\,\mathrm{Hom}(k^d, k^n)$ is a single point of $\mathbb{P}\,\mathrm{Hom}(k^d, \mathcal{S}_{d,n})$ if φ is injective, namely φ viewed as a map from k^d to its image $\varphi(k^d) \subseteq k^n$, whereas if $\dim \varphi(k^d) < d$ then the fiber in $\mathbb{P}\,\mathrm{Hom}(k^d, \mathcal{S}_{d,n})$ is in bijection with all d-dimensional subspaces $L \subseteq k^n$ containing $\varphi(k^d) \subseteq k^n$, since for each such $L \supseteq \varphi(k^d)$ we have the element of the fiber given by viewing φ as a map from k^d to L. In fact, $\mathbb{P}\,\mathrm{Hom}(k^d, \mathcal{S}_{d,n})$ is the iterated blow-up of $\mathbb{P}\,\mathrm{Hom}(k^d, k^n)$ along the locus of non-full rank maps, ordered in increasing order of rank. Note that the morphism to $Gr(d,n)$ is a $\mathbb{P}^{d^2-1}$-bundle; in particular, it is flat.

On the other hand, the rational S-quotient map (3) is resolved by the $\mathbb{P}^{m-1}$-bundle given by the projectivization of the total space the direct sum of the dual line bundles to the tautological bundles:

$$\mathbb{P}\left(\bigoplus_{i=1}^{m} \mathcal{O}(e_i)\right)$$

$$\mathbb{P}\,\mathrm{Hom}(k^d, k^n) - - - - - - - - - - - - \rightarrow \prod_{i=1}^{m} \mathbb{P}\,\mathrm{Hom}(k^d, k^{r_i})$$

Here $\mathcal{O}(e_j)$ denotes the pull-back of $\mathcal{O}(1)$ along the jth projection

$$\prod_{i=1}^{m} \mathbb{P}\,\mathrm{Hom}(k^d, k^{r_i}) \to \mathbb{P}\,\mathrm{Hom}(k^d, k^{r_j}) \cong \mathbb{P}^{r_j d - 1}.$$

One can see this as follows. The base locus for this map consists of matrices where any of the $r_i \times d$ blocks (corresponding to the diagonal subtorus action) are entirely zero, so to resolve this map we need to blow up this locus. Since it is a union of linear subspaces meeting transversely, this can be done one subspace at a time, in any order, and we thus reduce to the standard observation that the total space of $\mathcal{O}(1)$ on any projective space $\mathbb{P}^\ell$ is the blow-up of $\mathbb{A}^{\ell+1}$ at the origin.

Putting this together, we get the following commutative diagram:

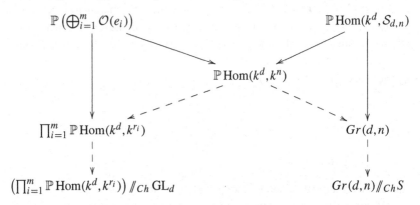

Here the vertical morphisms are both projective space bundles, the diagonal morphisms are birational, and the dashed arrows are all rational quotient maps—on the left by the torus first then GL_d, and on the right by GL_d first then the torus.

The rest of Thaddeus' argument now goes through essentially verbatim. The universal family of cycles on $\mathrm{Gr}(d,n)$ over the Chow quotient $\mathrm{Gr}(d,n) /\!/_{Ch} S \subseteq$ Chow$(\mathrm{Gr}(d,n))$ pulls back along the flat morphism to a family of cycles on $\mathbb{P}\mathrm{Hom}(k^d, \mathcal{S}_{d,n})$ over $\mathrm{Gr}(d,n) /\!/_{Ch} S$ with general fiber a $(\mathrm{GL}_d \times S)$-orbit closure. This family pushes forward along the birational morphism to an S-invariant family of cycles on $\mathbb{P}\mathrm{Hom}(k^d, k^n)$. The restriction of the cycles in this family to the complement of the base locus of the torus quotient map (3) pushes forward along this quotient map, a geometric quotient, and yields a family of cycles on $\prod_{i=1}^m \mathbb{P}\mathrm{Hom}(k^d, k^{r_i})$ over $\mathrm{Gr}(d,n) /\!/_{Ch} S$. Since $\mathrm{Gr}(d,n) /\!/_{Ch} S$ is reduced and the cycles over it in this last family all have the expected dimension, there is an induced morphism

$$\mathrm{Gr}(d,n) /\!/_{Ch} S \to \mathrm{Chow}\left(\prod_{i=1}^m \mathbb{P}\mathrm{Hom}(k^d, k^{r_i})\right)$$

by [16, Theorem 3.21]. A general point of this Chow quotient gets sent to a GL_d-orbit closure, so the image of this morphism is contained in the Chow quotient $\left(\prod_{i=1}^m \mathbb{P}\mathrm{Hom}(k^d, k^{r_i})\right) /\!/_{Ch} \mathrm{GL}_d$. On the other hand, the same argument applied symmetrically to other side of the above big commutative diagram yields a morphism between these Chow quotients in the other direction. Since these Chow quotients are separated varieties, to show that these morphisms are inverse to each other, and hence that the two Chow quotients are isomorphic, it suffices to show that they are inverse on open dense loci. For this we apply the naive argument discussed at the beginning of this proof, regarding commuting group actions, to see that indeed these maps identify generic orbit closures. □

The classical Gale duality is a bijection between sufficiently general configurations of n points in $\mathbb{P}^d$ and sufficiently general configurations of n points in $\mathbb{P}^{n-d-2}$, up to projectivity (see, for instance, [3, §4.1]). Kapranov showed [14, Corollary 2.3.14] that this extends to an involutive isomorphism of Chow quotients

$$(\mathbb{P}^{d-1})^n /\!\!/_{Ch} \mathrm{GL}_d \cong (\mathbb{P}^{n-d-2})^n /\!\!/_{Ch} \mathrm{GL}_{n-d-1},$$

obtained by applying his Chow-theoretic Gelfand–MacPherson correspondence to both sides of the isomorphism $\mathrm{Gr}(d,n) \cong \mathrm{Gr}(n-d,n)$ induced by orthogonal complement. An immediate corollary of our generalized Gelfand–MacPherson is the generalized Gale duality Corollary 1.4 stated in the introduction. Indeed, the orthogonal complement isomorphism $\mathrm{Gr}(d,n) \cong \mathrm{Gr}(n-d,n)$ is torus-equivariant so descends to an isomorphism $\mathrm{Gr}(d,n)/\!\!/_{Ch}S \cong \mathrm{Gr}(n-d,n)/\!\!/_{Ch}S$ of Chow quotients for any subtorus S, and applying our generalized Gelfand–MacPherson isomorphisms to both sides of this isomorphism provides our generalized Gale duality isomorphism.

Remark For parameters $m, d, d_1, \ldots, d_m$ such that

$$2d - m = \sum_{i=1}^{m} d_i,$$

our generalized Gale duality sends configurations of m parameterized linear subspaces of dimensions $d_1, \ldots, d_m$ in $\mathbb{P}^d$ to configurations of m parameterized linear subspaces of dimensions $d_1, \ldots, d_m$ in $\mathbb{P}^d$, so in this situation one could study "self-associated" configurations, generalizing the maximal torus case studied by Kapranov in [14, Paragraph (2.3.9)] (see also [6, §II] for another setting for self-association).

6 The Borel transfer principle and the Chen–Gibney-Krashen moduli space

Consider a connected unipotent group H, and suppose G is a reductive group containing H as a closed subgroup. The quotient G/H, where H acts on the right, is a quasi-affine variety (and if H is positive-dimensional then it is not affine); it admits a natural embedding in the affinization

$$(G/H)^{\text{aff}} := \text{Spec } \mathcal{O}_{G/H}(G/H) = \text{Spec } \mathcal{O}_G(G)^H$$

which is a scheme possibly of infinite type since the ring of invariants of a non-reductive group need not be finitely generated.

Example 6.1 Let

$$G := \text{Spec } k[x_{11}, x_{12}, x_{21}, x_{22}, (x_{11}x_{22} - x_{12}x_{21})^{-1}] \cong \text{GL}_2$$

be the affine group variety of 2×2 invertible matrices, and let $H := \text{Spec } k[s] \cong \mathbb{G}_a$ be the subgroup of unipotent matrices of the form $\begin{pmatrix} 1 & s \\ 0 & 1 \end{pmatrix}$. The quotient G/H is the quasi-affine variety $\mathbb{A}^2 \setminus \{0\}$, because the affinization is

$$\begin{aligned}
(G/H)^{\text{aff}} &= \text{Spec } k[x_{11}, x_{12}, x_{21}, x_{22}, (x_{11}x_{22} - x_{12}x_{21})^{-1}]^H \\
&= \text{Spec } k[x_{11}, x_{21}] \cong \mathbb{A}^2
\end{aligned}$$

but the image of the quotient morphism $G \to (G/H)^{\text{aff}}$ does not include the origin since a matrix where x_{11} and x_{21} are both zero is not invertible. In this case the affinization is of finite type.

Continue to let G and H be a reductive group and unipotent subgroup as above, and suppose now that X is an affine variety with an H-action that extends to a G-action. The classical Borel transfer principle states, in the language of (non-reductive) GIT, that there is an isomorphism

$$X /\!\!/ H \cong \left((G/H)^{\text{aff}} \times X\right) /\!\!/ G,$$

where G acts diagonally on this product, with the G-action on $(G/H)^{\text{aff}}$ induced by left-multiplication of G on itself, and the symbol "$/\!\!/$" simply means to take Spec of the ring of invariants [5, §5.1]. This allows one to replace a non-reductive invariant ring with a reductive invariant ring, though in the process one replaces the k-algebra being acted upon with one that need not be finitely generated. This is often a useful tradeoff as it means instead of studying the H-action of X, it suffices to study the typically simpler H-action on G together with the (again, typically simpler) G-action on X. For some elegant applications and illustrations of the classical Borel transfer principle the reader is encouraged to consult [11, Chapter 2], and for a globalization of the geometric formulation of the Borel transfer principle described in this paragraph to the case that X is projective see [5, §5.1].

The definition of a Chow quotient is perfectly valid for any algebraic group, not just reductive groups, so a natural question, which seems not to have appeared in the literature previously, is whether this global Borel transfer principle for GIT quotients extends to Chow quotients:

Question Let G be a reductive group containing a connected unipotent closed subgroup H, let $\overline{G/H}$ be a projective completion of the quotient, and let X be a projective variety with a G-action. Is there an isomorphism

$$X /\!\!/_{Ch} H \overset{?}{\cong} \left(\overline{G/H} \times X \right) /\!\!/_{Ch} G,$$

or at least are there reasonable hypotheses guaranteeing such an isomorphism?

The projective completion here is needed since Chow varieties are only defined for projective varieties. In what follows we show that a specific instance of this question is an open question about a Grassmannian Chow quotient first asked (in casual conversation) by Krashen.

Consider the diagonal subtorus action on $\mathrm{Gr}(d, n)$ defined by $\vec{r} = (d - 1, 1, \ldots, 1)$, so that S is the rank $n - d + 2$ torus that acts by rescaling the first $d - 1$ columns of a matrix together and the last $n - d + 1$ columns individually. By our generalized Gelfand–MacPherson isomorphism (Theorem 1.3) we have

$$\mathrm{Gr}(d, n) /\!\!/_{Ch} S \cong \left(\mathbb{P} \mathrm{Hom}(k^{d-1}, k^d) \times (\mathbb{P}^{d-1})^{n-d+1} \right) /\!\!/_{Ch} \mathrm{GL}_d, \quad (4)$$

a compactification of the configuration space of $n - d + 1$ points and a parameterized hyperplane in $\mathbb{P}^{d-1}$. On the other hand, the Chen–Gibney-Krashen moduli space $T_{d-1, n-d+1}$ is a compactification of the same configuration space [4], and Krashen's question is whether these are isomorphic. In [8] it is shown that $T_{d-1, n-d+1}$ is isomorphic to the normalization of the Chow quotient $(\mathbb{P}^{d-1})^{n-d+1} /\!\!/_{Ch} H$, where $H \cong \mathbb{G}_m^2 \rtimes \mathbb{G}_a^{d-1}$ is the non-reductive subgroup of GL_d fixing a hyperplane pointwise. Since this H-action extends to the standard GL_d-action, we can apply Question 6 and ask whether this non-reductive Chow quotient is isomorphic to the reductive Chow quotient $\left(\overline{\mathrm{GL}_d /H} \times (\mathbb{P}^{d-1})^{n-d+1} \right) /\!\!/ \mathrm{GL}_d$. The following lemma describes $\overline{\mathrm{GL}_d /H}$ and the induced group actions and implies that this reductive Chow quotient is precisely the one appearing in our generalized Gelfand–MacPherson correspondence, the right side of Equation (4), and hence as claimed that the Krashen question is a specific instance of Question 6:

$$
\begin{array}{ccc}
T_{d-1, n-d+1} & \overset{?}{=\!=\!=\!=\!=} & \mathrm{Gr}(d, n) /\!\!/_{Ch} S \\
\| & & \| \\
(\mathbb{P}^{d-1})^{n-d+1} /\!\!/_{Ch} H & \underset{?}{=\!=\!=} & \left(\overline{\mathrm{GL}_d /H} \times (\mathbb{P}^{d-1})^{n-d+1} \right) /\!\!/ \mathrm{GL}_d
\end{array}
$$

The left vertical equality (up to normalization) here is [8], the right vertical equality is the following lemma together with the Gelfand–MacPherson isomorphism, the top horizontal equality is the Krashen question, and the bottom horizontal equality is a special instance of Question 6.

Lemma 6.2 *For the right-multiplication action of H on GL_d, the quotient GL_d / H is isomorphic to the open subvariety of $\mathbb{P}\mathrm{Hom}(k^{d-1}, k^d)$ consisting of projective equivalence classes of full rank $d \times (d-1)$ matrices. The left-multiplication action of GL_d on itself descends to an action on this quotient corresponding, via this isomorphism, to left matrix multiplication.*

Certainly the most natural projective completion to take for the space of full rank matrices is its Zariski closure in the space of all matrices, hence $\overline{\mathrm{GL}_d / H} = \mathbb{P}\mathrm{Hom}(k^{d-1}, k^d)$.

Proof If we choose coordinates so that the fixed hyperplane is defined by the vanishing of the first coordinate, then $H \cong \mathbb{G}_m^2 \rtimes \mathbb{G}_a^{d-1}$ consists of matrices of the form

$$\begin{pmatrix} t_1 & 0 & \cdots & 0 \\ s_1 & t_2 & & 0 \\ \vdots & & \ddots & \\ s_{d-1} & 0 & \cdots & t_2 \end{pmatrix}$$

for $s_i \in k$ and $t_i \in k^\times$.

Since the additive action is normalized by the torus action, we can compute the quotient in stages:

$$\mathrm{GL}_d / H \cong (\mathrm{GL}_d / \mathbb{G}_a^{d-1}) / \mathbb{G}_m^2.$$

We claim $\mathrm{GL}_d / \mathbb{G}_a^{d-1}$ is the space of full rank $d \times (d-1)$ matrices. Indeed, by viewing

$$\mathrm{GL}_d \subseteq \mathrm{Hom}(k^d, k^d) \cong \mathbb{A}^{d^2}$$

as the affine open complement of the hypersurface $\det = 0$, the ring of invariants for the $\mathbb{G}_a^{d-1}$-action is generated by all entries of the matrix except for those of the first column. Thus the categorical quotient, in the category of affine varieties, is

$$\mathrm{GL}_d /\!/ \mathbb{G}_a^{d-1} \cong \mathrm{Hom}(k^{d-1}, k^d) \cong \mathbb{A}^{(d-1)d}.$$

However, similar to the situation in Example 6.1, since this is a non-reductive quotient the quotient morphism need not be surjective, and indeed in the present situation its image is manifestly the set of full rank matrices.

The residual $\mathbb{G}_m^2$-action on this space of full rank $d \times (d-1)$ matrices has the $\mathbb{G}_m$ factor corresponding to t_1 acting trivially and the $\mathbb{G}_m$ factor corresponding to t_2 acting by rescaling all entries equally, so the quotient by $\mathbb{G}_m^2$ is simply the projectivization. The assertion about the induced left-multiplication action of GL_d on this space of matrices follows immediately from our explicit description of the quotient in terms of invariants as the rightmost $d-1$ columns of a square $d \times d$ matrix of indeterminates. $\square$

References

[1] Alexeev, Valery. 2015. *Moduli of weighted hyperplane arrangements.* Advanced Courses in Mathematics. CRM Barcelona. Birkhäuser/Springer, Basel. Edited by Gilberto Bini, MartíLahoz, Emanuele Macrìand Paolo Stellari.

[2] Ardila, Federico, Benedetti, Carolina, and Doker, Jeffrey. 2010. Matroid polytopes and their volumes. *Discrete Comput. Geom.*, **43**(4), 841–854.

[3] Caminata, A., Giansiracusa, N., Moon, H.-B., and Schaffler, L. 2018. Equations for point configurations to lie on a rational normal curve. *Adv. in Math.*, **340**, 653–683.

[4] Chen, L., Gibney, A., and Krashen, D. 2009. Pointed trees of projective spaces. *J. Algebraic Geom.*, **18**(3), 477–509.

[5] Doran, Brent, and Kirwan, Frances. 2007. Towards non-reductive geometric invariant theory. *Pure Appl. Math. Q.*, **3**(1, Special Issue: In honor of Robert D. MacPherson. Part 3), 61–105.

[6] Eisenbud, David, and Popescu, Sorin. 2000. The projective geometry of the Gale transform. *J. Algebra*, **230**(1), 127–173.

[7] Fulton, William, and MacPherson, Robert. 1994. A compactification of configuration spaces. *Ann. of Math. (2)*, **139**(1), 183–225.

[8] Gallardo, Patricio, and Giansiracusa, Noah. 2018. Modular interpretation of a non-reductive Chow quotient. *Proc. Edinb. Math. Soc. (2)*, **61**(2), 457–477.

[9] Gelfand, I. M., Goresky, R. M., MacPherson, R. D., and Serganova, V. V. 1987. Combinatorial geometries, convex polyhedra, and Schubert cells. *Adv. in Math.*, **63**(3), 301–316.

[10] Giansiracusa, Noah, and Gillam, William Danny. 2014. On Kapranov's description of $\overline{M}_{0,n}$ as a Chow quotient. *Turkish J. Math.*, **38**(4), 625–648.

[11] Grosshans, Frank. 1997. *Algebraic homogeneous spaces and invariant theory.* Berlin New York: Springer.

[12] Hacking, Paul, Keel, Sean, and Tevelev, Jenia. 2006. Compactification of the moduli space of hyperplane arrangements. *J. Algebraic Geom.*, **15**(4), 657–680.

[13] Herzog, Jürgen, and Hibi, Takayuki. 2002. Discrete polymatroids. *J. Algebraic Combin.*, **16**(3), 239–268.

[14] Kapranov, M. M. 1993. *Chow quotients of Grassmannians. I.* Adv. Soviet Math., vol. 16. Amer. Math. Soc., Providence, RI.

[15] Kapranov, M. M., Sturmfels, B., and Zelevinsky, A. V. 1991. Quotients of toric varieties. *Math. Ann.*, **290**(4), 643–655.

[16] Kollár, János. 1996. *Rational curves on algebraic varieties*. Ergebnisse der Mathematik und ihrer Grenzgebiete. 3. Folge. A Series of Modern Surveys in Mathematics [Results in Mathematics and Related Areas. 3rd Series. A Series of Modern Surveys in Mathematics], vol. 32. Springer-Verlag, Berlin.

[17] Thaddeus, Michael. 1999. Complete collineations revisited. *Math. Ann.*, **315**(3), 469–495.

11

Quantum Kirwan for Quantum K-theory

E. González[a] and C. Woodward[b]

Abstract. For G a complex reductive group and X a smooth projective or convex quasi-projective polarized G-variety we construct a formal map in quantum K-theory

$$\kappa_X^G : QK_G^0(X) \to QK^0(X /\!\!/ G)$$

from the equivariant quantum K-theory $QK_G^0(X)$ to the quantum K-theory of the geometric invariant theory quotient $X /\!\!/ G$, assuming the quotient $X /\!\!/ G$ is a smooth Deligne-Mumford stack with projective coarse moduli space. As an example, we give a presentation of the (possibly bulk-shifted) quantum K-theory of any smooth proper toric Deligne-Mumford stack with projective coarse moduli space, generalizing the presentation for quantum K-theory of projective spaces due to Buch–Mihalcea [7] and (implicitly) of Givental–Tonita [18]. We also provide a wall-crossing formula for the K-theoretic gauged potential

$$\tau_X^G : QK_G^0(X) \to \Lambda_X^G$$

under variation of geometric invariant theory quotient, a proof of the invariance of τ_X^G under (strong) crepant transformation assumptions, and a proof of the abelian non-abelian correspondence relating τ_X^G and τ_X^T for $T \subset G$ a maximal torus.

1 Introduction

We aim to describe the behavior of quantum K-theory under the operation of geometric invariant theory quotient. Let X be a smooth projective G-variety

[a] University of Massachusetts, Boston
[b] Rutgers University, New Jersey
Partially supported by NSF grants DMS-1510518 and DMS-1711070

with polarization $\mathcal{L} \to X$. The geometric invariant theory (git) quotient $X /\!\!/ G$ is a proper smooth Deligne-Mumford stack with projective coarse moduli space. Let $K_G^0(X)$ denote the even topological or algebraic G-equivariant K-cohomology of X. The *Kirwan map* is the ring homomorphism in K-theory ("K-theoretic reduction")

$$\kappa_X^G \colon K_G^0(X) \to K^0(X /\!\!/ G), \quad [E] \mapsto [(E|X^{\mathrm{ss}})/G] \tag{1}$$

obtained by restricting a vector bundle E to the semistable locus X^{ss} and passing to the stack quotient. If X is projective then Kirwan showed that (1) is surjective in rational cohomology [39]. The analogous results in K-theory hold by work of Harada–Landweber [30, Theorem 3.1] and Halpern–Leistner [27, Corollary 1.2.3]. The Kirwan map can often be used to compute the K-theory of a git quotient. In particular, the Kirwan map allows a simple presentation of the K-theory of a smooth projective toric variety, equivalent to the presentation given in Vezzosi-Vistoli [55, Section 6.2].

A quantum deformation of the K-theory ring was introduced by Givental [17] and Y.P. Lee [41]. In this deformation, the tensor product of vector bundles is replaced by a certain push-pull over the moduli space of stable maps. The virtual fundamental class in cohomology is replaced by a *virtual structure sheaf* introduced in [41], and integrals over the moduli space of stable maps in K-theory are called *K-theoretic Gromov–Witten invariants*. For many purposes quantum K-theory is expected to be more natural than the quantum cohomology ring, which can be obtained as a limit; see for example [43]. In particular, the K-theoretic Gromov–Witten invariants are integers. Computations in quantum K-theory have been rather rare; even the quantum K-theory of projective space seems to have been computed only recently by Buch–Mihalcea [7].

We develop a quantum version of Kirwan's map in K-theory. As applications, we give presentations of the quantum K-theory of toric varieties, generalizing a computation of Buch–Mihalcea [7] in the case of projective spaces. Let

$$\Lambda_X^G \subset \mathrm{Map}(H_2^G(X, \mathbb{Q}), \mathbb{Q})$$

denote the equivariant Novikov ring associated to the polarization, with q^d denoting for q a formal variable the delta function at $d \in H_2^G(X, \mathbb{Q})$. The quantum K-theory product is defined by Givental-Lee [17, 41] as a pull-push over moduli spaces of stable maps; in order to make the products finite we define the equivariant quantum K-theory as the completion of $K_G \otimes \Lambda_X^G$ with respect to the ideal $I_G^X(c)$ of elements $E \in K_G \otimes \Lambda_X^G$ with $\mathrm{val}_q(E) > c$:

$$QK_G^0(X) = \lim_{n \leftarrow}(K_G^0(X) \otimes \Lambda_X^G)/I_G^X(c)^n. \tag{2}$$

The main result is the following:

Theorem 1.1 *Let G be a complex reductive group and X be a smooth polarized projective (or convex quasiprojective) G-variety with locally free git quotient $X /\!\!/ G$. There exists a canonical* Kirwan map *in quantum K-theory*

$$\kappa_X^G : QK_G^0(X) \to QK^0(X /\!\!/ G)$$

with the property that the linearization $D_\alpha \kappa_X^G$ is a homomorphism:

$$D_\alpha \kappa_X^G(\beta \star \gamma) = D_\alpha \kappa_X^G(\beta) \star D_\alpha \kappa_X^G(\gamma).$$

If $X /\!\!/ G$ is a free quotient then κ_X^G is surjective.[1]

The convexity assumption is satisfied if, for example, the variety is a vector space with a torus action such that all weights are properly contained in a half-space. For example, for toric varieties that may be realized as git quotients we explicitly compute the kernel of the map to obtain a presentation of the orbifold quantum K-theory at a point determined by the presentation, generalizing the presentation of ordinary K-theory of non-singular toric varieties due to Vezzosi-Vistoli [55, Section 6.2] and the case of quantum K-theory of projective spaces due to Buch–Mihalcea [7]: Let G be a complex torus and X a vector space with weight spaces $X_1, \ldots, X_k$ and weights $\mu_1, \ldots, \mu_k$ define the *completed equivariant quantum K-theory* $\widehat{QK}_G^0(X)$ to be the ring with generators $X_1^{\pm 1}, \ldots, X_k^{\pm 1}$ formally completed by the ideal generated by $(1 - X_j^{-1})$, $j = 1, \ldots, k$.[2] The K-theoretic Batyrev (or quantum K-theoretic Stanley–Reisner) ideal is the ideal $QKSR_X^G$ generated by the relations

$$\prod_{(\mu_j, d) \geq 0} (1 - X_j^{-1})^{\mu_j(d)} = q^d \prod_{(\mu_j, d) < 0} (1 - X_j^{-1})^{-\mu_j(d)}. \tag{3}$$

Theorem 1.2 *Suppose that G is a torus with Lie algebra $\mathfrak{g}$, that X is a G-vector space with weights $\mu_1, \ldots, \mu_k \in \mathfrak{g}_\mathbb{R}^\vee$ contained in an open half-space in $\mathfrak{g}_\mathbb{R}^\vee$, and that X is equipped with a polarization so that $X /\!\!/ G$ is a non-singular proper toric Deligne-Mumford stack with projective coarse moduli space. Let $T = (\mathbb{C}^\times)^k / G$ denote the residual torus. Then the quantum K-theory ring $QK^0(X /\!\!/ G)$ with bulk deformation $\kappa_X^G(0)$ is isomorphic to the quotient $\widehat{QK}_G^0(X)/QKSR_X^G$.*

---

[1] Computations suggest that the map κ_X^G might be surjective even for locally free quotients of proper free actions.

[2] In the case of orbifold quotients, a more complicated formal completion is necessary, see Definition 4.2.

Example 1.3 (Weighted projective spaces) Suppose that $G = \mathbb{C}^\times$ acts on $X = \mathbb{C}^k$ with weights $\mu_1, \ldots, \mu_k \in \mathbb{Z}$, so that $X /\!\!/ G$ is the weighted projective space $\mathbb{P}(\mu_1, \ldots, \mu_k)$. Then the T-equivariant quantum K-theory of $X /\!\!/ G$ has canonical presentation with generators and a single relation (in this case the formal completion is not necessary):

$$QK^0(X /\!\!/ G) \cong \frac{\Lambda_X^G[X_1^{\pm 1}, \ldots, X_k^{\pm 1}]}{\langle \prod_j (1 - X_j^{-1})^{\mu_j(d)} - q \rangle}. \tag{4}$$

In this case, the bulk deformation $\kappa_X^G(0)$ turns out to vanish, see Lemma 3.6 below, and X_j is the class of the line bundle associated to the weight μ_j.

Example 1.4 ($B\mathbb{Z}_2$) This is a sub-example of the previous Example 1.3; we include it to emphasize the importance of working over the equivariant Novikov ring. Suppose that $G = \mathbb{C}^\times$ acts on $X = \mathbb{C}$ with weight two. Then the quantum K-theory of $X /\!\!/ G = B\mathbb{Z}_2$ has generators $X^{\pm 1}$ with single relation

$$QK(B\mathbb{Z}_2) = \frac{\mathbb{Z}[X^{\pm 1}, q]}{\langle (1 - X^{-1})^2 - q \rangle}.$$

On the other hand, without the equivariant Novikov ring $K(B\mathbb{Z}_2; \mathbb{Z})$ is simply the group ring on $\mathbb{Z}_2$ via the identification of representations with their characters. Let

$$\delta_{\pm 1} \in K(B\mathbb{Z}_2; \mathbb{Z})$$

be the delta functions at the group elements $\pm 1 \in \mathbb{Z}_2$. Proposition 4.1 below shows that $(1 - X^{-1})$ maps to $\sqrt{q}\delta_{-1}$ under $D_0\kappa_X^G$. This implies the relation

$$(\sqrt{q}\delta_{-1})^2 - q = q\delta_{(-1)(-1)} - q = 0$$

since δ_1 is the identity. This matches with the relation

$$\delta_{-1}^2 = \delta_1$$

in $K(B\mathbb{Z}_2)$, the group algebra of $\mathbb{Z}_2$ since the product is given by convolution.

Since the relations are essentially the same as those in quantum cohomology, one obtains a generalization of the isomorphism between K-theory and cohomology of toric varieties induced by identifying cohomological and K-theoretic first Chern classes of divisors in Vezzosi-Vistoli [55, Section 6.2]: The quantum K-theory ring at bulk deformation $\kappa_X^G(0) \in QK^0(X /\!\!/ G)$ is canonically isomorphic to the quantum cohomology ring $QH(X /\!\!/ G)$

at bulk deformation $\kappa_X^{G,\,\mathrm{coh}}(0) \in QH(X /\!\!/ G)$ (where $\kappa_X^{G,\,\mathrm{coh}}: QH_G(X) \to QH(X /\!\!/ G)$ is the cohomological quantum Kirwan map) via a map defined on generators by

$$QK^0(X /\!\!/ G) \to QH(X /\!\!/ G), \quad D_\alpha \kappa_X^G(X_j - 1) \mapsto D_\alpha \kappa_X^{G,\,\mathrm{coh}}(c_1(X_j)).$$

In particular, the quantum K-theory of toric Deligne-Mumford stacks is generically semisimple.

We thank Ming Zhang for helpful comments and an anonymous referee for pointing out an important omission in the orbifold case.

2 Equivariant quantum K-theory

We recall the following basics of equivariant quantum K-theory, following Buch–Mihalcea [7] and Iritani–Milanov–Tonita [33]. Let X be a G-variety. The *equivariant K-homology group* $K_0^G(X)$ is the Grothendieck group of coherent G-sheaves on X, that is, the free Abelian group generated by isomorphism classes of G-equivariant coherent sheaves modulo relations whenever there exists an equivariant exact sequence: For G-equivariant sheaves $\mathcal{F}, \mathcal{F}', \mathcal{F}''$ we have an implication

$$0 \to \mathcal{F}' \to \mathcal{F} \to \mathcal{F}'' \to 0 \quad \implies \quad [\mathcal{F}] = [\mathcal{F}'] + [\mathcal{F}''].$$

The K-homology $K_G^0(X)$ is naturally a module over the *equivariant K-cohomology ring* $K_G^0(X)$ of G-equivariant vector bundles on X. Both the multiplicative structure of $K_G^0(X)$ and the module structure of $K_0^G(X)$ are given by tensor products. If X is non-singular then the map from $K_G^0(X)$ to $K_0^G(X)$ that sends a vector bundle to its sheaf of sections is an isomorphism.

Equivariant K-theory has the following functoriality properties. Given a G-equivariant morphism $f: X \to Y$ between varieties X, Y there is a ring homomorphism

$$f^*: K_G^0(Y) \to K_G^0(X), \quad [E] \mapsto [f^* E]$$

given by pull-back of bundles. If f is proper then there is a pushforward map

$$f_*: K_0^G(X) \to K_0^G(Y), \quad f_*[\mathcal{F}] \mapsto \sum_{i \geq 0} (-1)^i [R^i f_* \mathcal{F}].$$

This map is a homomorphism of $K_G(Y)$ modules by the projection formula.

The quantum product in K-theory is defined by incorporating contributions from moduli spaces of stable maps. Let X be a smooth projective G-variety. For integers $g, n \geq 0$ and a class $d \in H_2(X, \mathbb{Z})$ let

$$\overline{\mathcal{M}}_{g,n}(X,d) = \left\{ (u : C \to X, \underline{z} \in C^n) \,\middle|\, \begin{array}{c} \# \operatorname{Aut}(u, \underline{z}) < \infty, \\ g(C) = g, \ u_*[C] = d \end{array} \right\}$$

denote the moduli stack of stable maps to X with n markings, genus g, and homology class d. The evaluation maps are denoted

$$\operatorname{ev} = (\operatorname{ev}_1, \ldots, \operatorname{ev}_n) : \overline{\mathcal{M}}_{g,n}(X,d) \to X^n.$$

Recall that a *perfect obstruction theory $E^\bullet$ admitting a global resolution by vector bundles* on a stack $\mathcal{M}$ is a pair (E, ϕ) consisting of an object of the bounded derived category of coherent sheaves on $\mathcal{M}$ that can be presented as a two term complex

$$E = [E^{-1} \to E^0] \in D^{[-1,0]}(\operatorname{Coh}(\mathcal{M}))$$

of vector bundles, together with a morphism

$$\phi : E \to L_{\mathcal{M}}, \quad h^0(\phi) \text{ iso}, \quad h^{-1}(\phi) \text{ epi}$$

to the ($L^{\geq -1}$ truncation of the) cotangent complex of $L_{\mathcal{M}}$, satisfying that (c.f. [4], [21]) $h^0(\phi)$ is an isomorphism and $h^{-1}(\phi)$ is an epimorphism. The perfect obstruction theory defines the *virtual tangent bundle*

$$T_{\mathcal{M}}^{\operatorname{vir}} = \operatorname{Def} - \operatorname{Obs} \in K^0(\mathcal{M}), \quad \operatorname{Def} := [(E^0)^\vee], \quad \operatorname{Obs} := [(E^{-1})^\vee]$$

as in [10, p. 21]. That is, the virtual tangent bundle is the K-theoretic difference of the deformation space and the obstruction space. The virtual normal cone $C \hookrightarrow E_1 = (E^{-1})^\vee$ induces the *virtual structure sheaf* [41] as the derived tensor product

$$\mathcal{O}_{\mathcal{M}}^{\operatorname{vir}} := \mathcal{O}_{\mathcal{M}} \overset{L}{\underset{\mathcal{O}_{E_1}}{\bigotimes}} \mathcal{O}_C \in \operatorname{Ob}(D\operatorname{Coh}(\mathcal{M})),$$

whose class in $K(\mathcal{M})$ is

$$[\mathcal{O}_{\mathcal{M}}^{\operatorname{vir}}] = \sum_{i=0}^{\infty} (-1)^i \operatorname{Tor}_{\mathcal{O}_{E_1}}^i (\mathcal{O}_{\mathcal{M}}, \mathcal{O}_C) \in K(\mathcal{M}).$$

For any class $\alpha \in K(\mathcal{M})$ we define the *virtual Euler characteristic*

$$\chi^{\operatorname{vir}}(\mathcal{M}; \alpha) = \chi(\mathcal{M}; \alpha \otimes \mathcal{O}_{\mathcal{M}}^{\operatorname{vir}}) \in \mathbb{Z}$$

the Euler characteristic after twisting by $\mathcal{O}_{\mathcal{M}}^{\operatorname{vir}}$. These constructions admit equivariant generalizations, so that for any genus $g \geq 0$ and markings $n \geq 0$

with $\mathcal{M} = \overline{\mathcal{M}}_{g,n}(X,d)$ the *virtual structure sheaf* $\mathcal{O}_{\mathcal{M}}^{\text{vir}}$ is an object in the G-equivariant bounded derived category for $\overline{\mathcal{M}}_{g,n}(X,d)$ introduced by Y.P. Lee [41]. It defines a class $[\mathcal{O}^{\text{vir}}]$ in the equivariant K-theory of $\overline{\mathcal{M}}_{g,n}(X,d)$. For classes $\alpha_1, \ldots, \alpha_n \in K_G(X)$ and a class $\beta \in K(\overline{\mathcal{M}}_{g,n})$ define the equivariant K-theoretic Gromov–Witten invariants

$$\langle \alpha_1, \ldots, \alpha_n; \beta \rangle_{g,n,d} := \chi_G(\text{ev}_1^* \alpha_1 \otimes \cdots \otimes \text{ev}_n^* \alpha_n \otimes f^* \beta \otimes [\mathcal{O}^{\text{vir}}]), \quad (5)$$

where χ_G is the equivariant Euler characteristic. In fact, because X is smooth, one may replace the algebraic K-cohomology group above by the topological equivariant K-cohomology, that is, the Grothendieck group of equivariant complex vector bundles on X, as in [41, Section 4]. In this case the equivariant Euler characteristic is then replaced by a proper push-forward in topological K-theory. Define *descendant invariants* by

$$\langle \alpha_1 L^{d_1}, \ldots, \alpha_n L^{d_n}; \beta \rangle_{g,n,d} \in K_G(\text{pt})$$

defined by insertion d_i cotangent lines L at the i-th marking z_i.

The K-theoretic Gromov–Witten invariants can be organized into a potential as follows. Let $\Lambda_X \subset \text{Map}(H_2(X), \mathbb{Q})$ denote the *Novikov ring* associated to the ample line bundle $\mathcal{L} \to X$. The elements of the Novikov ring are formal combinations

$$\Lambda_X = \left\{ \sum_{d \in H_2(X)} c_d q^d \right\}$$

such that for any $E > 0$ the number of coefficients c_d with $(d, c_1(\mathcal{L})) < E$ is finite. Define the *equivariant quantum K-theory* as the completion (2). The K-theoretic *genus zero Gromov–Witten potential* with insertions is the formal function which we write informally

$$\mu_X : QK_G^0(X) \to QK_G^0(\text{pt}), \quad \alpha \mapsto \sum_{d \in H_2(X)} \sum_{n \geq 0} \langle \alpha, \ldots, \alpha; 1 \rangle_{0,n,d} \frac{q^d}{n!}; \quad (6)$$

what this means is that each Taylor coefficient of μ_X is well-defined. In the following expressions involving μ_X will be understood in this sense. For any element $\sigma \in QK_G^0(\text{pt})$ we denote by $\partial_\sigma \mu_X$ the differentiation of μ_X in the direction of σ. The *quantum K-theory pairing* at $\alpha \in QK_G^0(X)$ is for $\sigma, \gamma \in QK_G^0(X)$

$$B_\alpha(\sigma, \gamma) = \partial_1 \partial_\sigma \partial_\gamma \mu_X(\alpha) \in QK_G^0(\text{pt}) \quad (7)$$

where the identity in $QK_G^0(X)$ is the structure sheaf $\mathcal{O}_X$. This recovers the usual pairing $\chi(\sigma \otimes \gamma)$ when $\alpha = 0, q = 0$. Note that the corresponding pairing in quantum cohomology is the classical pairing; the existence of quantum corrections in the K-theoretic pairing is due to a modification in the contraction axioms [41, Section 3.7], this is an important new feature of quantum K-theory. The *quantum K-theory product* on $QK_G^0(X)$ with bulk deformation α is the formal product defined by

$$B_\alpha(\sigma \star_\alpha \gamma, \kappa) = \partial_\sigma \partial_\gamma \partial_\kappa \mu_X(\alpha).$$

As in the product in quantum cohomology, for each choice of $\alpha \in QK_G^0(X)$ we obtain a formal Frobenius algebra structure on $QK_G^0(X)$, by an argument of Givental [17]. Notably, the product does not satisfy a divisor axiom. However, quantum K-theory has better properties in other respects. For example, the small quantum K-theory (product at $\alpha = 0$) is defined over the integers, since the virtual Euler characteristics are virtual representations.

Later we will need a slight reformulation of the quantum-corrected inner product in (7). Let ev_{n+1}^d denote the restriction of the evaluation map

$$\mathrm{ev}_{n+1} \colon \overline{\mathcal{M}}_{0,n+1}(X) \to X$$

to $\overline{\mathcal{M}}_{0,n+1}(X, d)$. Define the (formal) *Maurer–Cartan map*

$$\mathcal{MC}_X^G \colon QK_G^0(X) \to QK_G^0(X)$$

$$\alpha \mapsto \sum_{n \geq 1, d \in H_2(X)} \mathrm{ev}_{n+2,*}^d \left(\mathrm{ev}_1^* \alpha \otimes \cdots \otimes \mathrm{ev}_n^* \alpha \otimes \mathrm{ev}_{n+1}^* 1 \right) \frac{q^d}{n!} \qquad (8)$$

where the push-forward is defined using the virtual structure sheaf. Then if $\underline{B}$ denotes the classical Mukai pairing we have

$$B_\alpha(\sigma, \gamma) = \underline{B}(D_\alpha \mathcal{MC}_X^G(\sigma), \gamma), \quad \underline{B}^{-1} B_\alpha = D_\alpha \mathcal{MC}_X^G,$$

where $D_\alpha \mathcal{MC}_X^G$ denotes the linearization of $\mathcal{MC}_X^G$ at α

$$\sigma \mapsto \sum_{n \geq 1, d \in H_2(X)} \mathrm{ev}_{n+2,*}^d \left(\mathrm{ev}_1^* \sigma \otimes \mathrm{ev}_2^* \alpha \otimes \cdots \otimes \mathrm{ev}_n^* \alpha \otimes \mathrm{ev}_{n+1}^* 1 \right) \frac{q^d}{n!}.$$

One can also consider twisted K-theoretic Gromov–Witten invariants as in Tonita [54]: Let

$$\overline{C}_{g,n}(X, d) \xrightarrow{\ e\ } X$$
$$\downarrow p$$
$$\overline{\mathcal{M}}_{g,n}(X, d).$$

denote the universal curve. If $E \rightarrow X$ is a G-equivariant vector bundle then the index class is defined by

$$\mathrm{Ind}(E) := [Rp_* e^* E] \in K^0(\overline{\mathcal{M}}_{g,n}(X,d)).$$

Its Euler class $[\mathrm{Eul}(\mathrm{Ind}(E))]$ is well-defined in $K^0_{\mathbb{C}^\times}(\overline{\mathcal{M}}_{g,n}(X,d))$ after localizing the equivariant parameter for the action of $\mathbb{C}^\times$ by scalar multiplication on the fibers of E at roots of unity. The genus $g = 0$ sum

$$\varphi^E_{n,d} \colon K^0_{G \times \mathbb{C}^\times}(X)^{\otimes n} \rightarrow K^{0,\mathrm{loc}}_{G \times \mathbb{C}^\times}(\mathrm{pt}),$$

$$(\alpha_1, \ldots, \alpha_n) \mapsto \chi_{G \times \mathbb{C}^\times}(\mathrm{ev}_1^* \alpha_1 \otimes \cdots \otimes \mathrm{ev}_n^* \alpha_n \otimes [\mathcal{O}^{\mathrm{vir}}] \otimes [\mathrm{Eul}(\mathrm{Ind}(E))])$$

$$(9)$$

produces the *E-twisted quantum K-theory*, again a Frobenius manifold.

An analog of the quantum connection in quantum K-theory was introduced by Givental [17]:With $m(\alpha)(\cdot) = \alpha \star \cdot$ quantum multiplication define a connection

$$\nabla^q_\alpha = (1 - z)\partial_\alpha + m(\alpha) \in \mathrm{End}(QK_{\mathbb{C}^\times}(X)).$$

By Givental [17] the quantum connection is flat. One of the goals of this paper is to give (somewhat non-explicit) formulas for its fundamental solutions.

3 Quantum K-theoretic Kirwan map

In this section we extend the definition of the quantum Kirwan map, defined in [57] to K-theory. Let G be a complex reductive group as in the introduction and let X be a smooth polarized projective G-variety with G-polarization, that is, ample G-line bundle, $\mathcal{L} \rightarrow X$. Suppose that G acts with finite stabilizers on the semistable locus, defined as the locus of points with non-vanishing invariant sections of some positive power of the polarization:

$$X^{\mathrm{ss}} = X^{\mathrm{ss}}(\mathcal{L}) = \{x \in X \mid \exists k > 0, s \in H^0(X, \mathcal{L}^{\otimes k})^G, s(x) \neq 0\} \subset X.$$

Equivalently, suppose that every orbit $Gx \subseteq X^{\mathrm{ss}}$ for $x \in X^{\mathrm{ss}}$ is closed. Denote the stack-theoretic quotient

$$X/\!\!/G := X^{\mathrm{ss}}/G$$

which is necessarily a smooth proper Deligne-Mumford stack with projective coarse moduli space. By definition X/G is the category whose objects

$$\mathrm{Ob}(X/G) = \{(P \rightarrow C, u \colon P \rightarrow X)\}$$

are pairs consisting of principal G-bundles P over some base C and equivariant maps $u: P \to X$, and whose morphisms are the natural commutative diagrams; X^{ss}/G is the sub-category of X/G whose objects (P, u) have the property that u takes values in the semi-stable locus X^{ss}.

3.1 Affine gauged maps

The quantum Kirwan map is defined by push-forward (in cohomology by integration) over moduli spaces of *affine gauged maps*.

Definition 3.1 (Affine gauged maps)   An affine gauged map to X/G is a datum $(C, \underline{z}, \lambda, u: C \to X/G)$ consisting of

- (Curve) a possibly-nodal projective curve $p: C \to S$ of arithmetic genus 0 over an algebraic space S;
- (Markings) sections $\underline{z} = (z_0, \dots, z_n: S \to C)$ disjoint from each other and the nodes;
- (One-form) a section $\lambda: C \to \mathbb{P}(\omega_{C/S} \oplus \mathbb{C})$ of the projective dualizing sheaf $\omega_{C/S}$;
- (Map) a map $u: C \to X/G$ to the quotient stack X/G;

satisfying the following conditions:

- (Scalings at markings) $\lambda(z_0) = \infty$ and $\lambda(z_i)$ is finite for $i = 1, \dots, n$;
- (Monotonicity) on any component $C_v \subset C$ on which $\lambda|C_v$ is non-constant, $\lambda|C_v$ has a single double pole, at the node $w \in C_v$ closest to z_0;
- (Map stability for infinity scaling) u takes values in the semistable locus $X /\!\!/ G$ on $\lambda^{-1}(\infty)$;
- (Bundle triviality for zero scaling) The bundle $u^*(X \to X/G)$ is trivializable on $\lambda^{-1}(0)$, or equivalently, u lifts to a map to X on $\lambda^{-1}(0)$.

The monotonicity assumption gives an affine structure near the double pole, thus the term *affine*. An affine gauged map given by a datum $(u: C \to X/G, z_0, \dots, z_n, \lambda)$ is *stable* if any component C_v of C on which u is trivializable has at least three special points, if $\lambda|C_v$ is zero or infinite, or two special points, if $\lambda|C_v$ is finite and non-zero. In the case that $X /\!\!/ G$ is only locally free, that is, has some finite but non-trivial stabilizers we also allow orbifold twistings at the nodes of C where $\lambda = \infty$, as in orbifold quantum cohomology. The *homology class* of an affine gauged map is the class $u_*[C] \in H_2^G(X, \mathbb{Q})$.

We introduce the following notation for moduli stacks. Let $\overline{\mathcal{M}}_n^G(\mathbb{A}, X)$ be the moduli stack of stable affine gauged maps to X and $\overline{\mathcal{M}}_n^G(\mathbb{A}, X, d)$ the locus of homology class d. Each $\overline{\mathcal{M}}_n^G(\mathbb{A}, X, d)$ is a proper Deligne-Mumford stack equipped with a perfect obstruction theory. The relative perfect obstruction theory on $\overline{\mathcal{M}}_n^G(\mathbb{A}, X)$ has complex dual to $Rp_*e^*T_{X/G}$, where p, e are maps from the universal curve $\overline{\mathcal{C}}_n^G(\mathbb{A}, X)$ as in the diagram

$$
\begin{array}{ccc}
\overline{\mathcal{C}}_n^G(\mathbb{A}, X) & \xrightarrow{\ e\ } & X/G \\
\ \downarrow p & & \\
\overline{\mathcal{M}}_n^G(\mathbb{A}, X). & &
\end{array}
$$

As in the construction of Y.P. Lee [41], the perfect obstruction theory determines a virtual structure sheaf $\mathcal{O}_{\mathcal{M}}^{\mathrm{vir}}$ in the bounded derived category of coherent sheaves on $\overline{\mathcal{M}}_n^G(\mathbb{A}, X)$. It defines a class $[\mathcal{O}_{\mathcal{M}}^{\mathrm{vir}}]$ in the rational K-theory of $\overline{\mathcal{M}}_n^G(\mathbb{A}, X)$. Let $\overline{I}_{X/G}$ denote the rigidified inertia stack of $X /\!\!/ G$ and

$$
\mathrm{ev} = (\mathrm{ev}_0, \mathrm{ev}_1, \ldots, \mathrm{ev}_n) \colon \overline{\mathcal{M}}_n^G(\mathbb{A}, X) \to \overline{I}_{X/G} \times (X/G)^n
$$

denote the evaluation maps at $z_0, \ldots, z_n$. If X is smooth projective then the moduli stack $\overline{\mathcal{M}}_n^G(\mathbb{A}, X)$ is proper. Properness also holds in certain other situations, such as if X is a vector space, G is a torus, and the weights of G are contained in an open half-space in $\mathfrak{g}^\vee$. For details on the proof of properness we refer the reader to [26].

The moduli stack of affine gauged maps also admits a forgetful map to a stack of domain curves. Denote by $\overline{\mathcal{M}}_n(\mathbb{A})$ the moduli stack $\overline{\mathcal{M}}_n^G(\mathbb{A}, X)$ in the case that X and G are points, which we call the stack of *affine scaled curves*. There is a forgetful morphism

$$
f \colon \overline{\mathcal{M}}_n^G(\mathbb{A}, X) \to \overline{\mathcal{M}}_n(\mathbb{A})
$$

defined by forgetting the morphism to X/G and collapsing all components that become unstable.

Example 3.2   The moduli stack $\overline{\mathcal{M}}_2(\mathbb{A})$ of twice-marked affine scaled curves is isomorphic to the projective line via the map

$$
\overline{\mathcal{M}}_2(\mathbb{A}) \to \mathbb{P}^1, \quad (C, z_0, z_1, z_2, \lambda) \mapsto \int_{z_1}^{z_2} \lambda;
$$

more precisely, the map above identifies the locus of affine scaled curves with irreducible domain with $\mathbb{C}^\times$. The compactification adds the two distinguished divisors

$$
D_{\{1,2\}}, D_{\{1\},\{2\}} \subset \overline{\mathcal{M}}_2(\mathbb{A})
$$

corresponding to loci where the markings $z_1, z_2 \in C$ are on the same component $C_v \subseteq C, v \in \text{Vert}(\Gamma)$ with zero scaling $\lambda|C_v = 0$ resp. different components $C_{v_1}, C_{v_2} \subset C, v_1 \neq v_2 \in \text{Vert}(\Gamma)$.

3.2 Affine gauged invariants and quantum Kirwan map

K-theoretic affine gauged Gromov–Witten invariants are defined as virtual Euler characteristics over the moduli stack of affine gauged maps. We introduce an equivariant version of the Novikov ring. Denote the equivariant polarization $\mathcal{L} \to X$. We denote the equivariant Novikov ring

$$\Lambda_X^G = \left\{ f(q) = \sum_{d \in H_2^G(X, \mathbb{Q})} c_d q^d \,\middle|\, \forall E, \# \text{Supp}^E(f) < \infty \right\} \tag{10}$$

where

$$\text{Supp}^E(f) = \left\{ d \in H_2^G(X) \,\middle|\, c_d \neq 0, \langle d, c_1^G(\mathcal{L}) \rangle < E \right\}.$$

The ring Λ_X^G depends on the (class of the) polarisation $\mathcal{L}$, but we omit $\mathcal{L}$ from the notation when it is clear which polarisation we are using. Redefine

$$QK_G^0(X) = \lim_{n \leftarrow} K_G^0(X) \otimes \Lambda_X^G / I_X^G(c)^n,$$

$$QK^0(X /\!\!/ G) = \lim_{n \leftarrow} K^0(X /\!\!/ G) \otimes \Lambda_X^G / I_X^G(c)^n;$$

in other words, from now on we work over the Novikov ring Λ_X^G. Let ev_0^d denote the restriction of

$$\text{ev}_0 \colon \overline{\mathcal{M}}_n^G(\mathbb{A}, X) \to \overline{I}_{X/G}$$

to $\overline{\mathcal{M}}_n^G(\mathbb{A}, X, d)$. Recall the formal map $\mathcal{MC}_{X/G}$ from $QK(X /\!\!/ G)$ to $QK(X /\!\!/ G)$ from (8). The map $\mathcal{MC}_{X/G}$ is formally invertible near 0. Its linearization at 0 is the identity modulo higher order terms, involving positive powers of q, hence it has a formal inverse $\mathcal{MC}_{X/G}^{-1}$ with $\mathcal{MC}_{X/G}^{-1}(0) = 0$.

Definition 3.3 The *quantum Kirwan map* in quantum K-theory with insertions $\beta_n \in K(\overline{\mathcal{M}}_{0,n})$ is the formal map

$$\kappa_X^G \colon QK_G^0(X) \to QK^0(X /\!\!/ G),$$

$$\alpha \mapsto \mathcal{MC}_{X/G}^{-1} \sum_{d \in H_2(X/G, \mathbb{Q}), n \geq 0} \frac{q^d}{n!} \text{ev}_{0,*}^d (\text{ev}_1^* \alpha \otimes \cdots \otimes \cdots \text{ev}_n^* \alpha \otimes f^* \beta_n). \tag{11}$$

The *linearized quantum Kirwan map* is obtained from the linearization of κ_X^G and correction terms arising from the quantum corrections in the inner product:

$$D_\alpha \kappa_X^G : QK_G^0(X) \to QK^0(X/\!\!/G),$$

$$\sigma \mapsto (D_{\kappa_X^G(\alpha)} \mathcal{MC}_{X/G})^{-1} \sum_{d,n} \frac{q^d}{(n-1)!} \, \mathrm{ev}_{0,*}^d (\mathrm{ev}_1^* \sigma \otimes \mathrm{ev}_2^* \alpha \otimes \cdots \mathrm{ev}_n^* \alpha).$$

$$(12)$$

Theorem 3.4 *Each linearization of κ_X^G is a $\star$-homomorphism:*

$$D_\alpha \kappa_X^G (\sigma \star_\alpha \gamma) = D_\alpha \kappa_X^G (\sigma) \star_{\kappa_X^G(\alpha)} D_\alpha \kappa_X^G (\gamma)$$

for any $\alpha \in QK_G^0(X)$.

Proof The proof is a consequence of an equivalence of divisor classes. As in the proof of associativity of quantum K-theory by Givental [17], consider the forgetful map

$$f_2 : \overline{\mathcal{M}}_n^G(\mathbb{A}, X) \to \overline{\mathcal{M}}_2(\mathbb{A}) \cong \mathbb{P}^1$$

forgetting all but the first and second markings and scaling. The inverse image $f_2^{-1}(\infty)$ consists of configurations

$$(u : C \to X/G, \lambda, \underline{z}) \in \mathrm{Ob}(\overline{\mathcal{M}}_n^G(\mathbb{A}, X))$$

where the first two incoming markings $z_1, z_2 \in C$ are on different components of the domain C is a union of boundary divisors:

$$f_2^{-1}(\infty) = \bigcup_\Gamma \overline{\mathcal{M}}_\Gamma^G(\mathbb{A}, X)$$

where Γ ranges over combinatorial type of colored tree with r colored vertices $v_1, \ldots, v_r \in \mathrm{Vert}(\Gamma)$ and one non-colored vertex $v_0 \in \mathrm{Vert}(\Gamma)$, with the edge labelled 1 attached to the first vertex v_1 and the edge labelled 2 attached to the second colored vertex v_2. We digress briefly to recall that if

$$D = \cup_{i=1}^n D_i$$

is a divisor with normal crossing singularities on a variety Y then the class of the structure sheaf $\mathcal{O}_D$ is

$$[\mathcal{O}_D] = \sum_{I \subset \{1, \ldots n\}} (-1)^{|I|} [\mathcal{O}_{D_I}] \in K(Y), \quad D_I = \bigcap_{i \in I} D_i.$$

The corresponding property for virtual structure sheaves in $\overline{\mathcal{M}}_{g,n}(X)$ is proved by Y.P. Lee [41, Proposition 11]. The intersection of any two strata

$\overline{\mathcal{M}}^G_{\Gamma_1}(\mathbb{A}, X)$, $\overline{\mathcal{M}}^G_{\Gamma_2}(\mathbb{A}, X)$ of codimension one is a stratum $\overline{\mathcal{M}}^G_{\Gamma_3}(\mathbb{A}, X)$ of lower codimension; there is an exact sequence of sheaves whose i-th term is the union of structure sheaves of strata of codimension i. Thus the structure sheaf of $f_2^{-1}(\infty)$ is identified in K-theory with the alternating sum of structure sheaves

$$[\mathcal{O}_{f_2^{-1}(\infty)}] = \sum_{k_1, k_2 \geq 0} (-1)^{k_1 + k_2} \sum_{\Gamma \in \mathcal{T}_\infty(k_1, k_2)} [\mathcal{O}_{\overline{\mathcal{M}}^G_\Gamma(\mathbb{A}, X, d)}]$$

where $\mathcal{T}_\infty(k_1, k_2)$ is the set of combinatorial types of affine scaled gauged maps with k_1, k_2 rational curves connecting the components containing z_1, z_2, with finite and non-zero scaling, and the component with infinite scaling containing z_0. On the other hand, the structure sheaf of $f_2^{-1}(0)$ is the alternating sum of structure sheaves

$$[\mathcal{O}_{f_2^{-1}(0)}] = \sum_{k \geq 0} (-1)^k \sum_{\Gamma \in \mathcal{T}_0(k)} [\mathcal{O}_{\overline{\mathcal{M}}^G_\Gamma(\mathbb{A}, X, d)}]$$

where $\mathcal{T}_0(k)$ is the set of combinatorial types with z_1, z_2 on one component, z_0 on another, and these two components related by a chain of k rational curves. The contribution of $\mathcal{T}_0(k)$ to the push-forward over $f_2^{-1}(0)$ is

$$(D_\alpha \mathcal{MC}_{X/G} - I)^k \sum_{d \in H_2^G(X, \mathbb{Q}), n \geq 1} \frac{q^d}{(n-1)!} \, \mathrm{ev}^d_{0,*}(\mathrm{ev}^*_1 \sigma \otimes \mathrm{ev}^*_2 \alpha \otimes \cdots \mathrm{ev}^*_n \alpha).$$

The inverse of the linearization of the map $\mathcal{MC}_{X/G}$,

$$D_\alpha \mathcal{MC}^{-1}_{X/G} = (I + (D_\alpha \mathcal{MC}_{X/G} - I))^{-1} = \sum_{k \geq 0} (-1)^k (D_\alpha \mathcal{MC}_{X/G} - I)^k.$$

$$(13)$$

Putting everything together and using (12) gives

$$(D_{\kappa^G_X(\alpha)} \mathcal{MC}_{X/G})(D_\alpha \kappa^G_X)(\sigma \star_\alpha \gamma)$$
$$= (D_{\kappa^G_X(\alpha)} \mathcal{MC}_{X/G})(((D_\alpha \kappa^G_X)\sigma) \star_{\kappa^G_X(\alpha)} ((D_\alpha \kappa^G_X)\gamma)).$$

Since $D_{\kappa^G_X(\alpha)} \mathcal{MC}_{X/G}$ is invertible, this implies the result. □

Remark (Inductive definition) For classes

$$\aleph_0 \in QK^0(X /\!\!/ G), \quad \aleph_1, \ldots, \aleph_n \in QK^0_G(X)$$

denote by

$$m_{n,d}(\aleph_0, \aleph_1, \ldots, \aleph_n)$$
$$:= \chi(\overline{\mathcal{M}}^G_n(\mathbb{A}, X, d), \, \mathrm{ev}^*_0 \aleph_0 \otimes \mathrm{ev}^*_1 \aleph_1 \otimes \ldots \mathrm{ev}^*_n \aleph_n \otimes [\mathcal{O}^{\mathrm{vir}}]) \in \mathbb{Z} \quad (14)$$

the virtual Euler characteristic. For classes

$$\aleph_0 \in QK^0(X/\!\!/ G), \quad \alpha, \aleph_1, \ldots, \aleph_k \in QK_G^0(X)$$

define

$$m_{k,d}^\alpha(\aleph_0, \ldots, \aleph_k) = \sum_{n \geq 0} \frac{1}{n!} m_{k+n,d}(\aleph_0, \ldots, \aleph_k, \alpha, \ldots, \alpha)$$

and similarly define $\varphi_{n,d}^{\kappa_X^G(\alpha)}$ by summing over all possible numbers of insertions of $\kappa_X^G(\alpha)$. Expanding the definition of the inner product we have (in topological K-theory)

$$\underline{B}(D_\alpha \kappa_X^G(\sigma), \gamma) = \sum_{d_0, \ldots, d_r, \aleph_1, \ldots, \aleph_r} (-1)^r q^d m_{2, d_0}^\alpha(\sigma, \aleph_1^\vee)$$

$$\left(\prod_{i=1}^{r-1} \varphi_{2, d_i}^{\kappa_X^G(\alpha)}(\aleph_i, \aleph_{i+1}^\vee) \right) \varphi_{2, d_r}^{\kappa_X^G(\alpha)}(\aleph_r, \gamma) \qquad (15)$$

where the sum is over all sequences of non-negative classes $(d_0, \ldots, d_r)$ such that $\sum d_i = d$ and $d_i > 0$ for $i > 0$ and $\aleph_1, \ldots, \aleph_r$ range over a basis for $QK(X/\!\!/ G)$. Equivalently, $D_\alpha \kappa_X^G(\sigma)$ can be defined by the inductive formula

$$\underline{B}(D_\alpha \kappa_X^G(\sigma), \gamma) = \sum_{d > 0} q^d m_{2, d}^\alpha(\sigma, \gamma)$$

$$- \sum_{\aleph, e > 0} q^e \underline{B}(D_\alpha \kappa_X^G(\sigma), \aleph) \varphi_{2, e}^{\kappa_X^G(\alpha)}(\aleph^\vee, \gamma). \qquad (16)$$

Definition 3.5 The *canonical bulk deformation* of $QK(X/\!\!/ G)$ is the value $\kappa_X^G(0) \in QK(X/\!\!/ G)$ of κ_X^G at 0.

Note that κ_X^G depends on the choice of presentation of $X/\!\!/ G$ as a git quotient. We give the following criterion for the canonical bulk deformation to vanish.

Lemma 3.6 *Suppose that the coarse moduli spaces of $\overline{\mathcal{M}}_0^G(\mathbb{A}, X, d)$ and $\mathcal{M}_{0,n}(X/\!\!/ G, d)$ are smooth, rational and connected with virtual fundamental sheaf equal to the usual structure sheaf. Suppose further that $H_2^G(X) \cong H_2(X/\!\!/ G)$ with $\overline{\mathcal{M}}_0^G(\mathbb{A}, X, d)$ non-empty iff $\mathcal{M}_{0,n}(X/\!\!/ G, d)$ is non-empty. Then $\kappa_X^G(0) = 0$.*

Proof Under the conditions in the Lemma, a result of Buch–Mihalcea [7, Theorem 3.1] implies that the push-forward of the structure sheaf over any moduli space of maps in the Lemma under any evaluation map is the structure sheaf of the target:

$$\mathrm{ev}^d_{0,*}[\mathcal{O}_{\overline{\mathcal{M}}^G_n(\mathbb{A},X,d)}] = [\mathcal{O}_{X/G}], \quad \mathrm{ev}^d_{1,*}[\mathcal{O}_{\overline{\mathcal{M}}_{0,n}(X/G,d)}] = [\mathcal{O}_{X/G}],$$

$$\forall d \in H^G_2(X) \cong H_2(X/\!/G), n \ge 0. \quad (17)$$

Thus

$$\sum_d q^d \, \mathrm{ev}^d_{0,*}[\mathcal{O}_{\overline{\mathcal{M}}^G_n(\mathbb{A},X,d)}] = \left(\sum_d q^d\right)[\mathcal{O}_{X/G}],$$

$$\sum_{d,n} q^d \, \mathrm{ev}^d_{1,*}[\mathcal{O}_{\overline{\mathcal{M}}_{0,1}(X,d)}] = \left(\sum_d q^d\right)[\mathcal{O}_{X/G}]$$

where both sums are over d such that $\overline{\mathcal{M}}^G_n(\mathbb{A},X,d), \overline{\mathcal{M}}_{0,n}(X/\!/G,d)$ are non-empty. It follows that

$$\mathcal{MC}_{X/G}(0) = \left(\sum_d q^d\right)[\mathcal{O}_{X/G}], \quad \kappa^G_X(0) = 0$$

as claimed.   □

Under the conditions of the lemma, one obtains a map of *small* quantum K-rings given by the linearized quantum Kirwan map

$$D_0\kappa^G_X \colon QK^0_G(X) \to QK^0(X/\!/G).$$

4 Quantum K-theory of toric quotients

In this section we use the K-theoretic quantum Kirwan map to give a presentation of the quantum K-theory of any smooth proper toric Deligne-Mumford stack with projective coarse moduli space. The case of projective spaces was treated in Buch–Mihalcea [7] and Iritani–Milanov–Tonita [33], and general toric varieties were treated in Givental [18]. The toric stacks we consider are obtained as git quotients for actions of tori on vector spaces. Let G be a torus with Lie algebra $\mathfrak{g}$. Let X be a vector space with a representation of G such that the weights

$$\mu_i \in \mathfrak{g}^\vee_\mathbb{R}, i = 1, \ldots, k$$

of the action are contained in the interior of a half-space in $\mathfrak{g}^\vee_\mathbb{R}$. For generic polarization, the git quotient $X/\!/G$ is smooth. We assume that $X/\!/G$ is non-empty, and for simplicity that the generic stabilizer is trivial. We have

$$QK_G^0(X) \cong QK_G^0(\text{pt}) = R(G)$$

where $R(G)$ denotes the Grothendieck group of finite-dimensional representations of G. Denote by $X_k \subset X$ the representation given by the k-th weight space. For any class $d \in H_2^G(X) \cong \mathfrak{g}_{\mathbb{Z}}$, define elements of $QK_G^0(X)$ by

$$\zeta_+(d) = \prod_{\mu_j(d) \geq 0} (1 - X_j^{-1})^{\mu_j(d)}, \quad \zeta_-(d) = q^d \prod_{\mu_j(d) \leq 0} (1 - X_j^{-1})^{-\mu_j(d)}.$$

Define the d-th *K-theoretic Batyrev element*

$$\zeta(d) = \zeta_+(d) - \zeta_-(d) \in QK_G^0(\text{pt}) \cong QK_G^0(X). \tag{18}$$

Proposition 4.1 *The kernel of $D_0\kappa_X^G$ contains the elements $\zeta(d), d \in H_2^G(X)$.*

Proof An argument using divided difference operators is given later in Example 5.6; here we give a geometric proof. The target X itself defines an element of $K_G^0(X)$ via pull-back under $X \to \text{pt}$. The pull-back $[\text{ev}_j^* X]$ is a class in $K(\overline{\mathcal{M}}_1^G(\mathbb{A}, X))$ for $j = 1, \ldots, n$. Define sections

$$\sigma_{i,j} : \overline{\mathcal{M}}_n^G(\mathbb{A}, X) \to \text{ev}_j^* X, \quad i \geq 0 \tag{19}$$

by composing the map $\overline{\mathcal{M}}_1^G(\mathbb{A}, X) \to \text{ev}_1^* X$ taking the i-th derivative of the map $u : C \to X/G$ at the marking z_j with the forgetful morphism $\overline{\mathcal{M}}_n^G(\mathbb{A}, X) \to \overline{\mathcal{M}}_1^G(\mathbb{A}, X)$. More precisely, suppose that $u : C \to X/G$ is given by a bundle $P \to C$ and a section $v : C \to P \times_G X$. In a local trivialization near z_j the section is given by a map $v : \mathbb{C} \to X$. Furthermore, the scaling λ on C necessarily pulls back to a non-zero scaling on $\mathbb{C}$, since there are no components of C with zero scaling. (There are no holomorphic curves in X, hence all curves with zero scaling are constant, and there is only one marking, hence no constant components with three special points and zero scaling.) Choose a coordinate z so that $\lambda = dz$. Let $\sigma_{i,j}([u]) \in \text{ev}_1^* X_j$ denote the i-th derivative of v at z_j with respect to the coordinate z.

We apply these canonical sections to the following Euler class computation. Each factor X_j defines a corresponding class $[X_j]$ in $K_G^0(X)$; we often omit the square brackets to simplify notation. Define bundles $E_\pm \to \overline{\mathcal{M}}_1^G(\mathbb{A}, X)$

$$E_\pm := \bigoplus_{\pm\mu_j(d) \geq 0} \text{ev}_1^* X_j^{\oplus \mu_j(d)}.$$

The Euler class of $E_\pm$ is

$$\text{Eul}(E)_\pm = \bigotimes_{\pm\mu_j(d) \geq 0} (1 - \text{ev}_1^* X_j)^{\otimes \mu_j(d)} \in K(\overline{\mathcal{M}}_1^G(\mathbb{A}, X)).$$

Given a section σ_j of $\mathrm{ev}_1^* X_j$ transverse to the zero section, there is a canonical isomorphism of the structure sheaf $\mathcal{O}_{\sigma_j^{-1}(0)}$ of $\sigma_j^{-1}(0)$ with

$$[\mathrm{Eul}(\mathrm{ev}_1^* X_j^\vee)] = [1 - \mathrm{ev}_1^* X_j^\vee] \in K(\overline{\mathcal{M}}_1^G(\mathbb{A}, X, d')).$$

This isomorphism is defined by the exact sequence

$$0 \to \mathrm{ev}_1^* X_j^\vee \to \mathcal{O} \to \mathcal{O}_{\sigma_j^{-1}(0)} \to 0.$$

Extending this by direct sums, any section $\sigma : \overline{\mathcal{M}}_1^G(\mathbb{A}, X, d') \to E_\pm$ transverse to the zero section defines an equality

$$[\mathcal{O}_{\sigma^{-1}(0)}] = [\mathrm{Eul}(E)_\pm^\vee] \in K(\overline{\mathcal{M}}_1^G(\mathbb{A}, X), d').$$

In particular, let σ denote the section of E_+ given by the derivatives

$$\sigma_{i,j}, i = 1, \ldots, d_j := \min(\mu_j(d), \mu_j(d'))$$

defined in (19). We construct a diagram

$$
\begin{array}{ccc}
\sigma^{-1}(0) & \xrightarrow{\ \iota\ } & \overline{\mathcal{M}}_1^G(\mathbb{A}, X, d') \\
\downarrow{\scriptstyle \delta} & & \\
\overline{\mathcal{M}}_1^G(\mathbb{A}, X, d' - d) & &
\end{array}
$$

as follows. The map ι is the inclusion. To construct δ, note that $\sigma^{-1}(0) \subset \overline{\mathcal{M}}_1^G(\mathbb{A}, X)$ consists of maps u whose j-th component u_j vanishes to order d_j at the marking z_1. Therefore, for any $[u] \in \sigma^{-1}(0)$ define a map of degree $d' - d$ by dividing by the j-th component of $u : C \to X/G$ on the component of C containing z_1 by $(z - z_1)^{d_j}$ on the component containing z_1, to obtain a map denoted $(z - z_1)^{-d} u$. The other components of C all map to $X /\!\!/ G$, and the action of $(z - z_1)^{-d}$ on the other components does not change the isomorphism class of u. It follows that there is a canonical map

$$\delta : \sigma^{-1}(0) \to \overline{\mathcal{M}}_1^G(\mathbb{A}, X, d' - d), \quad [u] \mapsto [u/(z - z_1)^d]. \tag{20}$$

The normal bundle to δ has Euler class the product of factors $(1 - X_j^{-1})^{\min(\mu_j(-d), \mu_j(d'-d))}$ over j such that $\mu_j(-d) \geq 0$.

The remaining factors are explained by the difference in obstruction theories. We denote by $p^{d'}$ the restriction of the projection p to maps of homology class d'. To compute the difference in classes we note that δ lifts to an inclusion of universal curves and (if $e^{d'}, e^{d'-d}$ denote the universal evaluation maps)

$$\iota^*[Rp_*^{d'}e^*T_{X/G}] - \delta^*[Rp_*^{d'-d}e^*T_{X/G}] = \iota^*[Rp_*^{d'} \bigoplus_j (\mathcal{O}_{z_1}(\mu_j(d')))]$$

$$- \delta^*[Rp^{d'-d} \bigoplus_j \mathcal{O}_{z_1}(\mu_j(d'-d))]$$

$$= \iota^*[E_+] - \rho^*[E_-].$$

Hence for any class $\alpha_0 \in K(X /\!\!/ G)$ we obtain

$$\chi^{\mathrm{vir}}(\overline{\mathcal{M}}_1^G(\mathbb{A}, X, d'), \mathrm{ev}_0^* \alpha_0 \otimes \mathrm{ev}^* \zeta_+(d))$$

$$= \chi^{\mathrm{vir}}(\overline{\mathcal{M}}_1^G(\mathbb{A}, X, d'-d), \mathrm{ev}_0^* \alpha_0 \otimes \mathrm{ev}^* \zeta_-(d)). \qquad (21)$$

That is,

$$m_{1,d'}(\alpha_0, \zeta_+(d)) = m_{1,d'-d}(\alpha_0, \zeta_-(d)).$$

By definition of the quantum Kirwan map this implies

$$D_0 \kappa_X^G(\zeta_+(d)) = q^d D_0 \kappa_X^G(\zeta_-(d)). \qquad \square$$

We wish to show that the elements in the lemma above generate, in a suitable sense, the kernel of the K-theoretic quantum Kirwan map. Define the *quantum K-theoretic Stanley–Reisner ideal* $QKSR_X^G$ to be the ideal in $QK_G^0(X)$ spanned by the K-theoretic Batyrev elements $\zeta(d), d \in H_2(X, \mathbb{Z})$ of (18). In general there are additional elements in the kernel of the linearized quantum Kirwan map. In order to remove these one must pass to a formal completion.

Definition 4.2   a. In the case that G acts freely on the semistable locus in X, for $l = (l_1, \dots, l_k) \in \mathbb{Z}_{\geq 0}^k$ define a filtration

$$QK_G^0(X)^{\geq l} := \prod_{j=1}^k (1 - X_j^{-1})^{l_j} QK_G^0(X) \subset QK_G^0(X).$$

b. More generally suppose that G acts on the semistable locus in X with finite stabilizers $G_x, x \in X^{\mathrm{ss}}$ and let

$$\mathcal{F}(X) = \bigcup_{j=1}^k \mathcal{F}_j(X) \subset \mathbb{C}$$

denote the set of roots of unity

$$\mathcal{F}_j(X) := \{g^{\mu_j} = \exp(2\pi i \mu_j(\xi)) \in \mathbb{C}^\times \mid g = \exp(\xi) \in G_x, x \in X^{\mathrm{ss}}\}$$

representing the roots of unity for the action of $g \in G_x, x \in X^{ss}$ on X_j.
Define

$$QK_G^0(X)^{\geq l} := \prod_{j=1}^{k} \prod_{\zeta \in \mathcal{F}_j(X)} (1 - \zeta X_j^{-1})^{l_j}.$$

Let $\widehat{QK}_G(X)$ denote the completion with respect to this filtration

$$\widehat{QK}_G(X) = \varprojlim_l QK_G^0(X)/QK_G^0(X)^{\geq l}.$$

Lemma 4.3 *For any d, n the K-theoretic Gromov–Witten invariants in the definition of κ_X^G vanish for $\alpha_0, \ldots, \alpha_n \in QK_G^0(X)^{\geq l}$ for $l_1 + \cdots + l_k$ sufficiently large.*

Proof We apply Tonita's virtual Riemann–Roch theorem [53]: If $\mathcal{M} = \overline{\mathcal{M}}_n^G(\mathbb{A}, X, d)$ embeds in a smooth global quotient stack and $I_{\mathcal{M}}$ its inertia stack then for any vector bundle V

$$\chi^{\mathrm{vir}}(\mathcal{M}, V) = \int_{[I_{\mathcal{M}}]^{\mathrm{vir}}} m^{-1} \mathrm{Ch}(V \otimes \mathcal{O}_{I_{\mathcal{M}}}^{\mathrm{vir}}) \mathrm{Td}(T I_{\mathcal{M}}) \mathrm{Ch}(\mathrm{Eul}(v_{\mathcal{M}}^{\vee}))$$

where $v_{\mathcal{M}}$ is the normal bundle of $I_{\mathcal{M}} \to \mathcal{M}$ and $m \colon I_{\mathcal{M}} \to \mathbb{Z}$ is the order of stabilizer function. The pullback $\mathrm{ev}_j^*(1 - \zeta X_j^{-1})$ restricts on a component of the inertia stack corresponding to an element $h \in G$ to the K-theory class $1 - \chi_j(h)^{-1} X_j^{-1}$, where χ_j is the character of X_j. The Chern character of each $1 - \chi_j(h)^{-1} \zeta X_j^{-1}$ has degree at least two for $\zeta = \chi_j(h)$. In this case if the sum $l_j, j \in I(h)$ of weights for which $\chi_j(h)$ is trivial is larger than the virtual dimension for the component of $I\mathcal{M}$ corresponding to h, the virtual integral over $I\mathcal{M}$ (which embeds in a smooth global quotient stack, see [57, Part 3, Proposition 9.14] and [24, Proposition 3.1 part (d)]) vanishes. $\square$

Lemma 4.3 implies that the quantum Kirwan map admits a natural extension to the formal completion:

$$\widehat{\kappa_X^G} \colon \widehat{QK}_G(X) \to QK(X /\!\!/ G).$$

Theorem 4.4 *(Theorem 1.2 from the Introduction.) The completed quantum Stanley–Reisner ideal is the kernel of the linearized quantum Kirwan map $D_0 \kappa_X^G$: We have an exact sequence*

$$0 \to \widehat{QKSR}_X^G \to \widehat{QK_G^0(X)} \to QK(X /\!\!/ G) \to 0.$$

Proof In [23, Theorem 2.6] we proved a version of Kirwan surjectivity for the cohomological quantum Kirwan map. The arguments given there hold equally well in rational topological K-theory as in cohomology. We address first the surjectivity of the right arrow in the sequence. By [23, Proposition 2.9] for d such that $\mu_j(d) > 0$ for all j we have

$$D_0 \kappa_X^G \left(\prod_i^k (1 - X_j^{-1})^{s \lceil \mu_i(d) \rceil} \right) = q^d [\mathcal{O}_{I_{X/G}(\exp(d))}] + h.o.t. \quad (22)$$

where *h.o.t.* denotes terms higher order in q. Since divisor intersections $[D_I] = \cap_{i \in I} [D_i]$ generate the cohomology $H(I_{X/G})$ of any $I_{X/G}$, the classes of their structure sheaves $[\mathcal{O}_{D_i}]$ generate the rational K-theory $K(I_{X/G})$. It follows that $\widehat{D_0 \kappa_X^G}$ is surjective.

To show exactness of the sequence, it suffices to show the equality of dimensions

$$\dim(\widehat{QK}_G(X)/QKSR_X^G) = \dim(QK(X /\!\!/ G)).$$

We recall the argument in the case that the generic stabilizer is trivial. Let $T = (\mathbb{C}^\times)^k / G$ denote the residual torus acting on $X /\!\!/ G$. The moment polytope of $X /\!\!/ G$ may be written

$$\Delta_{X/G} = \{ \mu \in \mathfrak{t}^\vee \mid (\mu, \nu_j) \geq c_j, j = 1, \ldots, k \}$$

where ν_j are normal vectors to the facets of $\Delta_{X/G}$, determined by the image of the standard basis vectors in $\mathbb{R}^k$ in $\mathfrak{t}$ under the quotient map, and c_j are constants determined by the equivariant polarization on X.

The quantum cohomology may be identified with the Jacobian ring of a *Givental potential* defined on the dual torus. Let $\Lambda = \exp^{-1}(1) \subset \mathfrak{t}$ denote the coweight lattice, and $\Lambda^\vee \subset \mathfrak{t}^\vee$ the weight lattice. The dual torus and *Givental potential* are

$$T^\vee = \mathfrak{t}^\vee / \Lambda^\vee, \quad W \colon T^\vee \to \mathbb{C}[q, q^{-1}], \quad y \mapsto \sum_{j=1}^k q^{c_j} y^{\nu_j}.$$

The quotient of $QK_G(X)$ by the Batyrev ideal maps to the ring $\mathrm{Jac}(W)$ of functions on $\mathrm{Crit}(W)$ by $(1 - X_j^{-1}) \mapsto y^{\nu_j}$. Denote by $\widehat{\mathrm{Crit}(W)}$ the intersection of $\mathrm{Crit}(W)$ with a product U of formal disks around $q = 0, X_j \in \mathcal{F}_j(X)$:

$$\widehat{\mathrm{Crit}(W)} = \mathrm{Crit}(W) \cap U.$$

Under the identification of the Jacobian ring, $\widehat{\mathrm{Crit}(W)}$ is the scheme of critical points $y(q)$ such that each y^{ν_j} approaches an element of $\mathcal{F}_j(X)$ as $q \to 0$. This definition differs from that in [23] in that we allow in theory critical points that converge to non-trivial roots of unity $\mathcal{F}_j(X)$ in the limit $q \to 0$.

To see that this gives the same definition as in [23] we must show that there are no families of critical points that converge to non-zero values of y^{ν_j} as

$q \to 0$. Let $y(q)$ be such a family. Necessarily the q-valuation $\mathrm{val}_q(y(q))$ of $y(q)$ lies in some face of the moment polytope. Since the moment polytope is simplicial, the normal vectors v_j of facets containing $\mathrm{val}_q(y(q))$ cannot be linearly dependent. Taking such v_j as part of a basis for the Lie algebra of the torus one may write $W(y) = y_1 + \cdots + y_k + h.o.t$ and taking partials with respect to $y_1, \ldots, y_k$ shows that these variables must vanish. Thus $y(q)$ converges to zero as $q \to 0$.

The dimension of $\widehat{\mathrm{Crit}}(W)$ can be computed using the toric minimal model program [23, Lemma 4.15]: under the toric minimal model program each flip changes the dimension of $\widehat{\mathrm{Crit}}(W)$ in the same as way as the dimension of $QK(X /\!\!/ G)$.

Similarly for a Mori fibration one has a product formula representing $\dim(QK(X /\!\!/ G))$ as the product of dimension of the base and fiber, and similarly for the Jacobian ring. It follows that $\dim(\widehat{\mathrm{Crit}}(W))$ is equal to $\dim(\widehat{QK}_G(X)/QKSR_X^G) = \dim(QK(X /\!\!/ G))$. $\qquad\square$

Remark a. The presentation above specializes to Vezzosi-Vistoli presentation [55, Theorem 6.4] by setting $q = 1$ in the case of smooth projective toric varieties. See Borisov-Horja [6] for the case of smooth Deligne-Mumford stacks.

b. The presentation above restricts Buch–Mihalcea presentation [7] in the case of projective spaces. In the case of projective (or more generally weighted projective spaces realized as quotients of a vector space by a $\mathbb{C}^\times$ action) we have $m_{0,d}(\alpha_0) = 1$ for any $d > 0$. This implies $\kappa_X^G(0) = 0$, by the arguments in Buch–Mihalcea [7]: The moduli stack is non-singular, has rational singularities, the evaluation map

$$\mathrm{ev}_0(d) \colon \overline{\mathcal{M}}_0^G(\mathbb{A}, X, d) \to X^{\exp(d)} /\!\!/ G$$

is surjective and has irreducible and rational fibers. By [7, Theorem 3.1], the push-forward of the structure sheaf $\mathcal{O}_{\overline{\mathcal{M}}_0^G(\mathbb{A}, X, d)}$ is a multiple of the structure sheaf on $X^{\exp(d)} /\!\!/ G$. It follows from the inductive formula (16) for $\kappa_X^G(0)$ that $\kappa_X^G(0)$ is the structure sheaf on $I_{X/G}$.

c. The linearized quantum Kirwan map has no quantum corrections in the case of circle group actions on vector spaces with positive weights. To see this, note that for $d > 0$ the push-forward of $\mathrm{ev}_1^*(1 - X_j)$ is the pushforward of the structure sheaf $\mathcal{O}_{\sigma_j^{-1}(0)}$ to $X /\!\!/ G$. By the argument in the previous item, this push-forward is equal to $[\mathcal{O}_{X/G}]$. For similar reasons, for $d > 0$ we have

$$m_{2,d}(\mathrm{ev}_1^*(1 - X_j), [\mathcal{O}_{\mathrm{pt}}]) = 1.$$

The formula (16) then implies that $D_0 \kappa_X^G (1 - X_j)$ has no quantum corrections, hence neither does $D_0 \kappa_X^G (X_j)$. It follows that in the presentation (4) the class X_j may be taken to be the line bundle associated to the weight μ_j on the weighted projective space $X /\!\!/ G = \mathbb{P}(\mu_1, \ldots, \mu_k)$. It seems to us at the moment that even in the case of Fano toric stacks one might have $\kappa_X^G (0) \neq 0$ and so the bulk deformation above may be non-trivial.

5 K-theoretic gauged Gromov–Witten invariants

In this section we define gauged K-theoretic Gromov–Witten invariants by K-theoretic integration over moduli stacks of Mundet-semistable maps to the quotient stack, and prove an adiabatic limit Theorem 5.4 relating the invariants.

5.1 The K-theoretic gauged potential

In our terminology, a gauged Gromov–Witten invariant is an integral over gauged maps, by which we mean maps to the quotient stack. Let C be a smooth projective curve.

Definition 5.1 (Gauged maps) A gauged map from C to X/G consists of

- (Curve) a nodal projective curve $\hat{C} \to S$ over an algebraic space S;
- (Markings) sections $z_0, \ldots, z_n : S \to \hat{C}$ disjoint from each other and the nodes;
- (One-form) a stable map $\hat{C} \to C$ of homology class $[C]$;
- (Map) a map $\hat{C} \to X/G$ to the quotient stack X/G, corresponding to a bundle $P \to \hat{C}$ and section $u : \hat{C} \to P(X) := (P \times X)/G$ which we required to be pulled back from a map $C \to BG$.

A gauged map is *stable* it satisfies a slope condition introduced by Mundet [46] which combines the slope conditions in Hilbert–Mumford and Ramanathan for G-actions and principal G-bundles respectively. Given a gauged map

$$(u : \hat{C} \to C \times X/G, z_0, \ldots, z_n)$$

let

$$\sigma : C \to P/R$$

be a parabolic reduction of P to a parabolic subgroup $R \subset P$. Let

$$\lambda \in \mathfrak{l}(P)^\vee$$

be a central weight of the Levi subgroup $L(P)$ of R. By twisting the bundle and section by the one-parameter subgroup $z^\lambda, z \in \mathbb{C}$ we obtain a family of gauged maps

$$u_\lambda : \hat{C} \times \mathbb{C}^\times \to C \times X/G.$$

By Gromov compactness the limit $z \to 0$ gives rise to an *associated graded* gauged map

$$u_\infty : \hat{C}_\infty \to X/G$$

equipped with a canonical infinitesimal automorphism

$$\lambda_\infty : \hat{C}_\infty \to \mathrm{aut}(P_\infty)$$

where P_∞ is the G-bundle corresponding to u_∞. The automorphism naturally acts on the determinant line bundle $\det \mathrm{aut}(P_\infty)$ as well as on the line bundle induced by the linearization $u_\infty^*(P_\infty(\mathcal{L}) \to P(X))$. The action on the first line bundle is given by a *Ramanathan weight* while the second is the *Hilbert–Mumford weight*

$$\lambda_\infty \cdot \delta_\infty = i\mu_\lambda^R(u)\delta_\infty \quad \lambda_\infty \cdot \tilde{u}_\infty = i\mu_\lambda^{HM}(u)\tilde{u}_\infty$$

for points δ_∞ resp. $\tilde{u}_\infty$ in the fiber of $\det \mathrm{aut}(P_\infty)$ resp. $u_\infty^*(P_\infty(\mathcal{L}) \to P(X))$. Let $\rho > 0$ be a real number. The *Mundet weight* is combination of the Ramanathan and Hilbert–Mumford weights with *vortex parameter* $\rho \in \mathbb{R}_{>0}$

$$\mu^M(\sigma, \lambda) = \rho\mu^R(\sigma, \lambda) + \mu^{HM}(\sigma, \lambda). \tag{23}$$

Definition 5.2 A gauged map $u : \hat{C} \to C \times X/G$ is *Mundet semistable* if

a. $\mu^M(\sigma, \lambda) \leq 0$ for all pairs (σ, λ) and
b. each component C_j of $\hat{C}$ on which u is trivializable (as a bundle with section) has at least three special points $z_i \in C_j$.

The map u is *stable* if, in addition, there are only finitely many automorphisms in $\mathrm{Aut}(u)$.

Gauged K-theoretic Gromov–Witten invariants are defined as virtual Euler characteristics over moduli stacks of Mundet-semistable gauged maps. Denote by $\overline{\mathcal{M}}^G(C, X)$ the moduli stack of Mundet semistable gauged maps. Assume that the semistable locus is equal to the stable locus, in which case $\overline{\mathcal{M}}^G(C, X)$ is a Deligne-Mumford stack with a perfect obstruction theory, proper for fixed numerical invariants [57]. Restriction to the sections defines an evaluation map

$$\mathrm{ev} : \overline{\mathcal{M}}^G(C, X) \to (X/G)^n.$$

Forgetting the map and stabilizing defines a morphism

$$f : \overline{\mathcal{M}}^G(C, X) \to \overline{\mathcal{M}}_n(C), \quad (C, u) \mapsto C^{\text{st}}$$

where $\overline{\mathcal{M}}_n(C)$ is the moduli stack of stable maps to C of class $[C]$. For classes $\alpha_1, \ldots, \alpha_n \in K_G^0(X)$, $\beta_n \in \overline{\mathcal{M}}_n(C)$ and $d \in H_2^G(X)$ we denote by

$$\tau_{X,n,d}^G(C, \alpha_1, \ldots, \alpha_n; \beta)$$
$$:= \chi^{\text{vir}}(\overline{\mathcal{M}}_n^G(C, X, d), \text{ev}_0^* \alpha_0 \otimes \text{ev}_1^* \alpha_1 \otimes \cdots \text{ev}_n^* \alpha_n) \otimes f^* \beta_n) \in \mathbb{Z} \quad (24)$$

the virtual Euler characteristic. Define the *gauged K-theoretic Gromov–Witten potential* as the formal sum

$$\tau_X^G : QK_G^0(X) \times K(\overline{\mathcal{M}}_n(C)) \to \Lambda_X^G,$$

$$(\alpha, \sigma) \mapsto \sum_{\substack{n \geq 0, d \in H_2^G(X, \mathbb{Z})}} \frac{q^d}{n!} \tau_{X,n,d}^G(C, \alpha, \ldots, \alpha; \beta_n). \quad (25)$$

In the case of domain the projective line the gauged potential can be further localized as follows. Let $C = \mathbb{P}^1$ be equipped with the standard $\mathbb{C}^\times$ action with fixed points $0, \infty \in \mathbb{P}^1$. Denote by z the equivariant parameter corresponding to the $\mathbb{C}^\times$-action. As in [57, Section 9] let $\overline{\mathcal{M}}_{n+1}^G(\mathbb{C}_+, X, d)^{\mathbb{C}^\times}$ denote the stack of $n+1$-marked Mundet-semistable gauged maps $P \to \mathbb{P}^1, u : \mathbb{P}^1 \to P(X)$ with the following properties: the data (P, u) are fixed up to automorphism by the $\mathbb{C}^\times$ action, and with one marking at 0 and the remaining markings mapping to components attached to 0, and the pair (P, u) is trivializable in a neighborhood of $\infty \in \mathbb{P}^1$. The moduli space $\overline{\mathcal{M}}_{n+1}^G(\mathbb{C}_-, X, d)^{\mathbb{C}^\times}$ is defined similarly but replacing 0 with ∞ and vice-versa. It is an observation of Givental (in a more restrictive setting) that the $\mathbb{C}^\times$-fixed locus in $\overline{\mathcal{M}}^G(\mathbb{P}, X, d)$ for sufficiently small vortex parameter ρ is naturally a union of fiber products

$$\overline{\mathcal{M}}_{n_-+n_+}^G(\mathbb{P}, X)^{\mathbb{C}^\times}$$
$$= \bigcup_{\substack{n_-+n_+=n \\ d_-+d_+=d}} \overline{\mathcal{M}}_{n_-+1}^G(\mathbb{C}_-, X, d_-)^{\mathbb{C}^\times} \times_{\overline{I}_{X/G}} \overline{\mathcal{M}}_{n_++1}^G(\mathbb{C}_+, X, d_+)^{\mathbb{C}^\times} \quad (26)$$

Indeed the bundle $P \to \mathbb{P}^1$ is given via the clutching construction by a transition map corresponding to an element $d \in \mathfrak{g}$ and the map u on $\mathbb{C}_\pm$ is given by an orbit of the one-parameter subgroup $u(z) = \exp(zd)x$ generated by d, for some $x \in X$.

Integration over the factors in this fiber product define *localized gauged graph potentials* $\tau_{X,\pm}^G$ as follows. The stack $\mathcal{M}^G(\mathbb{C}_\pm, X, d)^{\mathbb{C}^\times}$ has a natural equivariant perfect obstruction theory, as a fixed point stack in $\mathcal{M}^G(\mathbb{P}^1, X, d)$.

The perfect obstruction theory for $\mathcal{M}^G(\mathbb{C}_\pm, X, d)$ on the fixed locus splits in the *fixed* and *moving* parts. A perfect obstruction theory for $\mathcal{M}^G(\mathbb{C}_\pm, X, d)^{\mathbb{C}^\times}$ can be taken to be the fixed part. Let $N_\pm$ denote the virtual normal complex of $\mathcal{M}^G(\mathbb{C}_\pm, X, d)^{\mathbb{C}^\times}$ in $\mathcal{M}^G(\mathbb{P}, X, d)$. Define

$$\tau^G_{X,\pm} : QK^0_G(X) \to QK(X /\!\!/ G)[z^{\pm 1}, z^{\mp 1}]],$$

$$\alpha \mapsto \sum_{n \geq 0, d \in H_2^G(X, \mathbb{Q})} \frac{q^d}{n!} \operatorname{ev}^d_{\infty, *} \frac{\operatorname{ev}^* \alpha^{\otimes n}}{\operatorname{Eul}(N_\pm^\vee)}.$$

Example 5.3 The gauged graph potentials for toric quotients are q-hypergeometric functions described in Givental–Lee [19]. Let G be a torus acting on a vector space X is a vector space with weights $\mu_1, \dots, \mu_k$ and weight spaces $X_1, \dots, X_k$ with free quotient $X /\!\!/ G$. For any given class $\phi \in H_2^G(X, \mathbb{Z}) \cong \mathfrak{g}_\mathbb{Z}$, we have (omitting the classical map $K^G(X) \to K(X /\!\!/ G)$ from the notation)

$$\tau^G_{X,-}(0) = \sum_{d \in H_2^G(X)} q^d \frac{\prod_{j=1}^k \prod_{m=-\infty}^0 (1 - z^m X_j^{-1})}{\prod_{j=1}^k \prod_{m=-\infty}^{\mu_j(d)} (1 - z^m X_j^{-1})}. \tag{27}$$

Note that the terms with $X /\!\!/_d G = \emptyset$ contribute zero in the above sum since in this case the factor in the numerator $\prod_{\mu_j(d) < 0}(1 - X_j^{-1})$ vanishes.

Arbitrary values of the gauged potential can be computed as follows, using a result of Y.P. Lee [42] on Euler characteristics on the moduli spaces of genus zero marked curves. Since there are no non-constant holomorphic spheres in X, the evaluation maps $\operatorname{ev}_1, \dots, \operatorname{ev}_n$ are equal on $\overline{\mathcal{M}}^G_n(\mathbb{C}_\pm, X)^{\mathbb{C}^\times}$. Let

$$L_i \to \overline{\mathcal{M}}^G_n(\mathbb{C}_\pm, X), \quad (L_i)_{u: C \to X/G, \lambda, \underline{z}} = T^\vee_{z_i} C$$

denote the cotangent line at the i-th marked point. We compute the push-pull as follows: On the component of $\overline{\mathcal{M}}^G_n(\mathbb{C}_\pm, X)^{\mathbb{C}^\times}$ corresponding to maps of degree d the pushforward is given by

$$\operatorname{ev}^d_{\infty, *} \frac{\operatorname{ev}^* \alpha^n}{\mp(1 - z^{\pm 1})(1 - z^{\pm 1} L_{n+1})}$$

$$= \frac{\Psi_d(\alpha)^{\otimes n}}{(1 - z^{\pm 1})^2} \chi \left(\overline{\mathcal{M}}_{0, n+1}, \sum_d (L_{n+1} z^{\pm 1})^d \right) \tag{28}$$

where

$$(\Psi_d \alpha)(g) = \alpha(g z^d).$$

The integral (28) can be computed using a result of Y.P. Lee [42, Equation (3)] on Euler characteristics over $\overline{\mathcal{M}}_{0,n+1}$ (note the shift by 1 in the variable n to relate to Lee's conventions):

$$\chi\left(\overline{\mathcal{M}}_{0,n+1}, \sum_d ((z^{\pm 1}L_{n+1})^{\otimes d})\right) = (1 - z^{\pm})^{n-1}.$$

This implies that for $\alpha \in K_G^0(X)$

$$\tau_{X,-}^G(\alpha) = \sum_{d \in H_2^G(X)} q^d \exp\left(\frac{\Psi_d(\alpha)}{1 - z^{-1}}\right) \frac{\prod_{j=1}^k \prod_{m=-\infty}^0 (1 - X_j^{-1}z^m)}{\prod_{j=1}^k \prod_{m=-\infty}^{\mu_j(d)} (1 - X_j^{-1}z^m)}. \quad (29)$$

This is a version of Givental's K-theoretic *I-function*, see Givental–Lee [19] and Taipale [50].

5.2 The adiabatic limit theorem

In the limit that the linearization tends to infinity, the gauged Gromov–Witten invariants are related to the Gromov–Witten invariants of the quotient in K-theory. Let $\mathbb{C}^\times$ act on $\mathcal{M}_X^G(\mathbb{A}, X)$ via the weight 1 action on $\mathbb{A}$. The quantum Kirwan map then has a natural $\mathbb{C}^\times$-equivariant extension

$$\kappa_X^G : K_G^0(X) \to K_{\mathbb{C}^\times}(X /\!\!/ G)$$

defined by push-forward using the $\mathbb{C}^\times$-equivariant virtual fundamental sheaf. The following is a K-theoretic version of a result of Gaio–Salamon [15]:

Theorem 5.4 (Adiabatic Limit Theorem) *Suppose that C is a smooth projective curve and X a polarized projective G-variety such that stable= semistable for gauged maps from C to X of sufficiently small ρ. Then*

$$\tau_{X/G} \circ \kappa_X^G = \lim_{\rho \to 0} \tau_X^G : QK_G^0(X) \to \Lambda_X^G \quad (30)$$

in the following sense: For a class $\beta \in K(\overline{\mathcal{M}}_{n,1}(C))$ let

$$\sum_{k=1}^l \beta_\infty^k \otimes \beta_1^k \otimes \cdots \beta_r^k, \quad \beta_0$$

be its pullbacks to

$$K\left(\overline{\mathcal{M}}_r(C) \times \prod_{j=1}^r \overline{\mathcal{M}}_{|I_j|}(\mathbb{C})\right), \quad resp. \ K(\overline{\mathcal{M}}_n(C))$$

respectively. Then

$$\sum_{I_1 \cup \ldots \cup I_r = \{1,\ldots,n\}} \sum_{k=1}^{l} \tau^r_{X/G}(\alpha, \beta^k_\infty) \circ \kappa^{G, |I_j|}_X(\alpha, \beta^k_j) = \lim_{\rho \to 0} \tau^{G,n}_X(\alpha, \beta_0).$$

Similarly for the localized graph potentials (without insertions of classes on the source moduli spaces)

$$\tau_{X/G,\pm} \circ \kappa^G_X = \tau^G_{X,\pm} : QK^0_G(X) \to QK_{\mathbb{C}^\times}(X /\!\!/ G).$$

In other words, the diagram

(31)

commutes in the limit $\rho \to 0$.

Sketch of Proof   The proof is similar to that in the cohomology case in [57]: the proof only used an equivalence of divisor classes in the moduli stacks of scaled gauged maps. Let $\overline{\mathcal{M}}_{n,1}(\mathbb{P}^1)$ denote the moduli space *scaled* maps to $\mathbb{P}^1$, that is, the space of maps $\phi: C \to \mathbb{P}^1$ of class $[\mathbb{P}^1]$ equipped with sections λ of the projectivized relative dualizing sheaf from [57]; this means that some component C_0 mapped isomorphically onto $\mathbb{P}^1$ while the remaining components maps to points; either $\lambda|C_0$ is finite, in which case the remaining components $C_v \subset C$ have $\lambda|C_v = 0$, or $\lambda|C_0$ is infinite in which case there are a collection of bubble trees $C_1, \ldots, C_k \subset C$ attached to C_0 of the form described in 3.1. In particular, if there are no markings $n = 0$ then there are no bubble components and there exists an isomorphism

$$\overline{\mathcal{M}}_{0,1}(\mathbb{P}^1) \cong \mathbb{P}^1$$

corresponding to the choice of section of the projectivized trivial sheaf. Denote by

$$f_{0,0}: \overline{\mathcal{M}}_{0,n}(\mathbb{P}^1) \to \overline{\mathcal{M}}_{0,0}(\mathbb{P}^1) \cong \mathbb{P}^1 \tag{32}$$

the forgetful morphism forgetting the markings $z_1, \ldots, z_n$ but remembering the scaling λ. We have the relation

$$[\mathcal{O}_{\cup_{\mathcal{P}} D_{\mathcal{P}}}] = [\mathcal{O}_{f^{-1}_{0,0}(0)}] \in K(\overline{\mathcal{M}}_n(\mathbb{P}^1))$$

where

$$D_{\mathcal{P}} \cong \overline{\mathcal{M}}_r(\mathbb{P}^1) \times \prod_{i=1}^{r} \overline{\mathcal{M}}_{i_j}(\mathbb{A})$$

is a divisor corresponding to the unordered partition

$$\mathcal{P} = \{\mathcal{P}_1, \dots, \mathcal{P}_r\}, \quad |\mathcal{P}_j| = I_j$$

of the markings $\mathcal{P}$ in groups of size $i_1, \dots, i_r$. The class $\cup_{\mathcal{P}} D_{\mathcal{P}}$ is locally the union of prime divisors in a toric variety and the standard resolution gives the equality in K-theory

$$[\mathcal{O}_{\cup_{\mathcal{P}} D_{\mathcal{P}}}] = \sum_{\mathcal{P}_1, \dots, \mathcal{P}_k} (-1)^{k-1} [\mathcal{O}_{D_{\mathcal{P}_1} \cap \dots \cap D_{\mathcal{P}_k}}].$$

Two divisors $D_{\mathcal{P}_1}, \dots, D_{\mathcal{P}_k}$ intersect if and only the partitions $\mathcal{P}_1, \dots, \mathcal{P}_k$ have a common refinement (take a curve $(C, \underline{z})$ in the intersection and the partition determined by the equivalence class given by two markings are equivalent if they lie on the same irreducible component) and any type $\mathcal{M}_\Gamma(\mathbb{P}^1)$ in the intersection corresponds to some common refinement $\mathcal{P}$. Thus in the case of a non-empty intersection $D_{\mathcal{P}_1} \cap \dots \cap D_{\mathcal{P}_k}$ we may assume that each $\mathcal{P}$ refines each $\mathcal{P}_j$.

Define a moduli space of maps with scaling as follows. If $(C, \lambda, \underline{z})$ is a scaled curve and $u: C \to X/G$ a morphism we say that the data $(C, \lambda, \underline{z})$ is stable if either $\lambda|C_0$ is finite and is Mundet semistable or $\lambda|C_0$ is infinite and each bubble tree is stable in the sense of 3.1. By [26], the moduli stack of scaled gauged maps $\overline{\mathcal{M}}_{n,1}^G(\mathbb{P}, X)$ is proper with a perfect obstruction theory. Virtual Euler characteristics over $\overline{\mathcal{M}}_{n,1}^G(\mathbb{P}, X)$ define invariants

$$K_G^0(X) \otimes K(\overline{\mathcal{M}}_{n,1}(\mathbb{P}^1)) \to \mathbb{Z}$$

with the property that $\alpha \otimes [\mathcal{O}_{\mathrm{pt}}]$ maps to $\tau_G^X(\alpha)$. We have a natural forgetful morphism

$$f: \overline{\mathcal{M}}_{n,1}^G(\mathbb{P}, X) \to \overline{\mathcal{M}}_{0,1}(\mathbb{P}^1) \cong \mathbb{P}$$

and the inverse image of ∞ is the union of divisor classes corresponding to partitions according to which markings lie on which bubble tree.

The contributions of intersections of divisors correspond to the terms in the Taylor expansion of the inverse of the Maurer–Cartan map, using the *tree inversion formula* of Bass–Connell–Wright [3, Theorem 4.1]. Consider the Taylor expansion

$$\mathcal{MC}(\alpha) = \mathrm{Id} + \sum_{k_1,\ldots,k_r \geq 0} \frac{\mathcal{MC}_{(i_1,\ldots,i_r;)}\alpha_1^{i_1}\ldots\alpha_r^{i_r}}{i_1!\ldots i_r!},$$

$$\mathcal{MC}_{(i_1,\ldots,i_r)} \in QK(X/\!\!/G). \tag{33}$$

The tree formula for the formal inverse $\mathcal{MC}^{-1}$ of $\mathcal{MC}$ reads

$$\mathcal{MC}^{-1} = \sum_{\Gamma} |\mathrm{Aut}(\Gamma)|^{-1} \prod_{v \in \mathrm{Vert}(\Gamma)} (-\mathcal{MC}_v)$$

where the sum over $\{1, \ldots, \dim(QH(X/\!\!/G))\}$-labelled trees Γ,

$$\mathcal{MC}_v(\sigma_1,\ldots,\sigma_{k(v)}) = \sum_{i_1,\ldots,i_{k(v)}} \mathcal{MC}_{i_1(v),\ldots,i_r(v)}(\sigma_1,\ldots,\sigma_r)$$

is the $|v|$-th Taylor coefficient in $\mathcal{MC} - \mathrm{Id}$ for the labels incoming the vertex v considered as a symmetric polynomial in the entries; and composition is taken on the tensor algebra using the tree structure on Γ. For example, for the tree corresponding to the bracketing (12)3 the contribution of $\mathcal{MC}_2$ to $\mathcal{MC}^{-1}$ is $(-\mathcal{MC}_2)(-\mathcal{MC}_2 \otimes \mathrm{Id})$. See Wright [58] for the extension of [3] to power series, and also Kapranov–Ginzburg [16, Theorem 3.3.9]. The argument for the localized gauged potential is similar, by taking fixed point components for the $\mathbb{C}^\times$-action on $\overline{\mathcal{M}}^G(\mathbb{P},X)$. $\qquad\qquad\square$

5.3 Divided difference operators

A result of Givental and Tonita [20] shows the existence of a difference module structure on quantum K-theory for arbitrary target which gives rise to relations in the quantum K-theory. We follow the treatment in [33, Section 2.5, esp. 2.10-2.11]. Let

$$\lambda_1,\ldots,\lambda_k \in K(X)$$

be classes of nef line bundles corresponding to a basis of $H^2(X,\mathbb{Z})$ and $m(\lambda_i)$ the corresponding endomorphisms of $QK(X)$ given by multiplication. Define endomorphisms

$$\mathcal{E}_i = \tau(m(\lambda_i)^{-1} z^{q_i \partial_{q_i}} \tau^{-1}) \in \mathrm{End}(QK_{\mathbb{C}^\times}(X))$$

where τ is a fundamental solution to $\nabla_\alpha^q \tau = 0$. Define

$$\mathcal{E}_{i,\,\mathrm{com}} = \mathcal{E}_i|_{z=1} \in \mathrm{End}(QK(X)).$$

Then any difference operator annihilating the J-function defines a relation in the quantum K-theory: Following [33, Remark 2.11] define

$$\tilde{\tau}^G_{X,\pm} = \left(\prod_{i=1}^r \lambda_i^{-\ln(q_i)/\ln(z)} \right) \tau^G_{X,\pm}.$$

The results in [33, Section 2] give relations corresponding to operators annihilating the fundamental solution. Working with topological K-theory we define coordinates $t_1, \ldots, t_r$ by

$$\aleph = t_1 \aleph_1 + \cdots t_r \aleph_r, \quad \aleph_1, \ldots, \aleph_r \text{ a basis of } QK(X /\!\!/ G).$$

Theorem 5.5 *For any divided-difference operator*

$$\square \in \mathbb{Q}[z, z^{-1}][[q, t]]\langle z^{q_i \partial_{q_i}}, (1 - z^{-1})\partial_\alpha, \alpha \in QH(X /\!\!/ G)\rangle$$

(where angle brackets denote the sub-ring of differential operators generated by these symbols) we have

$$\square(z, q, t, z^{q_i \partial_{q_i}}, (1 - z^{-1})\partial_\alpha)\tilde{\tau}^G_{X,-} = 0 \implies \square(z, q, t, \mathcal{E}_i, \nabla^z_\alpha)1 = 0$$

and setting $z = 1$ gives rise to the relation involving quantum multiplication $m(\alpha)$ by α

$$\square(1, q, t, \mathcal{E}_{i, \text{com}}, m(\alpha)) = 0$$

in the quantum K-theory $QK(X /\!\!/ G)$.

Example 5.6 (Toric varieties) In the toric case, we recover some of the relations on quantum K-theory from Theorem 1.2 as follows. The operators

$$\mathcal{D}_{i,k} := 1 - X_i(z^{q_i \partial_{q_i}} + z^{-k}(1 - z^{-1})\partial_i)$$

satisfy

$$\left(\prod_{\mu_i(d) \geq 0} \prod_{k=0}^{\mu_i(d)-1} \mathcal{D}_{i,k} - q^d \prod_{\mu_i(d) < 0} \prod_{i=1}^{-\mu_i(d)} \mathcal{D}_{i,k} \right) \hat{\tau}^G_{X,-} = 0. \tag{34}$$

Using the expression (29) and $\Psi_d(X_i) = X_i z^{\mu_i(d)}$ we have

$$(1 - z)\partial_i \tau^G_{X,-}$$

$$= \sum_{d \in H_2^G(X)} q^d X_i z^{\mu_i(d)} \exp\left(\frac{\Psi_d(\alpha)}{1 - z^{-1}} \right) \frac{\prod_{j=1}^k \prod_{m=-\infty}^0 (1 - X_j^{-1} z^m)}{\prod_{j=1}^k \prod_{m=-\infty}^{\mu_j(d)} (1 - X_j^{-1} z^m)}.$$

$$\tag{35}$$

Hence

$$
\prod_{\mu_i(d)\geq 0} \prod_{k=0}^{\mu_i(d)-1} \mathcal{D}_{i,k} \tau_{X,-}^G(\alpha,q,z)
$$

$$
= \sum_{d\in H_2^G(X)} q^d \prod_{k=0}^{\mu_i(d)-1} (1 - X_i z^{\mu_i(d)-k}) \exp\left(\frac{\Psi_d(\alpha)}{1-z^{-1}}\right)
$$

$$
\frac{\prod_{j=1}^k \prod_{m=-\infty}^0 (1 - X_j^{-1} z^m)}{\prod_{j=1}^k \prod_{m=-\infty}^{\mu_j(d)} (1 - X_j^{-1} z^m)}. \tag{36}
$$

This equals

$$
q^d \prod_{\mu_i(d)<0} \prod_{k=1}^{-\mu_i(d)} \mathcal{D}_{i,k} \tau_{X,-}^G
$$

hence the relation (34). By 5.5 one obtains quantum Stanley–Reisner relations as in (3).

6 Wall-crossing in K-theory

In this section we recall results on the dependence of the K-theoretic invariants of the quotient on the choice of polarization due to Kalkman [36], in the case of cohomology, and Metzler [45], in the case of K-theory.

6.1 The master space and its fixed point loci

The geometric invariant theory quotients for two different polarizations may be written as quotients by a circle group action on a *master space*. Let $\mathcal{L}_\pm \to X$ two G-polarizations of X. Define

$$
\mathcal{L}_t := \mathcal{L}_-^{(1-t)/2} \otimes \mathcal{L}_+^{(1+t)/2}, \quad t\in(-1,1)\cap\mathbb{Q} \tag{37}
$$

the one-parameter family of rational polarizations given by interpolation.

Definition 6.1 The wall-crossing datum $(\mathcal{L}_-,\mathcal{L}_+)$ is *simple* if the following conditions are satisfied:

a. The only *singular value* is $t=0$, this is,

$$
X^{ss}(\pm) := X^{ss}(\mathcal{L}_t), \quad 0 < \pm t \leq 1
$$

is constant. We assume that stable equals semi-stable for $t \neq 0$, that is, there are no positive-dimensional stabilizer subgroups of points in the semi-stable locus.

b. The *strictly semistable set* $X^{\mathrm{ss}}(\mathcal{L}_0) \setminus (X^{\mathrm{ss}}(+) \cup X^{\mathrm{ss}}(-))$ is connected.

c. The only infinite stabiliser subgroup is a circle group:

$$G_x \cong \mathbb{C}^\times, \quad \forall x \in X^{\mathrm{ss}}(\mathcal{L}_0) \setminus (X^{\mathrm{ss}}(+) \cup X^{\mathrm{ss}}(-)), \quad \dim(G_x) > 0.$$

Given a simple wall-crossing, choose a one-parameter subgroup $\lambda : \mathbb{C}^\times \to G$, a connected component Z_λ of the semi-stable fixed points $X^{\mathrm{ss}}(\mathcal{L}_0)^\lambda$.

Definition 6.2 The *master space* introduced Dolgachev–Hu [13] and Thaddeus [51] is defined as follows. The projectivization $\mathbb{P}(\mathcal{L}_- \oplus \mathcal{L}_+) \to X$ of the direct sum $\mathcal{L}_- \oplus \mathcal{L}_+ \to X$ is itself a G-variety, and has a natural polarization given by the relative hyperplane bundle $\mathcal{O}_{\mathbb{P}(\mathcal{L}_- \oplus \mathcal{L}_+)}(1)$ of the fibers, with stalks

$$\mathcal{O}_{\mathbb{P}(\mathcal{L}_- \oplus \mathcal{L}_+)}(1)_{[l_-, l_+]} = \mathrm{span}(l_- + l_+)^\vee. \tag{38}$$

Let $\pi : \mathbb{P}(\mathcal{L}_- \oplus \mathcal{L}_+) \to X$ denote the projection. The group $\mathbb{C}^\times$ acts on $\mathbb{P}(\mathcal{L}_- \oplus \mathcal{L}_+)$ by rotating the fiber $w[l_-, l_+] = [l_-, wl_+]$. The space of sections of $\mathcal{O}_{\mathbb{P}(\mathcal{L}_- \oplus \mathcal{L}_+)}(k)$ has a decomposition under the natural $\mathbb{C}^\times$-action with eigenspaces given by the sections of

$$\pi^* \mathcal{L}_-^{k_-} \otimes \pi^* \mathcal{L}_+^{k_+}, \quad k_- + k_+ = k, k_\pm \geq 0.$$

The G-semistable locus in $\mathbb{P}(\mathcal{L}_- \oplus \mathcal{L}_+)$ is the union of loci of invariant eigensections and so the union of $\pi^{-1}(X^{\mathrm{ss},t})$ where $X^{\mathrm{ss},t} \subset X$ is the semistable locus for

$$\mathcal{L}_t = \mathcal{L}_-^{(1-t)/2} \otimes \mathcal{L}_+^{(1+t)/2}, t \in [-1, 1].$$

Let

$$\tilde{X} := \mathbb{P}(\mathcal{L}_- \oplus \mathcal{L}_+) /\!\!/ G$$

denote the geometric invariant theory quotient, by which we mean the stack-theoretic quotient of the semistable locus. The assumption on the action of the stabilizers implies that the action of G on the semistable locus in $\mathbb{P}(\mathcal{L}_- \oplus \mathcal{L}_+)$ is locally free, so that stable=semistable for $\mathbb{P}(\mathcal{L}_- \oplus \mathcal{L}_+)$. It follows that $\tilde{X}$ is a proper smooth Deligne-Mumford stack. The quotient $\tilde{X}$ contains the quotients of $\mathbb{P}(\mathcal{L}_\pm) \cong X$ with respect to the polarizations $\mathcal{L}_\pm$, that is, $X /\!\!/_\pm G$. Moreover the quotient $\tilde{X}^{\mathrm{ss}}(t)/\mathbb{C}^\times$ with respect to the $\mathbb{C}^\times$-linearisation $\mathcal{O}(t)$ is isomorphic to $X^{\mathrm{ss}}(\mathcal{L}_t)/G$. This ends the definition.

Next we recall from [24] the fixed point sets for the circle action on the master space.

Lemma 6.3 *The fixed point set $\tilde{X}^{\mathbb{C}^\times}$ is the union of the quotients $X /\!\!/_{\pm} G$ and the locus in $X^{\mathbb{C}^\times,\,ss}$ in $X^{\mathbb{C}^\times}$ that is semistable for some $t \in (-1, 1)$. The normal bundle of $X^{\mathbb{C}^\times,\,ss}$ in $\tilde{X}$ is isomorphic to the normal bundle in X, while the normal bundles of $X /\!\!/_{\pm} G$ are isomorphic to $\mathcal{L}_{\pm}^{\pm 1}$.*

Proof Any fixed point is a pair $[l_-, l_+]$ with a positive dimensional stabiliser under the action of $\mathbb{C}^\times$. Necessarily the points with $l_- = 0, l_+ = 0$ are fixed and they correspond to the quotients $X /\!\!/_{\pm} G$. However there are other fixed points when l_-, l_+ are both non-zero when the projection to X is fixed by a one-parameter subgroup, as we now explain. For any $\zeta \in \mathfrak{g}$, we denote by $G_\zeta \subset G$ the stabiliser of the line $\mathbb{C}\zeta$ under the adjoint action of G. If $[l] \in \tilde{X}^{\mathbb{C}^\times}$, with $l \in \mathbb{P}(\mathcal{L}_- \oplus \mathcal{L}_+)$ then $[l] \in \tilde{X}^\xi$, where ξ is a generator of the Lie algebra of $\mathbb{C}^\times$ and $\tilde{X}^\xi$ is the zero set of the vector field $\xi_{\tilde{X}}$ generated by ξ. Since $\tilde{X}$ is the quotient of $\mathbb{P}(\mathcal{L}_- \oplus \mathcal{L}_+)$ by G, if ξ_L denotes the vector field on $\mathbb{P}(\mathcal{L}_- \oplus \mathcal{L}_+)$ generated by ξ then $\xi_L(l) = \zeta_L(l)$ for some $\zeta \in \mathfrak{g}$. Since G acts locally freely ζ must be unique. Integrating gives $z \cdot l = z^\zeta l$ for all $z \in \mathbb{C}^\times$. By uniqueness of ζ, this holds for every point in the component of $\tilde{X}^\xi$ containing $[l]$. Thus for any fixed point $\tilde{x} \in \tilde{X}^{\mathbb{C}^\times}$ with $\tilde{x} = [l]$ for some $l \in \mathbb{P}(\mathcal{L}_- \oplus \mathcal{L}_+)$, there exists $\zeta \in \mathfrak{g}$ such that

$$\forall z \in \mathbb{C}^\times, \quad z \cdot l = z^\zeta l,$$

where $z \mapsto z^\zeta$ is the one-parameter subgroup generated by ζ. By the definition of semistability, the argument above and our wall-crossing assumption, any fixed point $\tilde{x} \in \tilde{X}^{\mathbb{C}^\times}$ is in the fibre over $x \in X$ that is 0-semistable and has stabilizer given by the one-parameter subgroup generated by $\zeta \in \mathfrak{g}$, that is, the weight of the one-parameter subgroup generated by ζ on $(\mathcal{L}_t)_x$ vanishes:

$$z^\zeta l = l, \quad \forall l \in (\mathcal{L}_t)_x.$$

Denote by $X^\zeta /\!\!/_0 (G_\zeta / \mathbb{C}_\zeta^\times)$ the quotient $[X^{ss}(\mathcal{L}_0)^\zeta / (G_\zeta / \mathbb{C}_\zeta^\times)]$. Conversely, taking any lift gives a morphism

$$\iota_\zeta : X^\zeta /\!\!/_0 (G_\zeta / \mathbb{C}_\zeta^\times) \to \tilde{X}^{\mathbb{C}^\times}.$$

The argument above shows that any (l_-, l_+) with both non-zero is in the image of some ι_ζ. The normal bundle $\nu_{\tilde{X}^{\mathbb{C}^\times}}$ of $\tilde{X}^{\mathbb{C}^\times}$ in $\tilde{X}$ restricted to the image of ι_ζ is isomorphic to the quotient of the normal bundle of $G \times_{G_\zeta} \mathbb{P}(\mathcal{L}_- \oplus \mathcal{L}_+)^{\xi+\zeta}$ by G, which in turn is isomorphic via projection to the quotient of the normal bundle of $\mathbb{P}(\mathcal{L}_- \oplus \mathcal{L}_+)^{\xi+\zeta}$ by $\mathfrak{g}/\mathfrak{g}_\zeta$, and then by the induced action of the smaller group $G_\zeta / \mathbb{C}_\zeta^\times$. $G_\zeta / \mathbb{C}_\zeta^\times$ acts canonically on the normal bundle $\nu_{X^\zeta} / (\mathfrak{g}/\mathfrak{g}_\zeta)$ and induces a bundle $(\nu_{X^\zeta} / (\mathfrak{g}/\mathfrak{g}_\zeta)) /\!\!/ (G_\zeta / \mathbb{C}_\zeta^\times)$ over $X^\zeta /\!\!/_0 (G_\zeta / C_\zeta^\times)$. The pull-back of the normal bundle $\iota_\zeta^*(\nu_{\tilde{X}^{\mathbb{C}^\times}})$ is canonically isomorphic to the quotient of $\nu_{X^\zeta}/(\mathfrak{g}/\mathfrak{g}_\zeta)$ by $G_\zeta/\mathbb{C}_\zeta^\times$, by an isomorphism that intertwines the action of $\mathbb{C}_\zeta^\times$ on $(\nu_{X^\zeta}/(\mathfrak{g}/\mathfrak{g}_\zeta))/\!\!/(G_\zeta/\mathbb{C}_\zeta^\times)$ with the action of $\mathbb{C}^\times$ on $\nu_{\tilde{X}^{\mathbb{C}^\times}}$. The final claim is left to the reader. □

Definition 6.4 We introduce the following notation. Denote by $X^{\zeta,0} \subset X^\zeta$ the fixed point loci $X^{\mathrm{ss}}(\mathcal{L}_0)^\zeta$ that are 0-semistable. Denote by $\nu_{X^{\zeta,0}}$ the $\mathbb{C}_\zeta^\times$-equivariant normal bundle of $X^{\zeta,0}$ modulo $\mathfrak{g}/\mathfrak{g}_\zeta$, quotiented by $G_\zeta/\mathbb{C}_\zeta^\times$.

A natural collection of K-theory classes on the master space is obtained by pull back. The projection $\mathbb{P}(\mathcal{L}_- \oplus \mathcal{L}_+) \to X$ is G-equivariant and $\mathbb{C}^\times$-invariant by construction. Consider the canonical map

$$\delta \colon K_G^0(X) \to K_{\mathbb{C}^\times}(\tilde{X}). \tag{39}$$

obtained by composition of the natural pull-back

$$K_G^0(X) \to K_{G \times \mathbb{C}^\times}(\mathbb{P}(\mathcal{L}_- \oplus \mathcal{L}_+))$$

with the Kirwan map

$$K_{G \times \mathbb{C}^\times}(\mathbb{P}(\mathcal{L}_- \oplus \mathcal{L}_+)) \to K_{\mathbb{C}^\times}(\mathbb{P}(\mathcal{L}_- \oplus \mathcal{L}_+)/\!\!/G) = K_{\mathbb{C}^\times}(\tilde{X}).$$

The composition of δ with the Kirwan map

$$\tilde{\kappa}_{X,t}^{\mathbb{C}^\times} \colon K_{\mathbb{C}^\times}(\tilde{X}) \to K(\tilde{X}/\!\!/_t \mathbb{C}^\times) = K(X/\!\!/_t G)$$

agrees with the pull-back to the $\mathcal{L}_t$-semistable locus $X^{\mathrm{ss}}(\mathcal{L}_t)$. It follows that

$$\tilde{\kappa}_{X,t}^{\mathbb{C}^\times} \circ \delta = \kappa_{X,t}^G \colon K_G^0(X) \to K(X/\!\!/_t G),$$

is the Kirwan map for the geometric invariant theory quotient of X with respect to the polarisation $\mathcal{L}_t$. In particular, $\delta(\alpha) \in K_{\mathbb{C}^\times}(\tilde{X})$ restricts to $\kappa_{X,\pm}^G \alpha$ on the two distinguished fixed point components $X/\!\!/_\pm G \subset \tilde{X}^{\mathbb{C}^\times}$.

The restrictions of these classes to fixed point sets are described as follows. After passing to a finite cover, we may assume that G_ζ splits as the product $G_\zeta/\mathbb{C}_\zeta^\times \times \mathbb{C}_\zeta^\times$. For any $\alpha \in K_G^0(X)$, the pull-back of $\tilde{\kappa}_{X,0}^{\mathbb{C}^\times}|_{\tilde{X}^{\mathbb{C}^\times}}(\alpha)$ under ι_ζ is equal to the image of α under the restriction map $K_G^0(X) \to K_{\mathbb{C}_\zeta^\times}(X^\zeta/\!\!/_0(G_\zeta/\mathbb{C}_\zeta^\times))$. Indeed $TX|_{X^\zeta}$ is the quotient of $T\mathbb{P}(\mathcal{L}_- \oplus \mathcal{L}_+)|_{\pi^{-1}(X^\zeta)}$ by $\mathbb{C}^\times$. Hence the action of G_ζ on $T\mathbb{P}(\mathcal{L}_- \oplus \mathcal{L}_+)|_{\pi^{-1}(X^\zeta)}$ induces an action of $G_\zeta/\mathbb{C}^\times$ on $TX|_{X^\zeta}$. After identifying $\mathbb{C}^\times \cong \mathbb{C}_\zeta^\times$, we have that $\iota_\zeta \circ \tilde{\kappa}_{X,0}^{\mathbb{C}^\times}$ is the pullback to $X^\zeta/\!\!/_0(G_\zeta/\mathbb{C}_\zeta^\times)$. For $\alpha \in K_G^0(X)$ we will denote by

$$\alpha_0 \in K_G(X^\zeta/\!\!/_0(G_\zeta/\mathbb{C}_\zeta^\times))$$

its restriction.

6.2 The Atiyah–Segal localization formula

In this section we review the Atiyah–Segal localization formula in equivariant K-theory [2], which expresses Euler characteristics as a sum over fixed point loci. Recall the definition of the Euler class in K-theory. Suppose Y is a smooth variety. For a vector bundle $E \to Y$ and a formal variable q denote the graded exterior power by

$$\bigwedge_q E = \sum_{k=0}^{\infty} q^k \bigwedge^k E \in K(Y)[q]. \tag{40}$$

The K-theoretic Euler class of E is defined by

$$\mathrm{Eul}(E) = \bigwedge_{-1} E^\vee = 1 - E^\vee + \bigwedge^2 E^\vee - \cdots \in K(Y).$$

Suppose now that $T = \mathbb{C}^\times$ acts trivially on Y, and that $a = \mathbb{C}_{(1)}$ is the weight 1 one-dimensional representation. Thus the equivariant K-theory of Y is given by

$$K_T(Y) = K(Y) \otimes K_T(\mathrm{pt}) = K(Y) \otimes \mathbb{Z}[z, z^{-1}].$$

If E is a coherent T-equivariant sheaf, its decomposition into isotypical components will be denoted

$$E = \bigoplus_{i=1}^{k} z^{\mu_i} E_i \tag{41}$$

where $\mu_i \in \mathbb{Z}$ is the weight of the action on E_i. The K-theoretical equivariant Euler class of E is given by

$$\mathrm{Eul}_T(E) = \prod_i \mathrm{Eul}(z^{\mu_i} E_i) \in K_T(Y).$$

The localization formula involves an integral over fixed point components with insertion of the inverted Euler class. Suppose that $T = \mathbb{C}^\times$ acts on X (non-trivially) and with fixed point set X^T. The previous paragraph discussion applies to the T-equivariant normal bundle ν_{X^T}, that is, the Euler class

$$\mathrm{Eul}_T(\nu_{X^T}) \in K_T(X^T) = K(X^T)[z, z^{-1}]$$

has been formally inverted through localisation. Denote by $K_T^{\mathrm{loc}}(X)$, the equivariant K-theory ring localized at the ideal of $R(T)$ generated by the Euler classes of representations $\mathbb{C}_{\mu_i}$ with weights $\mu_i, i = 1, \ldots, k$ (or more alternatively, one could localize at the roots of unity in the equivariant parameter).[3]

---

[3] In the orbifold case, one must localize at the roots of unity that appear in the denominators in the Riemann–Roch formula as in Tonita [53].

The *K-theoretic localisation formula* of Atiyah–Segal [2], see also [12, Chapter 5]) states that in $K_T^{\mathrm{loc}}(X)$

$$[\mathcal{O}_X] = [\iota_* \left(\mathcal{O}_{X^T} \otimes \mathrm{Eul}_T(\nu_{X^T}^{\vee})^{-1}\right)] = \left[\sum_{F \to X^T} \mathcal{O}_F \otimes \mathrm{Eul}_T(\nu_F^{\vee})^{-1}\right] \quad (42)$$

for the inclusion $\iota : X^T \to X$. Here the sum runs over all fixed components F of X^T. In terms of Euler characteristics we have

$$\chi(X; \mathcal{F}) = \sum_{F \to X^T} \chi(F; \mathcal{F} \otimes \mathrm{Eul}_T(\nu_F^{\vee})^{-1}) \in R(T).$$

The Atiyah–Segal localization formula implies a wall-crossing formula for K-theoretic integrals under variation of geometric invariant theory quotient due to Metzler [45]. Metzler's formula uses the following notion of *residue*, related to a power series expansion in a localized ring. For any T-equivariant locally free sheaf $\mathcal{F}$ of rank r on a variety Y with trivial T-action the class

$$\mathcal{F}_0 = \mathcal{F} - r\mathcal{O}_Y \in K_T(Y)$$

is nilpotent (c.f. [12, Prop. 5.9.5]). By taking exterior powers with notation as in (40) we have

$$\bigwedge\nolimits_q \mathcal{F} = \bigwedge\nolimits_q (\mathcal{F}_0 + r\mathcal{O}_Y) = (1 + q)^r \bigwedge\nolimits_{\frac{q}{1+q}} \mathcal{F}_0 \in K_T(Y)[q].$$

Using this formula on the weight μ_i bundle E_i gives

$$\mathrm{Eul}_T(E_i) = (1 - z^{-\mu_i})^{r_i}(1 + N_i)$$

where $r_i = \mathrm{rank}\, E_i$, and $N_i \in K(Y) \otimes \mathbb{Z}[z, z^{-1}]$ is a combination of nilpotent elements in $K(Y)$ whose coefficients are monomials of the form $\frac{z^{-\mu_i}}{1 - z^{-\mu_i}}$. Thus

$$\mathrm{Eul}_T(E) = \prod_i (1 - z^{-\mu_i})^{r_i}(1 + N_i), \quad (43)$$

where

$$N_i = \sum N_{k,i} s_k(z) \in K(Y) \otimes \mathbb{Z}[[z, z^{-1}]]. \quad (44)$$

The equation (44) is a finite sum where $N_{k,i}$ is a nilpotent element in $K(Y)$ and $s_k(z)$ is a rational function in z that has no pole at $z = 0$ nor $z = \infty$. Thus $\mathrm{Eul}_T(E)^{-1}$ has only poles at roots of unity. In particular the Euler classes can formally be inverted, since the leading term is invertible.

Example 6.5 We give the following example of formal power series associated to inverted Euler classes. Let $L \to Y$ be a T-equivariant line bundle of weight $\mu = \pm 1$. Then

$$\mathrm{Eul}_T(L) = 1 - z^{\mp 1} L^{\vee} \in K_T(Y).$$

This expression can be expanded, using the nilpotent element $L_0 = L^{\vee} - 1$, as

$$1 - z^{\mp 1} L^{\vee} = 1 - z^{\mp 1}(1 + L_0) = (1 - z^{\mp 1}) \left(1 + \frac{z^{\mp 1} L_0}{1 - z^{\mp 1}} \right).$$

This ends the example.

The residue of a K-theoretic class is a difference between the residues of the characters at zero and infinity. For any class $\alpha \in K_T^{\mathrm{loc}}(Y)$ there exist unique expansions

$$\alpha = \sum_{n \geq 0} \alpha_{0,n} z^n \in K(Y)[z^{-1}, z]], \quad \alpha_{0,n} \in K(Y)$$

and

$$\alpha = \sum_{n \geq 0} \alpha_{\infty,n} z^{-n} \in K(Y)[[z^{-1}, z], \quad \alpha_{\infty,n} \in K(Y).$$

The *residue* is the map

$$\mathrm{Resid} \colon K_T^{\mathrm{loc}}(Y) \to K(Y); \quad \alpha \mapsto \alpha_{\infty,0} - \alpha_{0,0} \tag{45}$$

assigning the difference of the constant coefficients in the power series expansions above.

Lemma 6.6 *Let $\alpha \in K_T^{\mathrm{loc}}(Y)$ be a class in the image of the inclusion $K_T(Y) \to K_T^{\mathrm{loc}}(Y)$. Then $\mathrm{Resid}\, \alpha = 0$.*

Proof In this case the coefficients $\alpha_{0,i} = \alpha_{\infty,i}$ are the coefficients of z^i in α, hence in particular $\alpha_{0,0} = \alpha_{\infty,0}$. $\square$

Example 6.7 Let Y be a point and $E = \mathbb{C}_{(1)}$ the representation with weight one. The inverted Euler class of E is defined formally by the associated series

$$\mathrm{Eul}(\mathbb{C}_{(1)})^{-1} = (1 - z^{-1})^{-1} = 1 + z^{-1} + z^{-2} + \cdots$$

$$\mathrm{Eul}(\mathbb{C}_{(1)})^{-1} = (1 - z^{-1})^{-1} = \frac{-1}{z^{-1}(1 - z)} = -z - z^2 - z^3 + \cdots.$$

It follows that the residue of the inverted Euler class is

$$\mathrm{Resid}(\mathrm{Eul}(\mathbb{C}_{(1)})) = 1 - 0 = 1 \in K(\mathrm{pt}) \cong \mathbb{Z}.$$

Similarly

$$\text{Resid}(\text{Eul}(\mathbb{C}_{(-1)})) = 0 - 1 = -1 \in K(\text{pt}) \cong \mathbb{Z}.$$

More generally for any line bundle L of weight ± 1, by Example 6.5 we have

$$\text{Resid Eul}(L)^{-1} = \text{Resid}\left(\frac{1}{1 - L^{\vee}}\right) = \pm 1. \tag{46}$$

Example 6.8 Suppose that T acts trivially on Y and that $E \to Y$ is a bundle with isotypical decomposition as in (41). Let

$$S_+, S_- \subset \{1, \ldots, k\}$$

denote the index sets for positive and negative weights μ_i respectively. Using (43), we have

$$\text{Eul}_T(E)^{-1} = \left(\prod_{i \in S_+}(1 - z^{-\mu_i})^{r_i}(1 + E_{+,i})\prod_{i \in S_-}(1 - z^{-\mu_i})^{r_i}(1 + E_{-,i})\right)^{-1}$$

where both $E_{\pm} \in K_T^{\text{loc}}(X)$ are as in Equation (44). To compute the residue, we may ignore the N_i terms. In this case the lowest order terms in the power series expansions in both z and z^{-1} have degree given by $\prod_{i \in S_+} \mu_i r_i$ and $\prod_{i \in S_-}(-\mu_i) r_i$ respectively. If both $S_{\pm}$ are non-empty, then Resid $\text{Eul}_T(E)^{-1} = 0$.

In general one can make different choices of residues, and the choice depends on the problem at hand, see [45, Section 4] for other examples.

6.3 The Kalkman–Metzler formula

The formulas of Kalkman and Metzler describe the wall-crossing of K-theoretic integrals under variation of geometric invariant theory quotient. We follow the same notations as in Section 6.1. Consider family of polarisations $\mathcal{L}_t \to X$ with a simple wall crossing at the singular time $t = 0$ with an associated master space $\tilde{X}$. Let

$$X^{\zeta,0} = X^{\text{ss}}(\mathcal{L}_0) \cap X^{\mathbb{C}^{\times}}$$

be the fixed point component of the one-parameter subgroup generated by ζ involved in the wall-crossing.

Definition 6.9   For an element $\zeta \in \mathfrak{g}$, define the *fixed point contribution*

$$\tau_{X,\zeta,0} \colon K_G^0(X) \to \mathbb{Z},$$

$$\alpha \mapsto \chi\left(X^{\zeta,0} /\!\!/ (G_\zeta/\mathbb{C}_\zeta^\times);\ \mathrm{Resid}\left(\frac{\alpha_0}{\mathrm{Eul}_{\mathbb{C}_\zeta^\times}(\nu_{X\zeta,0})}\right)\right). \qquad (47)$$

Here α_0 is the image of α under the map $K_G^0(X) \to K_{\mathbb{C}_\zeta^\times}(X^{\zeta,0})$.

A version of the following formula was proved by Metzler [45, Theorem 1.1] in the differential geometric setting of manifolds with circle actions.

Theorem 6.10 (Kalkman–Metzler wall-crossing)   *Let X be a smooth G-variety (projective or affine) and suppose that $\mathcal{L}_\pm \to X$ are polarizations inducing a simple wall-crossing. Then*

$$\tau_{X/\!+G} \kappa_{X,+}^G - \tau_{X/\!-G} \kappa_{X,-}^G = \tau_{X,\zeta,0}. \qquad (48)$$

Example 6.11   Our first example concerns the passage from projective space to the empty set by variation of git quotient. Consider $G = \mathbb{C}^\times$ acting on $X = \mathbb{C}^{n+1}$ diagonally with weight one. A polarisation is given by a choice of another character $\ell \in \mathbb{Z}$. For $\ell < 0$ the quotient is empty and for $\ell > 0$, $X/\!\!/G = \mathbb{P}^n$. Let $\mathbb{C}_{(k)}$ be the weight $k \geq 0$ representation. The sheaf $\mathcal{F} = \mathcal{O}_X \otimes \mathbb{C}_{(k)}$ descends to $\mathcal{O}(k) \to \mathbb{P}^n$. The fixed point set X^G is the single point $0 \in X$ whose normal bundle is identified with $\mathbb{C}^{n+1}$ itself. The inverted Euler class is $(1 - z^{-1})^{-(n+1)}$ and can be identified with the class of the symmetric algebra $\mathrm{Sym}(z^{-1}\mathbb{C}^{n+1})$. The fixed point contribution of the wall crossing formula (48) equals

$$\mathcal{O}(k)|_0 \otimes \mathrm{Sym}(z^{-1}\mathbb{C}^{n+1}) = z^k \mathrm{Sym}(z^{-1}\mathbb{C}^{n+1}).$$

The residue, corresponding to the z^0 coefficient is the degree k part $\mathrm{Sym}^k(\mathbb{C}^{n+1})$. The wall crossing formula now reads

$$\chi(\mathbb{P}^n, \mathcal{O}(k)) = \chi(X^G, \mathrm{Sym}^k(\mathbb{C}^{n+1})) = \dim \mathrm{Sym}^k(\mathbb{C}^{n+1}) = \binom{n+k}{k}.$$

Similarly, if one considers $\mathcal{O}(-k)$ and the identity

$$\mathrm{Sym}\, V = (-1)^{n+1} \det V^\vee \otimes \mathrm{Sym}\, V^\vee,$$

we get that the wall crossing contribution is

$$\mathrm{Resid}\, z^{-k} z^{n+1} \mathrm{Sym}(\mathbb{C}^{n+1}).$$

This is the degree $k - n - 1$ part $\mathrm{Sym}^{k-n-1}(\mathbb{C}^{n+1})$, hence

$$\chi(\mathbb{P}^n, \mathcal{O}(-k)) = (-1)^n \chi(X^G, \mathrm{Sym}^{k-n-1}(\mathbb{C}^{n+1})) = (-1)^n \binom{k-1}{k-n-1}.$$

Example 6.12 Our second example concerns the Cremona transformation. Let $X = (\mathbb{P}^1)^3$ with polarization

$$\mathcal{L} = \mathcal{O}_{\mathbb{P}^1}(c_1) \boxtimes \mathcal{O}_{\mathbb{P}^1}(c_2) \boxtimes \mathcal{O}_{\mathbb{P}^1}(c_3), \quad c_1 < c_2 < c_3.$$

Consider $G = \mathbb{C}^\times$ acting on each factor $\mathbb{P}^1$ by $t[z_0, z_1] = [z_0, tz_1]$ and the action on $\mathcal{O}_{\mathbb{P}^1}(n)$ so that the weights at the fibres over the fixed points $[1, 0]$, $[0, 1]$ are $n/2$, $-n/2$. We consider the family of polarizations $\mathcal{L}_t = \mathcal{L} \otimes \mathbb{C}_t$ obtained by shifting $\mathcal{L}$ by a trivial line bundle weight weight t. The singular values t are given by the weights $(\pm c_1 \pm c_2 \pm c_3)/2$ of the action on the fibre $\mathcal{L}_t|_p$ at the fixed points. Hence, there are eight chambers for t on which the geometric invariant theory quotient $X /\!/_t G$ is constant. Two have empty git quotients. In the first and last chamber, we have $X /\!/_t G \cong \mathbb{P}(\mathbb{C}^3)$ resp. $\mathbb{P}((\mathbb{C}^3)^\vee)$, while the six intermediate wall-crossings induce three blow-ups and three blow-downs involved in the classic Cremona transformation of $\mathbb{P}^2$.

We consider how the Euler characteristic of the structure sheaf changes under wall-crossing. Consider the structure sheaf $\mathcal{O}_X$, whose restriction to any of the chambers is again the respective structure sheaf $\mathcal{O}_t$. We now describe the change in the Euler characteristic $\chi(X /\!/_t G, \mathcal{O}_t)$ as we vary t under the wall-crossing from the chamber $t < -c_1 - c_2 - c_3$ to the first non-empty chamber $t \in (c_1 - c_2 - c_3, -c_1 + c_2 - c_3)$, corresponding to the wall defined by the fixed point $x = ([1, 0], [1, 0], [1, 0]) \in X$. Since all weights of the action on the tangent bundle at this fixed point $x \in X$ are 1, we have that the Euler class at this fixed point is $(1 - z^{-1})^3$. The wall crossing formula yields

$$\chi(\mathbb{P}^2, \mathcal{O}) = 0 + \chi\left(x, \text{Resid}\,\frac{\mathcal{O}|_x}{(1 - z^{-1})^3}\right) = 1$$

as expected.

The next wall-crossing, corresponds to the blow-up $\mathrm{Bl}(\mathbb{P}^2)$ of $\mathbb{P}^2$ at a fixed point. The normal bundle at this point has weights $-1, 1, 1$ and thus the wall crossing formula yields

$$\chi(\mathrm{Bl}(\mathbb{P}^2), \mathcal{O}) = \chi(\mathbb{P}^2, \mathcal{O}) + \chi\left(x, \text{Resid}\,\frac{\mathcal{O}|_x}{(1 - z)(1 - z^{-1})^2}\right)$$

$$= \chi(\mathbb{P}^2, \mathcal{O}) + 0 = 1$$

since the residue is zero each time there are contributions with both positive and negative weights (since the numerator is trivial) which is consistent, since the Euler characteristic $\chi(X, \mathcal{O}_X)$ is a birational invariant. In fact the invariance of the Euler characteristic $\chi(X /\!/ G, \mathcal{O}_{X/G})$ under git birational transformation follows from Example 6.8.

Proof of Theorem 6.10   Consider the master space $\tilde{X}$ associated to the wall-crossing. Let $\alpha \in K_G^0(X)$, and consider its image $\delta(\alpha) \in K_{\mathbb{C}^\times}(\tilde{X})$ of (39). Using K-theoretic localization (42) on the $\mathbb{C}^\times$-space $\tilde{X}$, and the identification of the fixed point components and normal bundles with those in the ambient space we obtain the relation

$$\delta(\alpha) = \iota_* \frac{\kappa_{X,-}^G(\alpha)}{\mathrm{Eul}_{\mathbb{C}^\times}(\nu_-)} + \iota_* \frac{\kappa_{X,+}^G(\alpha)}{\mathrm{Eul}_{\mathbb{C}^\times}(\nu_+)} + \iota_* \frac{\alpha_0}{\mathrm{Eul}_{\mathbb{C}^\times_\zeta}(\nu_{X^{\zeta,0}})},$$

where $\iota : F \to \tilde{X}^{\mathbb{C}^\times}$ is the inclusion of (each) fixed point components. By applying residues and Euler characteristics

$$\chi\left(\tilde{X}; \mathrm{Resid}\,\delta(\alpha)\right) = \chi\left(X/\!\!/_-G; \mathrm{Resid}\,\frac{\kappa_{X,-}^G(\alpha)}{\mathrm{Eul}_{\mathbb{C}^\times}(\nu_-)}\right)$$

$$+ \chi\left(X/\!\!/_+G; \mathrm{Resid}\,\frac{\kappa_{X,+}^G(\alpha)}{\mathrm{Eul}_{\mathbb{C}^\times}(\nu_+)}\right) + \chi\left(X^{\zeta,0}; \mathrm{Resid}\,\frac{\alpha_0}{\mathrm{Eul}_{\mathbb{C}^\times}(\nu_{X^{\zeta,0}})}\right).$$
$$(49)$$

The left hand side of (49) is zero by the definition of residue (c.f. Remark 6.6) . The group $\mathbb{C}^\times$ acts on $\nu_\pm$ with weights ∓ 1 on the normal bundles $\nu_\pm \to X/\!\!/_\pm G$ of $X/\!\!/_\pm G$ in $\tilde{X}$ respectively, since they are canonically identified with $(\mathcal{L}_+ \otimes \mathcal{L}_-)^{\pm 1}$. Hence one obtains

$$0 = -\tau_{X/\!/_+G}\kappa_{X,+}^G\alpha + \tau_{X/\!/_-G}\kappa_{X,-}^G\alpha + \chi\left(X^{\zeta,0}; \; \mathrm{Resid}\,\frac{\alpha_0}{\mathrm{Eul}_{\mathbb{C}^\times_\zeta}(\nu_{X^{\zeta,0}})}\right)$$

as claimed.   □

We remark that Kalkman–Metzler wall-crossing formula Theorem 6.10 can be generalised to more complicated wall-crossing, such as when there are multiple singular times and fixed components. The details are left to the reader.

6.4 The virtual wall-crossing formula

A virtual version of the Kalkman–Metzler formula follows from a virtual version of localization in K-theory proved by Halpern–Leistner [28, Theorem 5.7]. Let $\mathcal{X}$ be a Deligne-Mumford G-stack with coarse moduli space X. Let $\mathcal{L}_\pm \to \mathcal{X}$ be G-polarizations of $\mathcal{X}$. That is, $\mathcal{L}_\pm$ are G-line bundles which are ample on the coarse moduli space X. Let

$$\mathrm{Pic}_{\mathbb{Q}}(\mathcal{X}) = \mathrm{Pic}(\mathcal{X}) \otimes_{\mathbb{Z}} \mathbb{Q}$$

denote the rational Picard group and

$$\mathcal{L}_t := \mathcal{L}_-^{(1-t)/2} \otimes \mathcal{L}_+^{(1+t)/2} \in \mathrm{Pic}_{\mathbb{Q}}(\mathcal{X}), \quad t \in (-1, 1) \cap \mathbb{Q}. \tag{50}$$

The family (50) is a one-parameter family of rational polarizations given by interpolation. We will also assume, for simplicity that there is a simple wall-crossing:

Assumption 6.13 *There is only one singular value $t = 0$, such that the semistable loci*

$$\mathcal{X}^{\mathrm{ss}}(+) := \mathcal{X}^{\mathrm{ss}}(\mathcal{L}_t), \quad \mathcal{X}^{\mathrm{ss}}(-) := \mathcal{X}^{\mathrm{ss}}(\mathcal{L}_t)$$

are constant for $\pm 1 \leq t < 0$. We assume that stable equals semi-stable for $t \neq 0$. Moreover $\mathcal{X}^{\mathrm{ss}}(\mathcal{L}_0) \setminus (\mathcal{X}^{\mathrm{ss}}(+) \cup \mathcal{X}^{\mathrm{ss}}(-))$ is connected. Assume that the stabiliser G_x for $x \in \mathcal{X}^{\mathrm{ss}}(\mathcal{L}_0) \setminus (\mathcal{X}^{\mathrm{ss}}(+) \cup \mathcal{X}^{\mathrm{ss}}(-))$ is isomorphic to $\mathbb{C}^\times$.

Consider the case that the given stack has an equivariant perfect obstruction theory. The G-action on $\mathcal{X}$ induces a canonical morphism

$$a^\vee : L_{\mathcal{X}} \to \underline{\mathfrak{g}}^\vee \subset L_{\mathcal{X} \times G} = L_{\mathcal{X}} \times \underline{\mathfrak{g}}^\vee$$

that we call the *infinitesimal action*. Composing with the morphism $E \to L_{\mathcal{X}}$ gives a natural lift $\tilde{a}^\vee : E \to \underline{\mathfrak{g}}^\vee$. Let $\mathcal{X}^{\mathrm{ss}}$ be the semistable locus for some polarization and assume stable=semistable, that is all the stabilisers are finite. Let $E^{\mathrm{ss}}, L_{\mathcal{X}^{\mathrm{ss}}}$ etc. be the restrictions to the semistable locus.

Lemma 6.14 *The perfect obstruction theory $E^{\mathrm{ss}} \to L_{\mathcal{X}^{\mathrm{ss}}} := L_{\mathcal{X}}|_{\mathcal{X}^{\mathrm{ss}}}$ descends to a perfect obstruction theory on the quotient $\mathcal{X}^{\mathrm{ss}}/G$.*

Proof From the fibration $\pi : \mathcal{X}^{\mathrm{ss}} \to \mathcal{X}^{\mathrm{ss}}/G$ one obtains an exact triangle of cotangent complexes

$$L_{\mathcal{X}^{\mathrm{ss}}/G} \to L_{\mathcal{X}^{\mathrm{ss}}} \xrightarrow{a^\vee} \underline{\mathfrak{g}}^\vee \to L_{\mathcal{X}^{\mathrm{ss}}/G}[1], \tag{51}$$

thus we can consider $L_{\mathcal{X}^{\mathrm{ss}}} \to \underline{\mathfrak{g}}^\vee$ as the cotangent complex of $\mathcal{X}^{\mathrm{ss}}/G$. Let $\mathrm{Cone}(\tilde{a}^\vee)$ denote the mapping cone of $\tilde{a}^\vee$. Then the exact triangle

$$\mathrm{Cone}(\tilde{a}^\vee) \to E^{\mathrm{ss}} \xrightarrow{\tilde{a}^\vee} \underline{\mathfrak{g}}^\vee \to \mathrm{Cone}(\tilde{a}^\vee)[1]$$

admits a morphism to (51), in particular making $\mathrm{Cone}(\tilde{a}^\vee) \to L_{\mathcal{X}^{\mathrm{ss}}/G}$ into an obstruction theory with amplitude in $[-1, 1]$. By the assumption on the finite stabilizers, this obstruction theory is perfect. $\square$

The perfect obstruction theories on the stack induce perfect obstruction theories on the fixed point components. Assume again that $\mathcal{X}$ is equipped with a G action. For any $\zeta \in \mathfrak{g}$, consider the fixed point stacks $\mathcal{X}^\zeta$. The restriction of the perfect obstruction theory $E^\bullet|_{\mathcal{X}^\zeta}$ decomposes as

$$E^\bullet|_{\mathcal{X}^\zeta} = E^{\bullet,\mathrm{mov}} + E^{\bullet,\mathrm{fix}}$$

where $E^{\bullet,\mathrm{mov}}$ is the *moving part* and $E^{\bullet,\mathrm{fix}}$ the *fixed part*. By results of [21], $E^{\bullet,\mathrm{fix}}$ yields an equivariant perfect obstruction theory for $\mathcal{X}^\zeta$, which is compatible with that on $\mathcal{X}$. Denote by

$$a_\zeta^\vee : L_{\mathcal{X}/G_\zeta} \to (\mathfrak{g}_\zeta/\mathbb{C}\zeta)^\vee$$

the map given by the infinitesimal action of $G_\zeta/\mathbb{C}_\zeta^\times$ on $\mathcal{X}/G_\zeta$. Denote by

$$\nu_{\mathcal{X}^\zeta} \in \mathrm{Ob}(D^b \mathrm{Coh}(\mathcal{X}^\zeta/G_\zeta))$$

the *moving part* of the mapping cone $\mathrm{Cone}(a_\zeta)$; this is the conormal complex for the embedding $\mathcal{X}^\zeta/G_\zeta \to \mathcal{X}/G_\zeta$ except for the factor $\mathbb{C}\zeta$ of automorphisms of the fixed point set. Denote by $\mathcal{X}^{\zeta,0}$ the locus of $\mathcal{X}^\zeta$ which is $\mathcal{L}_0$-semistable and by $\nu_{\mathcal{X}^{\zeta,0}}$ the restriction of $\nu_{\mathcal{X}^\zeta}$ to $\mathcal{X}^{\zeta,0}$.

A virtual version of localization in K-theory for stacks is proved in Halpern–Leistner [28, Theorem 5.7]. A special case is the following: Assume that a torus T acts on $\mathcal{X}$ (e.g. the action of $\mathbb{C}^\times$ on a master space for wall-crossing) and let $\mathcal{X}^T$ denote the fixed locus. The virtual tangent space restricted to $\mathcal{X}^T$ decomposes into

$$\mathrm{Def} - \mathrm{Obs}|_{\mathcal{X}^T} = \mathrm{Def}^{\mathrm{fix}} - \mathrm{Obs}^{\mathrm{fix}} + \mathrm{Def}^{\mathrm{mov}} - \mathrm{Obs}^{\mathrm{mov}}$$

its fixed and moving parts consisting of T-modules with trivial and non-trivial weights. The K-class $\nu_{\mathcal{X}^T} = (\mathrm{Def}^{\mathrm{mov}} - \mathrm{Obs}^{\mathrm{mov}})$ is the virtual *normal bundle*. In the case $T = \mathbb{C}^\times$ this class splits into classes $\nu_{\mathcal{X}^T,\pm}$ corresponding to negative and positive weights and the inverted Euler class of $\nu_{\mathcal{X}^T}$ maybe defined as the tensor product

$$\mathrm{Eul}(\nu_{\mathcal{X}^T})^{-1} = \mathrm{Eul}(\nu_{\mathcal{X}^T,+})^{-1}\,\mathrm{Eul}(\nu_{\mathcal{X}^T,-})^{-1}$$

where as in [28, Footnote 24]

$$\mathrm{Eul}(\nu_{\mathcal{X}^T,-})^{-1} = \mathrm{Sym}(\nu_{\mathcal{X}^T,-})$$
$$\mathrm{Eul}(\nu_{\mathcal{X}^T,+})^{-1} = \mathrm{Sym}(\nu_{\mathcal{X}^T,+}^\vee) \otimes \det(\nu_{\mathcal{X}^T,+})[-\mathrm{rank}(\nu_{\mathcal{X}^T,+})]$$

is an infinite sum of bundles such that each weight component is finite; this suffices for the finiteness of the formulas below. The residue of such classes is defined as before in (45), by taking the difference in expansion as formal power series in z and z^{-1}.

Theorem 6.15 (Virtual localisation) *[28, 5.6,5.7] Suppose that $\mathcal{X}$ is the a proper global quotient of a quasiprojective scheme by a reductive group equipped with an equivariant $T = \mathbb{C}^\times$-action, and E is a T-equivariant sheaf on $\mathcal{X}$. Then the Euler characteristic of E is computed by*

$$\chi^{\mathrm{vir}}(\mathcal{X}, E) = \chi^{\mathrm{vir}}(\mathcal{X}^T, E \otimes \mathrm{Eul}_T(\nu_{\mathcal{X}_T})^{-1})$$
$$= \sum_{F \subset \mathcal{X}^T} \chi^{\mathrm{vir}}(F, E \otimes \mathrm{Eul}_T(\nu_F)^{-1})$$

for the inclusion $\iota : \mathcal{X}^T \to \mathcal{X}$. Here the sum runs over all fixed components F of $\mathcal{X}^T$ and the Euler class $\mathrm{Eul}_T(\nu_{\mathcal{X}_T})$ are defined as in 6.2, and it is invertible in localised equivariant K-theory.

Proof This is stated for quasi-smooth schemes Y in [28, Theorem 5.7]. The statement for quasi-smooth global quotient stacks $\mathcal{X} = Y/G$ by reductive groups G follows from the identification of the Euler characteristic $\chi(\mathcal{X}, E/G)$ on $\mathcal{X}$ with the invariant part of the Euler characteristic $\chi(Y, E)^G$ upstairs, where the Bialynicki–Birula decomposition satisfies the requirements of the stratification by [29, Section 1.4.1]. $\qquad\Box$

The formula (52) implies a formula for Euler characteristics as a sum over fixed point components, in which the fixed point contributions are of the following form. Let $\mathrm{Eul}_{\mathbb{C}^\times_\zeta}(\nu_{\mathcal{X}^{\zeta,0}})$ be the equivariant Euler class of the normal bundle in $K_{\mathbb{C}^\times_\zeta}(\mathcal{X}^{\zeta,0})$, Let $\tau_{\mathcal{X},\zeta,0}$ denote the equivariant virtual Euler characteristic twisted by $\mathrm{Eul}^{-1}_{\mathbb{C}^\times_\zeta}(\nu_{\mathcal{X}^{\zeta,0}})$ combined with the residue

$$\tau_{\mathcal{X},\zeta,0} : K_{\mathbb{C}^\times_\zeta}(\mathcal{X}^{\zeta,0}) \to \mathbb{Z}; \quad \sigma \mapsto \chi^{\mathrm{vir}}_{\mathbb{C}^\times_\zeta}\left(\mathcal{X}^{\zeta,0}; \mathrm{Resid}\, \frac{\sigma}{\mathrm{Eul}_{\mathbb{C}^\times_\zeta}(\nu_{\mathcal{X}^{\zeta,0}})}\right).$$

Let $\tau_{\mathcal{X}/\pm G}$ denote the virtual Euler characteristic

$$\tau_{\mathcal{X}/\pm G} : K(\mathcal{X}/\!\!/_\pm G) \to \mathbb{Z}, \quad \alpha \mapsto \chi^{\mathrm{vir}}(\mathcal{X}/\!\!/_\pm G; \alpha),$$

on the quotients as before. We have the following virtual Kalkman–Metzler wall-crossing formula:

Theorem 6.16 *Let $\mathcal{X}$ be a proper Deligne-Mumford global-quotient G-stack equipped with a G-equivariant perfect obstruction theory which admits a global resolution by vector bundles. Let $\mathcal{L}_\pm \to \mathcal{X}$ be G-line bundles whose associated wall-crossing is simple. Then*

$$\tau_{\mathcal{X}/+G}\,\kappa^G_{\mathcal{X},+} - \tau_{\mathcal{X}/-G}\,\kappa^G_{\mathcal{X},-} = \tau_{\mathcal{X},\zeta,0} \tag{52}$$

Example 6.17 (Wall-crossing over a nodal fixed point)   Suppose that $\mathcal{X} = \mathbb{P}^1 \cup \mathbb{P}^1$ is a nodal projective line with a single node $x_0 \in \mathcal{X}$, equipped with the standard $G = \mathbb{C}^\times$ action on each component, so that the weights of the action on the tangent spaces at the node are ± 1. Equip $\mathcal{X}$ with a polarization so that the weights are ± 1 at the smooth fixed points, and 0 at the nodal point $x_0 = \mathcal{X}^{\zeta,0}$. Then $\mathcal{X} /\!\!/_t G$ is a point for $t \in (-1,1)$, and is singular for $t = 0$. Since $\mathcal{X}$ is a complete intersection, $\mathcal{X}$ it has a perfect obstruction theory [4, Example before Remark 5.4] and the virtual wall-crossing formula of Theorem 6.16 applies. We examine the wall-crossing for the class $\mathcal{O}_\mathcal{X}$ at the singular value $t = 0$. The virtual normal complex at the nodal point is the quotient of $\mathbb{C}_1 \oplus \mathbb{C}_{-1}$, the sum of one-dimensional representations with weights $1, -1$, modulo their tensor product $\mathbb{C}_1 \otimes \mathbb{C}_{-1}$, which has weight $1 - 1 = 0$. Hence the normal complex has inverted Euler class

$$\mathrm{Eul}_{\mathbb{C}^\times}(\nu_{\mathcal{X}^{\zeta,0}})^{-1} = \frac{1}{(1 - z^{-1})(z - 1)}$$

whose residue is zero by Example 6.8. The Euler characteristics on the left and right hand sides are 1 while the wall-crossing term is

$$1 - 1 = \tau_{\mathcal{X}/_+ G} \, \kappa_{\mathcal{X},+}^G - \tau_{\mathcal{X}/_- G} \, \kappa_{\mathcal{X},-}^G$$
$$= \tau_{\mathcal{X},\zeta,0} = \chi^{\mathrm{vir}}\left(x_0, \mathrm{Eul}_{\mathbb{C}^\times}(\nu_{\mathcal{X}^{\zeta,0}})^{-1}\right) = 0$$

compatible with the wall-crossing formula.

The proof of Theorem 6.16 uses the construction of a master space for this set up. However, the same construction as before with small modifications applies.

Lemma 6.18   *There exists a proper Deligne-Mumford $\mathbb{C}^\times$-stack $\tilde{\mathcal{X}}$ equipped with a line bundle ample for the coarse moduli space whose git quotients $\tilde{\mathcal{X}} /\!\!/_t \mathbb{C}^\times$ are isomorphic to those $\mathcal{X} /\!\!/_t G$ of $\mathcal{X}$ by the action of G with respect to the polarization $\mathcal{L}_t$ and whose fixed point set $\tilde{\mathcal{X}}^{\mathbb{C}^\times}$ is given by the union*

$$\tilde{\mathcal{X}}^{\mathbb{C}^\times} = (\mathcal{X} /\!\!/_- G) \cup (\mathcal{X} /\!\!/_+ G) \cup \iota_\zeta (\mathcal{X}^\zeta /\!\!/_0 (G_\zeta / \mathbb{C}_\zeta^\times)) \qquad (53)$$

where ι_ζ is the natural map to $\tilde{\mathcal{X}}$ as before. Furthermore, $\tilde{\mathcal{X}}$ has a perfect obstruction theory admitting a global resolution by vector bundles with the property that the virtual normal complex of $\mathcal{X}^\zeta /\!\!/_0 (G_\zeta / \mathbb{C}_\zeta^\times)$ is isomorphic to the image of $\nu_{\mathcal{X}^\zeta} / (\mathfrak{g}/\mathbb{C}\zeta)$ under the quotient map $\mathcal{X}^\zeta \to \mathcal{X}^\zeta /\!\!/ (G_\zeta / \mathbb{C}_\zeta^\times)$, by an isomorphism that intertwines the action of $\mathbb{C}_\zeta^\times$ on $(\nu_{\mathcal{X}^\zeta} / (\mathfrak{g}/\mathbb{C}\zeta)) /\!\!/ (G_\zeta / \mathbb{C}_\zeta^\times)$ with the action of $\mathbb{C}^\times$ on $\nu_{\tilde{\mathcal{X}}^{\mathbb{C}^\times}}$.

Proof The construction of the master space is the same as in the smooth case, that is, the master space is the stack-theoretic quotient $\tilde{\mathcal{X}} = \mathbb{P}(\mathcal{L}_- \oplus \mathcal{L}_+) /\!/ G$. The action of G on the semistable locus in $\mathbb{P}(\mathcal{L}_- \oplus \mathcal{L}_+)$ is locally free by assumption. It follows that $\tilde{\mathcal{X}}$ is a proper Deligne-Mumford stack, and by Lemma 6.14 has a perfect obstruction theory induced from the natural obstruction theory on $\mathbb{P}(\mathcal{L}_- \oplus \mathcal{L}_+)$ given by considering it as a bundle over $\mathcal{X}$. The quotient $\tilde{\mathcal{X}}$ is such that $\tilde{\mathcal{X}} /\!/_t \mathbb{C}^\times$ is isomorphic to $\mathcal{X} /\!/_t G$ for $t \neq 0$. In fact it contains the quotients of $\mathbb{P}(\mathcal{L}_\pm) \cong \mathcal{X}$ with respect to the polarizations $\mathcal{L}_\pm$, that is, $\mathcal{X} /\!/_\pm G$.

The same argument as before describes the fixed point loci: they correspond to fixed point loci in $\mathbb{P}(\mathcal{L}_- \oplus \mathcal{L}_+)$ for one-parameter subgroups of $\mathbb{C}^\times \times G$. Given such a locus $\mathbb{P}(\mathcal{L}_- \oplus \mathcal{L}_+)^{\xi+\zeta}$, the pull-back of the virtual normal complex is by definition the moving part of $\mathrm{Cone}(\tilde{a}^\vee_{\mathbb{P}(\mathcal{L}_- \oplus \mathcal{L}_+)})$, where

$$\tilde{a}^\vee_{\mathbb{P}(\mathcal{L}_- \oplus \mathcal{L}_+)} \colon E_{\mathbb{P}(\mathcal{L}_- \oplus \mathcal{L}_+)} \to \underline{\mathfrak{g}}^\vee_\zeta$$

is the lift of the infinitesimal action of G_ζ. Consider the fibration $\pi \colon \mathbb{P}(\mathcal{L}_- \oplus \mathcal{L}_+) \to \mathcal{X}$. By definition $E_{\mathbb{P}(\mathcal{L}_- \oplus \mathcal{L}_+)}$ fits into an exact triangle

$$E_\mathcal{X} \to E_{\mathbb{P}(\mathcal{L}_- \oplus \mathcal{L}_+)} \to L_\pi \to E_\mathcal{X}[1].$$

Over the complement $\mathbb{P}(\mathcal{L}_- \oplus \mathcal{L}_+) \setminus (D_0 \cup D_\infty) \subset \mathbb{P}(\mathcal{L}_- \oplus \mathcal{L}_+)$ of the sections at zero and infinity we may identify $L_\pi \cong \mathbb{C}$ using the $\mathbb{C}^\times$-action on the fibers, by the assumption on the weights of the $\mathbb{C}_\zeta$ action on the fiber. Thus the projection to $\mathcal{X}$ identifies

$$\mathrm{Cone}(\tilde{a}^\vee_{\mathbb{P}(\mathcal{L}_- \oplus \mathcal{L}_+)} |_{\mathbb{P}(\mathcal{L}_- \oplus \mathcal{L}_+)^{\xi+\zeta}}) \to \pi^* \mathrm{Cone}(\tilde{a}^\vee_{\mathcal{X}^\zeta})$$

where

$$\tilde{a}^\vee_{\mathcal{X}^\zeta} \colon E_\mathcal{X} |_{\mathcal{X}^\zeta} \to (\underline{\mathfrak{g}_\zeta / \mathbb{C}\zeta})^\vee$$

is the lift of the infinitesimal action of $\mathfrak{g}_\zeta / \mathbb{C}\zeta$. Now the virtual normal complex is by definition the $\mathbb{C}^\times$-moving part of the perfect obstruction theory; the Lemma follows. $\qquad \square$

Proof of Theorem 6.16 For any equivariant class $\alpha \in K_G(\mathcal{X})$, its pullback to $\mathbb{P}(\mathcal{L}_- \oplus \mathcal{L}_+)$ descends to a class $\tilde{\alpha} \in K_G(\tilde{\mathcal{X}})$ whose restriction to $\mathcal{X} /\!/_\pm G$ is $\kappa^G_{X,\pm}(\alpha)$, and whose pullback under $X^{\zeta,0} \to \tilde{X}^{\mathbb{C}^\times}$ is $\iota_{X^\zeta,0}\alpha$. By virtual localisation (52) and the description of the fixed set, the normal bundles in Lemma 6.18

$$0 = \chi^{\mathrm{vir}}(\tilde{\mathcal{X}}; \mathrm{Resid}\,\delta(\alpha))$$

$$= \chi^{\mathrm{vir}}\left(\mathcal{X}\,/\!/_{-}G; \mathrm{Resid}\,\frac{\kappa^G_{\mathcal{X},-}(\alpha)}{\mathrm{Eul}_{\mathbb{C}^\times}(\nu_-)}\right) + \chi^{\mathrm{vir}}\left(\mathcal{X}\,/\!/_{+}G; \mathrm{Resid}\,\frac{\kappa^G_{\mathcal{X},+}(\alpha)}{\mathrm{Eul}_{\mathbb{C}^\times}(\nu_+)}\right)$$

$$+ \chi^{\mathrm{vir}}\left(\mathcal{X}^{\zeta,0}/(G_\zeta/\mathbb{C}^\times); \mathrm{Resid}\,\frac{\iota^*_{\mathcal{X}^{\zeta,0}}\alpha}{\mathrm{Eul}_{\mathbb{C}^\times_\zeta}(\nu_{\mathcal{X}^{\zeta,0}})}\right). \tag{54}$$

As in (6.7) we have

$$\mathrm{Resid}\,\chi\left(\mathcal{X}\,/\!/_{\mp}G, \frac{\kappa^G_{\mathcal{X},\mp}(\alpha)}{\mathrm{Eul}_{\mathbb{C}^\times}(\nu_\mp)}\right) = \pm\chi(\mathcal{X}\,/\!/_{\mp}G, \kappa^G_{\mathcal{X},\mp}(\alpha).)$$

Indeed, by definition the normal bundle ν_F at $\mathcal{X}\,/\!/_{\mp}G$ has virtual dimension one with positive resp. negative weight, and the inverted Euler class $\mathrm{Eul}(\nu)^{-1}$ is the symmetric product $\mathrm{Sym}(\nu)$ for the Bialynicki–Birula decomposition for positive weight, or $\mathrm{Sym}(\nu^\vee)\det(\nu^\vee)$ for the decomposition with negative weight. [4] For $\chi(\mathcal{X}\,/\!/_{+}G)$ the invariant part of the first is the trivial line bundle, while for the second the invariant part vanishes since the weights are positive so the difference in (45) of $\frac{\kappa^G_{\mathcal{X},\mp}(\alpha)}{\mathrm{Eul}_{\mathbb{C}^\times}(\nu_\mp)}$ is $\kappa^G_{\mathcal{X},\mp}(\alpha)$. For $\chi(\mathcal{X}\,/\!/_{-}G)$, the weight of ν is negative and $\mathrm{Sym}(\nu)$ appears in the Bialynicki–Birula decomposition for negative weight. Thus the residue is $\kappa^G_{\mathcal{X},\mp}(\alpha)$, which completes the proof.   □

7 Wall-crossing in quantum K-theory

The main result in this section, Theorem 7.5 below, relates the quantum K-theory pairings on both sides of a wall-crossing. Let $X\,/\!/_{\pm}G$ denote the associated quotient stacks $[X^{\mathrm{ss}}(\pm)/G]$ at times $t = \pm 1$ and let

$$\kappa^G_{X,\pm}\colon QK^0_G(X, \mathcal{L}_\pm) \to QK^0_{\mathbb{C}^\times}(X\,/\!/_{\pm}G)$$

denote the associated quantum Kirwan maps. Consider the graph potentials

$$\tau_{X/_{\pm}G}\colon QK^0_{\mathbb{C}^\times}(X\,/\!/_{\pm}G) \to \Lambda^G_{X,\mathcal{L}_\pm}.$$

---

[4] That these classes both agree with the inverted Euler class in localized K-theory follows by inspection from Riemann–Roch [53]. In fact, the agreement with the inverted Euler class is not necessary and one may take the difference in the Halpern–Leistner version of virtual localization for the Bialynicki–Birula decomposition and its opposite as the definition of residue.

Denote by

$$QK_G^{0,\,\text{fin}}(X) \subset QK_G^0(X, \mathcal{L}_-) \cap QK_G^0(X, \mathcal{L}_+)$$

the subset of classes of sums lying in both completions; for example, any finite sum lies in this intersection. We want to establish a formula for the difference

$$\tau_{X/_+ G}\, \kappa_{X,+}^G - \tau_{X/_- G}\, \kappa_{X,-}^G : QK_G^{0,\,\text{fin}}(X) \to \Lambda_X^G$$

as a sum of fixed point contributions given by gauged Gromov–Witten invariants with smaller structure groups $G_\lambda / \mathbb{C}_\lambda^\times$.

7.1 Master space for gauged maps and wall-crossing

We recall from [24, Proposition 3.1] the construction of master space whose quotients are the moduli stacks of Mundet stable gauged maps.

Proposition 7.1 (Existence of a master space) *Under assumptions of simple wall-crossing 6.13, for each equivariant degree $d \in H_2^G(X)$ there exists a proper Deligne-Mumford $\mathbb{C}^\times$-stack $\overline{\mathcal{M}}_n^G(C, X, \mathcal{L}_-, \mathcal{L}_+, d)$ with the following properties:*

a. *The stack $\overline{\mathcal{M}}_n^G(C, X, \mathcal{L}_-, \mathcal{L}_+, d)$ has a $\mathbb{C}^\times$-equivariant perfect obstruction theory, relative over the moduli stack $\overline{\mathfrak{M}}_n(C)$ of prestable maps to C of class $[C]$.*

b. *the git quotients of $\overline{\mathcal{M}}_n^G(C, X, \mathcal{L}_-, \mathcal{L}_+)$ by the $\mathbb{C}^\times$-action are isomorphic to the moduli stacks $\overline{\mathcal{M}}_n^G(C, X, \mathcal{L}_-^{(1-t)/2} \otimes \mathcal{L}_+^{(1+t)/2})$ for parameter $t \in (-1, 1)$;*

c. *the $\mathbb{C}^\times$-fixed substack includes $\overline{\mathcal{M}}_n^G(C, X, \mathcal{L}_-, d)$ and $\overline{\mathcal{M}}_n^G(C, X, \mathcal{L}_+, d)$; the other fixed point components are isomorphic to substacks of reducible elements of $\overline{\mathcal{M}}_n^G(C, X, \mathcal{L}_-^{(1-t)/2} \otimes \mathcal{L}_+^{(1+t)/2})$ for $t \in (-1, 1)$ consisting of gauged maps with $\mathbb{C}^\times$-automorphisms.*

d. *$\overline{\mathcal{M}}_n^G(C, X, \mathcal{L}_-, \mathcal{L}_+, d)$ admits an embedding in a non-singular Deligne-Mumford stack.*

For any fixed point component $F \subset \overline{\mathcal{M}}_n^G(C, X, \mathcal{L}_-, \mathcal{L}_+, d)$ denote by ν_F the normal complex, that is, the $\mathbb{C}^\times$-moving part of the perfect obstruction theory of Proposition 7.1 part (a). The following is a direct application of virtual wall crossing applied to the master space. Denote the evaluation map

$$\text{ev} : \overline{\mathcal{M}}_n^G(C, X, \mathcal{L}_-, \mathcal{L}_+, d) \to (X/G)^n.$$

Proposition 7.2 *For any class* $\alpha \in K_G^0(X)^n$

$$\chi^{\mathrm{vir}}\left(\overline{\mathcal{M}}_n^G(C,X,\mathcal{L}_+,d);\, \mathrm{ev}^*\,\alpha\right) - \chi^{\mathrm{vir}}\left(\overline{\mathcal{M}}_n^G(C,X,\mathcal{L}_-,d);\, \mathrm{ev}^*\,\alpha\right)$$

$$= \sum_F \chi^{\mathrm{vir}}\left(F;\, \mathrm{Resid}\,\frac{\iota_F^*\,\mathrm{ev}^*\,\alpha}{\mathrm{Eul}_{\mathbb{C}^\times}(\nu_F)}\right) \tag{55}$$

where F ranges over the fixed point components of $\mathbb{C}^\times$ on the moduli $\overline{\mathcal{M}}_n^G(C,X,\mathcal{L}_-,\mathcal{L}_+,d)$ that are not equal to $\overline{\mathcal{M}}_n^G(C,X,\mathcal{L}_\pm,d)$.

7.2 Reducible gauged maps

We analyze further the fixed point contributions in (55), which come from reducible gauged maps. Let X be a smooth projective G-variety. Let $Z \subset G$ a central subgroup. For any principal G-bundle $P \to C$, the right action of Z on P induces an action on the associated bundle $P(X)$, and so on the space of sections of $P(X)$. The fixed point set of Z on $P(X)$ is $P(X)^Z = P(X^Z)$, the associated bundle with fiber the fixed point set $X^Z \subset X$. The action of Z on the space of sections of $P(X)$ preserves Mundet semistability, since the parabolic reductions are invariant under the action and the Mundet weights are preserved. This induces an action of Z on $\overline{\mathcal{M}}_n^G(C,X)$.

Proposition 7.3 *Let $Z \subset G$ be a central subgroup. The fixed point locus for the action of Z on $\overline{\mathcal{M}}_n^G(C,X)$ is the substack whose objects are tuples*

$$(p\colon P \to C, u\colon \hat{C} \to P(X), \underline{z})$$

such that

a. *u takes values in $P(X^Z)$ on the principal component C_0;*
b. *for any bubble component $C_i \subset \hat{C}$ mapping to a point in C, u maps C_i to a one-dimensional orbit of Z on $P(X)$; and*
c. *any node or marking of $\hat{C}$ maps to the fixed point set $P(X^Z)$.*

We introduce notation for the various substacks of reducible maps. Let $\zeta \in \mathfrak{g}$ generate a one-parameter subgroup $\mathbb{C}_\zeta^\times \subset G$. Recall that G_ζ denotes the centralizer in G and so it contains $\mathbb{C}_\zeta^\times$ as a central subgroup. Let

$$\overline{\mathcal{M}}_n^{G_\zeta}(C,X,\mathcal{L}_t,\zeta,d)$$

denote the stack of $\mathcal{L}_t$-Mundet-semistable morphisms from C to X/G_ζ that are $\mathbb{C}_\zeta^\times$-fixed and take values in X^ζ on the principal component and in X/G_ζ

on the bubbles. Because these gauged maps correspond to the smaller group G_ζ, we call them as *reducible* gauged maps.

Each component of reducibles has an equivariant perfect obstruction theory. Recall that the obstruction theory for the moduli of gauged maps $\overline{\mathcal{M}}_n^G(C, X, d)$ is given by the complex $Rp_* e^* T_{X/G}^\vee$, which is relative with respect to $\overline{\mathfrak{M}}_n(C)$. The moduli $\overline{\mathcal{M}}_n^{G_\zeta}(C, X, \mathcal{L}_0, \zeta, d)$ is an Artin stack, and if every automorphism group is finite modulo $\mathbb{C}_\zeta^\times$, it is a proper Deligne-Mumford stack with a $\mathbb{C}^\times$-equivariant relatively perfect obstruction theory over $\overline{\mathfrak{M}}_n(C)$. This follows from the fact that the relative perfect obstruction theory for $\overline{\mathcal{M}}_n^{G_\zeta}(C, X, \mathcal{L}_t, \zeta, d)$ is pulled back from that on the $\mathbb{C}^\times$-fixed point set in the master space $\overline{\mathcal{M}}_n^G(C, X, \mathcal{L}_-, \mathcal{L}_+, d)^{\mathbb{C}^\times}$. This coincides with the $\mathbb{C}_\zeta^\times$-equivariant obstruction theory on the stack $\overline{\mathcal{M}}_n^{G_\zeta}(C, X, \mathcal{L}_0, \zeta, d)$ whose relative part is the *cone* with target the trivial bundle $\underline{\mathbb{C}_\zeta^\vee}$ with fiber the Lie algebra $\mathbb{C}_\zeta$ of $\mathbb{C}_\zeta^\times$

$$\mathrm{Cone}(Rp_* e^* T_{X/G}^\vee \to \mathbb{C}_\zeta^\vee)$$

given by the infinitesimal action of $\mathbb{C}_\zeta^\times$. The complex $Rp_* e^* T_{X/G}^\vee$ itself is not perfect because of the $\mathbb{C}_\zeta^\times$-automorphisms; taking the cone has the effect of cancelling this additional automorphisms. Denote by ν_0 the virtual (co)normal complex of the morphism

$$\overline{\mathcal{M}}_n^{G_\zeta}(C, X, \mathcal{L}_t, \zeta) \to \overline{\mathcal{M}}_n^G(C, X, \mathcal{L}_-, \mathcal{L}_+),$$

and as before, denote by

$$\mathrm{Eul}_{\mathbb{C}^\times}(\nu_0) \in K(\overline{\mathcal{M}}_n^G(C, X, \mathcal{L}_0, \zeta))$$

its Euler class.

Virtual Euler characteristics over the reducible gauged maps gives rise to the fixed point contributions in the wall-crossing formula: these are *ζ-fixed K-theoretic gauged Gromov–Witten invariants*. The ζ-fixed K-theoretic gauged Gromov–Witten invariants that appear in the wall-crossing formula involve further twists by the inverse of Euler classes of the virtual normal complex $\mathrm{Eul}_{\mathbb{C}^\times}(\nu_0)^{-1}$. Before we made this explicit, we need to allow a slightly larger coefficient ring. Denote by

$$\tilde\Lambda_X^G := \mathrm{Map}(H_2^G(X, \mathbb{Z}), \mathbb{Q})$$

the space of $\mathbb{Q}$-valued functions on $H_2^G(X, \mathbb{Z})$ (cf. Equation (10)). Note that $\tilde\Lambda_X^G$ has no ring structure extending that on Λ_X^G. The space $\tilde\Lambda_X^G$ can be viewed as the space of distributions in the quantum parameter q, and we use it as a master space interpolating Novikov parameters for the quotients with respect to $\mathcal{L}_t$ as t varies. Let $\tilde{\Lambda}^G_{X,\,\mathrm{fin}}$ denote the subspace of finite sums.

Definition 7.4   Let $X, G, \mathcal{L}_\pm, \zeta \in \mathfrak{g}$ as above, such that there is simple wall-crossing at the unique singular time $t = 0$ and such that $X^{\zeta,0}$ is non-empty. The *fixed point potential* associated to this data is the map

$$\tau_{X,\zeta,0} \colon QK^{0,\,\mathrm{fin}}_G(X) \to \tilde{\Lambda}^G_X$$

$$\alpha \mapsto \sum_{d \in H^G_2(X,\mathbb{Z})} \sum_{n \geq 0} \chi\left(\overline{\mathcal{M}}^G_n(C,X,\mathcal{L}_0,\zeta,d); \mathrm{Resid}\,\frac{\mathrm{ev}^*(\alpha,\ldots,\alpha)}{\mathrm{Eul}_{\mathbb{C}^\times}(\nu_0)}\right)\frac{q^d}{n!},$$

(56)

for $\alpha \in K^0_G(X)$, extended to $QK^{0,\,\mathrm{fin}}_G(X)$ by linearity. Here we omit the restriction map $K^{0,\,\mathrm{fin}}_G(X) \to K^0_{G_\zeta}(X)$ to simplify notation.

Remark   The fixed point potential $\tau_{X,\zeta,0}$ takes values in $\tilde{\Lambda}^G_X$ rather than in $\Lambda^G_X(\mathcal{L}_0)$ because Gromov compactness fails for gauged maps in the case that a central subgroup $\mathbb{C}^\times_\zeta$ acts trivially. Indeed, in this case, the energy $\langle d, c_1(L)\rangle$ of a gauged map of class d does not determine the isomorphism class of the bundle, since twisting by a character of $\mathbb{C}^\times_\zeta$ does not change the energy.

7.3 The wall-crossing formula

We may now prove the quantum version of the Kalkman–Metzler formula.

Theorem 7.5 (Wall-crossing for gauged potentials)   *Let X be a smooth G-variety. Suppose that $\mathcal{L}_\pm \to X$ are polarizations such that there is simple wall-crossing. Then the gauged Gromov–Witten potentials are related by*

$$\tau^G_{X,+} - \tau^G_{X,-} = \tau_{X,\zeta,0} \tag{57}$$

where the same Mundet semistability parameter should be used to define the potentials on both sides of the equation.

Proof of Theorem 7.5   The statement follows from virtual Kalkman–Metzler formula 6.16 applied to the master space $\overline{\mathcal{M}}^G_n(C,X,\mathcal{L}_-,\mathcal{L}_+)$ and the identification of fixed point contributions as reducible gauged maps described in Sections 7.1 and 7.2.   □

Combining Theorem 7.5 with the adiabatic limit (30) yields:

Theorem 7.6 (Quantum Kalkman–Metzler formula)   *Suppose that X is equipped with polarizations $\mathcal{L}_\pm$ so that the wall crossing is simple (the only* *singular polarisation is $\mathcal{L}_0$). Then the Gromov–Witten invariants of $X /\!/_{\pm} G$ are related by twisted gauged Gromov–Witten invariants with smaller subgroup $G_\zeta \subset G$*

$$\tau_{X/\!+G} \, \kappa_{X,+}^G - \tau_{X/\!-G} \, \kappa_{X,-}^G = \lim_{\rho \to 0} \tau_{X,\zeta,0}. \tag{58}$$

In other words, failure of the following square

$$
\begin{array}{ccccc}
QK_G(X,\mathcal{L}_-) & \longleftarrow & QK_G^{0,\,\mathrm{fin}}(X) & \longrightarrow & QK_G(X,\mathcal{L}_+) \\
{\scriptstyle \kappa_{X,-}^G}\downarrow & & & & \downarrow{\scriptstyle \kappa_{X,+}^G} \\
QK(X/\!/_- G) & & & & QK(X/\!/_+ G) \\
{\scriptstyle \tau_{X/\!-G}}\downarrow & & & & \downarrow{\scriptstyle \tau_{X/\!+G}} \\
\Lambda_{X,\mathcal{L}_-}^G & \longrightarrow & \tilde{\Lambda}_X^G & \longleftarrow & \Lambda_{X,\mathcal{L}_+}^G
\end{array}
\tag{59}
$$

to commute is measured by an explicit sum of wall-crossing terms given by the contribution of the fixed gauged potential. We remark that if the wall-crossing is not simple, the contributions on the right-hand side of the wall-crossing formula might come from several singular values $t \in (-1,1)$ as the polarisations $\mathcal{L}_t$ varies; however a simple modification of the argument above proves it as well.

8 Crepant wall-crossing

In this section we use the quantum Kalkman–Metzler formula to prove a version of the crepant transformation conjecture for K-theoretic Gromov–Witten invariants, under some rather strong assumptions on the weights involved in the wall-crossing. We assume the following symmetry condition on the weights involved in the wall-crossing. Suppose we have a birational transformation of git type

$$\phi : X /\!/_- G \dashrightarrow X /\!/_+ G$$

defined by a simple wall-crossing induced by two polarisations $\mathcal{L}_+, \mathcal{L}_-$ as in the previous sections. Suppose that for $\zeta \in \mathfrak{g}$, the fixed point component $X^{\zeta,0}$ is the one contributing to the wall-crossing term, and let $\nu_{X^{\zeta,0}} \to X^{\zeta,0}$ be its normal bundle in X. Let

$$\nu_{X^{\zeta,0}} = \bigoplus_j \nu_j$$

be the isotypical decomposition so that $\mathbb{C}_\zeta^\times$ acts on ν_j with weight μ_j. Note as before that all the $\mu_j \neq 0$. Let $r_j = \operatorname{rank} \nu_j$.

Definition 8.1   The birational transformation $\phi \colon X /\!\!/_- G \to X /\!\!/_+ G$ is *simply crepant* if the set of weights μ_i of the normal bundle of $X^{\zeta,0}$ in X is invariant under multiplication by -1, that is, whenever μ_j is a weight with multiplicity r_j then so is $-\mu_j$ with the same multiplicity.

If the wall-crossing is not simple, it is simply crepant if the condition in 8.1 holds for all fixed point components contributing to the wall-crossing terms.

We show invariance for the gauged potentials under crepant wall-crossing if a certain symmetrised version of the Euler characteristics are used. Let T be a torus acting on a Deligne-Mumford stack $\mathcal{X}$, endowed with a perfect obstruction theory. Suppose $x \in \mathcal{X}^T$ is an isolated fixed point. Locally the virtual tangent space

$$T_x^{\mathrm{vir}} := \operatorname{Def}_x - \operatorname{Obs}_x .$$

can be decomposed as

$$T_x^{\mathrm{vir}} = \bigoplus_i \mathbb{C}_{a_i} - \bigoplus_j \mathbb{C}_{b_j}$$

where a_i, b_j are the weights of the deformation and obstruction spaces respectively. Define

$$\widehat{\mathcal{O}}_{\mathcal{X}}^{\mathrm{vir}} := \mathcal{O}_{\mathcal{X},x}^{\mathrm{vir}} \otimes (K_{\mathcal{X}}^{\mathrm{vir}})^{1/2}, \quad K_{\mathcal{X}}^{\mathrm{vir}} := (\det T_{\mathcal{X}}^{\mathrm{vir}})^{-1}$$

where a square root can be defined in rational K-theory via the Chern character [14]. The resulting K-theoretic Gromov–Witten invariants obtained by replacing the virtual structure sheaf by this shift quantum K-theory at level $-1/2$ in the language of Ruan–Zhang [47]. At an isolated fixed point x we have

$$\widehat{\mathcal{O}}_x^{\mathrm{vir}} := \frac{\prod_j (b_j^{1/2} - b_j^{-1/2})}{\prod_i (a_i^{1/2} - a_i^{-1/2})}$$

where $a_i^{1/2}, b_j^{1/2}$ are formal, since they represent weights only after passing to a cover $\hat{T} \to T$.

The virtual localization formula may be re-written in terms of the shifted structure sheaves. Let $\hat{A}(\cdot)$ be the denominator of the A-hat genus, mapping $R(T)$ to the space of functions defined on some cover

$$\widehat{A}(a_1 + a_2) = \widehat{A}(a_1)\widehat{A}(a_2); \quad \widehat{A}(a) = \frac{1}{a^{1/2} - a^{-1/2}}.$$

where a is a weight (representation) of T. Define an extension to $\mathcal{F} \in K_T^0(X)$ by

$$\widehat{A}(\mathcal{F}) = \prod_j \widehat{A}(y_j),$$

where the product runs over the equivariant Chern roots $y_j \in K_T^0(X)$ of $\mathcal{F}$. Then localization (52) becomes

$$\widehat{\mathcal{O}}_{\mathcal{X}}^{\mathrm{vir}} = \iota_*(\widehat{A}(T_{\mathcal{X}^T}^{\mathrm{vir}}) \, \widehat{\mathcal{O}}_{\mathcal{X}^T}^{\mathrm{vir}}). \tag{60}$$

This can be made more explicit as follows. For each component $F \subset \mathcal{X}^T$ we have a decomposition

$$T_{\mathcal{X}}^{\mathrm{vir}}|_F = T_F^{\mathrm{vir}} + \nu_F$$

and therefore

$$(K_{\mathcal{X}}|_F)^{1/2} = K_F^{1/2}(\det \nu_F)^{-1/2}.$$

It follows that

$$\frac{\mathcal{O}_F \otimes (K_{\mathcal{X}}^{\mathrm{vir}}|_F)^{1/2}}{\mathrm{Eul}_T(\nu_F)} = \mathcal{O}_F \otimes (K_F^{\mathrm{vir}})^{1/2} \otimes \frac{(\det \nu_F)^{-1/2}}{\mathrm{Eul}_T(\nu_F)}. \tag{61}$$

By considering the decomposition

$$\nu_F = \bigoplus_i z^{\mu_i} \nu_{F,i}, \tag{62}$$

in isotypical components, we have

$$\mathrm{Eul}_T(\nu_F) = \prod_i \mathrm{Eul}_T(z^{\mu_i}\nu_{F,i}) = \prod_{i,j}(1 - z^{-\mu_i}x_{i,j}^{-1})$$

where $x_{i,j}$ are the Chern roots of $\nu_{F,i}$. Since $(\det \nu_F)^{1/2} = \prod_{i,j}(z^{\mu_i}x_{ij})^{1/2}$ we have

$$\mathrm{Eul}_T(\nu_F)^{-1}(\det \nu_F)^{-1/2} = \prod_{i,j}\widehat{A}(z^{\mu_i}x_{ij})^{-1} = \widehat{A}(\nu_F)^{-1}. \tag{63}$$

For our arguments below, we need to discuss the asymptotic behaviour of $\widehat{A}(\nu_F)$. Consider the decomposition of ν_F as in (62) and the Euler class expansion (43) for each of its isotypical components. Thus

$$\widehat{A}(\nu_F) = \frac{(\det \nu_F)^{-1/2}}{\prod_i(1 - z^{-\mu_i})^{r_i}(1 + N_i)}$$

with $N_i \in K(F) \otimes K_T^{\mathrm{loc}}(\mathrm{pt})$ as in (44). Therefore

$$\widehat{A}(\nu_F) = \prod_i \left(\frac{z^{-\mu_i/2}}{1 - z^{-\mu_i}} \right)^{r_i} \cdot O(z) = \prod_i \widehat{A}(\mathbb{C}_{\mu_i})^{r_i} \cdot O(z), \qquad (64)$$

where $\mathbb{C}_{\mu_i}$ is the representation with weight μ_i and $O(z)$ is a term that converges to 0 as $z^{\pm 1} \to 0$.

8.1 Symmetrised wall-crossing

We can define symmetric versions of the gauged K-theoretic Gromov–Witten potentials previously studied by considering Euler characteristics with respect to $\widehat{\mathcal{O}}^{\mathrm{vir}}$. In the following, we add a hat to any expression whose definition now uses $\widehat{\mathcal{O}}^{\mathrm{vir}}$ rather than $\mathcal{O}^{\mathrm{vir}}$. The proof of the quantum Kalkman formula in Theorem 7.6 relied on virtual localisation. If instead we use localisation for the symmetrised virtual structure sheaf we obtain the following:

$$\widehat{\tau}_{X/+G} \, \widehat{\kappa}_{X,+}^G - \widehat{\tau}_{X/-G} \, \widehat{\kappa}_{X,-}^G = \lim_{\rho \to 0} \widehat{\tau}_{X,\zeta,0}. \qquad (65)$$

The symmetrised virtual structure sheaves satisfy good properties under the action of the Picard stack on the locus of reducible maps. Let

$$\mathrm{Pic}(C) := \mathrm{Hom}(C, B\mathbb{C}^{\times})$$

denote the Picard stack of line bundles on C. The Lie algebra $\mathfrak{g}_\zeta$ has a distinguished factor generated by ζ, and using an invariant metric the weight lattice of $\mathfrak{g}_\zeta$ has a distinguished factor $\mathbb{Z}$ given by its intersection with the Lie algebra of $\mathbb{C}_\zeta^{\times}$. After passing to a finite cover, we may assume that $G_\zeta \cong (G_\zeta/\mathbb{C}_\zeta^{\times}) \times \mathbb{C}_\zeta^{\times}$. The Picard stack $\mathrm{Pic}(C)$ acts on the moduli stack of reducible gauged maps $\overline{\mathcal{M}}_n^{G_\zeta}(C, X, \mathcal{L}_0, \zeta)$ as follows. Recall that a reducible gauged map $(P, \widehat{C}, u)$, where $P \to C$ is a G-bundle and $u \colon \widehat{C} \to P(X)$ is ζ-fixed. The restriction of u to the principal component of C maps into the fixed point locus X^ζ. For Q an object of $\mathrm{Pic}(C)$ and $(P, \widehat{C}, u)$ an object of $\overline{\mathcal{M}}_n^{G_\zeta}(C, X, \mathcal{L}_0, \zeta)$ define

$$Q(P, \widehat{C}, u) := (P \times_{\mathbb{C}_\zeta^{\times}} Q, \widehat{C}, v) \qquad (66)$$

where the section v is defined as follows: We have an identification of bundles $(P \times_{\mathbb{C}_\zeta^{\times}} Q)(X^\zeta) \cong P(X^\zeta)$ since the action of $\mathbb{C}_\zeta^{\times}$ on X^ζ is trivial. Hence the principal component u_0, which is a section of $P(X^\zeta)$ induces the corresponding section v_0 of $(P \times_{\mathbb{C}_\zeta^{\times}} Q)(X^\zeta)$. Each bubble component of u maps into a fiber of $P(X)$, canonically identified with X up to the action of G_ζ.

So u induces the corresponding bubble map of v into a fiber of $(P \times_{\mathbb{C}_\zeta^\times} Q)(X)$, well-defined up to isomorphism. Note that if the degree of $(P, \hat{C}, u)$ is d the degree of $Q(P, \hat{C}, u)$ is $d + c_1(Q)$.

The Picard action preserves semistable loci in the large area limit. Indeed, because the Mundet weights $\mu_M(\sigma, \lambda)$ approach the Hilbert–Mumford weight $\mu_{HM}(\sigma, \lambda)$ as $\rho \to 0$, the limiting Mundet weight is unchanged by the shift by Q in the limit $\rho \to 0$ and so Mundet semistability is preserved. Thus for ρ^{-1} sufficiently large the action of an object Q of $\mathrm{Pic}(C)$ induces an isomorphism

$$\mathcal{S}^\delta : \overline{\mathcal{M}}_n^{G_\zeta}(C, X, \mathcal{L}_0, \zeta, d) \to \overline{\mathcal{M}}_n^{G_\zeta}(C, X, \mathcal{L}_0, \zeta, d + \delta) \qquad (67)$$

where $\delta = c_1(Q)$.

Lemma 8.2 (Action of the Picard stack on fixed loci) *The action of $Pic(C)$ in (67) induces isomorphisms of the relative obstruction theories on $\mathcal{M}_n^{G_\zeta}(C, X, \mathcal{L}_0, \zeta, d)$ preserving the restriction of symmetrised virtual structure sheaves $\hat{\mathcal{O}}_{\mathcal{M}_n(C, X, \mathcal{L}_0)}^{\mathrm{vir}}$, and preserving the class $\mathrm{ev}^* \alpha$ for any $\alpha \in K_G^0(X)^n$.*

Proof The action of $\mathrm{Pic}(C)$ lifts to the universal curves, denoted by the same notation. Since the relative part of the obstruction theory on $\overline{\mathcal{M}}_n^{G_\zeta}(C, X, \mathcal{L}_0, \zeta, d)$ is the $\mathbb{C}_\zeta^\times$-invariant part of $Rp_* e^* T_{X/G_\zeta}^\vee$ up to the factor $\mathbb{C}_\zeta$, the isomorphism preserves the relative obstruction theories and so the virtual structure sheaves. (Note that on the principal component, the invariant part is $Rp_* e^* T_{X^\zeta/G_\zeta}^\vee$ which is unchanged by the tensor product by $\mathbb{C}_\zeta^\times$-bundles, while on the bubble components $Rp_* e^* T_{X/G}^\vee$ is unchanged by the tensor product by Q.) Since the evaluation map is unchanged by pull-back by $\mathcal{S}^\delta$ (up to isomorphism given by twisting by Q), the class $\mathrm{ev}^* \alpha$ is preserved. $\square$

Theorem 8.3 (Wall-crossing for crepant birational transformations of git type) *Suppose that $X, G, \mathcal{L}_\pm$ define a simple wall-crossing, and C has genus zero. If the wall-crossings is simply crepant then*

$$\widehat{\tau}_{X/-G} \circ \widehat{\kappa}_{X,-}^G \underset{a.e.}{=} \widehat{\tau}_{X/+G} \circ \widehat{\kappa}_{X,+}^G$$

almost everywhere (a.e.) in the quantum parameter q.

The following remark explains precisely in what sense *a.e.* is used in Theorem 8.3.

Remark In the Schwartz theory of distributions (Hörmander [32]) denote by $\mathcal{T}(S^1)$ the space of *tempered distributions*. Fourier transform identifies $\mathcal{T}(S^1)$ with the space of functions on $\mathbb{Z}$ with polynomial growth. The variable q is a coordinate on the punctured plane $\mathbb{C}^\times$ and any formal power series in q, q^{-1}

defines a distribution on S^1, which is tempered if the coefficient of q^d has polynomial growth in d. In particular the series $\sum_{d\in\mathbb{Z}} q^d$ is the delta function at $q = 1$, and its Fourier transform is the constant function with value 1. Any distribution of the form $\sum_{d\in\mathbb{Z}} f(d)q^d$, for $f(d)$ polynomial, is a sum of derivatives of the delta function (since Fourier transform takes multiplication to differentiation) and so is almost everywhere zero.

We study the dependence of the fixed point contributions $\tau_{X,\zeta,d,0}$ with respect to the Picard action defined in (66). Suppose that Q is a $\mathbb{C}_\zeta^\times$-bundle of first Chern class the generator of $H^2(C)$, after the identification $\mathbb{C}_\zeta^\times \to \mathbb{C}^\times$. Denote the corresponding class in $H_2^{G_\zeta}(X^\zeta)$ by δ. Consider the action of the $\mathbb{Z}$-subgroup $\mathbb{Z}_Q \subset \mathrm{Pic}(C)$ generated by Q. The contribution of any component $\overline{\mathcal{M}}_n^{G_\zeta}(C,X,\mathcal{L}_0,\zeta,d)$ of class $d \in H_2^G(X)$ differs from that from the component induced by acting by $Q^{\otimes r}, r \in \mathbb{Z}_Q$, of class $d+r\delta$, by the ratio of symmetrised Euler classes of the moving parts of the virtual normal complexes

$$\widehat{A}((Rp_*e^*T_{X/G}^\vee)^+)\widehat{A}(S^{r\delta,*}(Rp_*e^*T_{X/G}^\vee)^+)^{-1} \tag{68}$$

As before, denote by ν_i be the subbundle of $\nu_{X^\zeta,0}$ of weight μ_i.

Lemma 8.4 *The $\widehat{A}$ classes relate by*

$$\widehat{A}((Rp_*e^*T_{X/G}^\vee)^+) = \widehat{A}(S^{r\delta,*}(Rp_*e^*T_{X/G}^\vee)^+)\left(\prod_i \widehat{A}(\nu_i)^{\mu_i}\right)^r.$$

Proof The Grothendieck–Riemann–Roch allows a computation of the Chern characters of the (representable) push-forwards. Consider the isotypical decomposition into $\mathbb{C}^\times$-bundles of the normal bundle to the fixed component $X^{\zeta,0}$ in X

$$\nu_{X^\zeta,0} = \bigoplus_{i=1}^{k} \nu_i$$

where $\mathbb{C}^\times$ acts on ν_i with non-zero weight $\mu_i \in \mathbb{Z}$. By the discussion above $e^*T_{X/G}^\vee$ is canonically isomorphic to $S^{r\delta}(e^*T_{X/G}^\vee)$ on the bubble components, since the G-bundles are trivial on those components. Because the pull-back complexes are isomorphic on the bubble components, the difference $(e^*T_{X/G}^\vee)^+ - S^{r\delta,*}(e^*T_{X/G}^\vee)^+$ is the pullback of the difference of the restrictions to the principal part of the universal curve, that is, the projection on the second factor

$$p_0\colon C \times \overline{\mathcal{M}}_n^{G_\zeta}(C,X,\mathcal{L}_0,\zeta) \to \overline{\mathcal{M}}_n^{G_\zeta}(C,X,\mathcal{L}_0,\zeta).$$

These restrictions are given by

$$(e^*T_{X/G})^{+,\,\mathrm{prin}} \cong \bigoplus_i e^* \nu_{X^{\zeta,t},i}$$

$$\mathcal{S}^{r\delta,*}(e^*T_{X/G})^{+,\,\mathrm{prin}} \cong \bigoplus_i e^* \nu_{X^{\zeta,t},i} \otimes (e_C^* \mathcal{Q} \times_{\mathbb{C}_{\zeta}^{\times}} \mathbb{C}_{r\mu_i})$$

where e_C is the map from the universal curve to C. The projection p_0 is a representable morphism of stacks given as global quotients. Let

$$\sigma : \overline{\mathcal{M}}_n^{G_{\zeta}}(C, X, \mathcal{L}_0, \zeta) \to C \times \overline{\mathcal{M}}_n^{G_{\zeta}}(C, X, \mathcal{L}_0, \zeta)$$

be a constant section of p_0, so that $c_1(\sigma^* e^* \nu_{X^{\zeta,0},i})$ is the "horizontal" part of $c_1(\nu_{X^{\zeta,0},i})$. By Grothendieck–Riemann–Roch for such stacks [52], [14]

$$\mathrm{Td}_{\mathcal{M}} \mathrm{Ch}(\mathcal{S}^{r\delta,*} \mathrm{Ind}(T_{X/G})^+) = p_{0,*}(\mathrm{Td}_{C \times \mathcal{M}} \mathrm{Ch}(\mathcal{S}^{r\delta,*} T_{X/G})^+)$$

$$= (1 - g) + \mathrm{Td}_{\mathcal{M}} \, p_{0,*} \, \mathrm{Ch}(\mathcal{S}^{r\delta,*} T_{X/G})^+)$$

$$= (1 - g) + \mathrm{Td}_{\mathcal{M}} \, p_{0,*} \sum_i \mathrm{Ch}(e^* \nu_{X^{\zeta,t},i}) \, \mathrm{Ch}((e_C^* \mathcal{Q} \times_{\mathbb{C}_{\zeta}^{\times}} \mathbb{C}_{r\mu_i}))$$

$$= (1 - g) + \mathrm{Td}_{\mathcal{M}} \, p_{0,*} \sum_i \mathrm{Ch}(e^* \nu_{X^{\zeta,t},i})(1 + r\mu_i \omega_C)$$

$$= p_{0,*}\left(\mathrm{Td}_{C \times \mathcal{M}} \mathrm{Ch}\left(\mathrm{Ind}(T_{X/G})^+ \oplus \bigoplus_i (\sigma^* e^* \nu_{X^{\zeta,t},i})^{\oplus r\mu_i}\right)\right)$$

$$= \mathrm{Td}_{\mathcal{M}} \mathrm{Ch}\left(\mathrm{Ind}(T_{X/G})^+ \oplus \bigoplus_i (\sigma^* e^* \nu_{X^{\zeta,0},i})^{\oplus r\mu_i}\right).$$

Hence

$$\mathrm{Ch}(\mathcal{S}^{r\delta,*} \mathrm{Ind}(T_{X/G})^+) = \mathrm{Ch}\left(\mathrm{Ind}(T_{X/G})^+ \oplus \bigoplus_{i=1}^{m} (\sigma^* e^* \nu_{X^{\zeta,0},i})^{\oplus r\mu_i}\right). \quad (69)$$

The equality of Chern characters above implies an isomorphism in rational topological K-theory. The difference in Euler classes (68) is therefore given by the Euler class of the last summand in (69)

$$\frac{\widehat{A}((Rp_* e^* T_{X/G}^{\vee})^+)}{\widehat{A}(\mathcal{S}^{r\delta,*}(Rp_* e^* T_{X/G}^{\vee})^+)} = \widehat{A}\left(\bigoplus_i (\sigma^* e^*(\nu_i))^{\oplus \mu_i r}\right)$$

$$= \left(\prod_i \widehat{A}(\nu_i)^{\mu_i}\right)^r.$$

$\square$

Proof of Theorem 8.3 Using the expansion of Euler classes as in (64) and (43), we have that by setting $r_i = \text{rank } \nu_i$ (on each component of the inertia stack, in the orbifold case)

$$\prod_i \widehat{A}(\nu_i)^{-\mu_i} = \prod_i (\zeta_i^{1/2} z^{\mu_i/2} - \zeta_i^{-1/2} z^{-\mu_i/2})^{-\mu_i r_i}(1+N)$$

where ζ_i are the roots of unity appearing in orbifold Riemann–Roch [53] and N is nilpotent. Let S_-, S_+ denote the indices for which μ_i is negative and respectively positive. Define

$$\Delta(z) := \frac{\prod_{j \in S_-} (\zeta_j^{1/2} z^{\mu_j/2} - \zeta_j^{-1/2} z^{-\mu_j/2})^{-\mu_j r_j}}{\prod_{i \in S_+} (\zeta_i^{1/2} z^{\mu_i/2} - \zeta_i^{-1/2} z^{-\mu_i/2})^{\mu_i r_i}}. \tag{70}$$

We can rewrite the difference

$$\left(\prod_i \widehat{A}(\nu_i)^{-\mu_i} \right)^r = (\Delta(z)(1+N))^r.$$

By the crepant wall-crossing assumption 8.1, the function $\Delta(z)$ is a constant, denoted Δ. Summing the terms from $\mathcal{S}^{r\delta}, r \in \mathbb{Z}$ we obtain that the wall-crossing contribution is

$$\sum_{r \in \mathbb{Z}} q^{d+\delta r} \widehat{\tau}_{X,\zeta,d+\delta r,0}(\alpha) = \sum_{r \in \mathbb{Z}} q^{d+r\delta} \cdot \chi_0(r) \tag{71}$$

where $\chi_0(r)$ is polynomial in r, since each N is nilpotent and the binomial coefficients from the expansion of $(1+N)^r$ are polynomial. Now

$$\sum_{r \in \mathbb{Z}} q^{\delta r} \in \mathcal{T}(S^1)$$

is a delta function and it vanishes almost everywhere in q^δ, see Remark 8.1. Since $\chi_0(r)$ is polynomial, (71) is the derivative of a delta function. Since

$$\widehat{\kappa}_X^{G,+} \widehat{\tau}_{X/+G} - \widehat{\kappa}_X^{G,-} \widehat{\tau}_{X/-G}$$

is a sum of wall-crossing terms of the type in (71), this completes the proof of Theorem 8.3. $\qquad \square$

9 Abelianization

In this section we compare the K-theoretic Gromov–Witten invariants of a git quotient with the quotient by a maximal torus, along the lines of the case of quantum cohomology investigated by Bertram–Ciocan-Fontanine–Kim [5] and Ciocan-Fontanine–Kim–Sabbah [8]. The analogous question for

K-theoretic I-functions of git quotients was already considered in Taipale [50] as well as Wen [56] and, around the same time as the first draft of this paper, Jockers, Mayr, Ninad, and Tabler [35].

Recall that the equivariant cohomology may be identified with the Weyl-invariant equivariant cohomology for the action of a maximal torus. We assume that G is a connected complex reductive group. Choose a maximal torus T and W its Weyl group. By a theorem of Harada–Landweber–Sjamaar [31, Theorem 4.9(ii), Section 6] if X is either a smooth projective G-variety or a G-vector space then restriction from the action of G to the action of the torus T defines an isomorphism onto the space of W-invariants

$$\text{Restr}_T^G : K_G^0(X) \cong K_T^0(X)^W$$

for either the topological or algebraic K-cohomology. Given a polarisation $\mathcal{L} \to X$ of the G action, consider the naturally induced T polarisation on X so that

$$X^{\text{ss},G}(\mathcal{L}) \subset X^{\text{ss},T}(\mathcal{L}).$$

We assume from now in this section that $QK_G^0(X)$ denotes the algebraic equivariant quantum K-cohomology. We relate the K-theoretic potentials of the two geometric invariant theory quotients $X /\!\!/ G$, and the *abelian* quotient $X /\!\!/ T$. Let $\nu_{\mathfrak{g}/\mathfrak{t}}$ denote the bundle over $X /\!\!/ T$ induced from the trivial bundle over X with fibre $\mathfrak{g}/\mathfrak{t}$. Consider the graph potential

$$\tau_{X/T} : QK_T^0(X) \to \Lambda_X^T$$

twisted by the Euler class of the index bundle associated to $\mathfrak{g}/\mathfrak{t}$:

$$\tau_{X/T}(\alpha, q) := \sum_{n \geq 0} \sum_{d \in H_2^G(X, \mathbb{Q})} \chi^{\text{vir}} \left(\overline{\mathcal{M}}_n(C, X /\!\!/ T, d); \text{ev}^* \, \alpha^n \, \text{Eul}(\text{Ind}(\mathfrak{g}/\mathfrak{t})) \right) \frac{q^d}{n!}.$$

$$(72)$$

Similarly T-gauged potential τ_X^T and the Kirwan map $\kappa_{X,T}$ will from now on denote the maps with the Euler twist above. The natural map $H_2^T(X, \mathbb{Q}) \to H_2^G(X, \mathbb{Q})$ induces a map of Novikov rings

$$\pi_T^G : \Lambda_X^T \to \Lambda_X^G, \quad \sum_{d \in H_2^T(X)} c_q q^d \mapsto \sum_{d \in H_2^G(X)} c_q q^{\pi(d)}.$$

By abuse of notation, denote again by

$$\text{Restr}: QK_G^0(X) \to QK_T^0(X)$$

the map induced by the restriction map above and the inclusion of the Weyl invariants $\Lambda_X^G \cong (\Lambda_X^T)^W \subset \Lambda_X^T$. As in Martin [44] the restriction map

$$\mathrm{Restr}_G^T : K(X/\!/T)^W \to K(X/\!/G)$$

is surjective and has kernel is the annihilator of $\mathrm{Eul}(\mathfrak{g}/\mathfrak{t})$, the set of classes that vanish when multiplied by $\mathrm{Eul}(\mathfrak{g}/\mathfrak{t})$. This map naturally extends to a map

$$\mathrm{Restr}_G^T : QK(X/\!/T)^W \to QK(X/\!/G)$$

by a similar definition on the twisted sectors. On the main sector Restr_G^T is given by restriction of a class

$$\alpha \in K(X/\!/T) = K(X^{\mathrm{ss}}(T)/T)$$

to $X^{\mathrm{ss}}(G)/T$ then followed by the identification of the Weyl invariant part with $K(X/\!/G)$ [44, Theorem A]. With these notations we have the following result.

Theorem 9.1 (Quantum Martin formula in quantum K-theory) *Let C be a smooth connected projective genus 0 curve, X a smooth projective or convex quasiprojective G-variety, and suppose that stable=semistable for T and G actions on X. The following equality holds on the topological quantum K-theory $QK_G^0(X)$:*

$$\tau_{X/G} \circ \kappa_{X,G} = |W|^{-1}\pi_T^G \circ \tau_{X/T}^{\mathfrak{g}/\mathfrak{t}} \circ \kappa_{X,T}^{\mathfrak{g}/\mathfrak{t}} \circ \mathrm{Restr}_T^G.$$

Similarly for (J-functions) localised graph potentials

$$\tau_{X/G,-} : QK_G^0(X) \to QK(X/\!/G)[z,z^{-1}]]$$
$$\tau_{X/T,-} : QK_T^0(X) \to QK(X/\!/T)[z,z^{-1}]]$$

we have

$$\tau_{X/G,-} \circ \kappa_{X,G} = \tau_{X,-}^G$$
$$= \mathrm{Restr}_G^T \circ \mathrm{Eul}(\mathfrak{g}/\mathfrak{t})^{-1}\tau_{X,-}^{T,\mathfrak{g}/\mathfrak{t}} \circ \mathrm{Restr}_T^G \qquad (73)$$
$$= \mathrm{Restr}_G^T \circ \mathrm{Eul}(\mathfrak{g}/\mathfrak{t})^{-1}\tau_{X/T,-}^{\mathfrak{g}/\mathfrak{t}} \circ \kappa_{X,T}^{\mathfrak{g}/\mathfrak{t}} \circ \mathrm{Restr}_T^G.$$

In particular commutativity of the following diagram holds:

$$
\begin{array}{ccc}
QK_G^0(X) & \xrightarrow{\mathrm{Restr}_T^G} & QK_T^0(X)^W \\
{\scriptstyle \kappa_{X,G}}\downarrow & & \downarrow{\scriptstyle \kappa_{X,T}} \\
QK(X/\!/G) & & QK(X/\!/T) \\
{\scriptstyle \tau_{X/G}}\downarrow & & \downarrow{\scriptstyle |W|^{-1}\tau_{X/T}} \\
\Lambda_X^G & \xleftarrow{\ \pi_T^G\ } & \Lambda_X^T
\end{array}
$$

Sketch of proof The argument is the same as that for cohomology in [25, Section 4]. In the case of projective target X, one can vary the vortex parameter $\rho \in \mathbb{R}_{>0}$ until one reaches the small-area chamber in which every bundle $P \to C$ appearing in the vortex moduli space is trivial (in genus zero). Indeed, for ρ^{-1} sufficiently large the Mundet weight is dominated by the Ramanathan weight, and this forces the bundle to be semistable of vanishing Chern class and so trivial. It follows that both $\overline{\mathcal{M}}^G(C, X, d)$ and $\overline{\mathcal{M}}^T(C, X, d)$ are quotients of open loci in the moduli stacks of stable maps $\overline{\mathcal{M}}_{0,n}(X, d)$ by G resp. T. In the small-area chamber abelianization holds for the localized potentials $\tau_{X,G}$ and $\tau_{X,T}$ by virtual non-abelian localization [28]: For sufficiently positive equivariant vector bundles V over $\overline{\mathcal{M}}_{0,n}(X, d)$ denote by $V /\!\!/ G$ the restriction to $\overline{\mathcal{M}}^G(C, X, d)$. Then

$$
\chi(\overline{\mathcal{M}}^G(C, X, d), V /\!\!/ G) = \chi(\overline{\mathcal{M}}_{0,n}(X, d), V)^G
$$
$$
= |W|^{-1} \chi(\overline{\mathcal{M}}_{0,n}(X, d), V \otimes \mathrm{Eul}(\mathfrak{g}/\mathfrak{t}))^T
$$
$$
= |W|^{-1} \chi(\overline{\mathcal{M}}^T(C, X, d), (V \otimes \mathrm{Eul}(\mathfrak{g}/\mathfrak{t})) /\!\!/ T)
$$

where the second equality holds by the Weyl character formula. If the stabilizers are at most one-dimensional then the wall-crossing formula of [24] implies that the variation in the gauged Gromov–Witten invariants τ_X^G with respect to the vortex parameter ρ is given by wall-crossing terms $\tau_{X,\zeta}^{G_\zeta}$ involving smaller-dimensional structure group given by the centralizers $G_\zeta, \zeta \in \mathfrak{g}$ of one-parameter subgroups generated by ζ: For any singular value ρ_0 and $\rho_\pm = \rho_0 \pm \epsilon$ for ϵ small we have

$$
\tau_{X,d}^G(\alpha, \rho_+) - \tau_{X,d}^G(\alpha, \rho_-) = \sum_\zeta \tau_{X,\zeta}^{G_\zeta}(\alpha, \rho_0) \tag{74}
$$

where $\tau_{X,\zeta}^{G_\zeta}$ is a moduli stack of ρ_0-semistable gauged maps fixed by the one-parameter subgroup generated by ζ as in Section 7. After possibly adding a parabolic structure the stabilizers of gauged maps that are ρ_0-semistable are one-dimensional and so the wall-crossing formula (74) holds. Furthermore, the fixed point components have structure group that reduces to $G_\zeta / \mathbb{C}_\zeta^\times$, which as such that objects in the fixed point components have trivial stabilizer. By induction we may assume that the abelianization formula holds for structure groups $G_\zeta / \mathbb{C}_\zeta^\times$ of lower dimension, and in particular for the invariants $\tau_{X,\zeta}^{G_\zeta}(\alpha, \rho_0)$. The result for other chambers $\rho \in (\rho_i, \rho_{i+1})$ holds by the wall-crossing formula Theorem 7.5 since, by the inductive hypothesis, the wall-crossing terms are equal. The conclusion for the git quotients then follows from the adiabatic limit theorem 5.4.

In the case of quasiprojective X we assume that G has a central factor $\mathbb{C}^\times \subset G$ and the moment map $\Phi \colon X \to \mathbb{R}$ for this factor on X is bounded from below. Then a similar wall-crossing argument obtained by varying the polarization $\lambda(t) \in H_G^2(X)$ from $\lambda(0) = \omega$ to to a chamber corresponding to an equivariant symplectic class $\lambda \in H_G^2(X)$ where $X /\!/_{\lambda(1)} G$ is empty, produces the same result [25]. Indeed the moduli space of gauged maps $\overline{\mathcal{M}}_n^G(\mathbb{P}^1, X)$ for the polarization $\lambda(1)$ is also empty, since for ρ small elements of $\overline{\mathcal{M}}_n^G(\mathbb{P}^1, X)$ must be generically semistable. On the other hand, the wall-crossing terms correspond to integrals over gauged maps whose structure group G_ζ is the centralizer of some one-parameter subgroup generated by a rational element $\zeta \in \mathfrak{t}$. By induction on the dimension of G_ζ we may assume that the wall-crossing terms for $\overline{\mathcal{M}}_n^G(\mathbb{P}^1, X)$ and $\overline{\mathcal{M}}_n^T(\mathbb{P}^1, X)$ are equal and we obtain the abelianization formula by induction.

For the localized potentials the same wall-crossing argument applied to the $\mathbb{C}^\times$-fixed point components $\overline{\mathcal{M}}_n^G(\mathbb{P}^1, X)^{\mathbb{C}^\times}$ and $\overline{\mathcal{M}}_n^T(\mathbb{P}^1, X)^{\mathbb{C}^\times}$ produces the abelianization formula

$$
\chi(\overline{\mathcal{M}}_n^G(\mathbb{P}^1, X)^{\mathbb{C}^\times}, V \otimes \mathrm{Eul}(\nu_G)^{-1})
$$
$$
= |W|^{-1} \chi(\overline{\mathcal{M}}_n^T(\mathbb{P}^1, X)^{\mathbb{C}^\times}, V \otimes \mathrm{Eul}(\mathrm{Ind}(\mathfrak{g}/\mathfrak{t}))) \tag{75}
$$

for any equivariant K-class V, where ν_G, ν_T are the normal bundles for the inclusion of the fixed point sets of the action of $\mathbb{C}^\times$. This formula holds as well after restricting to fixed point components with markings $z_1, \ldots, z_n$ mapping to $0 \in \mathbb{P}^1$ and z_{n+1} mapping to ∞, and taking V to be a class of the form $\mathrm{ev}_1^* \alpha \otimes \cdots \mathrm{ev}_n^* \alpha \otimes \mathrm{ev}_{n+1}^* \alpha_0$. One obtains the formula (with superscript class denoting the classical Kirwan map)

$$
\langle \tau_{X,-}^G(\alpha), \kappa_{X,G}^{\mathrm{class}}(\alpha_0) \rangle = |W|^{-1} \langle \tau_{X,-}^{T, \mathfrak{g}/\mathfrak{t}}(\mathrm{Restr}_T^G \alpha), \kappa_{X,T}^{\mathrm{class}}(\alpha_0) \rangle
$$
$$
= \langle \mathrm{Restr}_G^T \mathrm{Eul}(\mathfrak{g}/\mathfrak{t})^{-1} \tau_{X,-}^{T, \mathfrak{g}/\mathfrak{t}}(\mathrm{Restr}_T^G \alpha), \kappa_{X,G}^{\mathrm{class}}(\alpha_0) \rangle
$$

(the second by Martin's formula [44]) from which the localized abelianization formula (73) follows. □

Example 9.2 The fundamental solution in quantum K-theory for the Grassmannian $G(r, n)$ is studied in Taipale [50, Theorem 1], Wen [56], and Jockers, Mayr, Ninad, and Tabler [35]. Let

$$
X = \mathrm{Hom}(\mathbb{C}^r, \mathbb{C}^n), \quad G = GL(r).
$$

The group G acts on X by composition on the right: $gx = x \circ g$. Choose a polarization $\mathcal{L} = X \times \mathbb{C}$ corresponding a positive central character of G. The semistable locus is then

$$X^{ss} = \{x \in X \mid \text{rank}(x) = r\}$$

and the git quotient

$$X /\!\!/ G \cong G(r,n).$$

The torus $T = (\mathbb{C}^\times)^r$ is a maximal torus of G and the git quotient by the maximal torus is

$$X /\!\!/ T \cong (\mathbb{P}^{n-1})^r.$$

We claim that the localized gauged potential $\tau_{X,-}^G$ is the restriction of the Ind($\mathfrak{g}/\mathfrak{t}$)-twisted potential $\tau_{X,-}^{T,\mathfrak{g}/\mathfrak{t}}$ given by

$$\tau_{X,-}^{T,\mathfrak{g}/\mathfrak{t}}(\alpha,q,z) = \sum_d q^d \tau_{X,-,d}^{T,\mathfrak{g}/\mathfrak{t}}(\alpha,q,z) \prod_{1 \le i < j \le r} (1 - X_i X_j^\vee) X^{2\rho}$$

$$\tau_{X,-,d}^{T,\mathfrak{g}/\mathfrak{t}}(\alpha,q,z) := \sum_{d_1+\ldots+d_r=d} \exp\left(\frac{\Psi_d(\alpha)}{1-z^{-1}}\right) (-1)^{d(r-1)} z^{\langle d+\rho,d+\rho\rangle - \langle \rho,\rho\rangle}$$

$$\left(\frac{\prod_{i<j}(1 - X_i X_j^\vee z^{d_i - d_j})}{\prod_{i=1}^r \prod_{l=1}^{d_i}(1 - X_i z^l)^n} \right). \tag{76}$$

Without the factors $1 - X_i X_j^\vee z^{d_i-d_j}$ and $1 - X_i X_j^\vee$ the expression (76) in the Lemma would be the formula (29) for $\tau_{X,-}^T$ discussed previously in the toric setting. The additional factors are given by the Euler class of the index bundle

$$\text{Eul}(\text{Ind}(\mathfrak{g}/\mathfrak{t})) = \frac{\prod_{i<j} \prod_{k=0}^{d_j-d_i}(1 - X_i X_j^\vee z^k)}{\prod_{i<j} \prod_{k=1}^{d_j-d_i-1}(1 - X_j X_i^\vee z^k)} \tag{77}$$

$$= \frac{\prod_{i<j} \prod_{k=0}^{d_j-d_i}(1 - X_i X_j^\vee z^{-k})}{\prod_{i<j}(-1)^{d_j-d_i-2} \prod_{k=1}^{d_j-d_i-1} z^{-2k} X_j X_i^\vee (1 - X_i X_j^\vee z^k)}$$

$$= (-1)^{d(r-1)} z^{\sum_{i<j} -2(d_j-d_i)(d_j-d_i-1)/2}$$

$$\prod_{i<j} X^{2\rho}(1 - X_i X_j^\vee)(1 - X_i X_j^\vee z^{d_j-d_i}) \tag{78}$$

$$= (-1)^{d(r-1)} z^{\langle d,d\rangle + 2\langle d,\rho\rangle}$$

$$\prod_{i<j} X^{2\rho}(1 - X_i X_j^\vee)(1 - X_i X_j^\vee z^{d_j-d_i}) \tag{79}$$

$$= (-1)^{d(r-1)} z^{\langle d+\rho,d+\rho\rangle - \langle \rho,\rho\rangle}$$

$$\prod_{i<j} X^{2\rho}(1 - X_i X_j^\vee)(1 - X_i X_j^\vee z^{d_j-d_i}) \tag{80}$$

where $\langle \cdot, \cdot \rangle$ is the Killing form. Note the missing factor of $X^{2\rho}$ in [50, (20)]; this factor re-appears in [50, (31)] but without the powers of z.

As pointed out to us by M. Zhang, the additional factors arising from the ρ-shift in (77) vanish when one uses the level $-1/2$-theory introduced by Ruan–Zhang [47]. It would be interesting to know how the relations depend on the level structure, and whether at level $-1/2$ the relations can be found using the difference module structure in (5.5); see Jockers, Mayr, Ninad, and Tabler [35] for further developments.

References

[1] S. Agnihotri. Quantum Cohomology and the Verlinde Algebra. PhD thesis, Oxford University, 1995.

[2] Atiyah, M. F.; Segal, G. B. The index of elliptic operators. II. *Ann. of Math.* (2) 87 1968 531–545.

[3] H. Bass, E. H. Connell, and D. Wright. The Jacobian conjecture: Reduction of degree and formal expansion of the inverse. *Bull. Amer. Math. Soc. (N.S.)* Volume 7, Number 2 (1982), 287–330.

[4] K. Behrend and B. Fantechi. The intrinsic normal cone. *Invent. Math.*, 128(1): 45–88, 1997.

[5] A. Bertram, I. Ciocan-Fontanine, and B. Kim. Gromov-Witten invariants for abelian and nonabelian quotients. *J. Algebraic Geom.*, 17(2):275–294, 2008.

[6] L. A. Borisov and R. P. Horja. On the K-theory of smooth toric DM stacks. In *Snowbird lectures on string geometry*, volume 401 of *Contemp. Math.*, pages 21–42. Amer. Math. Soc., Providence, RI, 2006.

[7] A. S. Buch and L. C. Mihalcea. Quantum K-theory of Grassmannians. *Duke Math. J.*, 156(3):501–538, 2011.

[8] I. Ciocan-Fontanine, B. Kim, and C. Sabbah. The abelian/nonabelian correspondence and Frobenius manifolds. *Invent. Math.*, 171(2):301–343, 2008.

[9] H.L. Chang, Y.H. Kiem and J. Li. Torus localization formulas for cosection localized virtual cycles. *Adv. Math.*, 308, 964–986, 2017.

[10] T.H. Coates. Riemann–Roch theorems in Gromov–Witten theory. Thesis (Ph.D.)-University of California, Berkeley. 2003.

[11] J.-L. Colliot-Thélène and J.-J. Sansuc. Fibrès quadratiques et composantes connexes rèelles. *Math. Ann.*, 244(2):105–134, 1979.

[12] N. Chriss and V. Ginzburg. Representation theory and complex geometry. Reprint of the 1997 edition. Modern Birkhäuser Classics. Birkhäuser Boston, Inc., Boston, MA, 2010.

[13] I. V. Dolgachev and Y. Hu. Variation of geometric invariant theory quotients. *Inst. Hautes Études Sci. Publ. Math.*, (87):5–56, 1998. With an appendix by Nicolas Ressayre.

[14] D. Edidin. Riemann–Roch for Deligne-Mumford stacks. *A celebration of algebraic geometry*, Clay Math. Proc. (18): 241–266, 2013.

[15] A. R. P. Gaio and D. A. Salamon. Gromov-Witten invariants of symplectic quotients and adiabatic limits. *J. Symplectic Geom.*, 3(1):55–159, 2005.

[16] V. Ginzburg and M. Kapranov. Koszul duality for operads. *Duke Math. J.* 76 (1994), no. 1, 203–272.

[17] A. Givental. On the WDVV equation in quantum K-theory. *Michigan Math. J.*, 48:295–304, 2000. Dedicated to William Fulton on the occasion of his 60th birthday.

[18] A. Givental. Permutation-equivariant quantum K-theory V. Toric q-hypergeometric functions. Preprint, https://math.berkeley.edu/~giventh/perm/perm5.pdf

[19] A. Givental and Y.-P. Lee. Quantum K-theory on flag manifolds, finite-difference Toda lattices and quantum groups. *Invent. Math.*, 151(1):193–219, 2003.

[20] A. Givental and V. Tonita. The Hirzebruch–Riemann–Roch theorem in true genus-0 quantum K-theory. arXiv:1106.3136.

[21] T. Graber and R. Pandharipande. Localization of virtual classes. *Invent. Math.*, 135(2):487–518, 1999.

[22] R. Goldin, M. Harada, T. S. Holm, and T. Kimura. The full orbifold K-theory of abelian symplectic quotients. *J. K-Theory*, 8(2):339–362, 2011.

[23] E. Gonzalez and C. Woodward. Quantum cohomology and minimal model programs. *Adv. Math.* 353:591–646, 2019.

[24] E. Gonzalez and C. Woodward. A wall-crossing formula for Gromov–Witten invariants under variation of git quotient 1208.1727.

[25] E. Gonzalez and C. Woodward. Quantum Witten localization and abelianization for qde solutions. arXiv:0811.3358.

[26] E. González, P. Solis and C. Woodward. Properness for scaled gauged maps Journal of Algebra, 490, 104–157. 2017

[27] D. Halpern–Leistner. The derived category of a GIT quotient. *J. Amer. Math. Soc.* 28: 871–912, 2015.

[28] D. Halpern–Leistner. Θ-stratifications, Θ-reductive stacks, and applications. Algebraic geometry: Salt Lake City 2015. Proc. Sympos. Pure Math. 97, 349–379. Amer. Math. Soc., Providence, RI. 2018.

[29] D. Halpern–Leistner. On the structure of instability in moduli theory,

[30] Megumi Harada and Gregory D. Landweber. Surjectivity for Hamiltonian G-spaces in K-theory. *Trans. Amer. Math. Soc.*, 359(12):6001–6025 (electronic), 2007.

[31] M. Harada, G. D. Landweber, and R. Sjamaar. Divided differences and the Weyl character formula in equivariant K-theory. *Math. Res. Lett.*, 17(3):507–527, 2010.

[32] L. Hörmander. *The Analysis of Linear Partial Differential Operators I*, volume 256 of *Grundlehren der mathematischen Wissenschaften*. Springer-Verlag, Berlin-Heidelberg-New York, second edition, 1990.

[33] H. Iritani, T. Milanov, and V. Tonita. Reconstruction and Convergence in Quantum K-Theory via Difference Equations. *Int. Math. Res. Not. IMRN*, 2015(11):2887–2937, 2014.

[34] H. Iritani. An integral structure in quantum cohomology and mirror symmetry for toric orbifolds. *Adv. Math.*, 222(3):1016–1079, 2009.

[35] H. Jockers, P. Mayr, U. Ninad, A. Tabler. Wilson loop algebras and quantum K-theory for Grassmannians. . *J. High Energ. Phys.* 36, 2020.

[36] J. Kalkman. Cohomology rings of symplectic quotients. *J. Reine Angew. Math.*, 485:37–52, 1995.

[37] Y. Kawamata. D-equivalence and K-equivalence. *J. Differential Geom.*, 61(1):147–171, 2002.

[38] Y. -H. Kiem and J. Li. A wall crossing formula of Donaldson-Thomas invariants without Chern-Simons functional. *Asian J. Math.* 17:63–94, 2013.

[39] F. C. Kirwan. *Cohomology of Quotients in Symplectic and Algebraic Geometry*, volume 31 of *Mathematical Notes*. Princeton Univ. Press, Princeton, 1984.

[40] M. Kontsevich. Enumeration of rational curves via torus actions. In *The moduli space of curves (Texel Island, 1994)*, pages 335–368. Birkhäuser Boston, Boston, MA, 1995.

[41] Y.-P. Lee. Quantum K-theory. I. Foundations. *Duke Math. J.*, 121(3):389–424, 2004.

[42] Y.-P. Lee. A formula for Euler characteristics of tautological line bundles on the Deligne-Mumford moduli spaces. *Int. Math. Res. Not.*, (8):393–400, 1997.

[43] A. Okounkov. Lectures on K-theoretic computations in enumerative geometry. Geometry of moduli spaces and representation theory, 251–380, IAS/Park City Math. Ser., 24, Amer. Math. Soc., Providence, RI, 2017.

[44] S. Martin. Symplectic quotients by a nonabelian group and by its maximal torus. 2000. Preprint, arXiv:math/0001002.

[45] D. Metzler. Cohomological localization for manifolds with boundary *Int. Math. Res. Not.*, (24):1239–1274, 2002.

[46] I. Mundet i Riera. A Hitchin-Kobayashi correspondence for Kähler fibrations. *J. Reine Angew. Math.*, 528:41–80, 2000.

[47] Y. Ruan and M. Zhang. The level structure in quantum K-theory and mock theta functions. 2018. Preprint, arXiv:1804.06552.

[48] Y. Ruan and M. Zhang. Verlinde/Grassmannian Correspondence and Rank 2 δ-wall-crossing. 2018. Preprint, arXiv:1811.01377.

[49] P. Solis A complete degeneration of the moduli of G-bundles on a curve. 2013. Preprint, arXiv:1311.6847.

[50] K. Taipale. K-theoretic J-functions of type A flag varieties. *Int. Math. Res. Not. IMRN*, (16):3647–3677, 2013.

[51] M. Thaddeus. Geometric invariant theory and flips. *J. Amer. Math. Soc.*, 9(3): 691–723, 1996.

[52] B. Toen. Théorèmes de Riemann-Roch pour les champs de Deligne-Mumford. *K-Theory*, 18(1):33–76, 1999.

[53] V. Tonita. A virtual Kawasaki-Riemann-Roch formula. *Pacific J. Math.* 268: 249–255, 2014.

[54] V. Tonita. Twisted K-theoretic Gromov–Witten invariants. *Math. Ann.* 372:489–526, 2018.

[55] G. Vezzosi and A. Vistoli. Higher algebraic K-theory for actions of diagonalizable groups. *Invent. Math.*, 153(1):1–44, 2003.

[56] Y. Wen. K-Theoretic I-function of $V//_\theta G$ and Application. 2019. Preprint, arXiv: 1906.00775.

[57] C. T. Woodward. Quantum Kirwan morphism and Gromov–Witten invariants of quotients I, II, III. Transformation Groups 20 (2015) 507–556, 881–920, 21 (2016) 1–39.

[58] D.Wright. The tree formulas for reversion of power series. Journal of Pure and Applied Algebra 57 (1989) 191–211.

12

Toric Varieties and a Generalization of the Springer Resolution

William Graham[a]

Abstract. Let $\mathfrak{g}$ be a semisimple Lie algebra and $\mathcal{N}$ the nilpotent cone in $\mathfrak{g}$. The Springer resolution of $\mathcal{N}$ has played an important role in representation theory. The variety $\mathcal{N}$ is equal to $\operatorname{Spec} R(\mathcal{O}^{pr})$, where $\mathcal{O}^{pr}$ is the open nilpotent orbit in $\mathcal{N}$. This paper constructs and studies an analogue of the Springer resolution for the variety $\mathcal{M} = \operatorname{Spec} R(\widetilde{\mathcal{O}}^{pr})$, where $\widetilde{\mathcal{O}}^{pr}$ is the universal cover of $\mathcal{O}^{pr}$. The construction makes use of the theory of toric varieties. Using this construction, we provide new proofs of results of Broer and of Graham about $\mathcal{M}$. Finally, we show that the construction can be adapted to covers of an arbitrary nilpotent orbit in $\mathfrak{g}$.

Dedicated to William Fulton on the occasion of his 80th birthday

1 Introduction

Let G be a semisimple simply connected algebraic group with Lie algebra $\mathfrak{g}$, let $\mathcal{N}$ be the nilpotent cone in $\mathfrak{g}$, and let $\widetilde{\mathcal{N}}$ denote the cotangent bundle of the flag variety of G. There is a map $\mu \colon \widetilde{\mathcal{N}} \to \mathcal{N}$, called the Springer resolution, which plays an important role in representation theory. The purpose of this paper is to construct and study an analogous map $\widetilde{\mathcal{M}} \to \mathcal{M}$, where $\mathcal{M}$ is defined as follows. The nilpotent cone $\mathcal{N}$ has a dense G-orbit $\mathcal{O}^{pr}$, called the principal or regular nilpotent orbit, and $\mathcal{N} = \operatorname{Spec} R(\mathcal{O}^{pr})$, where $R(\mathcal{O}^{pr})$ denotes the ring of regular functions on $\mathcal{O}^{pr}$. The variety $\mathcal{M}$ is defined to be $\operatorname{Spec} R(\widetilde{\mathcal{O}}^{pr})$, where $\widetilde{\mathcal{O}}^{pr}$ is the universal cover of $\mathcal{O}^{pr}$. There is a commutative diagram

[a] University of Georgia

$$\begin{array}{ccc} \widetilde{\mathcal{M}} & \xrightarrow{\;\widetilde{\eta}\;} & \widetilde{\mathcal{N}} \\[2pt] \widetilde{\mu}\downarrow & & \downarrow\mu \\[2pt] \mathcal{M} & \xrightarrow{\;\eta\;} & \mathcal{N}. \end{array} \qquad\qquad (1)$$

Because the construction of $\widetilde{\mathcal{M}}$ makes use of toric varieties, the rich theory of these varieties can be used to obtain detailed information about $\widetilde{\mathcal{M}}$ and about the map $\widetilde{\mathcal{M}} \to \widetilde{\mathcal{N}}$.

The motivation for constructing $\widetilde{\mathcal{M}} \to \mathcal{M}$ is to explore what constructions involving the Springer resolution can yield in the setting of $\mathcal{M}$, which is in some sense richer than $\mathcal{N}$. One original motivation for considering covers of orbits came from the theory of Dixmier algebras, which are algebras related to the infinite-dimensional representations of $\mathfrak{g}$. More recently, work of Russell and Graham, Precup and Russell ([19], [8]) has uncovered a close connection in type A between the map $\widetilde{\mathcal{M}} \to \mathcal{M}$ and Lusztig's generalized Springer correspondence, which played an important role in the development of character sheaves. In this paper, we apply the commutative diagram above to give new proofs of a result of Broer (see [4]) showing that $\mathcal{M}$ is Gorenstein with rational singularities, and of a formula for the G-module decomposition of $R(\widetilde{\mathcal{O}})$ (see [9]).

The center Z of G acts trivially on $\widetilde{\mathcal{N}}$. To define $\widetilde{\mathcal{M}}$, we use the theory of toric varieties to modify the construction of $\widetilde{\mathcal{N}}$ and obtain a variety where Z acts faithfully. The variety $\mathcal{N}$ is isomorphic to $G \times^B \mathfrak{u}$, where B is a Borel subgroup of G with unipotent radical U, and $\mathfrak{u}$ is the Lie algebra of U. Using toric varieties, we construct a variety $\widetilde{\mathfrak{u}}$ on which Z acts faithfully, and define $\mathcal{M} = G \times^B \widetilde{\mathfrak{u}}$. Since $\widetilde{\mathfrak{u}}$ maps to $\mathfrak{u}$, we obtain the map $\widetilde{\eta}\colon \widetilde{\mathcal{M}} = G \times^B \widetilde{\mathfrak{u}} \to \widetilde{\mathcal{N}} = G \times^B \mathfrak{u}$.

The proof of Theorem 4.1 in [18] contains a related construction where a variety of the form $G \times^B \widetilde{V}$ maps to the closure of $\widetilde{\mathcal{O}}^{pr}$ in a representation of G. The variety $\widetilde{V}$ is described as the closure of a B-orbit in a representation, and it is not analyzed in detail in [18]. In our setting, because toric varieties are so well understood, we can obtain detailed results about $\widetilde{\mathfrak{u}}$ and hence about the geometry of $\widetilde{\mathcal{M}}$. This understanding is used in the results of this paper, as well as in the applications to the generalized Springer correspondence. Although $\widetilde{\mathcal{M}} \to \mathcal{M}$ is an analogue of the Springer resolution, it is not a resolution of singularities since $\widetilde{\mathcal{M}}$ is not in general smooth; it is locally the quotient of a smooth variety by a finite group (see Proposition 4.4). A genuine resolution of singularities of $\mathcal{M}$ can be constructed from $\widetilde{\mathcal{M}}$ using toric resolutions, and we use this in our proofs. Although a G-equivariant resolution of singularities exists by general theory, the commutative diagram above, as well as the detailed results about the map $\widetilde{\mathcal{M}} \to \widetilde{\mathcal{N}}$, are crucial to the applications.

The construction of this paper can be adapted to any G-equivariant covering of the principal orbit, but for simplicity, we consider only the universal cover. Also, in the final section of the paper, we show that the construction can be generalized to the setting of covers of any nilpotent orbit, not only the principal orbit (note that [18, Theorem 4.1] also applies to any nilpotent orbit cover). However, the construction is more difficult to study explicitly for general orbits, because the theory of toric varieties is not available, and further work would be required to obtain precise results about the geometry.

The contents of the paper are as follows. Section 2 contains notation and background results. Section 3 introduces and studies the toric varieties used in our main construction. In Section 4, we construct $\widetilde{\mathcal{M}}$ and the commutative diagram (1). Although $\widetilde{\mathcal{M}}$ is not in general smooth, we construct a resolution $\widehat{\mathcal{M}} \to \widetilde{\mathcal{M}}$, which is useful in proofs. Section 5 studies canonical and dualizing sheaves on $\widehat{\mathcal{M}}$ and $\widetilde{\mathcal{M}}$ and uses results about these sheaves to give new proofs of the results of Broer and of Graham (see [4] and [9]) mentioned above. The methods of this paper lead to a slightly different formula for the G-module decomposition of $R(\widetilde{\mathcal{O}}^{pr})$ than in [9], but they agree because of a fact about Weyl group conjugacy of certain weights (see Proposition 5.9). In Section 6 we apply the techniques of this paper to show that the closure in $\mathcal{M}$ of the B-orbit $B \cdot \widetilde{\nu}$ is not normal; here $\widetilde{\nu}$ is a certain element in the orbit cover $\widetilde{\mathcal{O}}^{pr}$. Finally, in Section 7, we show how this construction extends to other nilpotent orbits besides the principal orbit.

Acknowledgments: This paper is dedicated to William Fulton on the occasion of his 80th birthday. The ideas and instruction he provided while I was a postdoc continue to deeply influence my mathematical work, and I remain very grateful for his help. I would also like to thank David Vogan, who initiated me into the study of nilpotent orbits and their covers. Thanks also to James Humphreys for pointing out an omission in the proof of Proposition 5.9. Finally, I thank my collaborators Martha Precup and Amber Russell for our joint work uncovering the connection of $\widetilde{\mathcal{M}}$ with the generalized Springer correspondence.

2 Preliminaries

In this paper we work over the ground field $\mathbb{C}$. If X is a complex algebraic variety, $R(X)$ will denote the ring of regular functions on X, and $\kappa(X)$ the field of fractions of $R(X)$. If X smooth and of pure dimension n, Ω_X denotes the sheaf of top degree differential forms (i.e., differential n-forms) on X.

This is the sheaf of sections of the top exterior power of the cotangent bundle T_X^*. Since we will have no use in this paper for forms of less than top degree, we will generally omit the superscript and write $\bigwedge T_X^* = \bigwedge^n T_X^*$.

If H is a linear algebraic group, H_0 will denote its identity component. We denote the Lie algebra of an algebraic group by the corresponding fraktur letter. If H acts on X, the stabilizer in H of $x \in X$ is denoted by H^x. If V is a vector space, let V_X denote the vector bundle $V \times X \to X$; if V is a representation of H, then V_X is an H-equivariant vector bundle. The orbit $H \cdot x$ is open in its closure $\overline{H \cdot x}$ (see [14, Section 2.1]). We will make use of the following lemma.

Lemma 2.1 *Suppose $f : X \to Y$ is an H-equivariant map of irreducible varieties of the same dimension, and suppose that $H \cdot y$ is open in Y. Let $x \in f^{-1}(y)$. Then $f^{-1}(H \cdot y) = H \cdot x$.*

Proof Since $x \in f^{-1}(y)$, we have $H \cdot x \subseteq f^{-1}(H \cdot y)$. We prove the reverse inclusion. Since f maps $H \cdot x$ surjectively onto $H \cdot y$, which has the same dimension as Y and X, it follows that $\dim H \cdot x = \dim X$. Hence $\overline{H \cdot x} = X$, so $H \cdot x$ is open in X. The complement of $H \cdot x$ in X consists of orbits of smaller dimension than $\dim H \cdot y$, so no such orbit can map H-equivariantly to $H \cdot y$. Hence $H \cdot x \supseteq f^{-1}(H \cdot y)$, as desired. $\square$

2.1 Tori

If T is an algebraic torus with Lie algebra $\mathfrak{t}$, $\widehat{T}$ will denote the character group of T. This is a free abelian group which can be viewed as a subset of $\mathfrak{t}^*$. We generally use a Greek letter to denote an element of $\widehat{T}$ when viewed as an element of $\mathfrak{t}^*$, and exponential notation to denote the same element viewed as a function on T, e.g. λ and e^λ. Thus, $R(T)$ is the span of e^λ for λ in the lattice $\widehat{T}$. Under the action of T on $R(T)$, e^λ is a weight vector of weight $-\lambda$, since for $s, t \in T$, we have

$$(t \cdot e^\lambda)(s) = e^\lambda(t^{-1}s) = e^\lambda(t^{-1})e^\lambda(s) = e^{-\lambda}(t)e^\lambda(s).$$

Let $\mathbb{V}^* = \widehat{T} \otimes_{\mathbb{Z}} \mathbb{R}$, and $\mathbb{V}$ the dual real vector space (the notation is motivated by the conventions of [7, Section 1.2]). We can identify $\mathbb{V}^*$ with the real span of $\widehat{T}$ in $\mathfrak{t}^*$.

Throughout this paper Z will denote a finite subgroup of T, and $T_{ad} = T/Z$ the quotient torus. For most of this paper, T and T_{ad} will be maximal tori of the algebraic groups G and G_{ad}, and Z will be the center of G, but in Section 3.1, T and Z can be arbitrary. Write $\mathcal{P} = \widehat{T}$ for the character group of T and $\mathcal{Q} = \widehat{T}_{ad}$. We have $\mathcal{P} \supset \mathcal{Q}$, and the quotient $\mathcal{P}/\mathcal{Q}$ is isomorphic to the character group $\hat{Z}$ of Z. Since $\mathcal{P}$ and $\mathcal{Q}$ span the same real subspace of $\mathfrak{t}^*$, we can identify $\mathbb{V} = \mathcal{P} \otimes_{\mathbb{Z}} \mathbb{R} = \mathcal{Q} \otimes_{\mathbb{Z}} \mathbb{R}$.

2.2 Sheaves on mixed spaces

The main construction of this paper is an example of a "mixed space". By this we mean a scheme of the form $G \times^H S$, where $G \supset H$ are linear algebraic groups, H acts on the scheme S, and $G \times^H S = (G \times S)/H$ where H acts on $G \times S$ by the mixing action: $h(g,s) = (gh^{-1}, hs)$. In this subsection we discuss some results about mixed spaces which will be useful in studying canonical sheaves. We provide some proofs for lack of a convenient reference. We will use the following notation. If S is a scheme with an H-action, there is an equivalence between the category of H-equivariant coherent sheaves on S and the category of G-equivariant coherent sheaves on $G \times^H S$. We refer to this as an induction equivalence, and denote this equivalence by $\mathcal{I}_H^G$, so if $\mathcal{F}$ is an H-equivariant coherent sheaf on S, we denote by $\mathcal{I}_H^G \mathcal{F}$ the corresponding G-equivariant sheaf on $G \times^H S$.

The induction equivalence is compatible with tensor products and pullbacks of sheaves, as well as direct images and higher direct images. This can be seen as follows. Let p and q be the projections from $G \times S$ to (respectively) $G \times^H S$ and S. These projections are faithfully flat. The pullback p^* induces an equivalence of categories between G-equivariant sheaves on $G \times^H S$ and $G \times H$-equivariant sheaves on $G \times S$. The pullback q^* induces an equivalence of categories between H-equivariant sheaves on S and $G \times H$-equivariant sheaves on $G \times S$. See [21, Section 6]. The induction equivalence is characterized by the equation $p^* \mathcal{I}_H^G \mathcal{F} = q^* \mathcal{F}$. Compatibility of the induction equivalence with tensor products holds because pullback of sheaves commutes with tensor product; compatibility with pullbacks follows from the functoriality of pullbacks; the compatibility with (higher) direct images follows from the compatibility of higher direct images with flat pullback ([10, III, Prop. 9.3]).

Let $e^\lambda \colon H \to \mathbb{G}_m$ be a homomorphism, and let $\mathbb{C}_\lambda$ denote the corresponding 1-dimensional representation of H. Let $\mathcal{L}_\lambda$ denote the sheaf of sections of the line bundle $G \times^H \mathbb{C}_\lambda \to G/H$ on G/H. We abuse notation and view $\mathbb{C}_\lambda$ as an H-equivariant sheaf over a point; then $\mathcal{L}_\lambda = \mathcal{I}_H^G(\mathbb{C}_\lambda)$.

If $\mathcal{F}$ is an H-equivariant sheaf on S, we abuse notation and write $\mathcal{F} \otimes \mathbb{C}_\lambda$ for the H-equivariant sheaf $\mathcal{F} \otimes p_S^* \mathbb{C}_\lambda$, where p_S is the projection from S to a point. The sheaf $\mathcal{F} \otimes \mathbb{C}_\lambda$ can be identified with the sheaf $\mathcal{F}$, but with action twisted by $\mathbb{C}_\lambda$: precisely, if $s \in \mathcal{F}(U)$ for some open set U of S, then $s \otimes 1 \in (\mathcal{F} \otimes \mathbb{C}_\lambda)(U)$, and $h(s \otimes 1) = e^\lambda(h)((hs) \otimes 1)$. Given an H-equivariant map $S \to S'$, we have $f_*(\mathcal{F} \otimes \mathbb{C}_\lambda) = f_*\mathcal{F} \otimes \mathbb{C}_\lambda$. Let $\pi \colon G \times^H S \to G/H$ denote the projection.

Lemma 2.2   *With notation as above, we have*

$$\mathcal{I}_H^G(\mathcal{F} \otimes \mathbb{C}_\lambda) = (\mathcal{I}_H^G \mathcal{F}) \otimes \pi^* \mathcal{L}_\lambda$$

as G-equivariant sheaves on $G \times^H S$.

Proof   By definition, $\mathcal{I}_H^G(\mathcal{F} \otimes \mathbb{C}_\lambda) = \mathcal{I}_H^G(\mathcal{F} \otimes p_S^* \mathbb{C}_\lambda)$. We have

$$\mathcal{I}_H^G(\mathcal{F} \otimes p_S^* \mathbb{C}_\lambda) = \mathcal{I}_H^G(\mathcal{F}) \otimes \mathcal{I}_H^G(p_S^* \mathbb{C}_\lambda)$$
$$= \mathcal{I}_H^G(\mathcal{F}) \otimes \pi^* \mathcal{I}_H^G(\mathbb{C}_\lambda) = (\mathcal{I}_H^G \mathcal{F}) \otimes \pi^* \mathcal{L}_\lambda;$$

here, the first equality uses the compatibility of induction with tensor product, and the second equality uses the compatibility of induction with pullback.   □

We will make use of the following lemma about the canonical sheaf of a mixed space.

Lemma 2.3   *With notation as above, suppose S is smooth, and let $\pi \colon G \times^H S \to G/H$ be the projection. Then as G-equivariant sheaves on $G \times^H S$, we have*

$$\Omega_{G \times^H S} = \mathcal{I}_H^G \Omega_S \otimes \pi^* \Omega_{G/H}.$$

Proof   With p and q as above, we have an $G \times H$-equivariant exact sequence of vector bundles on $G \times S$, where the first map in the sequence is the composition $q^* TS \to T(G \times S) \to p^* T(G \times^H S)$:

$$0 \to q^*(TS) \to p^* T(G \times^H S) \to p^* \pi^* T(G/H) \to 0.$$

By the equivalence of categories discussed above, this exact sequence induces a G-equivariant exact sequence of vector bundles on $G \times^H S$:

$$0 \to V \to T(G \times^H S) \to \pi^* T(G/H) \to 0,$$

where $p^* V = q^*(TS)$. This implies that

$$\bigwedge T^*(G \times^H S) = \bigwedge V^* \otimes \pi^* \bigwedge T^*(G/H).$$

Therefore the sheaves of sections of these line bundles are equal; this is the statement of the lemma (note that the sheaf of sections of $\bigwedge V^*$ is $\mathcal{I}_H^G \Omega_S$).

□

2.3 Semisimple groups

For the rest of this paper, G will denote a complex semisimple simply connected algebraic group with center Z. The corresponding adjoint group is $G_{ad} = G/Z$.

We introduce some notation related to nilpotent elements which will be in effect for the remainder of the paper except for Section 7, since in that section we consider other nilpotent orbits besides the principal (that is, regular) nilpotent orbit. Let ν denote a principal nilpotent element in $\mathfrak{g}$; this means that the principal nilpotent orbit $G \cdot \nu = \mathcal{O}^{pr} \cong G/G^{\nu}$ is a dense open subvariety of $\mathcal{N}$. The universal cover of $\mathcal{O}^{pr}$ is $\widetilde{\mathcal{O}}^{pr} = G/G_0^{\nu}$. Write $\widetilde{\nu} \in \widetilde{\mathcal{O}}^{pr}$ for the coset $1 \cdot G_0^{\nu}$; then $G^{\widetilde{\nu}} = G_0^{\nu}$.

Choose a standard $\mathfrak{sl}_2$-triple $\{h, e, f\}$ such that $e = \nu$ is the nilpositive element. Write $\mathfrak{g}_i$ for the i-eigenspace of ad h on $\mathfrak{g}$; then $\mathfrak{g} = \oplus \mathfrak{g}_i$, and $\mathfrak{g}_i = 0$ if i is odd (in other words, ν is an even nilpotent element). Also, $\nu \in \mathfrak{g}_2$. Write $\mathfrak{g}_{\geq k} = \oplus_{i \geq k} \mathfrak{g}_i$ and $\mathfrak{g}_{>k} = \oplus_{i>k} \mathfrak{g}_i$. Let $\mathfrak{t} = \mathfrak{g}_0$, and let T denote the corresponding connected subgroup of G. Then T is a maximal torus of G, and $T_{ad} = T/Z$ is a maximal torus of G_{ad}. In this setting, $\mathcal{P} = \widehat{T}$ is the weight lattice of G, and $\mathcal{Q} = \widehat{T}_{ad}$ is the root lattice. Let $\mathfrak{u} = \mathfrak{g}_{>0}$; then $\mathfrak{b} = \mathfrak{t} + \mathfrak{u}$ is a Borel subalgebra of $\mathfrak{g}$. We have $\mathfrak{u}_i = \mathfrak{g}_i$ if $i > 0$. Let B and U be the subgroups of G whose Lie algebras are $\mathfrak{b}$ and $\mathfrak{u}$, respectively. By [2], we have $G^{\nu} = B^{\nu} = T^{\nu}U^{\nu}$. Because ν is principal, T^{ν} is equal to the center Z of G. We let W denote the Weyl group of T in G.

Let Φ denote the set of roots of $\mathfrak{t}$ in $\mathfrak{g}$, and let Φ^+ denote the positive root system whose elements are the weights of $\mathfrak{t}$ on $\mathfrak{u}$. Let $\alpha_1, \ldots, \alpha_r$ denote the simple roots, and let ξ denote the sum of the simple roots. We can choose vectors E_{α_i} in the root spaces $\mathfrak{g}_{\alpha_i}$ such that $\nu = \Sigma E_{\alpha_i}$ [15]. The vectors $E_{\alpha_1}, \ldots, E_{\alpha_r}$ form a basis of $\mathfrak{g}_2$. Let $E_{\alpha_1}^*, \ldots, E_{\alpha_r}^*$ denote the dual basis of $\mathfrak{g}_2^*$. Then $E_{\alpha_i}^*$ is a weight vector of weight $-\alpha_i$. The ring $R(\mathfrak{g}_2)$ is isomorphic to the polynomial ring $\mathbb{C}[E_{\alpha_1}^*, \ldots, E_{\alpha_r}^*]$. If we embed T_{ad} into $\mathfrak{g}_2$ via the map $t \mapsto t \cdot \nu$, then we obtain an inclusion $R(\mathfrak{g}_2) \hookrightarrow R(T_{ad})$ satisfying $E_{\alpha_i}^* \mapsto e^{\alpha_i}$.

3 The toric varieties $\mathcal{V}$ and $\mathcal{V}_{ad}$

The main construction of this paper uses toric varieties $\mathcal{V}$ and $\mathcal{V}_{ad}$. In this section we introduce these varieties and study some of their basic properties.

3.1 Toric varieties and finite quotients

In this section T is an arbitrary torus and Z is a finite subgroup of T. The relationship $T/Z = T_{ad}$ extends to a relationship between toric varieties for the tori T and T_{ad}.

Recall from Section 2.1 that we write $\mathcal{P} = \hat{T} \supset \mathcal{Q} = \hat{T}_{ad}$. If S is any subset of P, let $\mathbb{C}[S]$ be the subspace of $R(T)$ spanned by the e^λ for $\lambda \in S$. If S contains 0 and is closed under addition, then $\mathbb{C}[S]$ is a subring of $R(T)$.

Suppose σ is a rational polyhedral cone in $\mathbb{V}$, with dual cone $\sigma^\vee$ in $\mathbb{V}^*$. Corresponding to σ, there are affine toric varieties $X(\sigma)$ and $X_{ad}(\sigma)$, for the tori T and T_{ad}, respectively. We have semigroups $\sigma_{\mathcal{P}}^\vee = \sigma^\vee \cap \mathcal{P}$ and $\sigma_{\mathcal{Q}}^\vee = \sigma^\vee \cap \mathcal{Q}$ (in [7], the semigroup is denoted S_σ, but since we are considering two tori we use a different notation so that we can distinguish the corresponding semigroups). By definition, $R(X(\sigma)) = \mathbb{C}[\sigma^\vee \cap \mathcal{P}]$ and $R(X_{ad}(\sigma)) = \mathbb{C}[\sigma^\vee \cap \mathcal{Q}]$. (The notation χ^λ is used in [7] for what we denote as e^λ.)

The T-equivariant inclusion $R(X_{ad}(\sigma)) \hookrightarrow R(X(\sigma))$ induces a T-equivariant map $\pi \colon X(\sigma) \to X_{ad}(\sigma)$. The first part of the next proposition is essentially in [7, Section 2.2].

Proposition 3.1   *(1) The natural inclusion* $R(X_{ad}(\sigma)) \hookrightarrow R(X(\sigma))$ *has image equal to* $R(X(\sigma))^Z$, *so the corresponding map* $\pi \colon X(\sigma) \to X_{ad}(\sigma)$ *is the quotient by* Z.

(2) Under this map, the inverse image of T_{ad} *is* T. *Hence the inverse image of* $1 \in T_{ad}$ *is* $Z \subset T$.

Proof If $\lambda \in \mathcal{P}$ then any element Z takes e^λ to a multiple of itself, and e^λ is fixed by Z if and only if $\lambda \in \mathcal{Q}$. This implies that $R(X(\sigma))^Z$ is the span of e^λ for $\lambda \in \sigma_{\mathcal{Q}}^\vee$, which is exactly $R(X_{ad}(\sigma))$. This proves (1). Since the map is finite and T-equivariant, the unique orbit in $X(\sigma)$ mapping to the open orbit T_{ad} in $X_{ad}(\sigma)$ is the open orbit T. This proves (2). □

The previous proposition shows that the fibers of $\pi \colon X(\sigma) \to X_{ad}(\sigma)$ over the open T_{ad}-orbit are isomorphic to Z. We now describe the fibers of π over the other T_{ad}-orbits in $X_{ad}(\sigma)$. We need some facts about toric varieties, and in particular the description of orbits, from [7, Section 3.1]. The T-orbits in $X(\sigma)$ correspond to faces of σ. If τ is a face of σ, the orbit $\mathcal{O}(\tau)$ is isomorphic to the torus $T(\tau) = \operatorname{Spec} \mathbb{C}[\tau^\perp \cap \mathcal{P}]$. If $\tau = \{0\}$ then $T(\tau) = T$ and $\mathcal{O}(\tau)$ is the open T-orbit. The closure of $\mathcal{O}(\tau)$, denoted by $V(\tau)$, is an affine toric variety for the torus $T(\tau)$, and $R(V(\tau)) = \mathbb{C}[\tau^\perp \cap \sigma_{\mathcal{P}}^\vee]$. The embedding of $V(\tau)$ into $X(\sigma)$ corresponds to the map on rings of regular functions

$$\mathbb{C}[\sigma_{\mathcal{P}}^\vee] \to \mathbb{C}[\tau^\perp \cap \sigma_{\mathcal{P}}^\vee]$$

taking e^λ to e^λ if $\lambda \in \tau^\perp$, and 0 otherwise.

The analogous picture holds for the T_{ad}-toric variety $X_{ad}(\sigma)$; we use the subscript ad for the analogous definitions in this case. We have a commuting diagram of coordinate rings, where all maps are inclusions:

$$
\begin{array}{ccc}
\mathbb{C}[\tau^{\perp} \cap \mathcal{P}] & \longleftarrow & \mathbb{C}[\tau^{\perp} \cap \sigma_{\mathcal{P}}^{\vee}] \\
\uparrow & & \uparrow \\
\mathbb{C}[\tau^{\perp} \cap \mathcal{Q}] & \longleftarrow & \mathbb{C}[\tau^{\perp} \cap \sigma_{\mathcal{Q}}^{\vee}]
\end{array}
\tag{2}
$$

Applying Spec to this diagram yields the following diagram of tori and orbit closures:

$$
\begin{array}{ccc}
T(\tau) & \longrightarrow & V(\tau) \\
\downarrow & & \downarrow \\
T_{ad}(\tau) & \longrightarrow & V_{ad}(\tau).
\end{array}
\tag{3}
$$

Define $Z(\tau) = \ker(T(\tau) \to T_{ad}(\tau))$. This is a finite group; if $\tau = \{0\}$ then $Z(\tau) = Z$. The character group $\widehat{Z}(\tau)$ is given by $\widehat{Z}(\tau) = \widehat{T}(\tau)/\widehat{T}_{ad}(\tau) = (\tau^{\perp} \cap \mathcal{P})/(\tau^{\perp} \cap \mathcal{Q})$. Note that the groups $Z(\tau)$ and $\widehat{Z}(\tau)$ are isomorphic.

Proposition 3.2 *(a) The map $\mathcal{O}(\tau) \to \mathcal{O}_{ad}(\tau)$ is a covering map with fibers $Z(\tau)$.*

(b) The map $V(\tau) \to V_{ad}(\tau)$ is an isomorphism $\Leftrightarrow$ the map is an isomorphism on the open orbits, i.e., $\mathcal{O}(\tau) \to \mathcal{O}_{ad}(\tau)$ is an isomorphism.

Proof (a) holds because the orbits are isomorphic to tori, and the map of orbits corresponds to the map of tori. For (b), by (2), we see that the map $V(\tau) \to V_{ad}(\tau)$ is an isomorphism exactly when $\sigma^{\vee} \cap \tau^{\perp} \cap \mathcal{P} = \sigma^{\vee} \cap \tau^{\perp} \cap \mathcal{Q}$, and the map $\mathcal{O}(\tau) \to \mathcal{O}_{ad}(\tau)$ is an isomorphism exactly when $\tau^{\perp} \cap \mathcal{P} = \tau^{\perp} \cap \mathcal{Q}$. The implication ($\Leftarrow$) follows immediately from this. The implication ($\Rightarrow$) follows from the fact that if λ is in the relative interior of $\sigma^{\vee} \cap \tau^{\perp} \cap \mathcal{Q}$, the rings $\mathbb{C}[\tau^{\perp} \cap \mathcal{P}]$ and $\mathbb{C}[\tau^{\perp} \cap \mathcal{Q}]$ are obtained (respectively) from $\mathbb{C}[\tau^{\perp} \cap \sigma_{\mathcal{P}}^{\vee}]$ and $\mathbb{C}[\tau^{\perp} \cap \sigma_{\mathcal{Q}}^{\vee}]$ by adjoining the element $e^{-\lambda}$ (cf. [7], Section 1.1, and Sec. 1.2, Prop. 2). $\square$

3.2 The toric varieties $\mathcal{V}_{ad} = \mathfrak{u}_2$ and $\mathcal{V}$

We adopt the notation of Section 2.3. Let $\mathcal{V}_{ad}$ denote the vector space $\mathfrak{u}_2 = \mathfrak{g}_2$. The map $T_{ad} \to T_{ad} \cdot \nu$ embeds T_{ad} as a dense orbit in $\mathcal{V}_{ad}$, and $\mathcal{V}_{ad}$ is an affine toric variety for T_{ad}. The nilpotent element ν corresponds to the identity element $1 \in T_{ad}$. As noted in Section 2.3, $R(\mathcal{V}_{ad}) = \mathbb{C}[E_{\alpha_1}^*, \ldots, E_{\alpha_r}^*]$. When viewed as a function on T, $E_{\alpha_i}^*$ is the function e^{α_i}, which is a weight vector of weight $-\alpha_i$.

Let $\sigma^\vee \subset \mathbb{V}^*$ be the cone corresponding to $\mathcal{V}_{ad}$. The set of simple roots $\{\alpha_i\}$ is a basis of $\mathbb{V}^*$, and $\sigma^\vee$ is the cone generated by the elements of the basis $\{\alpha_i\}$ of simple roots. The dual cone σ is generated by the elements of the dual basis $\{v_i\}$ of $\mathbb{V}$.

Let $\mathcal{V}$ denote the affine toric variety for T defined using the cone σ, but using the lattice $\mathcal{P}$ instead of $\mathcal{Q}$. Thus, $R(\mathcal{V}) = \mathbb{C}[\sigma_\mathcal{P}^\vee]$ and $R(\mathcal{V}_{ad}) = \mathbb{C}[\sigma_\mathcal{Q}^\vee]$.

Since in general the cone $\sigma^\vee$ is not generated by part of a basis for $\mathcal{P}$ (see Example 3.4), the cone σ is not generated by part of a basis for $\mathcal{P}^\vee$. It follows by [7, Section 2.1] that the variety $\mathcal{V}$ is singular in general; we will see below that it is a quotient of a smooth variety by a finite group. Any toric variety has a resolution of singularities which is itself a toric variety (for the same torus); such a resolution can be obtained by taking the toric variety associated to an appropriate subdivision of the fan of the original toric variety (see [7, Section 2.6]). In subsequent sections, we will let $\widehat{\mathcal{V}} \to \mathcal{V}$ denote a resolution of singularities of $\mathcal{V}$ obtained in this way; note that such a resolution is proper.

In each coset of $\mathcal{P}$ mod $\mathcal{Q}$, we consider two distinguished elements:

(1) The minimal dominant weight λ_{dom}

(2) The element $\lambda_R = \Sigma a_\alpha \alpha$ with $0 \le a_\alpha < 1$.

Note that for the identity coset, our convention is that $\lambda_{dom} = \lambda_R = 0$.

It sometimes happens that the elements λ_{dom} and λ_R corresponding to the same coset are equal. Remarkably, they are always conjugate by the Weyl group (see Proposition 5.9). The letter R in the notation λ_R is chosen because these weights are related to $R(\mathcal{V}) = \mathbb{C}[\sigma_\mathcal{P}^\vee]$ by the following proposition (whose proof is immediate).

Proposition 3.3 *The semigroup $\sigma_\mathcal{P}^\vee$ is generated by the simple roots α and the λ_R. Hence $R(\mathcal{V})$ is a free $R(\mathcal{V}_{ad})$-module with basis given by the e^{λ_R}.* □

Example 3.4   For $\mathfrak{g} = \mathfrak{sl}_4$, the minimal dominant weights are $\lambda_1 = \frac{1}{4}(3\alpha_1 + 2\alpha_2 + \alpha_3)$, $\lambda_2 = \frac{1}{2}(\alpha_1 + 2\alpha_2 + \alpha_3)$, $\lambda_3 = \frac{1}{4}(\alpha_1 + 2\alpha_2 + 3\alpha_3)$. So $\lambda_{1,R} = \lambda_1$, $\lambda_{2,R} = \frac{1}{2}(\alpha_1 + \alpha_3)$, $\lambda_{3,R} = \lambda_3$. Note that under the usual identification of $\mathfrak{t}^*$ with the subspace of $\mathbb{C}^4$ where all coordinates sum to zero, and $\alpha_i = \epsilon_i - \epsilon_{i+1}$, we have $\lambda_2 = (\frac{1}{2}, \frac{1}{2}, -\frac{1}{2}, -\frac{1}{2})$ and $\lambda_{2,R} = (\frac{1}{2}, -\frac{1}{2}, \frac{1}{2}, -\frac{1}{2})$. We see that λ_2 and $\lambda_{2,R}$ are indeed conjugate by the Weyl group S_4, as asserted above. Write $x_i = e^{\alpha_i}$ and $w_i = e^{\lambda_{i,R}}$. Then

$$R(T_{ad}) = \mathbb{C}[x_1^{\pm 1}, x_2^{\pm 1}, x_3^{\pm 1}]$$
$$R(T) = \mathbb{C}[w_1^{\pm 1}, w_2^{\pm 1}, w_3^{\pm 1}].$$

The embedding $R(T_{ad}) \hookrightarrow R(T)$ is given by $x_1 \mapsto \frac{w_1 w_2}{w_3}, x_2 \mapsto \frac{w_1 w_3}{w_2^2}, x_3 \mapsto \frac{w_2 w_3}{w_1}$. This follows from the equations $\alpha_1 = \lambda_{1,R} + \lambda_{2,R} - \lambda_{3,R}; \alpha_2 = \lambda_{1,R} - 2\lambda_{2,R} + \lambda_{3,R}; \alpha_3 = -\lambda_{1,R} + \lambda_{2,R} + \lambda_{3,R}$. We have

$$R(\mathcal{V}_{ad}) = \mathbb{C}[x_1, x_2, x_3]$$

$$R(\mathcal{V}) = \mathbb{C}[w_1, w_2, w_3, x_1, x_2, x_3] = \mathbb{C}\left[w_1, w_2, w_3, \frac{w_1 w_2}{w_3}, \frac{w_1 w_3}{v_2^2}, \frac{w_2 w_3}{w_1}\right].$$

Note that when we write $R(\mathcal{V}) = \mathbb{C}[w_1, w_2, w_3, x_1, x_2, x_3]$, we are not asserting that $R(\mathcal{V})$ is a polynomial ring in the variables w_i and x_i; we are describing $R(\mathcal{V})$ as a subring of $R(T)$, where the x_i are expressed in terms of the w_i as described above.

Proposition 3.5 *There is a torus T_d with a surjective map $T_d \to T$ and finite kernel Z_d, and a T_d-toric variety $\mathcal{V}_d \cong \mathbb{C}^r$ such that $\mathcal{V}_d / Z_d \cong \mathcal{V}$. Hence $\mathcal{V}$ is an orbifold.*

Proof Let d be a positive integer such that $\mathcal{P}$ is contained in the lattice $\mathcal{Q}_d = \frac{1}{d}\mathcal{Q}$. (For example, if $\mathfrak{g} = \mathfrak{sl}_n$, then $d = n$ suffices.) Let T_d be the torus whose character group $\widehat{T}_d$ is $\mathcal{Q}_d$; let Z_d be the kernel of the surjective map $T_d \to T$. The character group $\widehat{Z}_d$ is isomorphic to $\mathcal{Q}_d / \mathcal{P}$. Let $\mathcal{V}_d$ be the affine toric variety for T_d corresponding to the cone $\sigma^\vee$. The semigroup $\sigma_{\mathcal{Q}_d}^\vee$ is generated by the elements $\frac{1}{d}\alpha_1, \ldots, \frac{1}{d}\alpha_r$. Since these elements form a basis for $\mathcal{Q}_d$, the toric variety $\mathcal{V}_d$ is isomorphic to $\mathbb{C}^r$. Moreover, by the arguments in Section 3.1, we have $\mathcal{V}_d / Z_d = \mathcal{V}$. $\square$

The fibers of the map $\pi \colon \mathcal{V} \to \mathcal{V}_{ad}$ can be calculated using Proposition 3.2. As noted above, the cone σ in $\mathbb{V}$ is generated by the v_i. The faces of σ correspond to subsets J of $\{1, \ldots, n\}$, the face τ_J being generated by the v_j with $j \in J$.

One can describe $\widehat{Z}(\tau_J) = (\tau_J^\perp \cap \widehat{T}) / (\tau_J^\perp \cap \widehat{T}_{ad})$ in terms of J. If J is empty then $\widehat{Z}(\tau) = \widehat{Z}$; in general there is a natural injection of $(\tau_J^\perp \cap \widehat{T}) / (\tau_J^\perp \cap \widehat{T}_{ad})$ into $\widehat{T}/\widehat{T}_{ad} = \widehat{Z}$. The image consists of the cosets $\lambda + \widehat{T}_{ad}$ which have nonempty intersection with $\tau_J^\perp$, i.e., those $\lambda + \widehat{T}_{ad}$ such that if $\lambda = \Sigma a_i \alpha_i$, then for all $j \in J$, we have $a_j \in \mathbb{Z}$. The cosets of $\widehat{T}_{ad}$ in $\widehat{T}$ have as representatives the minimal dominant weights; these and the corresponding a_i are listed in [13, Section 3.13]. From this list we can determine $\widehat{Z}(\tau)$, or equivalently $Z(\tau)$. We state the answer as the next proposition, using the numbering of simple roots from [13]. The statement of this proposition in an earlier version of this paper contained some errors; the correct statement was given by Russell in [19], to which we refer for the proof. This result also appears (with a small error in type D) in Section 2.4 of [20], using different notation and conventions.

Proposition 3.6 *Let $\mathcal{V}$ and $\mathcal{V}_{ad}$ be toric varieties associated to a simple Lie algebra as above, and let J be a nonempty subset of $\{1, \ldots, n\}$. The group $Z(J) = Z(\tau_J)$ is given as follows.*

A_n : $Z(J) = \mathbb{Z}/c\mathbb{Z}$ where $c = \gcd(J \cup \{n + 1\})$
B_n : $Z(J) = \mathbb{Z}/2\mathbb{Z}$ if all $j \in J$ are even
$\quad\quad$ $Z(J) = \{1\}$ otherwise
C_n : $Z(J) = \mathbb{Z}/2\mathbb{Z}$ if $n \notin J$
$\quad\quad$ $Z(J) = \{1\}$ otherwise
D_n : $Z(J) = \mathbb{Z}$ if $n - 1, n \notin J$ and all $j \in J$ are even
$\quad\quad$ $Z(J) = \mathbb{Z}/2\mathbb{Z}$ if $n - 1, n \notin J$ and not all $j \in J$ are even
$\quad\quad$ $Z(J) = \mathbb{Z}/2\mathbb{Z}$ if exactly one of $n - 1$ and n is in J,
$\quad\quad\quad\quad\quad\quad\quad\quad$ all $j \in J$ such that $j < n - 1$ are even,
$\quad\quad\quad\quad\quad\quad\quad\quad$ and $n = 4k + 2$ for some $k \geq 1$
$\quad\quad$ $Z(J) = \{1\}$ otherwise
E_6 : $Z(J) = \mathbb{Z}/3\mathbb{Z}$ if none of $1, 3, 5, 6$ are in J; otherwise $Z(J) = \{1\}$
E_7 : $Z(J) = \mathbb{Z}/2\mathbb{Z}$ if none of $2, 5, 7$ are in J; otherwise $Z(J) = \{1\}$.

4 The generalization of the Springer resolution

In this section we construct the variety $\widetilde{\mathcal{M}}$, and study its properties. In particular, we show that there is a commutative diagram

$$
\begin{array}{ccc}
\widetilde{\mathcal{M}} & \xrightarrow{\ \widetilde{\eta}\ } & \widetilde{\mathcal{N}} \\
\widetilde{\mu}\downarrow & & \downarrow\mu \\
\mathcal{M} & \xrightarrow{\ \eta\ } & \mathcal{N},
\end{array}
$$

where the map $\widetilde{\mu}$ is proper and an isomorphism over $\widetilde{\mathcal{O}}^{pr}$ (see Theorem 4.8).

4.1 The variety $\widetilde{\mathfrak{u}}$

By definition, $\mathcal{V}_{ad} = \mathfrak{u}_2 \cong \mathfrak{u}/\mathfrak{u}_{\geq 4}$. (Recall that $\mathfrak{u}_i = \mathfrak{g}_i$ for $i > 0$.) With the interpretation of $\mathcal{V}_{ad}$ as $\mathfrak{u}/\mathfrak{u}_{\geq 4}$, since the B-action on $\mathfrak{u}$ preserves $\mathfrak{u}_{\geq 4}$, the variety $\mathcal{V}_{ad}$ has a B-action. Under this action, the unipotent radical U acts trivially, and the projection $\mathfrak{u} \to \mathcal{V}_{ad}$ is B-equivariant. Consider the maps

$$\widehat{\mathcal{V}} \xrightarrow{\widehat{\rho}} \mathcal{V} \xrightarrow{\rho} \mathcal{V}_{ad}, \tag{4}$$

where, as in Section 3.2, $\widehat{\mathcal{V}}$ is a toric resolution of singularities of $\mathcal{V}$. These maps are T-equivariant, and become B-equivariant if we extend the T-actions on $\widehat{\mathcal{V}}$ and $\mathcal{V}$ to B-actions by requiring that U act trivially. We define $\widetilde{\mathfrak{u}} = \widetilde{\mathcal{V}} \times_{\mathcal{V}_{ad}} \mathfrak{u}$ and $\widehat{\mathfrak{u}} = \widehat{\mathcal{V}} \times_{\mathcal{V}_{ad}} \mathfrak{u}$. Consider the following diagram, where the squares are fiber squares:

$$
\begin{array}{ccccc}
\widehat{\mathfrak{u}} & \xrightarrow{\widehat{\sigma}} & \widetilde{\mathfrak{u}} & \xrightarrow{\sigma} & \mathfrak{u} \\
{\scriptstyle \widehat{p}_1}\Big\downarrow & & {\scriptstyle \widetilde{p}_1}\Big\downarrow & & {\scriptstyle p_1}\Big\downarrow \\
\widehat{\mathcal{V}} & \xrightarrow{\widehat{\rho}} & \mathcal{V} & \xrightarrow{\rho} & \mathcal{V}_{ad}.
\end{array}
\tag{5}
$$

In this diagram, we have identified $\mathcal{V}_{ad} \times_{\mathcal{V}_{ad}} \mathfrak{u}$ with $\mathfrak{u}$. The maps $\widehat{p}_1$, $\widetilde{p}_1$, and p_1 are projections onto the first factors of the fiber products. Under our identification of $\mathcal{V}_{ad} \times_{\mathcal{V}_{ad}} \mathfrak{u}$ with $\mathfrak{u}$, the map $\sigma : \widetilde{\mathfrak{u}} = \mathcal{V} \times_{\mathcal{V}_{ad}} \mathfrak{u} \to \mathfrak{u}$ is simply the projection on the second factor. We write $\gamma = \sigma \circ \widehat{\sigma} : \widehat{\mathfrak{u}} \to \mathfrak{u}$. The fiber products $\widehat{\mathfrak{u}}$ and $\widetilde{\mathfrak{u}}$ have B-actions coming from the B-actions on each factor, and the maps above are all B-equivariant. Also, $\widehat{\sigma}$ is proper, and σ is finite, since these properties hold for the maps of toric varieties in (4). As algebraic varieties, we have

$$\widetilde{\mathfrak{u}} \cong \mathcal{V} \times \mathfrak{u}_{\geq 4}, \quad \widehat{\mathfrak{u}} \cong \widehat{\mathcal{V}} \times \mathfrak{u}_{\geq 4}. \tag{6}$$

The reason for defining $\widetilde{\mathfrak{u}}$ and $\widehat{\mathfrak{u}}$ as fiber products as in (5), rather than by the formulas of (6), is that the fiber product definition allows us to equip $\widetilde{\mathfrak{u}}$ and $\widehat{\mathfrak{u}}$ with a B-action.

Extend the Z-action on $\mathcal{V}$ to a Z-action on $\widetilde{\mathfrak{u}}$ by $z \cdot (w, x) = (zw, x)$.

Proposition 4.1 (1) The actions of B and Z on $\widetilde{\mathfrak{u}}$ commute, and $\widetilde{\mathfrak{u}}/Z \cong \mathfrak{u}$ as B-varieties.

(2) $R(\widetilde{\mathfrak{u}})$ is the integral closure of $R(\mathfrak{u})$ in $\kappa(\widetilde{\mathfrak{u}})$.

Proof (1) The actions of B and Z on $\mathcal{V}$ and on $\mathfrak{u}$ commute (since the B-action on $\mathcal{V}$ comes from the T-action and $Z \subset T$, and the Z-action on $\mathfrak{u}$ is trivial). Hence the actions of B and Z on the fiber product $\widetilde{\mathfrak{u}} = \mathcal{V} \times_{\mathcal{V}_{ad}} \mathfrak{u}$ commute. To see that $\widetilde{\mathfrak{u}}/Z \cong \mathfrak{u}$ as B-varieties, observe that the varieties $\mathcal{V}$, $\mathcal{V}_{ad}$, $\widetilde{\mathfrak{u}}$ and $\mathfrak{u}$ are all affine, and $R(\mathfrak{u}) = R(\mathcal{V}) \otimes_{R(\mathcal{V}_{ad})} R(\mathfrak{u})$. The B-equivariant inclusion $R(\mathcal{V})^Z \to R(\mathcal{V})$ induces a B-equivariant map

$$R(\mathcal{V})^Z \otimes_{R(\mathcal{V}_{ad})} R(\mathfrak{u}) \to (R(\mathcal{V}) \otimes_{R(\mathcal{V}_{ad})} R(\mathfrak{u}))^Z = R(\widetilde{\mathfrak{u}})^Z. \tag{7}$$

This map is an isomorphism: indeed, $R(\mathfrak{u})$ is free over $R(\mathcal{V}_{ad})$, and given a basis $\{b_i\}$ for $R(\mathfrak{u})$ over $R(\mathcal{V}_{ad})$, and elements a_i of $R(\mathcal{V})$, the sum $\sum a_i \otimes b_i$

is Z-invariant if and only if each a_i is Z-invariant. Hence the corresponding map of affine varieties

$$(\mathcal{V} \times_{\mathcal{V}_{ad}} \mathfrak{u})/Z \to (\mathcal{V}/Z) \times_{\mathcal{V}_{ad}} \mathfrak{u}$$

is an isomorphism. Since $\mathcal{V}/Z = \mathcal{V}_{ad}$, this gives an isomorphism

$$\widetilde{\mathfrak{u}}/Z \to \mathcal{V}_{ad} \times_{\mathcal{V}_{ad}} \mathfrak{u} = \mathfrak{u},$$

proving (1).

(2) Since $\widetilde{\mathfrak{u}}/Z = \mathfrak{u}$, we have $R(\widetilde{\mathfrak{u}})^Z = R(\mathfrak{u})$. Hence $R(\widetilde{\mathfrak{u}})$ is integral over $R(\mathfrak{u})$. Since $\mathcal{V}$ is a toric variety, it is normal, hence so is $\widetilde{\mathfrak{u}} \cong \mathcal{V} \times \mathfrak{u}_{\geq 4}$. Hence $R(\widetilde{\mathfrak{u}})$ is integrally closed in $\kappa(\widetilde{\mathfrak{u}})$. It follows that $R(\widetilde{\mathfrak{u}})$ is the integral closure of $R(\mathfrak{u})$ in $\kappa(\widetilde{\mathfrak{u}})$. $\square$

4.2 The variety $\mathcal{M} = \operatorname{Spec} R(\widetilde{\mathcal{O}}^{pr})$

In this subsection, we recall some known facts we will need about $\mathcal{M}$ and $\mathcal{N}$, due to Brylinski and Kostant [5]. We include some proofs.

The principal nilpotent orbit $\mathcal{O}^{pr}$ is open in $\mathcal{N}$, and its complement has codimension 2. Since $\mathcal{N}$ is normal (by [16]), the restriction map $R(\mathcal{N}) \to R(\mathcal{O}^{pr})$ is an isomorphism. We identify $\mathcal{O}^{pr}$ with the homogeneous variety G/G^v by the map $G/G^v \to \mathcal{O}^{pr}, gG^v \mapsto g \cdot v$.

We define $\widetilde{\mathcal{O}}^{pr}$ to be the homogeneous variety G/G_0^v, and write $\widetilde{v}$ for the coset $1 \cdot G_0^v$ in $\widetilde{\mathcal{O}}^{pr}$. Let $\mathcal{M} = \operatorname{Spec} R(\widetilde{\mathcal{O}}^{pr})$. The inclusion $R(\mathcal{O}^{pr}) \hookrightarrow R(\widetilde{\mathcal{O}}^{pr})$ induces a map $\eta \colon \mathcal{M} \to \mathcal{N}$. Since the map

$$\widetilde{\mathcal{O}}^{pr} = G/G^{\widetilde{v}} \to \mathcal{O}^{pr} \cong G/G^v$$

is G-equivariant, the map $\eta : \mathcal{M} \to \mathcal{N}$ is as well. We claim that $\mathcal{M}/Z = \mathcal{N}$. Indeed, this equation is equivalent to $R(\mathcal{M})^Z = R(\mathcal{N})$, or equivalently, $R(\widetilde{\mathcal{O}}^{pr})^Z = R(\mathcal{O}^{pr})$. Since $R(\widetilde{\mathcal{O}}^{pr}) = R(G)^{G_0^v}$, and $G^v = ZG_0^v$, we have $R(\widetilde{\mathcal{O}}^{pr})^Z = R(G)^{G^v}$, proving the claim.

The variety $\widetilde{\mathcal{O}}^{pr}$ is quasi-affine (this is proved in the proof of Theorem 4.1 in [18]). This can be seen as follows (the argument of [18] can be simplified since we are only considering the principal orbit). Let V be a representation of G containing a highest weight vector v such that the stabilizer group Z^v is trivial. Then the map $G/G_0^v \to \mathfrak{g} \times V, gG_0^v \mapsto g \cdot (v, v)$ embeds $\widetilde{\mathcal{O}}^{pr}$ into $\mathfrak{g} \times V$, so $\widetilde{\mathcal{O}}^{pr}$ is quasi-affine. Let Y be the closure of $\widetilde{\mathcal{O}}^{pr}$ in $\mathfrak{g} \times V$. Since any orbit is open in its closure, $\widetilde{\mathcal{O}}^{pr}$ is open in Y. The projection $\mathfrak{g} \times V \to \mathfrak{g}$ maps Y to $\mathcal{N}$, and takes $\widetilde{\mathcal{O}}^{pr}$ to $\mathcal{O}^{pr}$.

Lemma 4.2 *There is a map* $\mathcal{M} \to Y$ *such that* $\mathcal{M}$ *is the normalization of* Y. *Hence there is an open embedding* $\tilde{\mathcal{O}}^{pr} = G/G_0^{\nu} \hookrightarrow \mathcal{M}$. *Moreover, the restriction map* $R(\mathcal{M}) \to R(\tilde{\mathcal{O}}^{pr})$ *is an isomorphism.*

Proof Let $\kappa(\tilde{\mathcal{O}}^{pr})$ be the function field of $\tilde{\mathcal{O}}^{pr}$. Since $\tilde{\mathcal{O}}^{pr}$ is quasi-affine, $\kappa(\tilde{\mathcal{O}}^{pr})$ is the fraction field of $R(\tilde{\mathcal{O}}^{pr}) = R(\mathcal{M})$; this is also the fraction field of $R(Y)$. Because $\tilde{\mathcal{O}}^{pr}$ is normal (being smooth), $R(\mathcal{M})$ is integrally closed in $\kappa(\tilde{\mathcal{O}}^{pr})$, so $\mathcal{M}$ is normal. Because $R(\mathcal{M})^Z = R(\mathcal{N})$, $R(\mathcal{M})$ is integral over $R(\mathcal{N})$ ([1, Ch. 5, Exer. 12]). The maps $\tilde{\mathcal{O}}^{pr} \to Y \to \mathcal{N}$ yield maps

$$R(\mathcal{N}) = R(\mathcal{O}^{pr}) \to R(Y) \to R(\mathcal{M}) = R(\tilde{\mathcal{O}}^{pr}).$$

Since the composition is injective, the map $R(\mathcal{N}) \to R(Y)$ is injective. Since $R(\mathcal{M})$ is integral over $R(\mathcal{N})$, it follows that $R(\mathcal{M})$ is integral over $R(Y)$. Since $R(\mathcal{M})$ is integrally closed, $R(\mathcal{M})$ is the integral closure of $R(Y)$ in its fraction field. Therefore the map $\mathcal{M} \to Y$ induced by the inclusion $R(Y) \to R(\mathcal{M})$ is the normalization map. Since the normalization map is an isomorphism over the locus of nonsingular points, it is an isomorphism over $\tilde{\mathcal{O}}^{pr}$. Hence we obtain an open embedding of $\tilde{\mathcal{O}}^{pr}$ into $\mathcal{M}$. The map $\mathcal{M} \to \mathcal{N}$ is finite and takes $\tilde{\mathcal{O}}^{pr}$ to $\mathcal{O}^{pr}$. Since the complement of $\mathcal{O}^{pr}$ in $\mathcal{N}$ has codimension 2, so does the complement of $\tilde{\mathcal{O}}^{pr}$ in $\mathcal{M}$. Since $\mathcal{M}$ is normal and the complement of $\tilde{\mathcal{O}}^{pr}$ has codimension 2, the restriction map $R(\mathcal{M}) \to R(\tilde{\mathcal{O}}^{pr})$ is an isomorphism. $\square$

By definition, $\tilde{\nu}$ is the coset $1 \cdot G^{\tilde{\nu}}$ in $\tilde{\mathcal{O}}^{pr}$, so, via the embedding of the lemma, we view $\tilde{\nu}$ as an element of $\mathcal{M}$. Under the composition $\mathcal{M} \to Y \to \mathcal{N}$, we have $\tilde{\nu} \mapsto (\nu, \nu) \mapsto \nu$. The following corollary (known to Brylinski and Kostant, who call $\mathcal{M}$ the normal closure of $\tilde{\mathcal{O}}^{pr}$) summarizes the situation.

Corollary 4.3 *There is a G-equivariant commutative diagram*

$$
\begin{array}{ccccc}
G/G^{\tilde{\nu}} & \xrightarrow{\;=\;} & \tilde{\mathcal{O}}^{pr} & \longrightarrow & \mathcal{M} \\
\downarrow & & \downarrow & & \downarrow{\scriptstyle \eta} \\
G/G^{\nu} & \xrightarrow{\;\cong\;} & \mathcal{O}^{pr} & \longrightarrow & \mathcal{N},
\end{array}
$$

where the horizontal maps into $\mathcal{M}$ *and* $\mathcal{N}$ *are open embeddings, and the vertical maps are quotients by the center* Z *of* G. $\square$

4.3 The generalized Springer resolution

We define $\widetilde{\mathcal{M}} = G \times^B \tilde{\mathfrak{u}}$ and $\widehat{\mathcal{M}} = G \times^B \hat{\mathfrak{u}}$. Since $\tilde{\mathcal{N}} = G \times^B \mathfrak{u}$, from our maps

$$\widehat{\mathfrak{u}} \xrightarrow{\widehat{\sigma}} \widetilde{\mathfrak{u}} \xrightarrow{\sigma} \mathfrak{u}$$

we obtain maps

$$\widehat{\mathcal{M}} \xrightarrow{\widehat{\eta}} \widetilde{\mathcal{M}} \xrightarrow{\widetilde{\eta}} \widetilde{\mathcal{N}}.$$

The map $\widehat{\eta}$ is proper (since $\widehat{\sigma}$ is proper) and $\widetilde{\eta}$ is finite (since σ is finite). In this section, we prove some properties of $\widetilde{\mathcal{M}}$, construct a map $\widetilde{\mathcal{M}} \to \mathcal{M}$, and study the properties of this map.

Let Z_d be the finite group defined in the proof of Proposition 3.5.

Proposition 4.4 *There is a covering of $\widetilde{\mathcal{M}}$ by open subvarieties W_i of the form $W_i = \widetilde{W}_i / Z_d$, where $\widetilde{W}_i$ is a smooth variety with an action of Z_d.*

Proof Because the group B is solvable, it is a special group, so the principal bundle $\pi : G \to G/B$ is locally trivial in the Zariski topology (see [6]). This means we can cover G/B with open subvarieties W'_i such that $W_i = \pi^{-1}(W'_i) \cong W'_i \times B$. Thus, $\widetilde{\mathcal{M}}$ is covered by the subvarieties

$$(W'_i \times B) \times^B \widetilde{\mathfrak{u}} \cong W'_i \times \widetilde{\mathfrak{u}} \cong W'_i \times V \times \mathfrak{u}_{\geq 4}.$$

By Proposition 3.5, $V = V_d/Z_d$, where $V_d \cong \mathbb{C}^r$ is smooth. The result follows. $\quad\square$

By the preceding proof, $\widetilde{\mathcal{M}}$ is locally the product of V and a smooth variety. Since V is not smooth in general, $\widetilde{\mathcal{M}}$ is not smooth in general.

Remark 4.5 The fibers of $\widetilde{\eta}$ can be described explicitly. For each subset J of the simple roots, one can define subsets $\widetilde{\mathcal{M}}_J$ and $\widetilde{\mathcal{N}}_J$ of $\widetilde{\mathcal{M}}$ and $\widetilde{\mathcal{N}}$ such that $\eta^{-1}(\widetilde{\mathcal{N}}_J) = \widetilde{\mathcal{M}}_J$, and the fibers of $\widetilde{\eta}$ over $\widetilde{\mathcal{N}}_J$ are isomorphic to the group $Z(J)$ calculated in Proposition 3.6.

In viewing V_{ad} as a T_{ad}-toric variety, we used the embedding $T_{ad} \hookrightarrow V_{ad}$ which takes 1 to v. We also denote by 1 the element in V corresponding to the identity in T under $T \hookrightarrow V$. Then, under $\rho : V \to V_{ad}$, we have $\rho(1) = 1$, and under $\sigma : \widetilde{\mathfrak{u}} \to \mathfrak{u}$, we have $\sigma(1, v) = v$.

Recall that $\widetilde{v} = 1 \cdot G_0^v \in \widetilde{\mathcal{O}}^{pr} = G/G_0^v$, and $G^{\widetilde{v}} = B^{\widetilde{v}} = B_0^v = U^v$.

Proposition 4.6 *(1) The stabilizer groups $B^{(1,v)}$ and $B^{\widetilde{v}}$ are equal.*

(2) Under the map $\sigma : \widetilde{\mathfrak{u}} \to \mathfrak{u}$, we have $\sigma^{-1}(B \cdot v) = B \cdot (1,v)$. Hence $B \cdot (1,v)$ is open in $\widetilde{\mathfrak{u}}$, and $\overline{B \cdot (1,v)} = \widetilde{\mathfrak{u}}$.

Proof We have $b = tu \in B^{(1,v)}$ if and only if $b \cdot (1,v) = (t, tuv) = (1,v)$. This holds if and only if $t = 1$ and $u \in U^v$, i.e. $b \in U^v = B^{\widetilde{v}}$. This proves (1).

For (2), the assertion $\sigma^{-1}(B \cdot v) = B \cdot (1, v)$ follows from Lemma 2.1. Since $B \cdot v$ is open in $\mathfrak{u}$, its inverse image $B \cdot (1, v)$ is open in $\widetilde{\mathfrak{u}}$. Since $\widetilde{\mathfrak{u}}$ is irreducible, it follows that $\overline{B \cdot (1, v)} = \widetilde{\mathfrak{u}}$. $\qquad\square$

There is a map $B \cdot (1, v) \to \mathcal{M}$ defined by $b \cdot (1, v) \mapsto b \cdot \widetilde{v}$. The equality of stabilizers $B^{(1,v)} = B^{\widetilde{v}}$ implies that this map yields an isomorphism $B \cdot (1, v) \to B \cdot \widetilde{v}$.

Proposition 4.7 *The map $B \cdot (1, v) \to \mathcal{M}$ extends to a map $\phi \colon \widetilde{\mathfrak{u}} \to \mathcal{M}$. We have a B-equivariant commutative diagram*

$$\begin{array}{ccc} \widetilde{\mathfrak{u}} & \xrightarrow{\ \phi\ } & \mathcal{M} \\ \sigma \downarrow & & \downarrow \eta \\ \mathfrak{u} & \longrightarrow & \mathcal{N}. \end{array} \qquad (8)$$

The map ϕ induces a map $\psi \colon \widetilde{\mathfrak{u}} \to \overline{B \cdot \widetilde{v}}$, which is the normalization map. We have $\psi^{-1}(B \cdot \widetilde{v}) = B \cdot (1, v)$, and ψ restricts to an isomorphism $B \cdot (1, v) \to B \cdot \widetilde{v}$.

Proof To show that the map extends, it is enough to show that the image of the pullback map $R(\mathcal{M}) \to R(B \cdot (1, v))$ lies in the subring $R(\widetilde{\mathfrak{u}})$. Consider the commutative diagram

$$\begin{array}{ccc} R(\widetilde{\mathfrak{u}}) \longrightarrow R(B \cdot (1, v)) \longleftarrow R(\mathcal{M}) \\ \uparrow \qquad\qquad \uparrow \qquad\qquad \uparrow \\ R(\mathfrak{u}) \longrightarrow R(B \cdot v) \longleftarrow R(\mathcal{N}). \end{array} \qquad (9)$$

Since $R(\mathcal{M})$ is integral over $R(\mathcal{N})$, the image of $R(\mathcal{M})$ in $R(B \cdot (1, v))$ is integral over the image of $R(\mathcal{N})$ in $R(B \cdot v)$. Since the map $B \cdot v \to \mathcal{N}$ extends to $\mathfrak{u} \to \mathcal{N}$, we know that the image of $R(\mathcal{N}) \to R(B \cdot v)$ lies in the subring $R(\mathfrak{u})$. We conclude that the image of $R(\mathcal{M})$ in $R(B \cdot (1, v))$ is integral over $R(\mathfrak{u})$. By Proposition 4.1, the integral closure of $R(\mathfrak{u})$ in $\kappa(\widetilde{\mathfrak{u}}) = \kappa(B \cdot (1, v))$ is $R(\widetilde{\mathfrak{u}})$, so the image of $R(\mathcal{M})$ lies in $R(\widetilde{\mathfrak{u}})$. This proves the first assertion of the proposition. The commutative diagram (8) of affine varieties follows from the commutative diagram (9) of rings of functions.

Since $B \cdot (1, v)$ is open in $\widetilde{\mathfrak{u}}$, the image of $\widetilde{\mathfrak{u}} \to \mathcal{M}$ lies in the closure of the image of $B \cdot (1, v)$, which is $\overline{B \cdot \widetilde{v}}$. Hence we obtain a birational map $\psi \colon \widetilde{\mathfrak{u}} \to \overline{B \cdot \widetilde{v}}$. To show that this is the normalization map, since $R(\widetilde{\mathfrak{u}})$ is integrally closed in $\kappa(\widetilde{\mathfrak{u}}) = \kappa(\overline{B \cdot \widetilde{v}})$, it suffices to show that $R(\widetilde{\mathfrak{u}})$ is integral over $R(\overline{B \cdot \widetilde{v}})$. To see this, observe that the map $\mathcal{M} \to \mathcal{N}$ takes $\overline{B \cdot \widetilde{v}}$ to $\mathfrak{u}$, so we have maps $R(\mathfrak{u}) \to R(\overline{B \cdot \widetilde{v}}) \to R(\widetilde{\mathfrak{u}})$. Since $R(\widetilde{\mathfrak{u}})$ is integral over $R(\mathfrak{u})$, the desired assertion follows.

The assertion $\psi^{-1}(B \cdot \widetilde{v}) = B \cdot (1, v)$ follows from Lemma 2.1. The fact that ψ restricts to an isomorphism $B \cdot (1, v) \to B \cdot \widetilde{v}$ follows from the equality $B^{(1,v)} = B^{\widetilde{v}}$ of stabilizer groups.

Finally, all the maps in the commutative diagram (8) are B-equivariant. Indeed, it was noted earlier that the map $\widetilde{u} \to u$ is B-equivariant and that the map $\mathcal{M} \to \mathcal{N}$ is G-equivariant. The horizontal maps are the extensions of maps from B-orbits; since the maps from B-orbits are B-equivariant by definition, all the maps on rings of functions are B-equivariant, and therefore the extensions of the maps from the B-orbits are B-equivariant as well. $\square$

The variety $\overline{B \cdot \widetilde{v}}$ is not normal in general (see Section 6).

Define $\widetilde{\mu} \colon \widetilde{\mathcal{M}} \to \mathcal{M}$ by $\widetilde{\mu}([g, \xi]) = g \cdot \phi(\xi)$, where ϕ is as in Proposition 4.7.

Theorem 4.8   *The map $\widetilde{\mu} \colon \widetilde{\mathcal{M}} \to \mathcal{M}$ is proper and is an isomorphism over $\widetilde{\mathcal{O}}^{pr}$. There is a commutative diagram*

$$
\begin{array}{ccc}
\widetilde{\mathcal{M}} & \xrightarrow{\ \widetilde{\eta}\ } & \widetilde{\mathcal{N}} \\
\widetilde{\mu} \downarrow & & \downarrow \mu \\
\mathcal{M} & \xrightarrow{\ \eta\ } & \mathcal{N},
\end{array}
$$

where all maps are G-equivariant. The horizontal maps in this diagram are quotients by Z.

Proof   The map $\widetilde{\mu}$ can be written as a composition:

$$\widetilde{\mathcal{M}} = G \times^B \widetilde{u} \to G \times^B \overline{B \cdot \widetilde{v}} \to G \times^B \mathcal{M} \xrightarrow{\cong} G/B \times \mathcal{M} \to \mathcal{M}.$$

The first map is finite since it is induced by the normalization map $\widetilde{u} \to \overline{B \cdot \widetilde{v}}$, which is finite. The second map is a closed embedding. The third map is the isomorphism which arises because $\mathcal{M}$ is a G-variety, not merely a B-variety; it is given by $[g, x] \mapsto (gB, g \cdot x)$. The fourth map is projection to the second factor. Since all of these maps are proper, so is their composition $\widetilde{\mu}$.

Since $\widetilde{u} \to \overline{B \cdot \widetilde{v}}$ is an isomorphism over the open orbit $B \cdot \widetilde{v}$, the map $G \times^B \widetilde{u} \to G \times^B \overline{B \cdot \widetilde{v}}$ is an isomorphism over the open subset $G \times^B B \cdot \widetilde{v}$. The map $G \times^B B \cdot (1, v) \to \mathcal{M}$ takes $G \times^B B \cdot (1, v)$ isomorphically onto the orbit $G \cdot \widetilde{v} = \mathcal{O}^{pr}$. It follows that $\widetilde{\mu}$ is an isomorphism over $\widetilde{\mathcal{O}}^{pr}$. Finally, the fact that the diagram (4.8) commutes follows from the commutativity of the diagram (8), and G-equivariance follows from the definitions of the maps.

As noted in Section 4.2, $\mathcal{M}/Z = \mathcal{N}$. To see that $\widetilde{\mathcal{M}}/Z = \widetilde{\mathcal{N}}$, note that the Z-action on $G \times^B \widetilde{u}$ is by

$$z \cdot (g, u) = (zg, u) = (gz, u) = (g, z \cdot u).$$

Since $\widetilde{\mathfrak{u}}/Z = \mathfrak{u}$ by Proposition 4.1, the fibers of the map $G \times^B \widetilde{\mathfrak{u}} \to G \times^B \mathfrak{u}$ are the orbits of Z, and the result follows. □

We have shown that $\widetilde{\mathcal{O}}^{pr}$ is a subvariety of $\mathcal{M}$; by Theorem 4.8, we can also view $\widetilde{\mathcal{O}}^{pr}$ as an open subvariety of $\widetilde{\mathcal{M}}$. Since $\widetilde{\mathcal{M}}$ is not in general smooth, $\widetilde{\mathcal{M}} \to \mathcal{M}$ is not a resolution of singularities of $\mathcal{M}$. However, the composition $\widehat{\mathcal{M}} \xrightarrow{\widehat{\eta}} \widetilde{\mathcal{M}} \xrightarrow{\widetilde{\mu}} \mathcal{M}$ is a resolution of the singularities of $\mathcal{M}$. Since $\widetilde{\mu} \circ \widehat{\eta}$ is G-invariant and an isomorphism over an open set, it is an isomorphism over $\widetilde{\mathcal{O}}^{pr}$.

Corollary 4.9 *We have isomorphisms*

$$R(\mathcal{M}) \xrightarrow{\widetilde{\mu}^*} R(\widetilde{\mathcal{M}}) \xrightarrow{\widehat{\eta}^*} R(\widehat{\mathcal{M}}) \xrightarrow{i^*} R(\widetilde{\mathcal{O}}^{pr}),$$

where the first two maps are pullbacks, and the third map is restriction, i.e., pullback via the inclusion $i : \widetilde{\mathcal{O}}^{pr} \to \widehat{\mathcal{M}}$.

Proof The maps $\widehat{\mathcal{M}} \to \widetilde{\mathcal{M}} \to \mathcal{M}$ are isomorphisms over $\widetilde{\mathcal{O}}^{pr}$, so $\widetilde{\mathcal{O}}^{pr}$ may be viewed as an open subvariety of $\mathcal{M}$, $\widetilde{\mathcal{M}}$, and $\widehat{\mathcal{M}}$. Since each of these are irreducible varieties, restriction to $\widetilde{\mathcal{O}}^{pr}$ is injective on regular functions. Thus, i^* and $\widehat{\eta}^* \circ i^*$ are injective. By Lemma 4.2, $\widetilde{\mu}^* \circ \widehat{\eta}^* \circ i^*$ is an isomorphism. It follows that each of the maps i^*, $\widehat{\eta}^*$, and $\widetilde{\mu}^*$ is an isomorphism. □

Example 4.10 We work out the construction of Theorem 4.8 in the case $G = SL(2) = Sp(2)$. We choose notation which will be compatible with Example 7.4 below. Let ϵ_i be the function on the space of 2×2 diagonal matrices defined by $\epsilon_i(\text{diag}(a_1, a_2)) = a_i$. The maximal torus $\mathfrak{t} \subset \mathfrak{g}$ consists of diagonal matrices of trace 0; the simple root α is usually written as $\alpha = \epsilon_1 - \epsilon_2$, but as an element of $\mathfrak{t}^*$, we can also write $\alpha = 2\epsilon_1$. Then the fundamental dominant weight is $\frac{1}{2}\alpha = \epsilon_1$. In this case, the principal nilpotent ν is a root vector $X_{2\epsilon_1}$. We have $\mathcal{V}_{ad} = \mathfrak{u} = \mathbb{C}_{2\epsilon_1}$ as a representation of T or B; then $\mathcal{V} = \mathbb{C}_{\epsilon_1}$, and the map $\mathcal{V} \cong \mathbb{C} \to \mathcal{V}_{ad} \cong \mathbb{C}$ is given by $z \mapsto z^2$. As is well-known, the variety $\mathcal{M}$ can be identified with $\mathbb{C}^2$, with the natural action of G. This can be seen as follows. Equip $\mathbb{C}^2$ with the symplectic form $dx_1 \wedge dx_2$. The action of G on $\mathbb{C}^2$ is Hamiltonian, with moment map $\eta: \mathbb{C}^2 \to \mathfrak{g}^*$. Using a G-invariant bilinear form (such as a constant multiple of the Killing form, with the constant chosen for convenience), we can identify $\mathfrak{g}^*$ with $\mathfrak{g}$ so that $\mu: \mathbb{C}^2 \to \mathfrak{g}$ is given in coordinates by

$$\mu : \begin{bmatrix} x_1 \\ x_2 \end{bmatrix} \mapsto c \begin{bmatrix} -x_1 x_2 & x_1^2 \\ x_2^2 & x_1 x_2 \end{bmatrix}.$$

Here c is a constant, and by choosing an appropriate multiple of the Killing form, we can assume $c = 1$. We see that μ takes $(1, 0)$ to the principal nilpotent $\nu = X_{2\epsilon_1}$, so $\mu \colon \mathbb{C}^2 \to \mathcal{N}$. Since μ is G-equivariant, and $\mathbb{C}^2$ and $\mathcal{N}$ both have dimension 2, we see that η restricts to a covering $\mathbb{C}^2 \setminus \{0\} \to \mathcal{O}^{pr}$. Since $\mathbb{C}^2 \setminus \{0\}$ is simply connected, we see that $\tilde{\mathcal{O}}^{pr} = \mathbb{C}^2 \setminus \{0\}$, so $\mathcal{M} = \mathrm{Spec}(\tilde{\mathcal{O}}^{pr}) = \mathbb{C}^2$, as claimed. We have

$$
\begin{array}{ccc}
\widetilde{\mathcal{M}} = G \times^B \mathbb{C}_{\epsilon_1} & \xrightarrow{\tilde{\eta}} & \widetilde{\mathcal{N}} = G \times^B \mathbb{C}_{2\epsilon_1} \\
\tilde{\mu} \downarrow & & \downarrow \mu \\
\mathcal{M} & \xrightarrow{\eta} & \mathcal{N}.
\end{array}
$$

The map $\widetilde{\mathcal{N}}$ is given by $\widetilde{\mathcal{N}} \colon [g, z] \mapsto [g, z^2]$. The vertical maps are given by $\tilde{\mu} \colon [g, z] \mapsto g \cdot \begin{bmatrix} z \\ 0 \end{bmatrix}$ and $\mu \colon [g, w] \mapsto g \cdot w X_{2\epsilon_1}$. Note that in this case, $\widetilde{\mathcal{N}} \cong T^*(G/B)$ is the vector bundle on $G/B = \mathbb{P}^1$ whose space of global sections is $\mathcal{O}_{\mathbb{P}^1}(-2)$. Also, $\widetilde{\mathcal{M}}$ is the tautological subbundle of the trivial bundle $\mathbb{C}^2 \times \mathbb{P}^1 \to \mathbb{P}^1$ whose space of global sections is $\mathcal{O}_{\mathbb{P}^1}(-1)$. This example generalizes to the minimal nilpotent orbit for the group $G = Sp(2n)$; see Example 7.4.

5 Canonical sheaves and applications

In [18] and [12], the canonical sheaves on orbits and resolutions are used in proving multiplicity formulas and proving that normalizations of orbit closures are Gorenstein with rational singularities. In this section we adapt these arguments to $\mathcal{M}$ using our resolution of singularities of $\mathcal{M}$. As observed earlier, the map $\widetilde{\mathcal{M}} \to \mathcal{M}$ is not a resolution of singularities, since $\widetilde{\mathcal{M}}$ is locally a quotient of a smooth variety by a finite group. Therefore, more background (e.g. an extension of the Grauert-Riemenschneider theorem to the orbifold situation) would be required to apply the arguments of [18] and [12] to the map $\widetilde{\mathcal{M}} \to \mathcal{M}$. In this paper, we circumvent this problem by making use of the auxiliary variety $\widehat{\mathcal{M}}$, which is smooth, and the composition $\widehat{\mathcal{M}} \to \widetilde{\mathcal{M}} \to \mathcal{M}$, which is a resolution of singularities.

The contents of the section are as follows. In Section 5.1 we prove a result of Broer [4, Cor. 6.3] that the variety $\mathcal{M}$ is Gorenstein with rational singularities (note that Broer's result applies in the setting of covers of arbitrary nilpotent orbits). The proof here is similar to the proof given by Hinich [12] for normalizations of orbit closures; we can use Hinich's argument because we have a resolution of $\mathcal{M}$ which also maps to $\widetilde{\mathcal{N}}$. (See Remark 5.3 for further discussion.) We then turn to canonical and dualizing sheaves, and describe the pushforward of the canonical sheaf $\Omega_{\widehat{\mathcal{M}}}$ on $\widehat{\mathcal{M}}$ under the map $h = \widetilde{\eta} \circ \widehat{\eta} : \widehat{\mathcal{M}} \to \widetilde{\mathcal{N}}$ (see Theorem 5.7), after some preliminary results about these sheaves on toric varieties and on the spaces $\widehat{u}$ and $\widetilde{u}$. Finally, in Section 5.4, we use this description to recover the formula for the G-module decomposition of $R(\widetilde{\mathcal{O}}^{pr})$ from [9]. Note that the formula from [9] is different from the formula that arises from the techniques of this paper; the equivalence of the formulas follows from Proposition 5.9. We remark that a description of $R(\widetilde{\mathcal{O}}^{pr})$ as a graded module was obtained by Broer (see [3]), refining the result of [9].

5.1 The Gorenstein property and rational singularities

Given an n-dimensional irreducible smooth variety X, recall that Ω_X^n denotes the sheaf of top degree differential forms (i.e., differential n-forms) on X. This is the sheaf of sections of $\bigwedge^n T^* X$. As in Section 2, since we will only consider forms of top degree, we write $\bigwedge T_X^* = \bigwedge^n T_X^*$ and $\Omega_X = \Omega_X^n$.

We recall our varieties and maps:

$$\widehat{\mathcal{M}} \xrightarrow{\widehat{\eta}} \widetilde{\mathcal{M}} \xrightarrow{\widetilde{\eta}} \widetilde{\mathcal{N}}$$
$$\widetilde{\mu} \downarrow \qquad\qquad \downarrow \mu$$
$$\mathcal{M} \xrightarrow{\eta} \mathcal{N}.$$

Write $\widehat{\mu}$ for the resolution of singularities $\widetilde{\mu} \circ \widehat{\eta} \colon \widehat{\mathcal{M}} \to \mathcal{M}$, and $h = \widetilde{\eta} \circ \widehat{\eta} \colon \widehat{\mathcal{M}} \to \widetilde{\mathcal{N}}$. The fact that $\widehat{\mathcal{M}}$ maps to $\widetilde{\mathcal{N}}$ as well as to $\mathcal{M}$ plays a significant role in the following proof.

Theorem 5.1 *The variety $\mathcal{M}$ is Gorenstein with rational singularities.*

Proof The map $\widehat{\mu} \colon \widehat{\mathcal{M}} \to \mathcal{M}$ is proper and birational, with $\widehat{\mathcal{M}}$ smooth and $\mathcal{M}$ normal. By a result of Hinich [12, Lemma 2.3], if there is a map $\varphi \colon \mathcal{O}_{\widehat{\mathcal{M}}} \to \Omega_{\widehat{\mathcal{M}}}$ such that the induced map

$$\widehat{\mu}_* \mathcal{O}_{\widehat{\mathcal{M}}} \to \widehat{\mu}_* \Omega_{\widehat{\mathcal{M}}} \tag{10}$$

is an isomorphism, then $\mathcal{M}$ is Gorenstein with rational singularities. Since $\mathcal{M}$ is an affine variety, to say that (10) is an isomorphism amounts to saying that the map

$$H^0(\varphi)\colon H^0(\widehat{\mathcal{M}},\mathcal{O}_{\widehat{\mathcal{M}}}) \to H^0(\widehat{\mathcal{M}},\Omega_{\widehat{\mathcal{M}}}) \tag{11}$$

is an isomorphism.

The variety $\widetilde{\mathcal{N}}$ is a holomorphic symplectic variety (since it is isomorphic to the cotangent bundle of G/B), and taking the top exterior power of the symplectic form gives a nowhere vanishing G-invariant section Ξ of $\Omega_{\widetilde{\mathcal{N}}}$. The pullback $\widehat{\Xi} = h^*(\Xi)$ of this section to $\widehat{\mathcal{M}}$ is a G-invariant section of $\Omega_{\widehat{\mathcal{M}}}$. Since the restriction of h to $\widetilde{\mathcal{O}}^{pr}$ is a covering map, the section $\widehat{\Xi}$ does not vanish at any point of the open set $\widetilde{\mathcal{O}}^{pr}$. Using $\widehat{\Xi}$, we define the map $\varphi\colon \mathcal{O}_{\widehat{\mathcal{M}}} \to \Omega_{\widehat{\mathcal{M}}}$ as the sheaf map which on any open set $\mathcal{U}$ takes $a \in O_{\widehat{\mathcal{M}}}(\mathcal{U})$ to $a\widehat{\Xi}|_U$. The induced map $H^0(\varphi)$ takes $a \in H^0(\widehat{\mathcal{M}},\mathcal{O}_{\widehat{\mathcal{M}}})$ to $a\,\widehat{\Xi}$. This map is injective because the section Ξ is not identically zero. To show that the map $H^0(\varphi)$ is surjective, suppose $\tau \in H^0(\widehat{\mathcal{M}},\Omega_{\widehat{\mathcal{M}}})$. Since Ξ is nowhere zero on $\widetilde{\mathcal{O}}^{pr}$, the quotient $\tau/\widehat{\Xi}$ defines an element r of $R(\widetilde{\mathcal{O}}^{pr})$. By Corollary 4.9, the element r is the restriction of an element (also denoted by r) of $R(\widehat{\mathcal{M}})$. Hence $\tau = r\,\widehat{\Xi} = H^0(\varphi)(r)$, so $H^0(\varphi)$ is surjective. Hence $H^0(\varphi)$ is an isomorphism, proving the theorem. $\qquad\qquad\square$

As a consequence of the proof, we obtain the following.

Corollary 5.2   *The map* $R(\mathcal{M}) \to H^0(\widehat{\mathcal{M}},\Omega_{\widehat{\mathcal{M}}})$ *taking* r *to* $r\,\widehat{\Xi}$ *is a G-module isomorphism.*

Proof   The map $r \mapsto r\,\widehat{\Xi}$ is the composition of the pullback map $R(\mathcal{M}) \to R(\widehat{\mathcal{M}})$ with the map (11). The pullback map is an isomorphism by Corollary 4.9, and the map (11) is an isomorphism by the proof of Theorem 5.1. Hence the map $r \mapsto r\,\widehat{\Xi}$ is an isomorphism; it is G-equivariant because $\widehat{\Xi}$ is G-invariant. $\qquad\qquad\square$

Remark 5.3   Broer's proof that $\mathcal{M}$ is Gorenstein with rational singularities makes use of the facts that $\mathcal{N}$ is Gorenstein with rational singularities, and that $\mathcal{M} \to \mathcal{N}$ is a finite map of normal varieties which is a covering map over an open set whose complement has codimension 2. By [4, Theorem 6.2], these hypotheses imply that $\mathcal{M}$ is Gorenstein with rational singularities. Using resolutions of singularities for normalizations of nilpotent orbit closures, Hinich proved that the normalization of any nilpotent orbit closure is Gorenstein with rational singularities. Thus (as noted earlier), Broer's result applies for arbitrary orbits, not only the principal orbit. In this paper, we do not make use of the fact that $\mathcal{N}$ is Gorenstein with rational singularities to deduce our result for $\mathcal{M}$. Rather, since we have constructed a resolution for $\mathcal{M}$, we apply the method of Hinich directly, using this resolution.

5.2 Canonical sheaves and toric varieties

We recall some facts about toric varieties from [7]. An affine toric variety $X(\sigma)$ is smooth if and only if σ is generated by part of a basis for $\mathcal{P}^\vee$. A general toric variety $X(\mathcal{F})$ is smooth if and only if this holds for each cone σ in the fan $\mathcal{F}$.

Suppose $\mathcal{F}'$ is a fan which is a subdivision of $\mathcal{F}$, i.e., $\mathcal{F}'$ and $\mathcal{F}$ have the same support, and every cone in $\mathcal{F}$ is a union of cones in $\mathcal{F}'$. There is a T-equivariant proper map $\pi: X(\mathcal{F}') \to X(\mathcal{F})$. Given any fan $\mathcal{F}$, one can choose a subdivision $\mathcal{F}'$ such that $X(\mathcal{F}')$ is smooth, and then π is a resolution of singularities. In this situation, $\omega_{X(\mathcal{F})} := \pi_* \Omega_{X(\mathcal{F}')}$ is independent of the subdivision chosen, and $R^i \pi_* \Omega_{X(\mathcal{F}')} = 0$ for $i > 0$ (see [7, Section 4.4]). On a complete toric variety, $\omega_{X(\mathcal{F})}$ is the dualizing sheaf. For the affine toric variety $X(\sigma)$, we view $\omega_{X(\sigma)}$ simply as an $R(X(\sigma))$-module. By [7, Section 4.4], it can be viewed as the $R(X(\sigma))$-submodule of $R(T)$ spanned by the e^μ such that μ is positive on all nonzero vectors in σ.

We now turn to the toric varieties $\mathcal{V}$ and $\mathcal{V}_{ad}$. Since these varieties are affine, we can view the sheaves $\omega_{\mathcal{V}}$ and $\Omega_{\mathcal{V}_{ad}}$ as (respectively) $R(\mathcal{V})$ and $R(\mathcal{V}_{ad})$-modules. The toric variety $\mathcal{V}_{ad}$ is the vector space $\mathfrak{u}_2$ spanned by the E_α, where α is simple, and the functions $x_i = e^{\alpha_i}$ give coordinates on $\mathcal{V}_{ad}$ (see Section 3.2). The sheaf $\Omega_{\mathcal{V}_{ad}}$ is the free $R(\mathcal{V}_{ad})$ module generated by $dx_1 dx_2 \cdots dx_r$, so it is spanned by the elements $x_1^{a_1} \cdots x_r^{a_r} dx_1 dx_2 \cdots dx_r$, where $a_i \geq 0$. Each of these elements is a T_{ad}-weight vector of weight $-\sum (a_i + 1)\alpha_i = -(\xi + \sum a_i \alpha_i)$. (Recall that $x_i = e^{\alpha_i}$ has weight $-\alpha_i$; the differential form $dx_1 dx_2 \cdots dx_r$ has weight $-\xi$, where, as in Section 2.3, ξ is the sum of the simple roots.) Thus, $\Omega_{\mathcal{V}_{ad}}$ is isomorphic to the $R(\mathcal{V}_{ad})$-submodule of $R(T_{ad})$, spanned by the e^μ, where $\mu = \sum (a_i + 1)\alpha_i$. We see that as a $R(\mathcal{V}_{ad})$-module, $\Omega_{\mathcal{V}_{ad}} = R(\mathcal{V}_{ad}) \otimes \mathbb{C}_{-\xi}$. Note that the cone σ of $\mathbb{V}$ is generated by the elements of the basis $\{v_i\}$ dual to the basis of simple roots, and $\mu(v_i) = a_i + 1$. That is, the elements μ are exactly those elements of the lattice Q whose values are positive on any nonzero element of σ, recovering the description of [7, Section 4.4].

We can phrase this in the language of vector bundles. Recall that if V is a vector space, then V_X denotes the vector bundle $V \times X \to X$. The top exterior power of the cotangent bundle of $\mathcal{V}_{ad} = \mathfrak{u}_2$ is the bundle $\bigwedge T^* \mathfrak{u}_2 = (\bigwedge \mathfrak{u}_2^*)_{\mathcal{V}_{ad}} = (\mathbb{C}_{-\xi})_{\mathcal{V}_{ad}}$, where the last equality holds since as a representation of T_{ad}, $\mathfrak{u}_2^* = \mathbb{C}_{-\xi}$. Therefore, as a T-equivariant vector bundle on $\mathcal{V}_{ad}$, $\bigwedge T^* \mathcal{V}_{ad} \cong (\mathbb{C}_{-\xi})_{\mathcal{V}_{ad}}$. This is a vector bundle which is trivial but not T-equivariantly trivial. The trivializing section has weight $-\xi$; it corresponds to the differential form $dx_1 dx_2 \cdots dx_r$.

Recall that in each coset of $\mathcal{P}$ mod $\mathcal{Q}$ there is a unique element $\lambda_R = \sum a_i \alpha_i$ such that $0 \le a_i < 1$. Write $\lambda_C = \xi - \lambda_R$; then $\lambda_C = \sum b_i \alpha_i$ is the unique element in the coset such that $0 < b_i \le 1$.

Consider the maps

$$\widehat{\mathcal{V}} \xrightarrow{\widehat{\rho}} \mathcal{V} \xrightarrow{\rho} \mathcal{V}_{ad}.$$

Proposition 5.4 *As an $R(T_{ad})$-module with T-action,*

$$(\rho \circ \widehat{\rho})_* \Omega_{\widehat{\mathcal{V}}} = \rho_*(\omega_{\mathcal{V}}) = \bigoplus_{\lambda_C} R(\mathcal{V}_{ad}) \otimes \mathbb{C}_{-\lambda_C} = \bigoplus_{\lambda_R} \Omega_{\mathcal{V}_{ad}} \otimes \mathbb{C}_{\lambda_R}. \quad (12)$$

Moreover, for $i > 0$,

$$R^i(\rho \circ \widehat{\rho})_* \Omega_{\widehat{\mathcal{V}}} = R^i \rho_*(\omega_{\mathcal{V}}) = 0. \quad (13)$$

Proof The equality $(\rho \circ \widehat{\rho})_* \Omega_{\widehat{\mathcal{V}}} = \rho_*(\omega_{\mathcal{V}})$ holds by definition of $\omega_{\mathcal{V}}$. The sheaf $\rho_*(\omega_{\mathcal{V}})$ corresponds to the $R(\mathcal{V}_{ad})$-module given by the space of global sections of $\omega_{\mathcal{V}}$. By [7, Section 4.4], this space corresponds to the set of e^μ for $\mu \in \mathcal{P}$ which are positive on any nonzero element of σ. This means that if $\mu = \sum b_i \alpha_i$, then each $b_i > 0$. Note that the b_i are rational numbers, but not necessarily integers.

Any $\mu = \sum c_i \alpha_i$ in $\mathcal{P}$, with each $c_i > 0$, can be written uniquely as $\mu = \lambda_C + \nu$ for some λ_C, where $\nu = \sum d_i \alpha_i$ and each d_i is a nonnegative integer. That is, $e^\mu = e^\nu \cdot e^{\lambda_C}$. We see that $\rho_*(\omega_{\mathcal{V}})$, viewed as an $R(\mathcal{V}_{ad})$-module, is equal to the $R(T_{ad})$-submodule of $R(T)$ given by $\oplus R(T_{ad}) \cdot e^{\lambda_C}$, where the sum is over all λ_C. Since e^{λ_C} is a T-weight vector of weight $-\lambda_C$, we see that as an $R(T_{ad})$-module with T-action, $\rho_*(\omega_{\mathcal{V}}) = \oplus R(\mathcal{V}_{ad}) \otimes \mathbb{C}_{-\lambda_C}$, proving the second equality of (12). Since $\Omega_{\mathcal{V}_{ad}} = R(\mathcal{V}_{ad}) \otimes \mathbb{C}_{-\xi}$, we have

$$\rho_*(\omega_{\mathcal{V}}) = \bigoplus \Omega_{\mathcal{V}_{ad}} \otimes \mathbb{C}_{\xi - \lambda_C} = \bigoplus \Omega_{\mathcal{V}_{ad}} \otimes \mathbb{C}_{\lambda_R}.$$

proving the third equality of (12).

To prove (13), consider the spectral sequence $R^i \rho_* \circ R^j \widehat{\rho}_* \Rightarrow R^{i+j}(\rho \circ \widehat{\rho})_*$. Since ρ is a finite map, for any coherent sheaf $\mathcal{F}$ on $\mathfrak{u}$, we have $R^i \rho_*(\mathcal{F}) = 0$ for $i > 0$. By [7], $R^i \widehat{\rho}_* \Omega_{\widehat{\mathcal{V}}} = 0$ for $i > 0$. We conclude that for $i > 0$, we have $R^i(\rho \circ \widehat{\rho})_* \Omega_{\widehat{\mathcal{V}}} = 0$, as desired. $\square$

5.3 Canonical sheaves on $\widehat{\mathfrak{u}}$ and $\widetilde{\mathfrak{u}}$

Recall our maps $\widehat{\mathfrak{u}} \xrightarrow{\widehat{\sigma}} \widetilde{\mathfrak{u}} \xrightarrow{\sigma} \mathfrak{u}$, with $\gamma = \sigma \circ \widehat{\sigma}$. As above, we let $\widehat{p}_1$, $\widetilde{p}_1$, and p_1 denote the projections of $\widehat{\mathfrak{u}}$, $\widetilde{\mathfrak{u}}$, and $\mathfrak{u}$ onto $\widehat{\mathcal{V}}$, $\mathcal{V}$, and $\mathcal{V}_{ad}$, respectively. We define $\omega_{\widetilde{\mathfrak{u}}} = \widehat{\sigma}_* \Omega_{\widehat{\mathfrak{u}}}$. In this section we describe $\sigma_* \omega_{\widetilde{\mathfrak{u}}} = \gamma_* \Omega_{\widehat{\mathfrak{u}}}$. The result is analogous to Proposition 5.4; the main issue in the proof is showing that the description obtained is B-equivariant (T-equivariance is clear).

As a variety,

$$\widehat{\mathfrak{u}} = \widehat{\mathcal{V}} \times \mathfrak{u}_{\geq 4}. \tag{14}$$

Since the tangent bundle $\mathfrak{u}_{\geq 4}$ is the trivial bundle $(\mathfrak{u}_{\geq 4})_{\mathfrak{u}_{\geq 4}}$, we have an identification

$$T\widehat{\mathfrak{u}} = \widehat{p}_1^* T\widehat{\mathcal{V}} \oplus (\mathfrak{u}_{\geq 4})_{\widehat{\mathfrak{u}}} \tag{15}$$

of vector bundles on $\widehat{\mathfrak{u}}$. Moreover, this identification is T-equivariant, since the decomposition (14) is T-equivariant. However, we cannot assert that the identifcation is B-equivariant, since (14) is not B-equivariant. Nevertheless, we have the following.

Lemma 5.5 *There are B-equivariant exact sequences of vector bundles on $\widehat{\mathfrak{u}}$ and $\mathfrak{u}$, respectively:*

$$0 \to (\mathfrak{u}_{\geq 4})_{\widehat{\mathfrak{u}}} \to T\widehat{\mathfrak{u}} \to \widehat{p}_1^* T\widehat{\mathcal{V}} \to 0,$$

and

$$0 \to (\mathfrak{u}_{\geq 4})_{\mathfrak{u}} \to T\mathfrak{u} \to p_1^* T\mathcal{V}_{ad} \to 0.$$

Proof We prove the lemma for the first sequence; the proof for the second is similar. The existence of the first exact sequence follows from the identification (15). We must prove that the maps involved are B-equivariant. Since the map $\widehat{p}_1 : \widehat{\mathfrak{u}} \to \widehat{\mathcal{V}}$ is B-equivariant, so is the induced map $T\widehat{\mathfrak{u}} \to \widehat{p}_1^* T\widehat{\mathcal{V}}$. The map $(\mathfrak{u}_{\geq 4})_{\widehat{\mathfrak{u}}} \to T\widehat{\mathfrak{u}}$ is T-equivariant, as follows from the T-invariance of (15), so it suffices to verify that the map is U-equivariant. Suppose $(\widehat{x}, q) \in \widehat{\mathcal{V}} \times \mathfrak{u}_{\geq 4} \cong \widehat{\mathfrak{u}}$. Let $x = \rho \circ \widehat{\rho} \circ \widehat{p}_1(\widehat{x}) \in \mathcal{V}_{ad}$. Let x' denote the element x, but viewed as an element of $\mathfrak{u}$ via our identification of $\mathfrak{u}_2$ as a subspace of $\mathfrak{u}$. Note that if $u \in U$, while we have $ux = x$, we may not have $ux' = x'$. We have

$$u \cdot (\widehat{x}, q) = (\widehat{x}, ux' - x' + uq).$$

Let $\xi \in \mathfrak{u}_{\geq 4}$, and let $\xi_{(\widehat{x},q)}$ denote the corresponding element in the fiber $(\widehat{\mathfrak{u}}_{\geq 4})_{\mathfrak{u},(\widehat{x},q)}$ The map $(\mathfrak{u}_{\geq 4})_{\widehat{\mathfrak{u}}} \to T\widehat{\mathfrak{u}}$ takes $\xi_{(\widehat{x},q)}$ to the vector $\xi_{(\widehat{x},q)}^+$ in $T_{(\widehat{x},q)}\widehat{\mathfrak{u}}$. Here $\xi_{(\widehat{x},q)}^+$ is the tangent vector which, when applied to a test function φ, yields

$$\xi_{(\widehat{x},q)}^+(\varphi) = \frac{d}{dt}\varphi(\widehat{x}, q + t\xi)|_{t=0}.$$

To prove U-equivariance of the map $(u_{\geq 4})_{\widehat{u}} \to T\widehat{u}$, it suffices to show that if $u \in U$, then

$$u_*(\xi^+_{(\widehat{x},q)}) = (u \cdot \xi)^+_{u \cdot (\widehat{x},q)}.$$

This follows from a direct calculation:

$$u_*(\xi^+_{(\widehat{x},q)})(\varphi) = \xi^+_{(\widehat{x},q)}(\varphi \circ u) = \frac{d}{dt}\varphi \circ u(\widehat{x}, q + t\xi)|_{t=0}$$

$$= \frac{d}{dt}\varphi(\widehat{x}, ux' - x' + uq + tu \cdot \xi)|_{t=0}$$

$$= (u \cdot \xi)^+_{u \cdot (\widehat{x},q)}(\varphi).$$

This completes the proof.   □

The analogue of Proposition 5.4 is the following.

Proposition 5.6 *As B-equivariant sheaves on* $\mathfrak{u}$,

$$\gamma_*\Omega_{\widehat{\mathfrak{u}}} = \sigma_*(\omega_{\widehat{\mathfrak{u}}}) = \bigoplus_{\lambda_R} \Omega_{\mathfrak{u}} \otimes \mathbb{C}_{\lambda_R}. \tag{16}$$

Also, for $i > 0$,

$$R^i\gamma_*\Omega_{\widehat{\mathfrak{u}}} = 0. \tag{17}$$

Proof The first equality of (16) follows from the definition of $\omega_{\widehat{\mathfrak{u}}}$, so we show $\gamma_*\Omega_{\widehat{\mathfrak{u}}} = \bigoplus_{\lambda_R} \Omega_{\mathfrak{u}} \otimes \mathbb{C}_{\lambda_R}$. The exact sequences of Lemma 5.5 imply that as B-equivariant sheaves, $\Omega_{\widehat{\mathfrak{u}}} = \widehat{p}_1^*\Omega_{\widehat{\mathcal{V}}} \otimes \bigwedge \mathfrak{u}^*_{\geq 4}$ and $\Omega_{\mathfrak{u}} = p_1^*\Omega_{\mathcal{V}_{ad}} \otimes \bigwedge \mathfrak{u}^*_{\geq 4}$. Therefore, with maps as in (5), we have

$$\gamma_*\Omega_{\widehat{\mathfrak{u}}} = \gamma_*\widehat{p}_1^*\Omega_{\widehat{\mathcal{V}}} \otimes \bigwedge \mathfrak{u}^*_{\geq 4} = \sigma_*\widehat{\sigma}_*\widehat{p}_1^*\Omega_{\widehat{\mathcal{V}}} \otimes \bigwedge \mathfrak{u}^*_{\geq 4}$$

$$= \sigma_*\widetilde{p}_1^*\widehat{\rho}_*\Omega_{\widehat{\mathcal{V}}} \otimes \bigwedge \mathfrak{u}^*_{\geq 4} = \sigma_*\widetilde{p}_1^*\omega_{\mathcal{V}} \otimes \bigwedge \mathfrak{u}^*_{\geq 4}$$

$$= p_1^*\rho_*\omega_{\mathcal{V}} \otimes \bigwedge \mathfrak{u}^*_{\geq 4}$$

$$= \bigoplus_{\lambda_R} p_1^*\Omega_{\mathcal{V}_{ad}} \otimes \bigwedge \mathfrak{u}^*_{\geq 4} \otimes \mathbb{C}_{\lambda_R} = \bigoplus_{\lambda_R} \Omega_{\mathfrak{u}} \otimes \mathbb{C}_{\lambda_R}.$$

This proves (16). To prove the vanishing result (17), it suffices to show that $R^i\gamma_*\widehat{p}_1^*\Omega_{\widehat{\mathcal{V}}} = 0$ for $i > 0$. Since $\gamma = \sigma \circ \widehat{\sigma}$, and σ is a finite map, it suffices (as in the proof of Proposition 5.4) to show that $R^i\widehat{\sigma}_*(\widehat{p}_1^*\Omega_{\widehat{\mathcal{V}}}) = 0$ for $i > 0$. In the left fiber square in (5), the maps $\widehat{p}_1$ and $\widetilde{p}_1$ are flat, so $R^i\widehat{\sigma}_*(\widehat{p}_1^*\Omega_{\widehat{\mathcal{V}}}) \cong \widetilde{p}_1^*R^i\widehat{\rho}_*(\Omega_{\widehat{\mathcal{V}}})$. As noted above, $R^i\widehat{\rho}_*(\Omega_{\widehat{\mathcal{V}}}) = 0$ for $i > 0$ by [7]; (17) follows.

                                                                        □

Recall our commutative diagram

$$\begin{array}{ccccc}
\widehat{\mathcal{M}} & \xrightarrow{\;\widehat{\eta}\;} & \widetilde{\mathcal{M}} & \xrightarrow{\;\widetilde{\eta}\;} & \widetilde{\mathcal{N}} \\
{\scriptstyle\widehat{\mu}}\big\downarrow & & & & \big\downarrow{\scriptstyle\mu} \\
& & \mathcal{M} & \xrightarrow{\;\eta\;} & \mathcal{N},
\end{array}$$

as well as the maps $h = \widetilde{\eta} \circ \widehat{\eta} \colon \widehat{\mathcal{M}} \to \widetilde{\mathcal{N}}$ and $\widehat{\mu} = \widetilde{\mu} \circ \widehat{\eta} \colon \widehat{\mathcal{M}} \to \mathcal{M}$.

Theorem 5.7 *Let* $\pi \colon \widetilde{\mathcal{N}} \to G/B$ *denote the projection, and* $\mathcal{L}_\lambda$ *the sheaf of sections of the line bundle* $G \times^B \mathbb{C}_\lambda$ *on* G/B. *As* G-*equivariant sheaves on* $\widetilde{\mathcal{N}}$, *we have*

$$h_* \Omega_{\widehat{\mathcal{M}}} = \bigoplus_{\lambda_R} \pi^* \mathcal{L}_{\lambda_R}.$$

Moreover,

$$R^i h_* \Omega_{\widehat{\mathcal{M}}} = 0$$

for $i > 0$.

Proof By definition, $\widehat{\mathcal{M}} = G \times^B \widehat{\mathfrak{u}}$, $\widetilde{\mathcal{M}} = G \times^B \widetilde{\mathfrak{u}}$ and $\widetilde{\mathcal{N}} = G \times^B \mathfrak{u}$. Let $\widehat{\pi}$, $\widetilde{\pi}$ and π, respectively, denote the projections of these spaces to G/B. By Lemma 2.3, we have

$$\Omega_{\widehat{\mathcal{M}}} = \mathfrak{I}_B^G \Omega_{\widehat{\mathfrak{u}}} \otimes \widehat{\pi}^* \Omega_{G/B}.$$

Since $\widehat{\pi} = \pi \circ h$, we have

$$h_* \Omega_{\widehat{\mathcal{M}}} = h_* \left(\mathfrak{I}_B^G \Omega_{\widehat{\mathfrak{u}}} \otimes h^* \pi^* \Omega_{G/B} \right) = h_* \left(\mathfrak{I}_B^G \Omega_{\widehat{\mathfrak{u}}} \right) \otimes \pi^* \Omega_{G/B}. \qquad (18)$$

The map h is $\mathfrak{I}_B^G \gamma$, where $\gamma = \sigma \circ \widehat{\sigma} \colon \widehat{\mathfrak{u}} \to \mathfrak{u}$. By the compatibility of $\mathfrak{I}_B^G$ with direct image, the right hand side is $\mathfrak{I}_B^G (\gamma_* \Omega_{\widehat{\mathfrak{u}}}) \otimes \pi^* \Omega_{G/B}$. By Proposition 5.6 and Lemma 2.3, this equals

$$\bigoplus_{\lambda_R} \mathfrak{I}_B^G (\Omega_{\mathfrak{u}} \otimes \mathbb{C}_{\lambda_R}) \otimes \pi^* \Omega_{G/B} = \bigoplus_{\lambda_R} \mathfrak{I}_B^G (\Omega_{\mathfrak{u}}) \otimes \pi^* \mathcal{L}_{\lambda_R} \otimes \pi^* \Omega_{G/B}$$

$$= \bigoplus_{\lambda_R} \Omega_{\widetilde{\mathcal{N}}} \otimes \pi^* \mathcal{L}_{\lambda_R}.$$

Since $\widetilde{\mathcal{N}}$ is isomorphic to the cotangent bundle of G/B, it has a G-invariant holomorphic symplectic form, whose top exterior power is a G-invariant nowhere vanishing section of $\Omega_{\widetilde{\mathcal{N}}}$. Hence, as G-equivariant sheaves, $\Omega_{\widetilde{\mathcal{N}}} \cong \mathcal{O}_{\widetilde{\mathcal{N}}}$. This proves the first equation of the theorem. Similarly, arguing as in (18), we have

$$R^i h_* \Omega_{\widehat{\mathcal{M}}} = \mathfrak{I}_B^G(R^i \gamma_* \Omega_{\widehat{\mathfrak{u}}}) \otimes \pi^* \Omega_{G/B}.$$

For $i > 0$, $R^i \gamma_* \Omega_{\widehat{\mathfrak{u}}} = 0$ by Proposition 5.6, so $R^i h_* \Omega_{\widehat{\mathcal{M}}} = 0$, as desired. $\qquad \square$

5.4 Multiplicity formulas

The main result of this section is a proof of a formula from [9] for the G-module decomposition of $R(\mathcal{M}) = R(\widetilde{\mathcal{O}}^{pr})$. The argument here is adapted from the argument given in [18], where a formula is given for the ring of functions on an arbitrary nilpotent orbit. This proof, which is different from the argument in [9], is possible because we have a resolution of $\mathcal{M}$ which also maps to $\widetilde{\mathcal{N}}$. The formula arising from the arguments of this paper is given in terms of the weights λ_R, while the formula in [9] is in terms of the weights λ_{dom}. The fact that the two formulas are equivalent follows from Proposition 5.9, which shows that the weights λ_{dom} and λ_R (belonging to the same coset of $\mathcal{P}$ mod $\mathcal{Q}$) are conjugate by the Weyl group W.

By definition, for any weight $\mu \in \mathcal{P}$, $\mathrm{Ind}_T^G(\mathbb{C}_\mu)$ is the space of global sections of the vector bundle $G \times^T \mathbb{C}_\mu \to G/T$ on the affine variety G/T. As a representation of G, $\mathrm{Ind}_T^G(\mathbb{C}_\mu)$ is a direct sum of finite dimensional irreducible representations of G. By Frobenius reciprocity, the multiplicity of an irreducible representation V in $\mathrm{Ind}_T^G(\mathbb{C}_\mu)$ equals the dimension of the μ-weight space of V. Since W-conjugate weights occur with the same multiplicity in V (see [13]), as representations of G, $\mathrm{Ind}_T^G(\mathbb{C}_\mu) = \mathrm{Ind}_T^G(\mathbb{C}_{w\mu})$ for any $w \in W$.

Recall that in each coset of $\mathcal{P}$ mod $\mathcal{Q}$ we have defined elements λ_R and λ_{dom}. For the identity coset $\mathcal{Q}$, $\lambda_R = \lambda_{dom} = 0$ (the convention of [9] was that $\lambda = 0$ was not considered as a weight of the form λ_{dom}, but listed separately).

Theorem 5.8 *As G-modules,*

$$R(\widetilde{\mathcal{O}}^{pr}) = R(\mathcal{M}) = \bigoplus_{\lambda_R} \mathrm{Ind}_T^G(\mathbb{C}_{\lambda_R}).$$

Proof The idea of the proof is to use the fact (see [18, Lemma 2.1]) that as a G-module, for any weight μ, we have

$$\chi(\widetilde{\mathcal{N}}, \pi^* \mathcal{L}_\mu) = \mathrm{Ind}_T^G(\mathbb{C}_\mu), \tag{19}$$

where $\pi \colon \widetilde{\mathcal{N}} \to G/B$ is the projection, and $\chi(\widetilde{\mathcal{N}}, \mathcal{F}) = \sum (-1)^i H^i(\widetilde{\mathcal{N}}, \mathcal{F})$ denotes the Euler characteristic. To apply this fact, we will express $R(\mathcal{M})$ in terms of global sections on $\widetilde{\mathcal{N}}$, and then use cohomology vanishing to identify the space of global sections with the Euler characteristic.

The argument is as follows. By Corollary 5.2, $R(\mathcal{M}) \cong H^0(\widehat{\mathcal{M}}, \Omega_{\widehat{\mathcal{M}}})$. Consider the composition

$$\widehat{\mathcal{M}} \xrightarrow{h} \widetilde{\mathcal{N}} \xrightarrow{\mu} \mathcal{N}.$$

The variety $\mathcal{N}$ is affine, and for any i, the sheaf $R^i(\mu \circ h)_*(\Omega_{\widehat{\mathcal{M}}})$ corresponds to the $R(\mathcal{N})$-module $H^i(\widehat{\mathcal{M}}, \Omega_{\widehat{\mathcal{M}}})$. In particular, $R(\mathcal{M})$ is the $R(\mathcal{N})$-module corresponding to the sheaf

$$(\mu \circ h)_*(\Omega_{\widehat{\mathcal{M}}}) = \mu_* \circ h_*(\Omega_{\widehat{\mathcal{M}}}) = \mu_*(\oplus_{\lambda_R} \pi^* \mathcal{L}_{\lambda_R}),$$

where the last equality follows from Theorem 5.7. Since $\mathcal{N}$ is affine, this sheaf corresponds to the $R(\mathcal{N})$-module $H^0(\widetilde{\mathcal{N}}, \oplus_{\lambda_R} \pi^* \mathcal{L}_{\lambda_R}) = \oplus_{\lambda_R} H^0(\widetilde{\mathcal{N}}, \pi^* \mathcal{L}_{\lambda_R})$. In light of the equality (19), to complete the proof, it suffices to show that for $i > 0$, $H^i(\widetilde{\mathcal{N}}, \pi^* \mathcal{L}_{\lambda_R}) = 0$.

Recall that $\widehat{\mu} = \widetilde{\mu} \circ \widehat{\eta} \colon \widehat{\mathcal{M}} \to \mathcal{M}$. We have $\mu \circ h = \eta \circ \widehat{\mu}$. Therefore,

$$R^i(\mu \circ h)_*(\Omega_{\widehat{\mathcal{M}}}) = R^i(\eta \circ \widehat{\mu})_*(\Omega_{\widehat{\mathcal{M}}}).$$

There is a spectral sequence

$$R^i \eta_* \circ R^j \widehat{\mu}_*(\Omega_{\widehat{\mathcal{M}}}) \Rightarrow R^{i+j}(\eta \circ \widehat{\mu})_*(\Omega_{\widehat{\mathcal{M}}}).$$

Since η is a finite map, the higher direct images $R^i \eta_*$ vanish. Since the map $\widehat{\mu}$ is proper and birational, the Grauert-Riemenschneider theorem implies that $R^j \widehat{\mu}_*(\Omega_{\widehat{\mathcal{M}}}) = 0$ for $j > 0$. We conclude that for $i > 0$, $R^i(\eta \circ \widehat{\mu})_*(\Omega_{\widehat{\mathcal{M}}}) = 0$, so $R^i(\mu \circ h)_*(\Omega_{\widehat{\mathcal{M}}}) = 0$. There is a spectral sequence

$$R^i \mu_* \circ R^j h_*(\Omega_{\widehat{\mathcal{M}}}) \Rightarrow R^{i+j}(\mu \circ h)_*(\Omega_{\widehat{\mathcal{M}}}).$$

By Theorem 5.7, $R^j h_* \Omega_{\widehat{\mathcal{M}}} = 0$ for $j > 0$. Hence, for $i > 0$, $(R^i \mu_*) h_*(\Omega_{\widehat{\mathcal{M}}}) = 0$. Since $\mathcal{N}$ is affine, the sheaf $(R^i \mu_*) h_*(\Omega_{\widehat{\mathcal{M}}})$ corresponds to the $R(\mathcal{N})$-module $H^i(\widetilde{\mathcal{N}}, h_*(\Omega_{\widehat{\mathcal{M}}}))$. Also, by Theorem 5.7, $h_* \Omega_{\widehat{\mathcal{M}}} = \oplus_{\lambda_R} \pi^* \mathcal{L}_{\lambda_R}$. We conclude that for all λ_R and for all $i > 0$, we have $H^i(\widetilde{\mathcal{N}}, \pi^* \mathcal{L}_{\lambda_R}) = 0$. This completes the proof. $\square$

Observe that the proof of Theorem 5.8 yielded the fact that for any λ_R and any $i > 0$, we have

$$H^i(\widetilde{\mathcal{N}}, \pi^* \mathcal{L}_{\lambda_R}) = 0. \tag{20}$$

Since we have chosen our Borel subalgebra to correspond to positive weight spaces (which is the opposite of the convention of [9]), general principles would suggest a vanishing theorem corresponding to negative weights. The equation above shows that the vanishing also holds for weights which are slightly nonnegative. In fact, more is true: one can show using a vanishing theorem of Hesselink (see [11]) that this vanishing holds for any W-conjugate of λ_{dom} (see [9, Theorem 1.3]). See [3, Theorem 2.4] for a further generalization.

The equivalence of Theorem 5.8 with the multiplicity result of [9] is a consequence of the following proposition.

Proposition 5.9 *The weights* λ_{dom} *and* λ_R *(in a fixed coset of* $\mathcal{P}$ *mod* $\mathcal{Q}$*) are* W*-conjugate.*

Proof This can be proved by explicit calculation. For example, consider type $A_{n-1} = \mathfrak{sl}_n$. In this case, $\mathfrak{t}$ can be identified as the subspace of $\mathbb{C}^n$ defined by the equation $x_1 + \cdots + x_n = 0$; the inner product $((x_1, \ldots, x_n), (y_1, \ldots, y_n)) = \sum x_i y_i$ gives an identification $\mathfrak{t} \simeq \mathfrak{t}^*$. We take as simple roots $\alpha_i = \epsilon_i - \epsilon_{i+1}$. The fundamental dominant weights $\lambda_1, \ldots, \lambda_{n-1}$ are all minuscule; in coordinates,

$$\lambda_i = \frac{1}{n}(n-i, \ldots, n-i, -i, \ldots, -i),$$

where the first i entries are equal to $n - i$. Suppose $\lambda_{dom} = \lambda_i$, and write

$$\lambda_R = (c_1, \ldots, c_n) = \sum a_k \alpha_k.$$

Then each c_k satisfies $nc_k \equiv -i \pmod{n}$, and $\sum c_k = 0$. The conditions $0 \le a_k < 1$ imply that $-n < nc_k < n$ for all k; therefore each nc_k is either $-i$ or $n - i$. The condition $\sum c_k = 0$ implies that exactly i of the nc_k must equal $n - i$. It follows that the coordinates $(c_1, \ldots, c_n)$ are a permutation of $\frac{1}{n}(n-i, \ldots, n-i, -i, \ldots, -i)$, so λ_R is W-conjugate to λ_{dom}. This proves the result for A_{n-1}; proofs for the other classical groups are similar but easier.

The exceptional groups with nonzero minimal dominant weights are types E_6 and E_7. For E_6, using the notation of [13], the minuscule weights are λ_1 and λ_6. These weights are interchanged by a diagram automorphism, so it suffices to consider the weight λ_1, which is given by

$$\lambda_1 = \frac{1}{3}(4\alpha_1 + 3\alpha_2 + 5\alpha_3 + 6\alpha_4 + 4\alpha_5 + 2\alpha_6).$$

Let $\lambda_{dom} = \lambda_1$; then

$$\lambda_R = \frac{1}{3}(\alpha_1 + 2\alpha_3 + \alpha_5 + 2\alpha_6).$$

If the inner product (μ, α_i) is 1, then the corresponding simple reflection s_i takes μ to $\mu - \alpha_i$. Using this fact repeatedly, we see that $w = s_4 s_5 s_2 s_4 s_3 s_1$ takes λ_{dom} to λ_R. The proof for E_7 is similar. $\square$

Using this, we recover the multiplicity formula of [9].

Corollary 5.10 *As G-modules,*

$$R(\widetilde{\mathcal{O}}^{pr}) = R(\mathcal{M}) = \bigoplus_{\lambda_{dom}} \mathrm{Ind}_T^G(\mathbb{C}_{\lambda_{dom}}).$$

Proof If λ_R and λ_{dom} are in the same coset of $\mathcal{P}$ mod $\mathcal{Q}$, then they are W-conjugate by Proposition 5.9, so $\mathrm{Ind}_T^G(\mathbb{C}_{\lambda_R}) = \mathrm{Ind}_T^G(\mathbb{C}_{\lambda_{dom}})$. $\qquad \square$

6 Nonnormality of a *B*-orbit closure

The methods of this paper, together with a result of [9], allow us to show that the variety $\mathcal{M}$ differs from $\mathcal{N}$ in that the subvariety $\overline{B \cdot \widetilde{v}}$ of $\mathcal{M}$ is not normal in general. In contrast, $\overline{B \cdot v} = \mathfrak{u}$ is isomorphic to affine space, so it is a normal subvariety of $\mathcal{N}$.

By Proposition 4.7, there is a map $\phi : \widetilde{\mathfrak{u}} \to \mathcal{M}$ which yields a map $\psi : \widetilde{\mathfrak{u}} \to \overline{B \cdot \widetilde{v}}$; the map ψ is the normalization map. Via pullback by ψ, we can identify $R(\overline{B \cdot \widetilde{v}})$ with a subring of $R(\widetilde{\mathfrak{u}})$. The two rings are equal if and only if $\overline{B \cdot \widetilde{v}}$ is normal. We will describe $R(\overline{B \cdot \widetilde{v}})$ as a subring of $R(\widetilde{\mathfrak{u}})$, and thus determine when $\overline{B \cdot \widetilde{v}}$ is normal. In particular, if $\mathfrak{g}$ is simple, this occurs only when $\mathfrak{g}$ is of type A_1 or A_2.

As a variety, $\widetilde{\mathfrak{u}} = \mathcal{V} \times \mathfrak{u}_{\geq 4}$, so we have an identification $R(\widetilde{\mathfrak{u}}) = R(\mathcal{V}) \otimes R(\mathfrak{u}_{\geq 4})$. The ring $R(\mathcal{V}_{ad})$ is the subring of $R(T)$ generated by the e^α as α runs over the simple roots. The ring $R(\mathcal{V})$ is the subring of $R(T)$ generated by the e^{λ_R} and $R(\mathcal{V}_{ad})$. Let S be the subring of $R(\mathcal{V})$ which is generated by the $e^{\lambda_{dom}}$ and $R(\mathcal{V}_{ad})$. The main result of this section is the following.

Theorem 6.1 *With notation as above, $R(\overline{B \cdot \widetilde{v}})$ is the subring $S \otimes R(\mathfrak{u}_{\geq 4})$ of $R(\widetilde{\mathfrak{u}}) = R(\mathcal{V}) \otimes R(\mathfrak{u}_{\geq 4})$.*

Proof We have

$$R(\mathcal{M}) \to R(\overline{B \cdot \widetilde{v}}) \to R(\widetilde{\mathfrak{u}}),$$

so the image of $R(\overline{B \cdot \widetilde{v}})$ in $R(\widetilde{\mathfrak{u}})$ is the same as the image of $R(\mathcal{M})$. Write A for the image of $R(\mathcal{M})$ in $R(\widetilde{\mathfrak{u}})$; we want to show that $A = S \otimes R(\mathfrak{u}_{\geq 4})$. As a ring, A is generated by the images of generators of $R(\mathcal{M})$. The commutative diagram (8) yields a commutative diagram of rings of regular functions:

$$
\begin{array}{ccc}
R(\widetilde{\mathfrak{u}}) & \longleftarrow & R(\mathcal{M}) \\
\uparrow & & \uparrow \\
R(\mathfrak{u}) & \longleftarrow & R(\mathcal{N})
\end{array}
$$

Each minuscule representation occurs exactly once in $R(\mathcal{M})$, and as a module for $R(\mathcal{N})$, $R(\mathcal{M})$ is generated by 1 and the minuscule representations in $R(\mathcal{M})$ (see [9, Cor. 3.4]). The image of $R(\mathcal{N})$ in $R(\widetilde{\mathfrak{u}})$ under the compositions in the above commutative diagram is the image of $R(\mathfrak{u}) = R(\mathcal{V}_{ad}) \otimes R(\mathfrak{u}_{\geq 4})$ in $R(\widetilde{\mathfrak{u}})$. To complete the proof we consider the minuscule representations in $R(\mathcal{M})$ and describe their images in $R(\widetilde{\mathfrak{u}})$.

Suppose V is minuscule with highest weight $\lambda = \lambda_{dom}$. Each weight of V occurs with multiplicity 1; choose a basis $\{v_\mu\}$ of V, where v_μ lies in the μ-weight space of V. Let w_0 be the longest element of the Weyl group W. We claim the following. The weight vector $v_{w_0\lambda}$ maps to a multiple of the function $e^{-w_0\lambda} = e^{-w_0\lambda} \otimes 1 \in R(\widetilde{\mathfrak{u}})$ (which is a weight vector of weight $w_0\lambda$); if $\mu \neq w_0\lambda$, v_μ maps to 0 in $R(\widetilde{\mathfrak{u}})$. The theorem is a consequence of the claim and the preceding discussion, noting that $-w_0\lambda$ is itself a minimal dominant weight.

Consider the commutative diagram

$$
\begin{array}{ccc}
R(\mathcal{M}) & \xrightarrow{\;\cong\;} & R(G/G^{\widetilde{v}}) \\
\downarrow & & \downarrow \\
R(\overline{B \cdot \widetilde{v}}) & \longrightarrow & R(B/B^{\widetilde{v}}) \supset R(\widetilde{\mathfrak{u}}).
\end{array}
\tag{21}
$$

The top arrow is induced by the map $G/G^{\widetilde{v}} = \widetilde{\mathcal{O}}^{pr} \to G \cdot \widetilde{v} \hookrightarrow \mathcal{M}$. The bottom map is the restriction map to $B \cdot \widetilde{v} \cong B/B^{\widetilde{v}}$; the inclusion $R(\widetilde{\mathfrak{u}}) \subset R(B/B^{\widetilde{v}})$ follows from the identifications $B/B^{\widetilde{v}} \cong B/B^{(1,v)} \cong B \cdot (1, v) \subset \widetilde{\mathfrak{u}}$ (recall that $B^{\widetilde{v}} = B^{(1,v)}$ by Proposition 4.6). We know that under these identifications, the image of the map $R(G/G^{\widetilde{v}}) \to R(B/B^{\widetilde{v}})$ lies in the subset $R(\widetilde{\mathfrak{u}})$ (cf. Proposition 4.7). The function $e^\mu \in R(\widetilde{\mathfrak{u}})$ corresponds to the function $r \in R(B/B^{\widetilde{v}})$ satisfying $r(b) = e^\mu(b)$, where as usual we extend the function e^μ from T to B by requiring $e^\mu(tu) = e^\mu(t)$ for $t \in T, u \in U$.

In light of the commutative diagram (21), to prove the claim, it suffices to show that (in the setting of the claim) if $r \in R(G/G^{\widetilde{v}})$ corresponds to $v_\mu \in V$, then $r(b) = 0$ if $\mu \neq w_0\lambda$, and $r(b) = e^{-w_0\lambda}(b)$ if $\mu = w_0\lambda$.

Let $\{f_\nu\}$ denote the basis of the dual representation V^* which is dual to the basis $\{v_\mu\}$ of V; then f_ν is a weight vector of weight $-\nu$. The representation V^* is also minuscule, with highest weight vector $\{f_{w_0\lambda}\}$ of weight $-w_0\lambda$.

Because V occurs once in $R(\mathcal{M}) = R(G/G^{\widetilde{v}})$, there is a unique (up to scaling) G-invariant map $V \to R(G/G^{\widetilde{v}})$. We can describe this map using Frobenius reciprocity. Suppose H is a subgroup of G. If V_2 is a representation of H, then by definition, $\mathrm{Ind}_H^G V_2$ is the space of maps $r: G \to V_2$ satisfying $r(gh) = h^{-1}r(g)$ for all $g \in G, h \in H$, so if $V_2 = \mathbb{C}$ is the trivial representation, $\operatorname{Ind}_H^G V_2 = R(G/H)$. If V_1 is a representation of G, and V_2 a representation of H, Frobenius reprocity gives a bijection $\operatorname{Hom}_H(\operatorname{Res}_H^G V_1, V_2) \rightarrow \operatorname{Hom}_G(V_1, \operatorname{Ind}_H^G V_2)$. If this bijection takes φ to Φ, then the relation between φ and Φ is as follows. If $v \in V_1$, $\Phi(v) \in \operatorname{Ind}_H^G V_2$ satisfies

$$\Phi(v)(g) = \varphi(g^{-1}v). \tag{22}$$

In our situation, Frobenius reciprocity gives an isomorphism

$$\operatorname{Hom}_{G^v}(\operatorname{Res}_{G^v}^G V, \mathbb{C}) \cong \operatorname{Hom}_G(V, R(G/G^v)). \tag{23}$$

Since V occurs once in $R(G/G^v)$, the dimension of the right hand side of (23) is 1. Hence $\dim \operatorname{Hom}_{G^{\tilde{v}}}(V, \mathbb{C}) = \dim(V^*)^{G^{\tilde{v}}} = 1$. The highest weight vector $f_{w_0\lambda}$ is in $(V^*)^{G^{\tilde{v}}}$, so it must span this space. Let $\varphi = f_{w_0\lambda}$, and let Φ be the corresponding element of $\operatorname{Hom}_G(V, R(G/G^v))$. By (22), $\Phi(v_\mu)(b) = f_{w_0\lambda}(b^{-1}v_\mu)$. Now, $f_{w_0\lambda}(v_\mu)$ is 0 unless v_μ is the lowest weight vector $v_{w_0\lambda}$ of V. Also, since $b^{-1} \in B$, $b^{-1}v_\mu$ is a linear combination of v_ν with $\nu \geq \mu$. Hence, $\Phi(v_\mu)(b) = f_{w_0\lambda}(b^{-1}v_\mu) = 0$ unless $\mu = w_0\lambda$, and $\Phi(v_{w_0\lambda})(b) = f_{w_0\lambda}(b^{-1}v_{w_0\lambda}) = e^{w_0\lambda}(b^{-1}) = e^{-w_0\lambda}(b)$. This proves the claim. The theorem follows. $\qquad\square$

Corollary 6.2 $\overline{B \cdot \tilde{v}}$ *is normal if and only if for all minimal dominant weights* λ_{dom}, *when* λ_{dom} *is expressed as a sum of simple roots, each coefficient is less than 1. In particular, if G is simple, $\overline{B \cdot \tilde{v}}$ is normal if and only if G is of type* A_1 *or* A_2.

Proof By Theorem 6.1, $\overline{B \cdot \tilde{v}}$ is normal if and only if in each coset of $\mathcal{P}$ mod $\mathcal{Q}$, we have $\lambda_{dom} = \lambda_R$, which occurs exactly when the expression for λ_{dom} as a sum of simple roots has each coefficient less than 1. For simple G, this only occurs when G is of type A_1 or A_2, as can be seen from the expressions for the minimal dominant weights in terms of simple roots given in [13] (see Exercise 13 of Section 13.4, and Table 1 of Section 13.2). $\qquad\square$

7 Other nilpotent orbits

In this section we extend the construction of $\widetilde{\mathcal{M}}$ to covers of other nilpotent orbits besides the principal nilpotent orbit. However, because we cannot apply the theory of toric varieties, the picture that is obtained is less complete than in the case of the principal orbit, and in order to extend the applications of our construction to the setting of general orbits, further study would be required. The example of the minimal nilpotent orbit for the group $G = Sp(2n)$ is worked out at the end of this section; this example generalizes the case of the principal nilpotent orbit in $G = Sp(2) = SL(2)$ (in this case, the principal orbit is the minimal orbit).

The construction in the proof of Theorem 4.1 of [18], discussed in the introduction, is related to the construction here, but it is somewhat different. In [18], the variety $\widetilde{V}$ is the closure of a P-orbit in a representation of G, and the map takes $G \times^P \widetilde{V}$ to the closure of the G-orbit in the representation. Here, the variety $\widetilde{V}$ is different; the construction in this paper of $\widetilde{V}$ is modeled on the construction of $\widetilde{\mathfrak{u}}$ given in previous sections, and the map is to $\widetilde{\mathcal{M}}_{\mathcal{O}} =$ Spec $R(\widetilde{\mathcal{O}})$ (which is the normalization of the closure of the G-orbit considered in [18]).

Note also that in some cases there exist natural resolutions of singularities not given by the construction of this paper, which are well-suited to applications. For example, as pointed out by the referee, the universal cover of the subregular nilpotent orbit in Sp(4) (more precisely, Spec of the functions on this cover) has a Springer-type resolution via the cotangent bundle of a minimal parabolic subgroup.

The meaning of the notation in this section is somewhat different than in other sections because we are dealing with arbitrary nilpotent orbits. We let v denote any nonzero nilpotent element of $\mathfrak{g}$. As in Section 2.3, choose a standard $\mathfrak{sl}_2$-triple $\{h, e, f\}$ such that $e = v$ is the nilpositive element. Write $\mathfrak{g}_i$ for the i-eigenspace of ad h on $\mathfrak{g}$; then $\mathfrak{g} = \oplus \mathfrak{g}_i$. Let $\mathfrak{l} = \mathfrak{g}_0$, $\mathfrak{u}_P = \mathfrak{g}_{\geq 1}$, and $\mathfrak{p} = \mathfrak{l} + \mathfrak{u}_P$; let L, U_P, and P denote the corresponding subgroups of G. Then P is a parabolic subgroup of G with Levi factor L. Let $V = \mathfrak{g}_{\geq 2} \subseteq \mathfrak{u}_P$. If v an even nilpotent element (that is, if $\mathfrak{g}_i = 0$ for i odd), as in the case when v is principal, then $V = \mathfrak{u}_P$. By [2, Prop. 2.4], we have $G^v = P^v = L^v U^v$. The group L^v is reductive; it is the centralizer in G of the Lie subalgebra generated by the standard triple. Since U^v is a unipotent group, it is connected. Write C_v for the component group $G^v/G_0^v = L^v/L_0^v$.

Write $\mathcal{O} = G \cdot v \cong G/G^v$ for the orbit of v in $\mathfrak{g}$, and let $\mathcal{N}_{\mathcal{O}} = $ Spec $R(\mathcal{O})$. Then $\mathcal{N}_{\mathcal{O}}$ is the normalization of $\overline{\mathcal{O}}$, the closure of $\mathcal{O}$ in $\mathfrak{g}$. The map $\mathcal{N}_{\mathcal{O}} \to \overline{\mathcal{O}}$ is an isomorphism over $\mathcal{O}$, so we will view $\mathcal{O}$ as a subset of $\mathcal{N}_{\mathcal{O}}$. Write $\widetilde{\mathcal{N}}_{\mathcal{O}} = G \times^P V$. The map $G \times^P V \to \overline{\mathcal{O}}$ defined by $[g, v] \to g \cdot v$ is a resolution of singularities and an isomorphism over $\mathcal{O}$ (see the proof of Theorem 3.1 in [18]). This map factors through the normalization $\mathcal{N}_{\mathcal{O}}$ of $\overline{\mathcal{O}}$, so we obtain maps

$$\widetilde{\mathcal{N}}_{\mathcal{O}} \xrightarrow{\mu} \mathcal{N}_{\mathcal{O}} \longrightarrow \overline{\mathcal{O}}.$$

The universal cover of $\mathcal{O}$ is $\widetilde{\mathcal{O}} = G/G_0^v$. Write $\widetilde{v}$ for the identity coset $1 \cdot G_0^v$ in $\widetilde{\mathcal{O}}$; then $G^{\widetilde{v}} = P^{\widetilde{v}} = G_0^v$. We write $\mathcal{M}_{\mathcal{O}} = $ Spec $R(\widetilde{\mathcal{O}})$. Since $\widetilde{\mathcal{O}}$ is normal (in fact, smooth), $\mathcal{M}_{\mathcal{O}}$ is a normal variety. The variety $\widetilde{\mathcal{O}}$ is quasi-affine (see the proof of Theorem 4.1 in [18]). The results of Brylinski and Kostant [5] from Section 4.2 apply to this situation

$$
\begin{array}{ccc}
\tilde{\mathcal{O}} & \longrightarrow & \mathcal{M}_\mathcal{O} \\
\downarrow & & \downarrow \\
\mathcal{O} & \longrightarrow & \mathcal{N}_\mathcal{O},
\end{array}
$$

where the horizontal maps are open embeddings, and the vertical maps are quotients by C_ν. Moreover, the restriction maps $R(\mathcal{N}_\mathcal{O}) \to R(\mathcal{O})$ and $R(\mathcal{M}_\mathcal{O}) \to R(\tilde{\mathcal{O}})$ are isomorphism.

Because L^ν and L_0^ν are reductive, the varieties L/L^ν and L/L_0^ν are affine, so their function fields are the fields of fractions of $R(L/L^\nu)$ and $R(L/L_0^\nu)$, respectively. We define $\mathcal{V}_{ad} = \mathfrak{g}_2$. Kostant (see [15, Section 4.2]) proved that the orbit $L \cdot \nu \cong L/L^\nu$ is open in $\mathcal{V}_{ad}$. Kostant observed [17] that the complement $L \cdot \nu$ in $\mathcal{V}_{ad}$ is the zero set of a single function f; this follows because L/L^ν is an affine subvariety of $\mathcal{V}_{ad}$. (In fact, Kostant proved more about f, but we do not need this here.) We conclude that $R(L/L^\nu) = R(\mathcal{V}_{ad})[\frac{1}{f}]$.

Kostant also proved that $P \cdot \nu = L \cdot \nu + \mathfrak{g}_{\geq 3}$ ([15, Section 4.2]). Hence $P \cdot \nu$ is open in V. Let A denote the integral closure of $R(\mathcal{V}_{ad})$ in the function field $\kappa(L/L_0^\nu)$ and let $\mathcal{V} = \operatorname{Spec} R(A)$, so $R(\mathcal{V}) = A$. The ring $R(\mathcal{V})$ is stable under the action of C_ν, and $R(\mathcal{V})^{C_\nu} = R(\mathcal{V}_{ad})$ (see [1, Ch. 5, Ex. 14]). Hence C_ν acts on $\mathcal{V}$, and $\mathcal{V}/C_\nu = \mathcal{V}_{ad}$.

Lemma 7.1 *We have an L-equivariant commutative diagram*

$$
\begin{array}{ccc}
L/L_0^\nu & \longrightarrow & \mathcal{V} \\
\downarrow & & \downarrow \\
L/L^\nu & \longrightarrow & \mathcal{V}_{ad},
\end{array}
\tag{24}
$$

where the horizontal maps are open embeddings, and the vertical maps are quotients by C_ν.

Proof Let K denote the function field $\kappa(L/L_0^\nu)$. The ring $R(L/L_0^\nu)$ is integrally closed in K, since L/L_0^ν is smooth, hence normal. Also, $R(L/L_0^\nu)$ is integral over $R(L/L^\nu)$, since $R(L/L_0^\nu)^{C_\nu} = R(L/L^\nu)$. Hence $R(L/L_0^\nu)$ is the integral closure in K of $R(L/L^\nu)$. On the other hand, $R(L/L^\nu) = R(\mathcal{V}_{ad})[\frac{1}{f}]$. By definition, $R(\mathcal{V})$ is the integral closure of $R(\mathcal{V}_{ad})$ in K. The compatibility of integral closure with localization ([1, Prop. 5.12]) implies that the integral closure of $R(L/L^\nu)$ in K is $R(\mathcal{V})[\frac{1}{f}]$. Hence $R(L/L_0^\nu) = R(\mathcal{V})[\frac{1}{f}]$, so

L/L_0^ν is an open subvariety of $\mathcal{V}$. The commutative diagram (24) follows from considering the corresponding diagram of rings of regular functions (the arrows are reversed). In the diagram (24), the left vertical and bottom horizontal maps are L-equivariant by construction. The right vertical map is L-equivariant because it corresponds to the inclusion $R(\mathcal{V}_{ad}) \to R(\mathcal{V})$, and $R(\mathcal{V})$ is stable under L since it is the integral closure of the ring $R(\mathcal{V}_{ad})$, on which L acts. The top horizontal arrow is L-equivariant since it corresponds to the inclusion $R(\mathcal{V}) \hookrightarrow R(L/L_0^\nu)$, and both $R(\mathcal{V})$ and $R(L/L_0^\nu)$ are stable under L.   □

We will abuse notation and write 1 for the image of the identity coset $1 \cdot L/L_0^\nu$ in $\mathcal{V}$.

We now define a variety $\widetilde{V}$ which is analogous to the variety $\widetilde{u}$ defined when we studied the principal nilpotent orbit. Recall that $V = \mathfrak{g}_{\geq 2}$. We can identify $\mathcal{V}_{ad} = \mathfrak{g}_2 \cong V/\mathfrak{g}_{\geq 3}$; with this identification, $\mathcal{V}_{ad}$ has an action of P, where the subgroup U_P acts trivially. We extend the L-action on V to a P-action by requiring that U_P act trivially; then the map $V \to \mathcal{V}_{ad}$ is P-equivariant. The projection $V \to \mathcal{V}_{ad}$ is P-equivariant. We define

$$\widetilde{V} = V \times_{\mathcal{V}_{ad}} V.$$

Because both factors in the fiber product have P-actions, and the maps are P-equivariant, the variety $\widetilde{V}$ has a P-action. We have $P^{(1,\nu)} = P^{\widetilde{\nu}} = P_0^\nu$. Since the orbit $P \cdot \nu$ is open in V, the orbit $P \cdot (1,\nu) \cong P \cdot \widetilde{\nu}$ is open in $\widetilde{V}$. The analogue of Proposition 4.7 holds in this setting.

Proposition 7.2 *The map $P \cdot (1,\nu) \to \mathcal{M}_\mathcal{O}$ extends to a map $\phi \colon \widetilde{V} \to \mathcal{M}_\mathcal{O}$. We have a commutative diagram*

$$
\begin{array}{ccc}
\widetilde{V} & \longrightarrow & \mathcal{M}_\mathcal{O} \\
\downarrow & & \downarrow \\
V & \longrightarrow & \mathcal{N}_\mathcal{O}.
\end{array}
\tag{25}
$$

Moreover, the resulting map $\psi \colon \widetilde{V} \to \overline{P \cdot \widetilde{\nu}}$ is the normalization map. We have $\psi^{-1}(P \cdot \widetilde{\nu}) = P \cdot (1,\nu)$, and ψ restricts to an isomorphism $P \cdot (1,\nu) \to P \cdot \widetilde{\nu}$.

The proof is essentially the same as the proof of Proposition 4.7; we omit the details.

Define $\widetilde{\mathcal{M}_\mathcal{O}} = G \times^P \widetilde{V}$. There is a G-equivariant map $\widetilde{\mu} \colon \widetilde{\mathcal{M}_\mathcal{O}} \to \mathcal{M}_\mathcal{O}$ defined by $\widetilde{\mu}([g,\xi]) = g \cdot \phi(\xi)$. Using Proposition 7.2, we obtain the analogue of Theorem 4.8 for an arbitrary nilpotent orbit.

Theorem 7.3 *Let $\mathcal{O} = G \cdot v$ be a nonzero nilpotent orbit in $\mathfrak{g}$. The map $\tilde{\mu} \colon \widetilde{\mathcal{M}_{\mathcal{O}}} \to \mathcal{M}_{\mathcal{O}}$ is proper and is an isomorphism over $\tilde{\mathcal{O}}$. There is a commutative diagram*

$$
\begin{array}{ccc}
\widetilde{\mathcal{M}_{\mathcal{O}}} & \xrightarrow{\;\tilde{\eta}\;} & \widetilde{\mathcal{N}_{\mathcal{O}}} \\
\tilde{\mu}\downarrow & & \downarrow\mu \\
\mathcal{M}_{\mathcal{O}} & \xrightarrow{\;\eta\;} & \mathcal{N}_{\mathcal{O}}.
\end{array}
$$

The horizontal maps in this diagram are quotients by C_v.

We omit the proof, which is similar to the proof of Theorem 4.8.

The variety $\widetilde{\mathcal{M}_{\mathcal{O}}}$ is normal, but for a general nilpotent orbit we do not know much else about it, since we do not know much about the variety $\mathcal{V}$. However, $\mathcal{V}$ can be viewed as a partial compactification of the homogeneous variety L/L_0^v, and perhaps the Luna–Vust theory of such compactifications can be applied to obtain more information.

Example 7.4 In the case where $\mathcal{O}$ is the minimal nilpotent orbit for $G = Sp(2n)$, the construction of this section yields a generalization of Example 4.10. We sketch the argument. Take $v = X_{2\epsilon_1}$. It is well-known that the Hamiltonian action of G on $\mathbb{C}^{2n}$ yields a moment map $\eta \colon \mathbb{C}^{2n} \to \overline{\mathcal{O}}$. This map restricts to a double cover over the orbit $\mathcal{O}$, so $\mathcal{M}_{\mathcal{O}} = \mathbb{C}^{2n}$. In this example, we have $\mathfrak{g}_i = 0$ for $i > 2$. The Levi factor L is isomorphic to $\mathbb{C}^* \times Sp(2n-2)$, and the character ϵ_1 generates the character group of the $\mathbb{C}^*$ factor of L. The group G acts transitively on $\mathbb{P}^{2n-1}$, and the stabilizer of the point $[1, 0, \ldots, 0] \in \mathbb{P}^{2n-1}$ is P, so $G/P \cong \mathbb{P}^{2n-1}$. If $a \in \mathbb{Z}$, write $\mathbb{C}_{a\epsilon_1}$ for the representation of L on which $\ell = (t, g)$ acts by $e^{a\epsilon_1}(t)$. Then, just as in the case $G = Sp(2)$, we have $\mathcal{V}_{ad} = \mathbb{C}_{2\epsilon_1}$ and $\mathcal{V} = \mathbb{C}_{\epsilon_1}$. Moreover, since $\mathfrak{g}_i = 0$ for $i > 2$, we have $\widetilde{\mathcal{V}} = \mathcal{V}$ and $\mathcal{V} = \mathcal{V}_{ad}$, and the unipotent radical U_P acts trivially on both $\widetilde{\mathcal{V}}$ and $\mathcal{V}$. The diagram (7.3) becomes

$$
\begin{array}{ccc}
\widetilde{\mathcal{M}_{\mathcal{O}}} = G \times^P \mathbb{C}_{\epsilon_1} & \xrightarrow{\;\tilde{\eta}\;} & \widetilde{\mathcal{N}_{\mathcal{O}}} = G \times^P \mathbb{C}_{2\epsilon_1} \\
\tilde{\mu}\downarrow & & \downarrow\mu \\
\mathcal{M}_{\mathcal{O}} & \xrightarrow{\;\eta\;} & \mathcal{N}_{\mathcal{O}}.
\end{array}
$$

The description of the maps is similar to Example 4.10. As in that example, $\widetilde{\mathcal{N}}$ is the vector bundle on $G/P = \mathbb{P}^{2n-1}$ whose space of global sections is $\mathcal{O}_{\mathbb{P}^{2n-1}}(-2)$, and $\widetilde{\mathcal{M}}$ is the tautological bundle of the trivial bundle $\mathbb{C}^{2n} \times \mathbb{P}^{2n-1} \to \mathbb{P}^{2n-1}$, whose space of global sections is $\mathcal{O}_{\mathbb{P}^{2n-1}}(-1)$.

References

[1] Atiyah, Michael F. and I. G. Macdonald, I.G. 1969. Introduction to commutative algebra. Addison-Wesley Series in Mathematics.

[2] Barbasch, Dan and Vogan, David. A, Jr., 1985. Unipotent representations of complex semisimple groups. *Ann. of Math.* (2), **121**, 41–110.

[3] Broer, Bram. 1993. Line bundles on the cotangent bundle of the flag variety. *Invent. Math.*, **113**, no. 1, 1–20.

[4] Broer, Abraham. 1998. Decomposition varieties in semisimple Lie algebras. *Canad. J. Math.*, **50**, 929–971.

[5] Brylinski, Ranee and Kostant, Bertram. 1994. Nilpotent orbits, normality and Hamiltonian group actions. *J. Amer. Math. Soc.* **7** (2), 269–298.

[6] Chevalley, Claude. 1958. Séminaire C. Chevalley; 2e année: 1958. Anneaux de Chow et applications. Secrétariat mathématique, 11 rue Pierre Curie, Paris, 1958.

[7] Fulton, William, 1993. Introduction to toric varieties. Annals of Mathematics Studies, vol. 131, Princeton University Press, Princeton, NJ, 1993, The William H. Roever Lectures in Geometry.

[8] Graham, William, Precup, Martha, and Russell, Amber. 2020. A new approach to the generalized Springer correspondence. arXiv:2002.12480.

[9] Graham, William A. 1992 Functions on the universal cover of the principal nilpotent orbit. *Invent. Math.*, **108** (1992), 15–27.

[10] Hartshorne, Robin. 1997. Algebraic geometry. Springer-Verlag, New York-Heidelberg, 1977, Graduate Texts in Mathematics, No. 52.

[11] Hesselink, Wim H. 1976. Cohomology and the resolution of the nilpotent variety. *Math. Ann.*, **223** (1976), 249–252.

[12] Hinich, V. 1991. On the singularities of nilpotent orbits. *Israel J. Math.*, **73** (1991), 297–308.

[13] Humphreys, James E. 1972. Introduction to Lie algebras and representation theory. Springer-Verlag, New York. Graduate Texts in Mathematics, Vol. 9.

[14] Jantzen, Jens Carsten. 2004. Nilpotent orbits in representation theory. Lie theory, Progr. Math., vol. 228, Birkhäuser Boston, Boston, MA, pp. 1–211.

[15] Kostant, Bertram. 1959. The principal three-dimensional subgroup and the Betti numbers of a complex simple Lie group. *Amer. J. Math.*, **81**, 973–1032.

[16] Kostant, Bertram. 1963. Lie group representations on polynomial rings. *Amer. J. Math.*, **85**, 327–404.

[17] Kostant, Bertram. 1990. Private Communication. 1990.

[18] McGovern, William M. 1989. Rings of regular functions on nilpotent orbits and their covers. *Invent. Math.*, **97** 209–217.

[19] Russell, Amber. 2012. Graham's variety and perverse sheaves on the nilpotent cone. Ph.D. thesis, Louisiana State University, 2012.

[20] Sommers, Eric. 1997. Nilpotent orbits and the affine flag manifold. Ph.D. thesis, Massachusetts Institute of Technology, 1997.

[21] Thomason, R. W. 1987. Algebraic K-theory of group scheme actions. Algebraic topology and algebraic K-theory (Princeton, N.J., 1983), Ann. of Math. Stud., vol. 113, Princeton Univ. Press, Princeton, NJ, 1987, pp. 539–563.

13

Toric Surfaces, Linear and Quantum Codes Secret Sharing and Decoding

Johan P. Hansen[a]

In genuine gratefulness and with warmest regard I dedicate this work to W. Fulton. For more than forty years Fulton's knowledge, insight and unrelenting creativity has served as an inspiration for me.

Abstract. Toric varieties and their associated toric codes, as well as determination of their parameters with intersection theory, are presented in the two dimensional case.

Linear Secret Sharing Schemes with strong multiplication are constructed from toric varieties and codes by the J.L. Massey construction.

Asymmetric Quantum Codes are obtained from toric codes by the A.R. Calderbank P.W. Shor and A.M. Steane construction of stabilizer codes from linear codes containing their dual codes.

Decoding of a class of toric codes is presented.

1 Introduction

We present error-correcting codes obtained from toric varieties of dimension n resulting in long codes of length q^n over the finite ground field $\mathbb{F}_q$. In particular we apply the construction to certain toric surfaces. The code parameters are estimated using intersection theory.

Linear secret sharing schemes from error-correcting codes can be constructed by the method of J. L. Massey. We apply his method to two classes of toric codes giving ideal schemes with strong multiplication with respect to certain adversary structures.

Quantum error-correcting codes can be constructed by the method of A. R. Calderbank, P. W. Shor and A. M. Steane – the (CSS) method. Our construction

[a] Aarhus University, Denmark
Johan Hansen passed away in January 2020. The version of his article included in this volume appears with the permission of the family.

of toric codes is suitable for making quantum codes and extends similar results obtained by A. Ashikhmin, S. Litsyn and M.A. Tsfasman obtained from Goppa codes on algebraic curves.

We utilized the inherent multiplicative structure on toric codes to decode a class of toric codes.

1.1 Error correcting codes

Codes are used in communication and storage of information.

The message is divided into blocks and extra information is appended before transmission allowing the receiver to correct errors.

Let $\mathbb{F}_q$ be a finite field with q elements. A word of length n in the alphabet $\mathbb{F}_q$ is a vector

$$\mathbf{c} = (c_1, c_2, \ldots, c_n) \in \mathbb{F}_q^n.$$

The Hamming weight $w(\mathbf{c})$ is the number of non-zero coordinates in $\mathbf{c}$. The Hamming distance $d(\mathbf{c}_1, \mathbf{c}_2)$ between two words is the Hamming weight $w(\mathbf{c}_1 - \mathbf{c}_2)$.

A linear code is a $\mathbb{F}_q$-linear subspace $C \subseteq \mathbb{F}_q^n$. The dimension of the code is $k = \dim_{\mathbb{F}_q} C$ and the minimum distance $d(C)$ of the code C is the minimal Hamming distance $d(\mathbf{c}_1, \mathbf{c}_2) = w(\mathbf{c}_1 - \mathbf{c}_2)$ between two code words $\mathbf{c}_1, \mathbf{c}_2 \in C$ with $\mathbf{c}_1 \neq \mathbf{c}_2$.

A linear code C can correct errors in t coordinates or less, if and only if $t < \frac{d(C)}{2}$.

For general presentations of the theory of error-correcting codes, see [41] and [33].

Example 1.1 (Reed–Solomon code)   Let $x_1, x_2, \ldots, x_n \in \mathbb{F}_q$ be n distinct elements and let $0 < k \leq n$.

To the word $(a_0, a_1, \ldots, a_{k-1}) \in \mathbb{F}_q^k$ of length k we associate the polynomial

$$f(X) = a_0 + a_1 X + \cdots + a_{k-1} X^{k-1} \in \mathbb{F}_q[X]$$

and upon evaluation the Reed–Solomon code word

$$\bigl(f(x_1), f(x_2), \ldots, f(x_n)\bigr) \in \mathbb{F}_q^n.$$

The Reed–Solomon code $C_{n.k} \subseteq \mathbb{F}_q^n$ is the subspace of all Reed–Solomon code words coming from all polynomials $f(X) \in \mathbb{F}_q[X]$ with $\deg f(X) < k \leq n$.

The Reed–Solomon code $C_{n.k} \subseteq \mathbb{F}_q^n$ has dimension $\dim_{\mathbb{F}_q}(C_{n,k}) = k$, minimum distance $d(C_{n.k}) = n - k + 1$ and corrects $t < \frac{d(C_{n,k})}{2}$ errors.

Example 1.2 (Goppa code) Let X be a projective algebraic curve defined over the finite field $\mathbb{F}_q$. Let D be a $\mathbb{F}_q$-rational divisor on X, and let $L(D)$ be the $\mathbb{F}_q$-rational functions f on X with $\operatorname{div} f + D \geq 0$. Let $P_1, P_2, \ldots, P_n \in X(\mathbb{F}_q)$ be $\mathbb{F}_q$-rational points on X, none in the support of D.

A $\mathbb{F}_q$–rational function $f \in L(D)$ gives rise to a code word

$$\big(f(P_1), f(P_2), \ldots, f(P_n)\big) \in \mathbb{F}_q^n .$$

The Goppa code $C \subseteq \mathbb{F}_q^n$ is the subspace of all Goppa code words coming from all $f \in L(D)$, see [23].

Good codes are constructed from curves with a large numbers of rational points compared to their genus. Using Shimura curves (modular curves) one find a sequence of codes which beats the Gilbert–Varshamov bound, see [63]. In [62] curves are presented from the function field point of view and in [5] good curves are constructed via Galois towers of function fields. Some Deligne–Lusztig curves with large automorphism groups have the maximal number of rational points compared to their genus, see [24], [30] and [29].

2 Toric varieties

The toric codes presented in Section 3 are obtained by evaluating certain $\mathbb{F}_q$-rational functions in $\mathbb{F}_q$-rational points on toric varieties defined over the finite field $\mathbb{F}_q$.

For the general theory of toric varieties we refer to [21], [17], and [45]. Here we will recall the parts of the theory needed for determination of the parameters of the toric codes.

2.1 Polytopes, normal fans and support functions

Let $M \simeq \mathbb{Z}^r$ be a free $\mathbb{Z}$-module of rank r over the integers $\mathbb{Z}$. Let $\square$ be an integral convex polytope in $M_{\mathbb{R}} = M \otimes_{\mathbb{Z}} \mathbb{R}$, i.e., a compact convex polyhedron such that the vertices belong to M.

Let $N = \operatorname{Hom}_{\mathbb{Z}}(M, \mathbb{Z})$ be the dual lattice with canonical $\mathbb{Z}$ – bi-linear pairing

$$\langle -, - \rangle \colon M \times N \to \mathbb{Z}.$$

Let $M_{\mathbb{R}} = M \otimes_{\mathbb{Z}} \mathbb{R}$ and $N_{\mathbb{R}} = N \otimes_{\mathbb{Z}} \mathbb{R}$ with canonical $\mathbb{R}$ - bi-linear pairing

$$\langle -, - \rangle \colon M_{\mathbb{R}} \times N_{\mathbb{R}} \to \mathbb{R}.$$

The r-dimensional *algebraic torus* $T_N \simeq (\overline{\mathbb{F}}_q{}^*)^r$ is defined by $T_N :=$ $\mathrm{Hom}_{\mathbb{Z}}(M, \overline{\mathbb{F}}_q{}^*)$. The multiplicative character $\mathbf{e}(m)$, $m \in M$ is the homomorphism

$$\mathbf{e}(m) \colon T \to \overline{\mathbb{F}}_q{}^*$$

defined by $\mathbf{e}(m)(t) = t(m)$ for $t \in T_N$. Specifically, if $\{n_1, \ldots, n_r\}$ and $\{m_1, \ldots, m_r\}$ are dual $\mathbb{Z}$-bases of N and M and we denote $u_j := \mathbf{e}(m_j)$, $j = 1, \ldots, r$, then we have an isomorphism $T_N \simeq (\overline{\mathbb{F}}_q{}^*)^r$ sending t to $(u_1(t), \ldots, u_r(t))$. For $m = \lambda_1 m_1 + \cdots + \lambda_r m_r$ we have

$$\mathbf{e}(m)(t) = u_1(t)^{\lambda_1} \cdot \cdots \cdot u_r(t)^{\lambda_r}. \tag{1}$$

Given an r-dimensional integral convex polytope $\square$ in $M_{\mathbb{R}}$, the support function $h_{\square} \colon N_{\mathbb{R}} \to \mathbb{R}$ is defined as $h_{\square}(n) := \inf\{< m, n > \mid m \in \square\}$ and the polytope $\square$ can be reconstructed from the support function

$$\square = \square_h = \{m \in M \mid\ < m, n > \geq h_{\square}(n) \quad \forall n \in N\}. \tag{2}$$

The support function $h_{\square}$ is piecewise linear in the sense that $N_{\mathbb{R}}$ is the union of a non-empty finite collection of strongly convex polyhedral cones in $N_{\mathbb{R}}$ such that $h_{\square}$ is linear on each cone.

A fan is a collection Δ of strongly convex polyhedral cones in $N_{\mathbb{R}}$ such that every face of $\sigma \in \Delta$ is contained in Δ and $\sigma \cap \sigma' \in \Delta$ for all $\sigma, \sigma' \in \Delta$.

The *normal fan* Δ is the coarsest fan such that $h_{\square}$ is linear on each $\sigma \in \Delta$, i.e., such that for all $\sigma \in \Delta$ there exists $l_{\sigma} \in M$ such that

$$h_{\square}(n) = \langle l_{\sigma}, n \rangle \quad \forall n \in \sigma. \tag{3}$$

The 1-dimensional cones $\rho \in \Delta$ are generated by unique primitive elements $n(\rho) \in N \cap \rho$ such that $\rho = \mathbb{R}_{\geq 0} n(\rho)$.

In the two-dimensional case we can upon refinement of the normal fan assume that two successive pairs of $n(\rho)$'s generate the lattice and we obtain *the refined normal fan*.

The *toric surface* $X_{\square}$ associated to the refined normal fan Δ of $\square$ is

$$X_{\square} = \bigcup_{\sigma \in \Delta} U_{\sigma} \tag{4}$$

where U_{σ} is the $\overline{\mathbb{F}}_q$-valued points of the affine scheme $\mathrm{Spec}(\overline{\mathbb{F}}_q[S_{\sigma}])$, i.e.,

$$U_{\sigma} = \{u \colon S_{\sigma} \to \overline{\mathbb{F}}_q \mid u(0) = 1,\ u(m + m') = u(m)u(m') \quad \forall m, m' \in S_{\sigma}\},$$

where S_σ is the additive sub-semi-group of M

$$S_\sigma = \{m \in M | \langle m, y \rangle \geq 0 \quad \forall y \in \sigma\}.$$

The *toric surface* $X_\square$ of (4) is irreducible, non-singular and complete as it is constructed from the refined normal fan.

If $\sigma, \tau \in \Delta$ and τ is a face of σ, then U_τ is an open subset of U_σ. Obviously, $S_0 = M$ and $U_0 = T_N$ such that the algebraic torus T_N is an open subset of $X_\square$.

The torus T_N *acts algebraically* on $X_\square$. On $u \in U_\sigma$ the action of $t \in T_N$ is obtained as

$$(tu)(m) := t(m)u(m) \qquad m \in S_\sigma$$

such that $tu \in U_\sigma$ and U_σ is T_N-stable. The set of orbits of this action is in one-to-one correspondence with Δ. For each $\sigma \in \Delta$ let

$$\text{orb}(\sigma) := \{u : M \cap \sigma \to \overline{\mathbb{F}}_q^* | \ u \text{ is a group homomorphism}\}.$$

Then $\text{orb}(\sigma)$ is a T_N orbit in $X_\square$. Define $V(\sigma)$ to be the closure of $\text{orb}(\sigma)$ in $X_\square$.

A Δ-linear support function h gives rise to the Cartier divisor D_h. Let $\Delta(1)$ be the 1-dimensional cones in Δ, then

$$D_h := - \sum_{\rho \in \Delta(1)} h(n(\rho)) V(\rho).$$

In particular

$$D_m = \text{div}(\mathbf{e}(-m)) \qquad m \in M.$$

Lemma 2.1 *Let h be a Δ-linear support function with associated Cartier divisor D_h and convex polytope $\square_h$ defined in (2). The vector space $H^0(X, O_X(D_h))$ of global sections of $O_X(D_h)$, i.e., rational functions f on $X_\square$ such that $\text{div}(f) + D_h \geq 0$, has dimension $\#(M \cap \square_h)$ (the number of lattice points in $\square_h$) and has $\{\mathbf{e}(m) | \ m \in M \cap \square_h\}$ as a basis, see (1).*

2.2 Polytopes, Cartier divisors and Intersection theory

For a fixed line bundle $\mathcal{L}$ on a variety X, given an effective divisor D such that $\mathcal{L} = O_X(D)$, the fundamental question to answer is: How many points from a fixed set $\mathcal{P}$ of rational points are in the support of D. This question is treated in general in [31] using intersection theory, see [22]. Here we will apply the same methods when X is a toric surface.

For a Δ-linear support function h and a 1-dimensional cone $\rho \in \Delta(1)$, we will determine the intersection number $(D_h; V(\rho))$ between the Cartier divisor

D_h and $V(\rho) = \mathbb{P}^1$. The cone ρ is the common face of two 2-dimensional cones $\sigma', \sigma'' \in \Delta(2)$. Choose primitive elements $n', n'' \in N$ such that

$$n' + n'' \in \mathbb{R}\rho$$
$$\sigma' + \mathbb{R}\rho = \mathbb{R}_{\geq 0}n' + \mathbb{R}\rho$$
$$\sigma'' + \mathbb{R}\rho = \mathbb{R}_{\geq 0}n'' + \mathbb{R}\rho.$$

Lemma 2.2 *For any $l_\rho \in M$, such that h coincides with l_ρ on ρ, let $\overline{h} = h - l_\rho$. Then the intersection number is*

$$intersection(D_h; V(\rho)) = -\left(\overline{h}(n') + \overline{h}(n'')\right).$$

In the 2-dimensional non-singular case, let $n(\rho)$ be a primitive generator for the 1-dimensional cone ρ. There exists an integer a such that

$$n' + n'' + an(\rho) = 0,$$

the orbit closure $V(\rho)$ is itself a Cartier divisor and the above gives the self-intersection number

$$(V(\rho); V(\rho)) = a.$$

Lemma 2.3 *Let D_h be a Cartier divisor and let $\square_h$ be the polytope associated to h, see (2). Then*

$$(D_h; D_h) = 2\mathrm{vol}_2(\square_h),$$

where vol_2 is the normalized Lebesgue-measure.

3 Toric Codes

In this section we will present the construction of toric codes and the derivation of their parameters, see [25], [26] and [27].

Definition 3.1 Let $M \simeq \mathbb{Z}^r$ be a free $\mathbb{Z}$-module of rank r over the integers $\mathbb{Z}$.

For any subset $U \subseteq M$, let $\mathbb{F}_q[U]$ be the linear span in $\mathbb{F}_q[X_1^{\pm 1}, \ldots, X_r^{\pm 1}]$ of the monomials

$$\{X^u = X_1^{u_1} \cdot \cdots \cdot X_r^{u_r} | u = (u_1, \ldots, u_r) \in U\}.$$

This is an $\mathbb{F}_q$-vector space of dimension equal to the number of elements in U.

Let $T(\mathbb{F}_q) = (\mathbb{F}_q^*)^r$ be the $\mathbb{F}_q$-rational points on the torus and let $S \subseteq T(\mathbb{F}_q)$ be any subset. The linear map that evaluates elements in $\mathbb{F}_q[U]$ at all the points in S is denoted by π_S:

$$\pi_S : \mathbb{F}_q[U] \to \mathbb{F}_q{}^{|S|}$$
$$f \mapsto (f(P))_{P \in S} \, .$$

With this notation $\pi_{\{P\}}(f) = f(P)$.

The toric code is the image $C_U = \pi_S(\mathbb{F}_q[U])$.

Remark 3.2 ($r = 1$, Reed–Solomon code)   Consider the special case where $M \simeq \mathbb{Z}$, $U = \square = [0, k-1] \subseteq M_\mathbb{R} = M \otimes_\mathbb{Z} \mathbb{R}$ and $S = T(\mathbb{F}_q) = \mathbb{F}_q{}^*$.

The toric code $C_\square$ associated to $\square$ is the linear code of length $n = (q-1)$ presented in Example 1.1 with $S = \{x_1, \dots, x_n\}$.

Remark 3.3 ($r = 2$)   Consider the special case where $M \simeq \mathbb{Z}^2$, $U = \square \subseteq M_\mathbb{R} = M \otimes_\mathbb{Z} \mathbb{R}$ is an integral convex polytope and $S = T(\mathbb{F}_q) = \mathbb{F}_q{}^* \times \mathbb{F}_q{}^*$.

Let $\xi \in \mathbb{F}_q$ be a primitive element. For any i such that $0 \le i \le q - 1$ and any j such that $0 \le j \le q - 1$, we let $P_{ij} = (\xi^i, \xi^j) \in S = \mathbb{F}_q{}^* \times \mathbb{F}_q{}^*$. Let m_1, m_2 be a $\mathbb{Z}$-basis for M. For any $m = \lambda_1 m_1 + \lambda_2 m_2 \in M \cap \square$, let $\mathbf{e}(m)(P_{ij}) := (\xi^i)^{\lambda_1}(\xi^j)^{\lambda_2}$.

The toric code $C_\square$ associated to $\square$ is the linear code of length $n = (q-1)^2$ generated by the vectors

$$\{(\mathbf{e}(m)(P_{ij}))_{i=0,\dots,q-1;\, j=0,\dots,q-1} \mid m \in M \cap \square\}.$$

The toric codes are evaluation codes on the points of toric varieties.

For each $t \in T(\mathbb{F}_q) = (\mathbb{F}_q^*)^r$ we evaluate the rational functions in $\mathrm{H}^0(X, O_X(D_h))$

$$\mathrm{H}^0(X, O_X(D_h)) \to \overline{\mathbb{F}}_q{}^*$$
$$f \mapsto f(t).$$

Let $\mathrm{H}^0(X, O_X(D_h))^{\mathrm{Frob}}$ denote the rational functions in $\mathrm{H}^0(X, O_X(D_h))$ that are invariant under the action of the Frobenius. Evaluating at all points in $T(\mathbb{F}_q)$, we obtain the code $C_\square$:

$$\mathrm{H}^0(X, O_X(D_h))^{\mathrm{Frob}} \to C_\square \subset (\mathbb{F}_q{}^*)^{\sharp T(\mathbb{F}_q)}$$
$$f \mapsto (f(t))_{t \in T(\mathbb{F}_q)} \, ,$$

as in Definition 3.1.

3.1 Hirzebruch surfaces and associated toric codes

The Hirzebruch surfaces and associated toric codes will be used in (4.3.1) to obtain Linear Secret Sharing Schemes. The toric surface constructed from then following polytope is a Hirzebruch surface, that is a $\mathbb{P}^1$ bundle over $\mathbb{P}^1$, see [21].

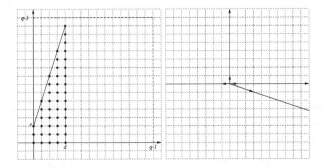

Figure 1 The polytope giving the Hirzebruch surface with vertices $(0,0),(d,0)$, $(d, e + rd),(0,e)$. The associated (refined) normal fan with generators of the 1-dimensional cones.

Let d, e, r be positive integers and let $\square$ be the polytope in $M_{\mathbb{R}}$ with vertices $(0,0),(d,0),(d,e+rd),(0,e)$, see Figure 1. Assume that $d < q - 1$, that $e < q - 1$ and that $e + rd < q - 1$. We have that $n(\rho_1) = \binom{1}{0}$, $n(\rho_2) = \binom{0}{1}$, $n(\rho_3) = \binom{-1}{0}$ and $n(\rho_4) = \binom{r}{-1}$. Let σ_1 be the cone generated by $n(\rho_1)$ and $n(\rho_2)$, σ_2 be the cone generated by $n(\rho_2)$ and $n(\rho_3)$, σ_3 the cone generated by $n(\rho_3)$ and $n(\rho_4)$ and σ_4 the cone generated by $n(\rho_4)$ and $n(\rho_1)$. The support function is:

$$h_{\square}\binom{n_1}{n_2} = \begin{cases} \binom{0}{0} \cdot \binom{n_1}{n_2} & \text{if } \binom{n_1}{n_2} \in \sigma_1, \\ \binom{d}{0} \cdot \binom{n_1}{n_2} & \text{if } \binom{n_1}{n_2} \in \sigma_2, \\ \binom{d}{e+rd} \cdot \binom{n_1}{n_2} & \text{if } \binom{n_1}{n_2} \in \sigma_3, \\ \binom{0}{e} \cdot \binom{n_1}{n_2} & \text{if } \binom{n_1}{n_2} \in \sigma_4. \end{cases}$$

Also

$$D_h := -\sum_{\rho \in \Delta(1)} h(n(\rho)) V(\rho) = d\, V(\rho_3) + e\, V(\rho_4)$$

and

$$\dim H^0(X, O_X(D_h)) = (d+1)(e+1) + r\frac{d(d+1)}{2}.$$

Theorem 3.4 *Let d, e, r be positive integers and let $\square$ be the polytope in $M_{\mathbb{R}}$ with vertices $(0,0),(d,0),(d,e+rd),(0,e)$, see Figure 1. Assume that $d < q - 1$, that $e < q - 1$ and that $e + rd < q - 1$. The toric code $C_{\square}$ has*

(i) length equal to $(q-1)^2$

(ii) dimension equal to $\#(M \cap \square) = (d+1)(e+1) + r\frac{d(d+1)}{2}$ (the number of lattice points in $\square$) and

Table 1. *Intersection numbers for divisors on Hirzebruch surfaces, calculated using Lemma 2.2 and Lemma 2.3.*

	$V(\rho_1)$	$V(\rho_2)$	$V(\rho_3)$	$V(\rho_4)$
$V(\rho_1)$	$-r$	1	0	1
$V(\rho_2)$	1	0	1	0
$V(\rho_3)$	0	1	r	1
$V(\rho_4)$	1	0	1	0

(iii) minimal distance equal to

$$\min\{(q-1-d)(q-1-e),(q-1)(q-1-e-rd)\}\,.$$

Proof Let $m_1 = (1,0) \in M_\mathbb{R}$. The $\mathbb{F}_q$-rational points of the torus $T \simeq \overline{\mathbb{F}}_q{}^* \times \overline{\mathbb{F}}_q{}^*$ belong to the $q-1$ lines on $X_\square$ given by $\prod_{\eta \in \mathbb{F}_q^*}(\mathbf{e}(m_1) - \eta) = 0$. Let $0 \neq f \in \mathrm{H}^0(X, O_X(D_h))$ and assume that f is zero along precisely a of these lines. As $\mathbf{e}(m_1) - \eta$ and $\mathbf{e}(m_1)$ have the same divisors of poles, they have equivalent divisors of zeroes, so

$$(\mathrm{div}(\mathbf{e}(m_1) - \eta))_0 \sim (\mathrm{div}(\mathbf{e}(m_1)))_0.$$

Therefore

$$\mathrm{div}(f) + D_h - a(\mathrm{div}(\mathbf{e}(m_1)))_0 \geq 0$$

or equivalently,

$$f \in \mathrm{H}^0(X, O_X(D_h - a(\mathrm{div}(\mathbf{e}(m_1)))_0)).$$

This implies that $a \leq d$ according to Lemma 2.1.

On any of the other $q - 1 - a$ lines, the number of zeroes of f is according to [31] at most the intersection number

$$(D_h - a(\mathrm{div}(\mathbf{e}(m_1)))_0; (\mathrm{div}(\mathbf{e}(m_1)))_0)\,. \tag{5}$$

The intersection number of (5) is calculated using Table 1 as

$$(\mathrm{div}(\mathbf{e}(m_1)))_0 = V(\rho_1) + r V(\rho_4)\,.$$

We get

$$\left(D_h - a(\mathrm{div}(\mathbf{e}(m_1)))_0; (\mathrm{div}(\mathbf{e}(m_1)))_0\right) = e + (d-a)r\,.$$

As $0 \leq a \leq d$, the total number of zeroes for f is at most

$$a(q-1) + (q-1-a)(e+(d-a)r)$$
$$\leq \max\{d(q-1) + (q-1-d)e, (q-1)(e+dr)\}.$$

This implies in both cases that the evaluation map

$$H^0(X, O_X(D_h))^{\mathrm{Frob}} \to C_\square \subset (\mathbb{F}_q)^{|T(\mathbb{F}_q)|}$$
$$f \mapsto (f(t))_{t \in T(\mathbb{F}_q)}$$

is injective and that the dimensions and the lower bounds for the minimal distances of the toric codes are as claimed.

To see that the lower bounds for the minimal distances are in fact the true minimal distances, we exhibit codewords of minimal weight.

Let $b_1, \ldots, b_{e+rd} \in \mathbb{F}_q^*$ be pairwise different elements. Then the function

$$x^d(y - b_1) \cdots \cdots (y - b_{e+rd}) \in H^0(X, O_X(D_h))^{\mathrm{Frob}}$$

evaluates to zero at the $(q-1)(e+rd)$ points

$$(x, b_j), \quad x \in \mathbb{F}_q^*, \quad j = 1, \ldots, e+rd$$

and gives a codeword of weight $(q-1)^2 - (q-1)(e+rd) = (q-1)(q-1-(e+rd))$. On the other hand, we let $a_1, \ldots, a_d \in \mathbb{F}_q^*$ be pairwise different elements and let $b_1, \ldots, b_e \in \mathbb{F}_q^*$ be pairwise different elements. Then the function

$$(x - a_1) \cdots \cdots (x - a_d)(y - b_1) \cdots \cdots (y - b_e) \in H^0(X, O_X(D_h))^{\mathrm{Frob}}$$

evaluates to zero at the $d(q-1) + (q-1)e - de$ points

$$(a_i, y), (x, b_j), \quad x, y \in \mathbb{F}_q^*, \quad i = 1, \ldots e, j = 1, \ldots, d$$

and gives a codeword of weight $(q-1-d)(q-1-e)$.   □

3.2 Some toric surfaces $X_{a,b}$ and their associated toric codes $C_{a,b}$

We present some toric surfaces and associated toric codes which we use in (4.3.2) to construct Linear Secret Sharing Schemes with strong multiplication and in Section 5 to construct quantum codes.

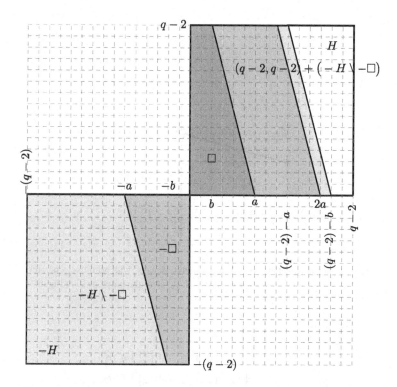

Figure 2 The convex polytope H with vertices $(0,0), (q-2,0), (q-2,q-2), (0,q-2)$ and the convex polytope $\Box$ with vertices $(0,0), (a,0), (b,q-2), (0,q-2))$ are shown. Also their opposite convex polytopes $-H$ and $-\Box$, the complement $-H \setminus -\Box$ and its translate $(q-2,q-2) + (-H \setminus -\Box)$ are depicted. Finally the convex hull of the reduction modulo $q-1$ of the Minkowski sum $U + U$ of the lattice points $U = \Box \cap M$ in $\Box$, is rendered. It has vertices $(0,0), (2a,0), (a+b,q-2)$ and $(0,q-2)$.

Let a,b be positive integers $0 \le b \le a \le q-2$, and let $\Box$ be the polytope in $M_\mathbb{R}$ with vertices $(0,0), (a,0), (b,q-2), (0,q-2)$ rendered in Figure 2 and with normal fan as in Figure 3.

The primitive generators of the 1-dimensional cones are

$$n(\rho_1) = \begin{pmatrix} 1 \\ 0 \end{pmatrix}, n(\rho_2) = \begin{pmatrix} 0 \\ 1 \end{pmatrix}, n(\rho_3) = \begin{pmatrix} \frac{-(q-2)}{\gcd(a-b,q-2)} \\ \frac{-(a-b)}{\gcd(a-b,q-2)} \end{pmatrix}, n(\rho_4) = \begin{pmatrix} 0 \\ -1 \end{pmatrix}.$$

For $i = 1, \ldots, 4$, the 2-dimensional cones σ_i are shown in Figure 3. The faces of σ_1 are $\{0, \rho_1, \rho_2\}$, the faces of σ_2 are $\{0, \rho_2, \rho_3\}$, the faces of σ_3 are $\{0, \rho_3, \rho_4\}$ and the faces of σ_4 are $\{0, \rho_4, \rho_1\}$.

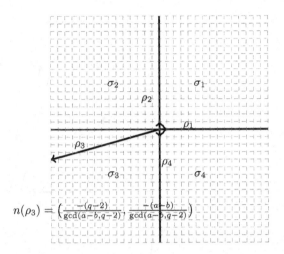

$$n(\rho_3) = \left(\frac{-(q-2)}{\gcd(a-b,q-2)}, \frac{-(a-b)}{\gcd(a-b,q-2)} \right)$$

Figure 3 The normal fan and its 1-dimensional cones ρ_i, with primitive generators $n(\rho_i)$, and 2-dimensional cones σ_i for $i = 1, \ldots, 4$ of the polytope $\square$ in Figure 2.

The support function of $\square$ is

$$h\binom{n_1}{n_2} = \begin{cases} \binom{0}{0} \cdot \binom{n_1}{n_2} & \text{if } \binom{n_1}{n_2} \in \sigma_1, \\ \binom{a}{0} \cdot \binom{n_1}{n_2} & \text{if } \binom{n_1}{n_2} \in \sigma_2, \\ \binom{b}{q-2} \cdot \binom{n_1}{n_2} & \text{if } \binom{n_1}{n_2} \in \sigma_3, \\ \binom{0}{q-2} \cdot \binom{n_1}{n_2} & \text{if } \binom{n_1}{n_2} \in \sigma_4. \end{cases} \qquad (6)$$

The related toric surface is in general singular as $\{n(\rho_2), n(\rho_3)\}$ and $\{n(\rho_3), n(\rho_4)\}$ are not bases for the lattice M. We can desingularize by subdividing the cones σ_2 and σ_3, however, our calculations will only involve the cones σ_1 and σ_2, so we refrain from that.

Theorem 3.5 *Assume a, b are integers with $0 \le b \le a \le q - 2$.*

Let $\square$ be the polytope in $M_{\mathbb{R}}$ with vertices $(0,0)$, $(a,0)$, $(b, q-2)$, $(0, q-2)$ rendered in Figure 2, and let $U = M \cap \square$ be the lattice points in $\square$.

(i) *The maximum number of zeros of $\pi_{T(\mathbb{F}_q)}(f)$ for $f \in \mathbb{F}_q[U]$ is less than or equal to*

$$(q-1)^2 - (q-1-a).$$

(ii) *The toric code $C_\square$ has*

    (a) *length equal to $(q-1)^2$,*

(b) *dimension equal to* #$(M \cap \square) = \frac{(q-1)(a+b+1)+\gcd(a-b,q-2)+1}{2}$ *(the number of lattice points in $\square$) and*

(c) *minimal distance greater than or equal to* $q - 1 - a$.

Proof Let $m_1 = (1,0)$. The $\mathbb{F}_q$-rational points of $T \simeq \overline{\mathbb{F}}_q^* \times \overline{\mathbb{F}}_q^*$ belong to the $q - 1$ lines on $X_\square$ given by

$$\prod_{\eta \in \mathbb{F}_q^*} (\mathbf{e}(m_1) - \eta) = 0.$$

Let $0 \neq f \in \mathrm{H}^0(X, O_X(D_h))$. Assume that f is zero along precisely c of these lines.

As $\mathbf{e}(m_1) - \eta$ and $\mathbf{e}(m_1)$ have the same divisors of poles, they have equivalent divisors of zeroes, so

$$(\mathbf{e}(m_1) - \eta)_0 \sim (\mathbf{e}(m_1))_0.$$

Therefore

$$\mathrm{div}(f) + D_h - c(\mathbf{e}(m_1))_0 \geq 0$$

or equivalently

$$f \in \mathrm{H}^0(X, O_X(D_h - c(\mathbf{e}(m_1))_0)).$$

This implies that $c \leq a$ according to Lemma 2.1.

On any of the other $q - 1 - c$ lines the number of zeroes of f is at most the intersection number

$$(D_h - c(\mathbf{e}(m_1))_0; (\mathbf{e}(m_1))_0).$$

This number can be calculated using Lemma 2.2 as $(\mathbf{e}(m_1))_0 = V(\rho_1)$ and

$$n(\rho_2) + n(\rho_4) + 0 \cdot n(\rho_1) = 0,$$

so

$$(\mathbf{e}(m_1))_0; (\mathbf{e}(m_1))_0) = 0. \tag{7}$$

We get from (6) and (7) that

$$(D_h - c(\mathbf{e}(m_1))_0; (\mathbf{e}(m_1))_0)$$
$$= (D_h; (\mathbf{e}(m_1))_0) - c(\mathbf{e}(m_1))_0; (\mathbf{e}(m_1))_0)$$
$$= -h_\square \binom{0}{1} - h_\square \binom{0}{-1} = q - 2.$$

As $0 \le c \le a$, we conclude the total number of zeroes for f is at most

$$c(q-1) + (q-1-c)(q-2) \le a(q-1) + (q-1-a)(q-2)$$
$$= (q-1)^2 - (q-1-a)$$

proving (i).

The claims in (ii) follows from (i) counting the number of lattice points.

□

3.3 Dual toric code

Let $M \simeq \mathbb{Z}^r$ be a free $\mathbb{Z}$-module of rank r over the integers $\mathbb{Z}$. Let $U \subseteq M$ be a subset, let $v \in M$ and consider translation $v + U := \{v + u | u \in U\} \subseteq M$.

Lemma 3.6 *Translation induces an isomorphism of vector spaces*

$$\mathbb{F}_q[U] \to \mathbb{F}_q[v + U]$$
$$f \mapsto f^v := X^v \cdot f.$$

We have that

(i) *The evaluations of $\pi_{T(\mathbb{F}_q)}(f)$ and $\pi_{T(\mathbb{F}_q)}(f^v)$ have the same number of zeroes on $T(\mathbb{F}_q)$.*

(ii) *The minimal number of zeros on $T(\mathbb{F}_q)$ of evaluations of elements in $\mathbb{F}_q[U]$ and $\mathbb{F}_q[v + U]$ are the same.*

(iii) *For $v = (v_1, \ldots, v_r)$ with v_i divisible by $q - 1$, the evaluations $\pi_S(f)$ and $\pi_S(f^v)$ are the same for any subset S of $T(\mathbb{F}_q)$.*

The lemma and generalizations have been used in several articles classifying toric codes, e.g., [36].

As an immediate consequence of (iii) above, see also [47, Theorem 3.3], we have:

Corollary 3.7 *Let $U \subseteq M$ be a subset and let*

$$\bar{U} := \{(\bar{u}_1, \ldots, \bar{u}_r) | \bar{u}_i \in \{0, \ldots, q-2\} \text{ and } \bar{u}_i \equiv u_i \bmod q - 1\}$$

be its reduction modulo $q - 1$. Then $\pi_S(\mathbb{F}_q[U]) = \pi_S(\mathbb{F}_q[\bar{U}])$ for any subset $S \subseteq T(\mathbb{F}_q)$.

Proposition 3.9 exhibits the dual code of the toric code $C = \pi_S(\mathbb{F}_q[U])$ defined in Definition 3.1.

Let $U \subseteq M$ be a subset, define its opposite as $-U := \{-u | u \in U\} \subseteq M$. The opposite maps the monomial X^u to X^{-u} and induces by linearity an isomorphism of vector spaces

$$\mathbb{F}_q[U] \to \mathbb{F}_q[-U]$$
$$X^u \mapsto \hat{X}^u := X^{-u}$$
$$f \mapsto \hat{f}.$$

On $\mathbb{F}_q^{|T(\mathbb{F}_q)|}$, we have the inner product

$$(a_0, \ldots, a_n) \star (b_0, \ldots, b_n) = \sum_{l=0}^{n} a_l b_l \in \mathbb{F}_q,$$

with $n = |T(\mathbb{F}_q)| - 1$.

Lemma 3.8 *Let $f, g \in \mathbb{F}_q[M]$ and assume $f \neq \hat{g}$, then*

$$\pi_{T(\mathbb{F}_q)}(f) \star \pi_{T(\mathbb{F}_q)}(g) = 0.$$

With this inner product we obtain the following proposition, e.g. [10, Proposition 3.5] and [48, Theorem 6].

Proposition 3.9 *Let*

$$H = \{0, 1, \ldots, q - 2\} \times \cdots \times \{0, 1, \ldots, q - 2\} \subset M .$$

Let $U \subseteq H$ be a subset. Then we have

(i) For $f \in \mathbb{F}_q[U]$ and $g \notin \mathbb{F}_q[-H \setminus -U]$, we have that $\pi_{T(\mathbb{F}_q)}(f) \star \pi_{T(\mathbb{F}_q)}(g) = 0$.

(ii) The orthogonal complement to $\pi_{T(\mathbb{F}_q)}(\mathbb{F}_q[U])$ in $\mathbb{F}_q^{|T(\mathbb{F}_q)|}$ is

$$\pi_{T(\mathbb{F}_q)}(\mathbb{F}_q[-H \setminus -U]),$$

i.e., the dual code of $C = \pi_{T(\mathbb{F}_q)}(\mathbb{F}_q[U])$ is $\pi_{T(\mathbb{F}_q)}(\mathbb{F}_q[-H \setminus -U])$.

Examples are shown in Figures 4 and 2.

4 Secret Sharing Schemes from toric varieties and codes

Secret sharing schemes were introduced in [8] and [49] and provide a method to split a *secret* into several pieces of information (*shares*) such that any large enough subset of the shares determines the secret, while any small subset of shares provides no information on the secret.

J. L. Massey presented in [43] a method for constructing linear secret sharing schemes from error-correcting codes. We apply his method to toric codes.

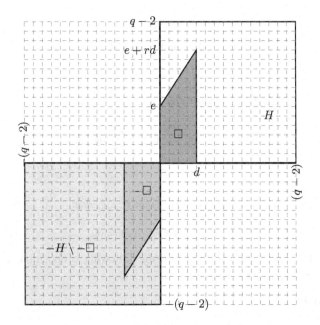

Figure 4 Hirzebruch surfaces. The convex polytope H with vertices $(0,0)$, $(q-2,0)$, $(q-2,q-2)$, $(0,q-2)$. The convex polytope $\square$ with vertices $(0,0)$, $(d,0)$, $(d,e+rd)$, $(0,e)$ and their opposite convex polytopes $-H$ and $-\square$. Also the (non-convex) polytope $-H \setminus -\square$ is depicted.

The schemes obtained are ideal and the number of players is $(q-1)r-1$ for any positive integer r. Examples of schemes which are quasi-threshold and have strong multiplication with respect to certain adversary structures are also presented.

Secret sharing schemes have found applications in cryptography, when the schemes have certain algebraic properties.

Linear secret sharing schemes (LSSS) are schemes where the secrets s and their associated shares $(a_1, \ldots, a_n)$ are elements in a vector space over $\mathbb{F}_q$. The schemes are called *ideal* if the secret s and the shares a_i are elements in that ground field $\mathbb{F}_q$. Specifically, if $s, \tilde{s} \in \mathbb{F}_q$ are two secrets with share vectors $(a_1, \ldots a_n)$, $(\tilde{a}_1, \ldots \tilde{a}_n) \in \mathbb{F}_q{}^n$, then the share vector of the secret $s + \lambda \tilde{s} \in \mathbb{F}_q$ is $(a_1 + \lambda \tilde{a}_1, \ldots, a_n + \lambda \tilde{a}_n) \in \mathbb{F}_q{}^n$ for any $\lambda \in \mathbb{F}_q$.

The reconstruction threshold of the linear secret sharing scheme is the smallest integer r such that any set of at least r of the shares $a_1, \ldots, a_n$ determines the secret s. The *privacy threshold* is the largest integer t such that no set of t (or fewer) elements of the shares $a_1, \ldots, a_n$ determines the secret s. The scheme is said to have *t-privacy*.

An ideal linear secret sharing scheme is said to have *multiplication* if the product of the shares determines the product of the secrets. It has *t-strong multiplication* if it has *t-privacy* and has multiplication for any subset of $n - t$ shares obtained by removing any t shares.

The properties of multiplication was introduced in [18]. Such schemes with multiplication can be utilized in the domain of multiparty computation (MPC), see [14], [7], [19] and [13].

4.1 Basic definitions and concepts – Linear Secret Sharing Schemes (LSSS)

This section presents basic definitions and concepts pertaining to linear secret sharing schemes as introduced in [43], [18], [15] and [16].

Let $\mathbb{F}_q$ be a finite field with q elements.

An *ideal linear secret sharing scheme* $\mathcal{M}$ over a finite field $\mathbb{F}_q$ on a set $\mathcal{P}$ of n players is given by a positive integer e, a sequence $V_1, \ldots, V_n$ of 1-dimensional linear subspaces $V_i \subset \mathbb{F}_q^e$ and a non-zero vector $u \in \mathbb{F}_q^e$.

An *adversary structure* $\mathcal{A}$, for a secret sharing scheme $\mathcal{M}$ on the set of players $\mathcal{P}$, is a collection of subsets of $\mathcal{P}$, with the property that subsets of sets in $\mathcal{A}$ are also sets in $\mathcal{A}$. In particular, the *adversary structure* $\mathcal{A}_{t,n}$ consists of all the subsets of size at most t of the set $\mathcal{P}$ of n players, and the *access structure* $\Gamma_{r,n}$ consists of all the subsets of size at least r of the set $\mathcal{P}$ of n players.

For any subset A of players, let $V_A = \sum_{i \in A} V_i$ be the $\mathbb{F}_q$-subspace spanned by all the V_i for $i \in A$.

The *access structure* $\Gamma(\mathcal{M})$ of $\mathcal{M}$ consists of all the subsets B of players with $u \in V_B$, and $\mathcal{A}(\mathcal{M})$ consists of all the other subsets A of players, that is $A \notin \Gamma(\mathcal{M})$.

A linear secret sharing scheme $\mathcal{M}$ is said to *reject* a given adversary structure $\mathcal{A}$, if $\mathcal{A} \subseteq \mathcal{A}(\mathcal{M})$. Therefore $A \in \mathcal{A}(\mathcal{M})$ if and only if there is a linear map from $\mathbb{F}_q^e$ to $\mathbb{F}_q$ vanishing on V_A, while non-zero on u.

The scheme $\mathcal{M}$ works as follows. For $i = 1, \ldots n$, let $v_i \in V_i$ be bases for the 1-dimensional vector spaces. Let $s \in \mathbb{F}_q$ be a *secret*. Choose at random a linear morphism $\phi : \mathbb{F}_q^e \to \mathbb{F}_q$, subject to the condition $\phi(u) = s$, and let $a_i = \phi(v_i)$ for $i = 1, \ldots, n$ be the *shares*

$$\phi: \mathbb{F}_q^e \to \mathbb{F}_q$$
$$u \mapsto s$$
$$v_i \mapsto a_i, i = 1, \ldots, n.$$

Then

- the shares $\{a_i = \phi(v_i)\}_{i \in A}$ determine the secret $s = \phi(u)$ uniquely if and only if $A \in \Gamma(\mathcal{M})$,
- the shares $\{a_i = \phi(v_i)\}_{i \in A}$ reveal no information on the secret $s = \phi(u)$, i.e., when $A \in \mathcal{A}(\mathcal{M})$.

Definition 4.1   Let $\mathcal{M}$ be a linear secret sharing scheme.

The *reconstruction threshold* of $\mathcal{M}$ is the smallest integer r so that any set of at least r of the shares $a_1, \ldots, a_n$ determines the secret s, i.e., $\Gamma_{r,n} \subseteq \Gamma(\mathcal{M})$.

The *privacy threshold* is the largest integer t so that no set of t (or less) elements of the shares $a_1, \ldots, a_n$ determine the secret s, i.e., $\mathcal{A}_{t,n} \subseteq \mathcal{A}(\mathcal{M})$. The scheme $\mathcal{M}$ is said to have *t-privacy*.

Definition 4.2   An ideal linear secret sharing scheme $\mathcal{M}$ has the *strong multiplication property* with respect to an adversary structure $\mathcal{A}$ if the following holds.

(i) $\mathcal{M}$ rejects the adversary structure $\mathcal{A}$.
(ii) Given two secrets s and $\tilde{s}$. For each $A \in \mathcal{A}$, the products $a_i \cdot \tilde{a}_i$ of all the shares of the players $i \notin A$ determine the product $s \cdot \tilde{s}$ of the two secrets.

4.2 Secret sharing schemes from toric codes using the construction of J. L. Massey

Linear secret sharing schemes obtained from linear codes were introduced by James L. Massey in [43] and were generalized in [16, Section 4.1]. Specifically, a linear secret sharing scheme with n players is obtained from a linear C code of length $n + 1$ and dimension k with privacy threshold $t = d' - 2$ and reconstruction threshold $r = n - d + 2$, where d is the minimum distance of the code and d' the minimum distance of the dual code.

This method of toric varieties also applies to construct algebraic geometric ideal secret sharing schemes (LSSS) defined over a finite ground field $\mathbb{F}_q$ with q elements. In a certain sense our construction resembles that of [15], where LSSS schemes were constructed from algebraic curves. However, the methods of obtaining the parameters are completely different.

The linear secret sharing schemes we obtain are *ideal* and the number of players can be of the magnitude q^r for any positive integer r.

The thresholds and conditions for strong multiplication are derived from estimates on the maximal number of zeroes of rational functions obtained via the cohomology and intersection theory on the underlying toric variety. In particular, we focus on toric surfaces.

4.3 The construction of Linear Secret Sharing Schemes (LSSS) from Toric Varieties

With notation as in Definition 3.1.

Definition 4.3 Let $S \subseteq T(\mathbb{F}_q)$ be any subset so that $P_0 \in S$. The linear secret sharing schemes (LSSS) $\mathcal{M}(U)$ with *support* S and $n = |S| - 1$ players is obtained as follows:

- Let $s_0 \in \mathbb{F}_q$ be a *secret* value. Select $f \in \mathbb{F}_q[U]$ at random, such that $\pi_{\{P_0\}}(f) = f(P_0) = s_0$.
- Define the n shares as the evaluations

$$\pi_{S \setminus \{P_0\}}(f) = (f(P))_{P \in S \setminus \{P_0\}} \in \mathbb{F}_q^{|S|-1} = \mathbb{F}_q^n.$$

Theorem 4.4 *Let $\mathcal{M}(U)$ be the linear secret sharing schemes of Definition 4.3 with $(q-1)^r - 1$ players.*

Let $r(U)$ and $t(U)$ be the reconstruction and privacy thresholds of $\mathcal{M}(U)$ as defined in Definition 4.1.

Then

$$r(U) \geq (\text{the maximum number of zeros of } \pi_{T(\mathbb{F}_q)}(f)) + 2$$

$$t(U) \leq (q-1)^r - (\text{the maximum number of zeros of } \pi_{T(\mathbb{F}_q)}(g)) - 2$$

for some $f \in \mathbb{F}_q[U]$ and for some $g \in \mathbb{F}_q[-H \setminus -U]$, where

$$\pi_{T(\mathbb{F}_q)} : \mathbb{F}_q[U] \to \mathbb{F}_q^{|T(\mathbb{F}_q)|}$$

$$f \mapsto \pi_{T(\mathbb{F}_q)}(f) = (f(P))_{P \in T(\mathbb{F}_q)}$$

$$\pi_{T(\mathbb{F}_q)} : \mathbb{F}_q[-H \setminus -U] \to \mathbb{F}_q^{|T(\mathbb{F}_q)|}$$

$$g \mapsto \pi_{T(\mathbb{F}_q)}(g) = (g(P))_{P \in T(\mathbb{F}_q)}.$$

Proof The minimal distance of an evaluation code and the maximum number of zeros of a function add to the length of the code.

The bound for $r(U)$ is based on the minimum distance d of the code $C = \pi_{T(\mathbb{F}_q)}(\mathbb{F}_q[U]) \subseteq \mathbb{F}_q^{|T(\mathbb{F}_q)|}$, the bound for $t(U)$ is based on the minimum distance d' of the dual code $C' = \pi_{T(\mathbb{F}_q)}(\mathbb{F}_q[-H \setminus -U] \subseteq \mathbb{F}_q^{|T(\mathbb{F}_q)|}$, using Proposition 3.9 to represent the dual code as an evaluation code.

The codes have length $|T(\mathbb{F}_q)|$, hence,

$$r(U) \geq |T(\mathbb{F}_q)| - d + 2$$
$$= (\text{the maximum number of zeros of } \pi_{T(\mathbb{F}_q)}(f)) + 2$$

$$t(U) \leq d' - 2$$
$$= |T(\mathbb{F}_q)| - (\text{the maximum number of zeros of } \pi_{T(\mathbb{F}_q)}(g)) - 2.$$

The results follow from the construction [43, Section 4.1]. $\qquad\square$

Theorem 4.5 *Let $U \subseteq H \subset M$ and let $U + U = \{u_1 + u_2 | u_1, u_2 \in U\}$ be the Minkowski sum. Let*

$$\pi_{T(\mathbb{F}_q)} : \mathbb{F}_q[U + U] \to \mathbb{F}_q^{|T(\mathbb{F}_q)|}$$

$$h \mapsto \pi_{T(\mathbb{F}_q)}(h) = (h(P))_{P \in T(\mathbb{F}_q)}.$$

The linear secret sharing schemes $\mathcal{M}(U)$ of Definition 4.3 with $n = (q - 1)^r - 1$ players, has strong multiplication with respect to $\mathcal{A}_{t,n}$ for $t \leq t(U)$, where $t(U)$ is the adversary threshold of $\mathcal{M}(U)$, if

$$t \leq n - 1 - (\text{the maximal number of zeros of } \pi_{T(\mathbb{F}_q)}(h)) \quad (8)$$

for all $h \in \mathbb{F}_q[U + U]$.

Proof For $A \in \mathcal{A}_{t,n}$, let $B := T(\mathbb{F}_q) \setminus (\{P_0\} \cup A)$ with $|B| = n - t$ elements. For $f, g \in \mathbb{F}_q[U]$, we have that $f \cdot g \in \mathbb{F}_q[U + U]$. Consider the linear morphism

$$\pi_B : \mathbb{F}_q[U + U] \to \mathbb{F}_q^{|B|} \quad (9)$$

$$h \mapsto (h(P))_{P \in B}. \quad (10)$$

evaluating at the points in B.

By assumption $h \in \mathbb{F}_q[U + U]$ can have at most $n - t - 1 < n - t = |B|$ zeros, therefore h cannot vanish identically on B, and we conclude that π_B is injective. Consequently, the products $f(P) \cdot g(P)$ of the shares $P \in B$ determine the product of the secrets $f(P_0) \cdot g(P_0)$, and the scheme has strong multiplication by Definition 4.2. $\qquad \square$

To determine the product of the secrets from the product of the shares amounts to decoding the linear code obtained as the image in (9), see Section 6.

4.3.1 Hirzebruch surfaces and their associated LSSS

Let d, e, r be positive integers and let $\square$ be the polytope in $M_{\mathbb{R}}$ with vertices $(0, 0), (d, 0), (d, e + rd), (0, e)$ with refined normal fan rendered in Figure 1. We obtain the following result as a consequence of Theorem 4.4 and the bounds obtained in Theorem 3.4 on the number of zeros of rational functions on Hirzebruch surfaces.

Theorem 4.6 *Let $\square$ be the polytope in $M_{\mathbb{R}}$ with vertices $(0, 0)$, $(d, 0)$, $(d, e + rd)$, $(0, e)$. Assume that $d \leq q - 2$, $e \leq q - 2$ and that $e + rd \leq q - 2$. Let $U = M \cap \square$ be the lattice points in $\square$.*

Let $\mathcal{M}(U)$ be the linear secret sharing schemes of Definition 4.3 with support $T(\mathbb{F}_q)$ and $(q - 1)^2 - 1$ players.

Then the number of lattice points in □ is

$$|U| = |(M \cap \square)| = (d+1)(e+1) + r\frac{d(d+1)}{2}.$$

The maximal number of zeros of a function $f \in \mathbb{F}_q[U]$ on $T(\mathbb{F}_q)$ is

$$\max\{d(q-1) + (q-1-d)e, (q-1)(e+dr)\}$$

and the reconstruction threshold as defined in Definition 4.1 of $\mathcal{M}(U)$ is

$$r(U) = 1 + \max\{d(q-1) + (q-1-d)e, (q-1)(e+dr)\}.$$

Remark 4.7 The polytope $-H \setminus -U$ is not convex, so our method using intersection theory does not determine the privacy threshold $t(U)$. It would be interesting to examine the methods and results of [36], [54], [37], [47], [6], [38], [53], [40], and [39] and for toric codes in this context.

4.3.2 Toric surfaces $X_{a,b}$ and codes $C_{a,b}$ – LSSS with strong multiplication

Let a, b be positive integers, $0 \le b \le a \le q-2$, and let □ be the polytope in $M_{\mathbb{R}}$ with vertices $(0,0)$, $(a,0)$, $(b, q-2)$, $(0, q-2)$ rendered in Figure 2 and with normal fan depicted in Figure 3. The corresponding toric surfaces and toric codes were studied in Section 3.2.

Under these assumptions the polytopes □, $-H \setminus -\square$ and $\square + \square$ are convex and we can use intersection theory on the associated toric surface to bound the number of zeros of functions and thresholds.

Theorem 4.8 *Assume a, b are integers with $0 \le b \le a \le q-2$.*

Let □ be the polytope in $M_{\mathbb{R}}$ with vertices $(0,0)$, $(a,0)$, $(b, q-2)$, $(0, q-2)$ rendered in Figure 2, and let $U = M \cap \square$ be the lattice points in □.

Let $\mathcal{M}(U)$ be the linear secret sharing schemes of Definition 4.3 with support $T(\mathbb{F}_q)$ and $n = (q-1)^2 - 1$ players.

(i) *The maximal number of zeros of $\pi_{T(\mathbb{F}_q)}(f)$ for $f \in \mathbb{F}_q[U]$ is less than or equal to*

$$(q-1)^2 - (q-1-a).$$

(ii) *The reconstruction threshold as defined in Definition 4.1 satisfies*

$$r(U) \le 1 + (q-1)^2 - (q-1-a).$$

(iii) *The privacy threshold as defined in Definition 4.1 satisfies*

$$t(U) \ge b - 1.$$

(iv) Assume $2a \leq q-2$. The secret sharing scheme has t-strong multiplication for

$$t \leq \min\{b - 1, (q - 2 - 2a) - 1\}.$$

Proof (i) reproduces Theorem 3.5 (i). According to Theorem 3.5 and Theorem 4.4, we have the inequality of (ii)

$$r(U) \leq 1 + (q - 1)^2 - (q - 1 - a).$$

We obtain (iii) by using the result in (i) on the polytope $(q - 2, q - 2) + (-H \setminus -\square)$ with vertices $(0,0)$, $(q-2-b,0)$, $(q-2-a,q-2)$ and $(q-2,q-2)$. The maximum number of zeros of $\pi_{T(\mathbb{F}_q)}(g)$ for $g \in \mathbb{F}_q[-H \setminus -U]$ is by Lemma 3.6 and the result in (i) less than or equal to $(q - 1)^2 - (q - 1 - (q - 2 - b)) = (q - 1)^2 - 1 - b$ and (iii) follows from Theorem 4.4.

To prove (iv) assume $t \leq (q - 2 - 2a) - 1$ and $t \leq b - 2$. We will use Theorem 4.5.

Consider the Minkowski sum $U + U$ and let $V = \overline{U + U}$ be its reduction modulo $q - 1$ as in Corollary 3.7. Under the assumption $2a \leq q - 2$, we have that $V = \overline{U + U}$ is the lattice points of the integral convex polytope with vertices $(0,0)$, $(2a,0)$, $(2b,q - 2)$ and $(0,q - 2)$.

By the result in (i) the maximum number of zeros of $\pi_{T(\mathbb{F}_q)}(h)$ for $h \in \mathbb{F}_q[V]$ is less than or equal to $(q - 1)^2 - (q - 1 - 2a)$. As the number of players is $n = (q-1)^2 - 1$, the right hand side of (8) of Theorem 4.5 is at least $(q - 2 - 2a) - 1$, which by assumption is at least t.

By assumption $t \leq b - 1$ and from (iii) we have that $b - 1 \leq t(U)$. We conclude that $t \leq t(U)$. $\square$

The bound in (i) of Theorem 4.8 can be improved for q sufficiently large. The polytope $\square$ includes the rectangle with vertices $(0,0)$, $(b,0)$, $(b,q - 2)$, $(0,q-2)$, which is the maximal Minkowski-reducible sub-polygon. The results of [36] and [54] give that the maximal number of zeroes of $\pi_{T(\mathbb{F}_q)}(f)$ for $f \in \mathbb{F}_q[U]$ cannot be larger than $(q - 1)^2 - (q - 1) + b$, with [54] giving the best bounds on q.

5 Asymmetric Quantum Codes on Toric Surfaces

A source on Quantum Computation and Quantum Information is [44].

5.1 Introduction

Our construction in Section 3 of toric codes is suitable for constructing quantum codes by the method of A. R. Calderbank, P. W. Shor and A.

M. Steane – the (CSS) method. Our constructions extended similar results obtained by A. Ashikhmin, S. Litsyn and M.A. Tsfasman in [4] from Goppa codes on algebraic curves.

Works of [50] and [59], [55] initiated the study and construction of quantum error-correcting codes. [12], [51] and [58] produced stabilizer codes (CSS) from linear codes containing their dual codes. For details see for example [3], [11] and [60].

Asymmetric quantum error-correcting codes are quantum codes defined over biased quantum channels: qubit-flip and phase-shift errors may have equal or different probabilities. The code construction is the CSS construction based on two linear codes. The construction appeared originally in [20], [32] and [61]. We present new families of toric surfaces, toric codes and associated asymmetric quantum error-correcting codes.

In [28] results on toric quantum codes are obtained by constructing a dualizing differential form for the toric surfaces.

A different approach to quantum codes defined on toric surfaces was originally presented in [34]. In [9] and [1] different codes are constructed by this method.

5.2 Asymmetric Quantum Codes

In the q-dimensional complex vector space $\mathbb{C}^q$ we use the Dirac notation in which the vectors are denoted by $|v\rangle$, called a ket, where v is a vector in $\mathbb{C}^q$. Let $|\beta_j\rangle$ with $i = 1, \ldots, q$ be a basis of $\mathbb{C}^q$.

Represent $|w\rangle$ and $\langle v|$ as the vectors

$$|w\rangle = \begin{bmatrix} w_1 \\ w_2 \\ \vdots \\ w_q \end{bmatrix} \quad \text{and} \quad \langle v| = \begin{pmatrix} \overline{v_1}, & \overline{v_2}, & \ldots & \overline{v_q} \end{pmatrix}.$$

Then the inner product $\langle v \mid w \rangle$ is the matrix product of the row vector $\langle v|$ times the column vector $|w\rangle$ independent of the choice of basis.

Let $|x\rangle$, $x \in \mathbb{F}_q$ be an orthonormal basis for $\mathbb{C}^q$. For $a, b \in \mathbb{F}_q$, the unitary operators $X(a)$ and $Z(b)$ on $\mathbb{C}^q$ are defined by

$$X(a)|x\rangle = |x + a\rangle, \qquad Z(b)|x\rangle = \omega^{\operatorname{tr}(bx)}|x\rangle,$$

where $\omega = \exp(2\pi i/p)$ is a primitive p-th root of unity and tr is the trace operation from $\mathbb{F}_q$ to $\mathbb{F}_p$.

More generally let $\mathcal{H}$ be the Hilbert space $\mathcal{H} = \mathbb{C}^{q^n} = \mathbb{C}^q \otimes \mathbb{C}^q \otimes \cdots \otimes \mathbb{C}^q$ with inner product extended from above.

For $\mathbf{a} = (a_1, \ldots, a_n) \in \mathbb{F}_q^n$ and $\mathbf{b} = (b_1, \ldots, b_n) \in \mathbb{F}_q^n$

$$X(\mathbf{a}) = X(a_1) \otimes \cdots \otimes X(a_n)$$
$$Z(\mathbf{b}) = Z(b_1) \otimes \cdots \otimes Z(b_n)$$

are the tensor products of n error operators.

With

$$\mathbf{E}_x = \{X(\mathbf{a}) = \bigotimes_{i=1}^{n} X(a_i)|\ \mathbf{a} \in \mathbb{F}_q^n, a_i \in \mathbb{F}_q\},$$

$$\mathbf{E}_z = \{Z(\mathbf{b}) = \bigotimes_{i=1}^{n} Z(b_i)|\ \mathbf{b} \in \mathbb{F}_q^n, b_i \in \mathbb{F}_q\}$$

the error groups $\mathbf{G}_x$ and $\mathbf{G}_z$ are

$$\mathbf{G}_x = \{\omega^c \mathbf{E}_x = \omega^c X(\mathbf{a})|\ \mathbf{a} \in \mathbb{F}_q^n, c \in \mathbb{F}_p\},$$
$$\mathbf{G}_z = \{\omega^c \mathbf{E}_z = \omega^c Z(\mathbf{b})|\ \mathbf{b} \in \mathbb{F}_q^n, c \in \mathbb{F}_p\}.$$

It is assumed that the groups $\mathbf{G_x}$ and $\mathbf{G_z}$ represent the qubit-flip and phase-shift errors.

Definition 5.1 (Asymmetric quantum code)   A q-ary asymmetric quantum code Q, denoted by $[[n, k, d_z/d_x]]_q$, is a q^k dimensional subspace of the Hilbert space $\mathbb{C}^{q^n}$ and can control all bit-flip errors up to $\lfloor \frac{d_x - 1}{2} \rfloor$ and all phase-flip errors up to $\lfloor \frac{d_z - 1}{2} \rfloor$. The code Q detects $(d_x - 1)$ qubit-flip errors as well as detects $(d_z - 1)$ phase-shift errors.

Let C_1 and C_2 be two linear error-correcting codes over $\mathbb{F}_q$, and let $[n, k_1, d_1]_q$ and $[n, k_2, d_2]_q$ be their parameters. For the dual codes $C_i^\perp$, we have $\dim C_i^\perp = n - k_i$ and if $C_1^\perp \subseteq C_2$ then $C_2^\perp \subseteq C_1$.

Lemma 5.2 *Let C_i for $i = 1, 2$ be linear error-correcting codes with parameters $[n, k_i, d_i]_q$ such that $C_1^\perp \subseteq C_2$ and $C_2^\perp \subseteq C_1$. Let*

$$d_x = \min\{\text{wt}(C_1 \setminus C_2^\perp), \text{wt}(C_2 \setminus C_1^\perp)\}$$
$$d_z = \max\{\text{wt}(C_1 \setminus C_2^\perp), \text{wt}(C_2 \setminus C_1^\perp)\}.$$

Then there is an asymmetric quantum code with parameters $[[n, k_1 + k_2 - n, d_z/d_x]]_q$. The quantum code is pure to its minimum distance, meaning that if $\text{wt}(C_1) = \text{wt}(C_1 \setminus C_2^\perp)$, then the code is pure to d_x, also if $\text{wt}(C_2) = \text{wt}(C_2 \setminus C_1^\perp)$, then the code is pure to d_z.

This construction is well-known, see for example [3], [11], [50], [59], [56], [57], [2]. The error groups $\mathbf{G_x}$ and $\mathbf{G_z}$ can be mapped to the linear codes C_1 and C_2.

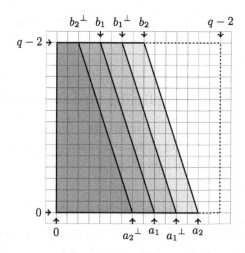

Figure 5 The polytope $\square_{a_i,b_i}$ is the polytope with vertices $(0,0),(a_i,0)$, $(b_i, q-2),(0,q-2)$. The polytopes giving the dual toric codes have vertices $(0,0),(a_i^\perp = q-2-b_i, 0),(b_i^\perp = q-2-a_i, q-2),(0,q-2)$.

5.2.1 Asymmetric Quantum Codes from Toric Codes

Let a, b be positive integers, $0 \le b \le a \le q-2$, and let $\square_{a,b}$ be the polytope in $M_\mathbb{R}$ with vertices $(0,0)$, $(a,0)$, $(b, q-2)$, $(0, q-2)$. The corresponding toric surfaces and toric codes were studied in Section 3.2.

From the results in Section 3.3 we conclude that the dual code $C_{a,b}^\perp$ is the toric code associated to the polytope $\square_{a^\perp, b^\perp}$ with vertices $(0,0)$, $(a^\perp, 0)$, $(b^\perp, q-2)$, $(0, q-2)$ where $b^\perp = q-2-a$ and $a^\perp = q-2-b$.

For $a_1 \le a_2$ we have the inclusions of polytopes $\square_{a_2^\perp, b_2^\perp} \subseteq \square_{a_1^\perp, b_1^\perp}$, see Figure 5, and corresponding inclusions of the associated toric codes

$$C_{a_2,b_2}^\perp \subseteq C_{a_1,b_1}, \quad C_{a_1,b_1}^\perp \subseteq C_{a_2,b_2}.$$

The nested codes given by the construction of Lemma 5.2 and the discussion above give rise to an asymmetric quantum code Q_{b_1,b_2}.

Theorem 5.3 (Asymmetric quantum codes Q_{a_1,b_1,a_2,b_2}) *Let $\mathbb{F}_q$ be a field with q elements and let $0 \le b_i \le a_i \le q-2$ for $i=1,2$.*

Then there is an asymmetric quantum code Q_{a_1,a_2,b_1,b_2} with parameters $[[(q-1)^2, k_1 + k_2 - (q-1)^2, d_z/d_x]]_q$, *where*

$$k_i = \frac{(q-1)(a_i + b_i + 1) + \gcd(a_i - b_i, q-2) + 1}{2}, \quad i = 1,2,$$

$$d_z \ge q-1-a_1,$$

$$d_x \ge q-1-a_2.$$

Proof The parameters and claims follow directly from Lemma 5.2 and Theorem 3.5. $\qquad\square$

6 Toric Codes, Multiplicative Structure and Decoding

We utilize the inherent multiplicative structure on toric codes to *decode* toric codes, resembling the decoding of Reed–Solomon codes and decoding by error correcting pairs, see [46], [35] and [42].

6.1 Multiplicative structure

In the notation of Section 3 let $\square$ and $\tilde{\square}$ be polyhedra in $\mathbb{R}^r$ and let $\square + \tilde{\square}$ denote their Minkowski sum. Let $U = \square \cap \mathbb{Z}^r$ and $\tilde{U} = \tilde{\square} \cap \mathbb{Z}^r$. The map

$$\mathbb{F}_q[U] \oplus \mathbb{F}_q[\tilde{U}] \to \mathbb{F}_q[U + \tilde{U}]$$
$$(f, g) \mapsto f \cdot g.$$

induces a multiplication on the associated toric codes

$$C_\square \oplus C_{\tilde{\square}} \to C_{\square + \tilde{\square}}$$
$$(c, \tilde{c}) \mapsto c \star \tilde{c}$$

with coordinatewise multiplication of the codewords – the *Schur* product.

Our goal is to use the multiplicative structure to correct t errors on the toric code $C_\square$. This is achieved choosing another toric code $C_{\tilde{\square}}$ that helps to reduce error-correcting to a *linear* problem.

Assume from now on:

(1) $|\tilde{U}| > t$, where $\tilde{U} = \tilde{\square} \cap \mathbb{Z}^2$.
(2) $d(C_{\square + \tilde{\square}}) > t$, where $d(C_{\square + \tilde{\square}})$ is the minimum distance of $C_{\square + \tilde{\square}}$.
(3) $d(C_{\tilde{\square}}) > n - d(C_\square)$, where $d(C_\square)$ and $d(C_{\tilde{\square}})$ are the minimum distances of $C_\square$ and $C_{\tilde{\square}}$.

6.2 Error-locating

Let the received word be $y(P) = f(P) + e(P)$ for $P \in T(\mathbb{F}_q)$, with $f \in \mathbb{F}_q[U]$ and error e of Hamming-weight at most t with support $T \subseteq T(\mathbb{F}_q)$, such that $|T| \le t$.

From (1), it follows that there is a $g \in \mathbb{F}_q[\tilde{U}]$, such that $g_{|T} = 0$ – an *error-locator*. To find g, consider the linear map:

$$\mathbb{F}_q[\tilde{U}] \oplus \mathbb{F}_q[U + \tilde{U}] \to \mathbb{F}_q{}^n \tag{11}$$

$$(g, h) \mapsto \big(g(P)y(P) - h(P)\big)_{P \in T(\mathbb{F}_q)}. \tag{12}$$

As $y(P) - f(P) = 0$ for $P \notin T$ (recall that the support of the error e is $T \subseteq T(\mathbb{F}_q)$), we have that $g(P)y(P) - (g \cdot f)(P) = 0$ for all $P \in T(\mathbb{F}_q)$. That is $(g, h = g \cdot f)$ is in the kernel of (11).

Lemma 6.1 *Let (g, h) be in the kernel of (11). Then $g|T = 0$ and $h = g \cdot f$.*

Proof

$$e(P) = y(P) - f(P), \quad P \in T(\mathbb{F}_q). \tag{13}$$

Coordinatewise multiplication yields by (11)

$$\begin{aligned}
g(P)e(P) &= g(P)y(P) - g(P)f(P) \\
&= h(P) - g(P)f(P)
\end{aligned}$$

for $P \in T(\mathbb{F}_q)$. The left hand side has Hamming weight at most t, the right hand side is a code word in $C_{\square + \tilde{\square}}$ with minimal distance strictly larger than t by assumption (2). Therefore both sides equal 0. $\qquad\square$

6.3 Error-correcting

Lemma 6.2 *Let (g, h) be in the kernel of (11) with $g|T = 0$ and $g \neq 0$. There is a unique f such that $h = g \cdot f$.*

Proof As in the above proof, we have

$$g(P)y(P) - g(P)f(P) = 0, \quad P \in T(\mathbb{F}_q). \tag{14}$$

Let $Z(g)$ be the zero-set of g with $T \subseteq Z(g)$. For $P \notin Z(g)$, we have $y(P) = f(P)$ and there are at least $d(C_{\tilde{\square}}) > n - d(C_{\square})$ such points by (3). This determines f uniquely as it is determined by the values in $n - d(C_{\square})$ points. $\qquad\square$

Example 6.3 Let $\square$ be the convex polytope with vertices $(0, 0)$, $(a, 0)$ and $(0, a)$. Let $\tilde{\square}$ be the convex polytope with vertices $(0, 0)$, $(b, 0)$ and $(0, b)$. Their Minkowski sum $\square + \tilde{\square}$ is the convex polytope with vertices $(0, 0)$, $(a + b, 0)$ and $(0, a + b)$, see Figure 6.

We have that

$$n = (q - 1)^2, \quad |\tilde{\square}| = \frac{(b + 1)(b + 2)}{2}, \quad d(C_{\square}) = (q - 1)(q - 1 - a),$$

$$d(C_{\tilde{\square}}) = (q - 1)(q - 1 - b) \quad \text{and} \quad d(C_{\square + \tilde{\square}}) = (q - 1)(q - 1 - (a + b))$$

for the associated codes over $\mathbb{F}_q$, see [25], [26] and [27].

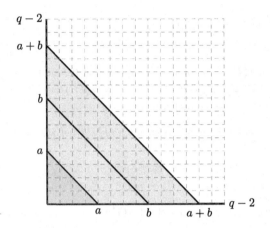

Figure 6 The convex polytope $\square$ with vertices $(0,0), (a,0)$ and $(0,a)$. The convex polytope $\tilde{\square}$ with vertices $(0,0), (b,0)$ and $(0,b)$). Their Minkowski sum $\square + \tilde{\square}$ having vertices $(0,0), (a+b,0)$ and $(0,a+b)$.

Let $q = 16$, $a = 4$ and $b = 8$. Then $n = 225$, $|\tilde{\square}| = 45$, $d(C_\square) = 165$, $d(C_{\tilde{\square}}) = 105$ and $d(C_{\square + \tilde{\square}}) = 45$.

As $d(C_{\tilde{\square}}) = 105 > 60 = n - d(C_\square)$, the procedure corrects t errors with

$$t < \min\{d(C_{\square + \tilde{\square}}), |\tilde{\square}|\} = 45 .$$

Remark 6.4 Error correcting pairs. [46] and [35] introduced the concept of error-correcting pairs for a linear code, see also [42]. Specifically for a linear code $C \subseteq \mathbb{F}_q^n$ a t-error correcting pair consists of two linear codes $A, B \subseteq \mathbb{F}_q^n$, such that

$$(A \star B) \perp C, \quad \dim_{\mathbb{F}_q} A > t, \quad d(B^\perp) > t, \quad d(A) + d(C) > n. \quad (15)$$

Here $A \star B = \{a \star b \mid a \in A, b \in B\}$ and $\perp$ denotes orthogonality with respect to the usual inner product. They described the known decoding algorithms for decoding t or fewer errors in this framework.

Also the decoding in the present paper can be described in this framework, taking $C = C_\square, A = C_{\tilde{\square}}$ and $B = (C \star A)^\perp$ using Proposition 3.9.

Acknowledgements. I would like to thank the anonymous reviewers for their valuable remarks and constructive comments, which significantly contributed to the quality of this paper.

References

[1] Albuquerque, Clarice Dias, Palazzo, Jr., Reginaldo, and Silva, Eduardo Brandani. 2010. Construction of new toric quantum codes. Pages 1–10 of: *Finite fields: theory and applications.* Contemp. Math., vol. 518. Amer. Math. Soc., Providence, RI.

[2] Aly, S. A., and Ashikhmin, A. 2010 (Jan). Nonbinary quantum cyclic and subsystem codes over asymmetrically-decohered quantum channels. Pages 1–5 of: *2010 IEEE Information Theory Workshop on Information Theory (ITW 2010, Cairo).*

[3] Ashikhmin, A., and Knill, E. 2001. Nonbinary quantum stabilizer codes. *IEEE Transactions on Information Theory,* **47**(7), 3065–3072.

[4] Ashikhmin, A., Litsyn, S., and Tsfasman, M.A. 2001. Asymptotically good quantum codes. *Physical Review A – Atomic, Molecular, and Optical Physics,* **63**(3), 1–5.

[5] Bassa, Alp, Beelen, Peter, Garcia, Arnaldo, and Stichtenoth, Henning. 2014. Galois towers over non-prime finite fields. *Acta Arith.,* **164**(2), 163–179.

[6] Beelen, Peter, and Ruano, Diego. 2009. The Order Bound for Toric Codes. Pages 1–10 of: Bras-Amorós, Maria, and Høholdt, Tom (eds), *Applied Algebra, Algebraic Algorithms and Error-Correcting Codes.* Lecture Notes in Computer Science, vol. 5527. Springer Berlin Heidelberg.

[7] Ben-Or, Michael, Goldwasser, Shafi, and Wigderson, Avi. 1988. Completeness Theorems for Non-Cryptographic Fault-Tolerant Distributed Computation (Extended Abstract). *In:* [52].

[8] Blakley, G.R. 1979. Safeguarding cryptographic keys. Pages 313–317 of: *Proceedings of the 1979 AFIPS National Computer Conference.* Monval, NJ, USA: AFIPS Press.

[9] Bombin, H., Andrist, Ruben S., Ohzeki, Masayuki, Katzgraber, Helmut G., and Martin-Delgado, M. A. 2012. Strong Resilience of Topological Codes to Depolarization. *Phys. Rev. X,* **2**(Apr), 021004.

[10] Bras-Amorós, Maria, and O'Sullivan, Michael E. 2008. Duality for some families of correction capability optimized evaluation codes. *Adv. Math. Commun.,* **2**(1), 15–33.

[11] Calderbank, A. Robert, Rains, Eric M., Shor, P. W., and Sloane, Neil J. A. 1998. Quantum error correction via codes over GF(4). *IEEE Trans. Inform. Theory,* **44**(4), 1369–1387.

[12] Calderbank, A.R., and Shor, P.W. 1996. Good quantum error-correcting codes exist. *Physical Review A – Atomic, Molecular, and Optical Physics,* **54**(2), 1098–1105.

[13] Cascudo, I. 2010. *On Asymptotically Good Strongly Multiplicative Linear Secret Sharing.* Ph.D. thesis, University of Oviedo.

[14] Chaum, David, Crépeau, Claude, and Damgård, Ivan. 1988. Multiparty Unconditionally Secure Protocols (Extended Abstract). *In:* [52].

[15] Chen, Hao, and Cramer, Ronald. 2006. Algebraic Geometric Secret Sharing Schemes and Secure Multi-Party Computations over Small Fields. Pages 521–536 of: Dwork, Cynthia (ed), *Advances in Cryptology - CRYPTO 2006.* Lecture Notes in Computer Science, vol. 4117. Springer Berlin Heidelberg.

[16] Chen, Hao, Cramer, Ronald, Goldwasser, Shafi, de Haan, Robbert, and Vaikuntanathan, Vinod. 2007. Secure computation from random error correcting codes. Pages 291–310 of: *Advances in cryptology—EUROCRYPT 2007.* Lecture Notes in Comput. Sci., vol. 4515. Springer, Berlin.

[17] Cox, David A., Little, John B., and Schenck, Henry K. 2011. *Toric varieties.* Graduate Studies in Mathematics 124. Providence, RI: American Mathematical Society (AMS). xxiv, 841 p.

[18] Cramer, Ronald, Damgård, Ivan, and Maurer, Ueli. 2000. General Secure Multiparty Computation from any Linear Secret-Sharing Scheme. Pages 316–334 of: Preneel, Bart (ed), *Advances in Cryptology - EUROCRYPT 2000.* Lecture Notes in Computer Science, vol. 1807. Springer Berlin Heidelberg.

[19] Cramer, Ronald, Damgård, Ivan, and Nielsen, Jesper Buus. 2015. *Secure Multiparty Computation and Secret Sharing.* Cambridge University Press.

[20] Evans, Z. W. E., Stephens, A. M., Cole, J. H., and Hollenberg, L. C. L. 2007. Error correction optimisation in the presence of X/Z asymmetry. *ArXiv e-prints*, Sept.

[21] Fulton, William. 1993. *Introduction to toric varieties.* Annals of Mathematics Studies, vol. 131. Princeton, NJ: Princeton University Press. The William H. Roever Lectures in Geometry.

[22] Fulton, William. 1998. *Intersection theory.* Second edn. Ergebnisse der Mathematik und ihrer Grenzgebiete. 3. Folge. A Series of Modern Surveys in Mathematics [Results in Mathematics and Related Areas. 3rd Series. A Series of Modern Surveys in Mathematics], vol. 2. Springer-Verlag, Berlin.

[23] Goppa, V. D. 1988. *Geometry and codes.* Mathematics and its Applications (Soviet Series), vol. 24. Kluwer Academic Publishers Group, Dordrecht. Translated from the Russian by N. G. Shartse.

[24] Hansen, Johan P. 1992. Deligne-Lusztig varieties and group codes. Pages 63–81 of: *Coding theory and algebraic geometry (Luminy, 1991).* Lecture Notes in Math., vol. 1518. Springer, Berlin.

[25] Hansen, Johan P. 1998. Toric Surfaces and Codes. Pages 42–43 of: *Information Theory Workshop.* IEEE.

[26] Hansen, Johan P. 2000. Toric surfaces and error-correcting codes. Pages 132–142 of: Buchmann, J., Hoeholdt, T., Stichtenoth, H., and Tapia-Recillas, H. (eds), *Coding theory, cryptography and related areas.* Springer.

[27] Hansen, Johan P. 2002. Toric Varieties Hirzebruch Surfaces and Error-Correcting Codes. *Applicable Algebra in Engineering, Communication and Computing,* **13**(4), 289–300.

[28] Hansen, Johan P. 2013. Quantum codes from toric surfaces. *IEEE Trans. Inform. Theory,* **59**(2), 1188–1192.

[29] Hansen, Johan P., and Pedersen, Jens Peter. 1993. Automorphism groups of Ree type, Deligne-Lusztig curves and function fields. *J. Reine Angew. Math.,* **440**, 99–109.

[30] Hansen, Johan P., and Stichtenoth, Henning. 1990. Group codes on certain algebraic curves with many rational points. *Appl. Algebra Engrg. Comm. Comput.,* **1**(1), 67–77.

[31] Hansen, Sren Have. 2001. Error-Correcting Codes from Higher-Dimensional Varieties. *Finite Fields and Their Applications,* **7**(4), 530 – 552.

[32] Ioffe, Lev, and Mézard, Marc. 2007. Asymmetric quantum error-correcting codes. *Phys. Rev. A*, **75**(Mar), 032345.

[33] Justesen, Jom, and Høholdt, Tom. 2017. *A Course in Error-Correcting Codes (EMS Textbooks in Mathematics)*. European Mathematical Society.

[34] Kitaev, A Yu. 1997. Quantum computations: algorithms and error correction. *Russian Mathematical Surveys*, **52**(6), 1191–1249.

[35] Kötter, Ralf. 1992. A unified description of an error locating procedure for linear codes. Pages 113–117 of: *Proceedings of Algebraic and Combinatorial Coding Theory*. Voneshta Voda.

[36] Little, John, and Schenck, Hal. 2006. Toric surface codes and Minkowski sums. *SIAM J. Discrete Math.*, **20**(4), 999–1014 (electronic).

[37] Little, John, and Schwarz, Ryan. 2007. On toric codes and multivariate Vandermonde matrices. *Appl. Algebra Engrg. Comm. Comput.*, **18**(4), 349–367.

[38] Little, John B. 2013. Remarks on generalized toric codes. *Finite Fields Appl.*, **24**, 1–14.

[39] Little, John B. 2017a. Corrigendum to "Toric codes and finite geometries" [Finite Fields Appl. 45 (2017) 203–216] [MR3631361]. *Finite Fields Appl.*, **48**, 447–448.

[40] Little, John B. 2017b. Toric codes and finite geometries. *Finite Fields Appl.*, **45**, 203–216.

[41] MacWilliams, F. J., and Sloane, N. J. A. 1977. *The theory of error-correcting codes. II*. North-Holland Publishing Co., Amsterdam-New York-Oxford. North-Holland Mathematical Library, Vol. 16.

[42] Márquez-Corbella, Irene, and Pellikaan, Ruud. 2016. A characterization of MDS codes that have an error correcting pair. *Finite Fields Appl.*, **40**, 224–245.

[43] Massey, James L. 2001. Some applications of code duality in cryptography. *Mat. Contemp.*, **21**, 187–209. 16th School of Algebra, Part II (Portuguese) (Brasília, 2000).

[44] Nielsen, Michael A., and Chuang, Isaac L. 2000. *Quantum Computation and Quantum Information*. Cambridge University Press.

[45] Oda, Tadao. 1988. *Convex bodies and algebraic geometry*. Springer.

[46] Pellikaan, Ruud. 1992. On decoding by error location and dependent sets of error positions. *Discrete Math.*, **106/107**, 369–381. A collection of contributions in honour of Jack van Lint.

[47] Ruano, Diego. 2007. On the parameters of r-dimensional toric codes. *Finite Fields Appl.*, **13**(4), 962–976.

[48] Ruano, Diego. 2009. On the structure of generalized toric codes. *J. Symbolic Comput.*, **44**(5), 499–506.

[49] Shamir, Adi. 1979. How to Share a Secret. *Commun. ACM*, **22**(11), 612–613.

[50] Shor, Peter W. 1995. Scheme for reducing decoherence in quantum computer memory. *Phys. Rev. A*, **52**(Oct), R2493–R2496.

[51] Shor, Peter W. 1996. Fault-tolerant quantum computation. Pages 56–65 of: *37th Annual Symposium on Foundations of Computer Science (Burlington, VT, 1996)*. Los Alamitos, CA: IEEE Comput. Soc. Press.

[52] Simon, Janos (ed). 1988. *Proceedings of the 20th Annual ACM Symposium on Theory of Computing, May 2–4, 1988, Chicago, Illinois, USA*. ACM.

[53] Soprunov, Ivan. 2015. Lattice polytopes in coding theory. *J. Algebra Comb. Discrete Struct. Appl.*, **2**(2), 85–94.

[54] Soprunov, Ivan, and Soprunova, Jenya. 2008/09. Toric surface codes and Minkowski length of polygons. *SIAM J. Discrete Math.*, **23**(1), 384–400.

[55] Steane, A. M. 1996a. Error correcting codes in quantum theory. *Phys. Rev. Lett.*, **77**(5), 793–797.

[56] Steane, A. M. 1996b. Simple quantum error-correcting codes. *Phys. Rev. A*, **54**(Dec), 4741–4751.

[57] Steane, A. M. 1999a. Quantum Reed-Muller codes. *IEEE Transactions on Information Theory*, **45**(5), 1701–1703.

[58] Steane, A.M. 1999b. Enlargement of Calderbank-Shor-Steane quantum codes. *IEEE Transactions on Information Theory*, **45**(7), 2492–2495.

[59] Steane, Andrew. 1996c. Multiple-particle interference and quantum error correction. *Proc. Roy. Soc. London Ser. A*, **452**(1954), 2551–2577.

[60] Steane, Andrew M. 1998. Quantum error correction. Pages 184–212 of: *Introduction to quantum computation and information*. World Sci. Publ., River Edge, NJ.

[61] Stephens, Ashley M., Evans, Zachary W. E., Devitt, Simon J., and Hollenberg, Lloyd C. L. 2008. Asymmetric quantum error correction via code conversion. *Phys. Rev. A*, **77**(Jun), 062335.

[62] Stichtenoth, Henning. 2009. *Algebraic function fields and codes*. Second edn. Graduate Texts in Mathematics, vol. 254. Springer-Verlag, Berlin.

[63] Tsfasman, M. A., Vlăduţ, S. G., and Zink, Th. 1982. Modular curves, Shimura curves, and Goppa codes, better than Varshamov-Gilbert bound. *Math. Nachr.*, **109**, 21–28.

Printed in the United States
by Baker & Taylor Publisher Services